FANTASTIC VOYAGE

LIVE LONG ENOUGH TO LIVE FOREVER

RAY KURZWEIL
AND
TERRY GROSSMAN
M.D.

A PLUME BOOK

PLUME
Published by Penguin Group
Penguin Group (USA) Inc., 375 Hudson Street, New York, New York 10014, U.S.A.
Penguin Group (Canada), 90 Eglinton Avenue East, Suite 700, Toronto, Ontario, Canada M4P 2Y3
(a division of Pearson Penguin Canada Inc.)
Penguin Books Ltd., 80 Strand, London WC2R 0RL, England
Penguin Ireland, 25 St. Stephen's Green, Dublin 2, Ireland (a division of Penguin Books Ltd.)
Penguin Group (Australia), 250 Camberwell Road, Camberwell, Victoria 3124, Australia
(a division of Pearson Australia Group Pty. Ltd.)
Penguin Books India Pvt. Ltd., 11 Community Centre, Panchsheel Park, New Delhi – 110 017, India
Penguin Books (NZ), 67 Apollo Drive, Rosedale, North Shore 0632, New Zealand
(a division of Pearson New Zealand Ltd.)
Penguin Books (South Africa) (Pty.) Ltd., 24 Sturdee Avenue, Rosebank,
Johannesburg 2196, South Africa

Penguin Books Ltd., Registered Offices: 80 Strand, London WC2R 0RL, England

Published by Plume, a member of Penguin Group (USA) Inc. This is an authorized reprint of a hardcover edition published by Rodale. For information address Rodale, 733 Third Avenue, 15th Floor, New York, NY 10017-3204.

First Plume Printing, October 2005
10

Illustrations designed by Ray Kurzweil and Terry Grossman, M.D., and illustrated by Laksman Frank

CIP data is available.
ISBN 1-57954-954-3 (hc.)
ISBN 978-0-452-28667-2 (pbk.)

Printed in the United States of America

PUBLISHER'S NOTE
Neither the publisher nor the author is engaged in rendering professional advice or services to the individual reader. The ideas, procedures, and suggestions contained in this book are not intended as a substitute for consulting with your physician. All matters regarding your health require medical supervision. Neither the author nor the publisher shall be liable or responsible for any loss or damage allegedly arising from any information or suggestion in this book.

While the author has made every effort to provide accurate telephone numbers and Internet addresses at the time of publication, neither the publisher nor the author assumes any responsibility for errors, or for changes that occur after publication. Further, publisher does not have any control over and does not assume any responsibility for author or third-party websites or their content.

To the memory of my father, Fredric, who could have been alive today if he had had the knowledge in this book.—*Ray*

To the memory of my grandfather, Jacob Light, who lived 104 healthy years without seeming to need any health advice at all.—*Terry*

CONTENTS

ACKNOWLEDGMENTS

Ray—I would like to express appreciation to my mother, Hannah, who gave me the confidence that I could solve any challenge, and to my wife, Sonya, and my kids, Ethan and Amy, who give me the motivation to live my life to the fullest.

Terry—I would like to express special thanks to my parents, Louis and Irene, who provided me a world of possibilities; my children, Abraya and Samuel, for whom unlimited possibilities lie ahead; and to Karen, who has already done the impossible.

We would both like to thank the many talented and devoted people who assisted us in bringing this project to life.

At Rodale: Stephanie Tade, whose enthusiasm, inspired leadership, and discerning ideas made this book possible; Chris Potash, whose insightful editing made a significant contribution; Kelly Schmidt, for coordinating the many details of book production; Rachelle V. S. Laliberte, for her expert copyediting; Louise Braverman, Cathy Gruhn, and Meghan Phillips, for their publicity work; and Amy Rhodes, vice president and publisher of Rodale Trade Books, for her continued support of the book.

Loretta Barrett, our literary agent, whose keen guidance and efforts established our wonderful partnership with Rodale.

Martine Rothblatt, whose dedication to health and longevity and enthusiastic support of this project have greatly enhanced the book and our experience of creating it.

Amara Angelica, whose tireless and insightful efforts led our research team. Amara also used her outstanding editing skills to help us articulate the complex issues in this book. Elizabeth Collins, who conducted thorough and extensive research on technology projects and provided extensive assistance with the endnotes and glossary. Celia Black-Brooks, who provided key insights on how to shape our message.

Laksman Frank, who created the attractive diagrams and images from our descriptions. Phil Cohen, who created the cardiovascular images in chapter 15.

Our research staff, Sarah Black, Tom Garfield, Emily Brown, and Daniel Pentlarge, who helped find and organize extensive research materials.

Nanda Barker-Hook, who helped manage the extensive logistics of the research and editorial processes.

Ken Linde and Matt Bridges, who provided computer systems support to keep our intricate work flow progressing smoothly.

Denise Scutellaro, Joan Walsh, Maria Ellis, and Bob Beal for doing the accounting on our complicated project.

Our peer expert readers:

Karen Kurtak, L.Ac., Dipl. Ac., for her contributions and ideas regarding the importance of organic foods and environmental toxins; Glenn S. Rothfeld, M.D., who carefully reviewed many of the chapters; Aubrey de Grey, for his expert review of the manuscript; Brian Grossman, M.D., D.V.M., regarding vitamin and mineral supplementation; Joel Miller, M.D., for reviews and contributions to our discussion of stress and balance; Melvin E. Clouse, M.D., for his thorough review of the chapter on heart disease; George Juetersonke, D.O., and Lee Light, M.D., who did a wonderful and thorough job of providing detailed review of several chapters; Robert Rowen, M.D., editor of *Second Opinion* newsletter; Drs. Jeff and Greg Stock, Patrick Hanaway, M.D., of Genova Diagnostics, Lynn Parry, M.D., John McNamara, M.D., Rob Lyons, M.D., Steve Parcell, N.D., and Bob Litman, R. Ph., who provided detailed reviews of specific chapters and material. A special thank-you also to Paul Dragul, M.D., for his support and inspiration.

Our lay readers, who provided keen insight to help shape the material: Aaron Kleiner, Martine Rothblatt, Mike Weiner, Nanda Barker-Hook, Sarah Black, Regina Mandl, Irene Grossman, and Robert Toale.

The above individuals provided many ideas and corrections that we were able to make thanks to their expert guidance. For any mistakes that remain, the authors take sole responsibility.

1

YOU CAN LIVE LONG ENOUGH TO LIVE FOREVER

Do not go gentle into that good night,
Old Age should burn and rave at close of day;
Rage, rage against the dying of the light.

—Dylan Thomas

"I don't want to achieve immortality through my work. I want to achieve immortality through not dying."

—Woody Allen

Written at the height of the Cold War, Isaac Asimov's 1966 science-fiction thriller *Fantastic Voyage* shifted the public's fascination from space travel to an even more fascinating journey—inside the human body. In the novel, scientists on "our side" as well as the unnamed "other side" have developed a miniaturization technology that promises victory for whoever can perfect it first. However, the technology has a fatal flaw: the miniaturization wears off quickly.

Professor Benes has figured out a breakthrough that overcomes this limitation, but before he has a chance to communicate his crucial insight, he falls into a coma, with a potentially fatal blood clot in his brain. Against a backdrop of international intrigue, our side sends in a submarine with a team of five people using the still time-limited miniaturization technology to travel inside Benes's body and destroy the blood clot.

The team includes pilot Owens, who helms the submarine *Proteus* (now blood cell–size); Duvall, a brilliant neurosurgeon in charge of the medical mission; Peterson, the beautiful surgical assistant (played by Raquel Welch in the highly successful movie version); Michaels, a human-circulatory expert; and Grant, the mission leader from central intelligence. In the course of the drama, readers and moviegoers are treated to a genuinely fantastic voyage through the human body as the intrepid crew battles enormous white blood cells, insidious antibodies, annoying platelets, and a myriad of other threats as they struggle to achieve their goal before the miniaturization catastrophically wears off.

The metaphor of *Fantastic Voyage* fits our book on several levels. First, we hope to treat you, our readers, to a fantastic voyage through the human body. Our understanding of the complex processes underlying life, disease, and aging has progressed enormously since 1966. We now have an unprecedented ability to comprehend our biology at the level of the tiniest molecular structures. We also have the opportunity to vastly extend our longevity, improve our well-being, and expand our ability to experience the world around us.

Asimov's fascination with miniaturization was prophetic. We are now in the early stages of a profound revolution in which we are indeed shrinking

Proteus **as Prototype**

A team at UCLA headed by biomedical engineer Carlo Montemagno is building a blood cell–size "submarine" intended for critical medical maneuvers inside the human body.[1] "In living systems, molecules perform repetitive functions the way machines do," Montemagno explains. "Some molecules take matter or information and move it from one location to another, while others filter and pump. I look at how to take pieces of these molecular machines and engineer them into hybrid devices. That means devices that are living and nonliving; they incorporate all the functionality you find in living systems but are artificial and engineered." The team has already created what it calls a nanocopter, with a propeller made of nickel and a motor the size of a virus that uses the body's own ATP (adenosine triphosphate, a complex molecule that stores energy) for power.

Virologist Peixuan Guo at Purdue University has created a remotely guided motorized nanomotor made from viral RNA and DNA and powered by the body's own chemical fuels. Guo has already guided his device inside cells to destroy the hepatitis virus.[2]

Another team at the University of California at Irvine is using a $2.9 million National Institutes of Health grant to develop a microscopic vessel that would be remotely piloted by surgeons through the esophagus, stomach, small intestine, and colon to find tiny tumors and perform immediate biopsies.[3] Ultimately, it will be able to destroy the tumors it finds.

The results of this technology revolution will go far beyond mere health maintenance to include a vast expansion of our human potential.

our technology down to the molecular level. We actually are developing blood cell–size submarines called nanobots (robots whose key features are measured in nanometers, or billionths of a meter) to be sent into the human body on vital health missions. Although we won't literally be shrinking ourselves to ride inside these nanobots, as in Asimov's imagined tale (at least not in the next several decades), we will be able to place ourselves in virtual-reality environments and see out of the eyes of these tiny robots. We will be able to control their movements as if we were inside, just as soldiers today remotely control intelligent weapons systems.

IMMORTALITY IS WITHIN OUR GRASP

Do we have the knowledge and the tools today to live forever? If all science and technology development suddenly stopped, the answer would have to be no. We do have the means to dramatically slow disease and the aging process far more than most people realize, but we do not yet have all the techniques we need to indefinitely extend human life. However, it is clear that far from halting, the pace of scientific and technological discovery is accelerating.

According to models that Ray has created, our paradigm-shift rate—the rate of technical progress—is doubling every decade, and the capability (price performance, capacity, and speed) of specific information technologies is doubling every year.[4] So the answer to our question is actually a definitive yes—the knowledge exists, if aggressively applied, for you to slow aging and disease processes to such a degree that you can be in good health and good spirits when the more radical life-extending and life-enhancing technologies become available over the next couple of decades.

Longevity expert and gerontologist Aubrey de Grey uses the metaphor of maintaining a house to explain this key concept. How long does a house last? The answer obviously depends on how well you take care of it. If you do nothing, the roof will spring a leak before long, water and the elements will invade, and eventually the house will disintegrate. But if you proactively take care of the structure, repair all damage, confront all dangers, and rebuild or renovate parts from time to time using new materials and technologies, the life of the house can essentially be extended without limit.

The same holds true for our bodies and brains. The only difference is that while we fully understand the methods underlying the maintenance of a house, we do not yet fully understand all of the biological principles of life.

But with our rapidly increasing comprehension of the human genome, the proteins expressed by the genome (proteome), and the biochemical processes and pathways of our metabolism, we are quickly gaining that knowledge. We are beginning to understand aging, not as a single inexorable progression but as a group of related biological processes. Strategies for reversing each of these aging progressions using different combinations of biotechnology techniques are emerging. Many scientists, including the authors of this book, believe that we will have the means to stop and even reverse aging within the next two decades. In the meantime, we can slow each aging process to a crawl using the methods outlined in this book.

In this way, the goal of extending longevity can be taken in three steps, or Bridges. This book is intended to serve as a guide to living long enough in good health and spirits—Bridge One—to take advantage of the full development of the biotechnology revolution—Bridge Two. This, in turn, will lead to the nanotechnology-AI (artificial intelligence) revolution—Bridge Three—which has the potential to allow us to live indefinitely.

This, then, is the premise of our book and the case we will make throughout: the knowledge of how to maintain our biological "house" and extend its longevity and vitality without limit is close at hand. We will tell you how to use the extensive knowledge that we do have today to remain healthy as the reverse engineering (decoding and understanding the principal methods) of our biology proceeds.

THE 21ST CENTURY IS WORTH LIVING TO EXPERIENCE

Most of our conceptions of human life in the 21st century will be turned on their head. Not the least of these is the expectation expressed in the adage about the inevitability of death and taxes. We'll leave the issue of the future of taxes to another book,[5] but belief in the inevitability of death and how this perspective will soon change is very much the primary theme of this book. As we succeed in understanding the genome and the proteome, many dramatic advances in treating disease and even reversing aging will emerge. The first two decades of the 21st century will be a golden era of biotechnology.

Many experts believe that within a decade we will be adding more than a year to human life expectancy every year. At that point, with each passing year, your remaining life expectancy will move further into the future. (Aubrey de Grey believes that we will successfully stop aging in mice—who share 99 percent of our genetic code—within 10 years, and that human therapies to halt and reverse aging will follow 5 to 10 years after that.) A small minority of older boomers will make it past this impending critical threshold. You can be among them. The authors of this book are of this generation and are intent on living through this threshold era in good health and spirits. Unfortunately, most of our fellow baby boomers remain oblivious to the hidden degenerative processes inside their bodies and will die unnecessarily young.

As interesting as the first two decades of this century are likely to be, subsequent decades should lead to even more dramatic changes. Ray has spent several decades studying and modeling technology trends and their impact on society. Perhaps his most profound observation is that the rate of change is itself accelerating. This means that the past is not a reliable guide to the future. The 20th century was not 100 years of progress at *today's* rate but, rather, was equivalent to about 20 years, because we've been speeding up to current rates of change. And we'll make another 20 years of progress at today's rate, equivalent to that of the entire 20th century, in the next 14 years. And then we'll do it again in just 7 years. Because of this exponential growth, the 21st century will equal 20,000 years of progress at today's rate of progress—1,000 times greater than what we witnessed in the 20th century, which itself was no slouch for change.

The result will be profound changes in every facet of our lives, from our health and longevity to our economy and society, even our concepts of who we are and what it means to be human. Within a couple of decades we will have the knowledge to revitalize our health, expand our experiences—such as full-immersion virtual reality incorporating all of the senses, augmented reality, and enhanced human intelligence and capability—and expand our horizons.

As we peer even further into the 21st century, nanotechnology will enable us to rebuild and extend our bodies and brains and create virtually any product from mere information, resulting in remarkable gains in prosperity. We will develop means to vastly expand our physical and mental capabilities by directly interfacing our biological systems with human-created technology.

Although human ability to take command of the course of life and death is controversial, we believe that the ability to broaden our horizons is a unique and desirable attribute of our species. And we certainly believe that it is worth the effort to remain healthy and vital today to experience this remarkable century ahead.

A Program That Is Important for All Ages

The longevity program that we lay out in this book is valuable for members of every generation:

- For young adults in their 20s and 30s, now is a good time to implement a healthy course. It's much easier to avoid disease progression than reverse damage later.
- For the youngest half of the baby boomers, now in their 40s, this is a critical time when processes of morbidity pick up speed.
- Older boomers now in their 50s are a pivotal group, living on the cusp. They are the last generation in which the vast majority will die in the more or less "old-fashioned" way, generally from debilitating progressive conditions that severely interfere with quality of life.
- For the senior generation, people 60 and older, it's not too late to reverse decades of damage and significantly extend longevity and vitality.

A DECADES-LONG MARCH TO HEALTH —OR DISEASE

The leading causes of death—heart disease, cancer, stroke, respiratory disease, kidney disease, liver disease, and diabetes[6]—do not appear out of the blue. They are the end result of processes that are decades in the making. To help you understand how long-standing imbalances in the metabolic processes underlying life functions can lead to disease, we have developed Ray & Terry's Longevity Program, which is laid out over the course of this book. (Our program is Bridge One, as mentioned above; Bridges Two and Three are detailed in chapter 2.) The advice we offer on how to keep your body optimally healthy—from what to put into it ("Food and Water," chapter 4) to how to fine-tune it ("Stress and Balance," chapter 23)—will enable you to determine your own specific health status and teach you how to take effective corrective action when necessary. Our program does require time and commitment to implement, but the rewards are considerable:

- Significant gains in how you feel, including the alleviation of various discomforts, improved gastrointestinal functioning, reversal of fatigue, improvements in sleep, enhanced mood, and attaining your optimal weight
- A greatly improved sense of well-being and increased levels of energy

- The comfort of knowing that you're on a path toward long-term health while significantly reducing the risk of chronic diseases such as heart disease, stroke, type 2 diabetes, and cancer

Conventional medical care is geared toward dealing with long-term degenerative processes only after they erupt into advanced clinical disease, but by this time it is often too late. It's like approaching a cliff but walking backward. You need to recognize that you're getting closer to the edge and stop. Once you fall off, it's difficult to do anything about it. This is what *Fantastic Voyage* is all about: to provide the knowledge and the specific steps to take, sooner rather than later, to extend your life, your vitality, and your well-being.

WHO IS THE ENEMY?

It is wise to consider the process of reversing and overcoming the dangerous progression of disease as a war. As in any war, if the enemy is at the gates—or worse, inside the gates—it's important to mobilize all the means of intelligence and weaponry that can be harnessed. That's why we'll advocate that key dangers be attacked on multiple fronts. For example, we'll discuss 10 approaches that should be practiced concurrently for preventing heart disease, particularly for people with elevated risk factors.

But if fighting disease and extending longevity and vitality is a war, who is the enemy? At the top of the list we should put ourselves. Of course, health issues get our attention the moment clinical disease strikes, but most people fail to focus on prevention and health enhancement in a timely manner *before* the onset of overt symptoms. Unfortunately, the medical profession is oriented toward detecting and treating these conditions only *after* they reach the point of crisis (symptom-control medicine), so most people receive limited guidance on disease prevention from their health professionals. You should not wait for others to show you the path to healing; the only person who can take responsibility for your health is you.

Our second enemy is the disease process itself. Our bodies evolved when it was not beneficial to the survival of the species for people to live beyond their child-rearing years and compete for the tribe's or community's limited food and other resources. Only a century and a half ago, life expectancy was 37 years.[7] If we want to remain vital for as long as possible, we cannot simply rely on the natural order that biological evolution has given us.

The third enemy is an increasingly vocal body of opinion that opposes extending human longevity on the basis that it supposedly violates the essence of human nature. Author Francis Fukuyama, for example, considers research that might extend human longevity beyond its current fourscore years to be immoral.[8] Opposition to certain biological technologies such as stem cell research is delaying vital therapies for a wide range of diseases. We should note that we don't consider these thinkers themselves to be our adversaries but, rather, their regressive ideas. The essence of the human species is to extend and expand our boundaries. Ultimately, such opposition will end up being mere stones in a torrent of innovation, with the continued flow of progress passing around these barriers. But even minor delays will result in the suffering and death of millions of people.

PUBLIC HEALTH RECOMMENDATIONS ARE COMPROMISED AT THE START

Many people believe that public health recommendations, such as the Department of Agriculture's Food Pyramid, represent our best collective wisdom.[9] People typically then go on to compromise (weaken) these recommendations further to meet their own priorities and circumstances without realizing that the recommendations come already pre-

Out of Sight, Out of Mind?

A significant number of cardiac patients receive the first warning that something is wrong when they suffer a heart attack, a third of which are fatal and another third of which cause permanent heart damage. Atherosclerosis, the buildup of cholesterol-laden deposits in the arteries, is rarely detected until it produces either a heart attack or, if the patient is "lucky," angina pain or an abnormal stress test. At that point, coronary artery disease is already advanced. More than 1,250,000 Americans suffer heart attacks each year, while 710,000 die of heart disease,[10] with most of these deaths following long periods of debilitation.

Hypertension, or high blood pressure, although easily diagnosed, has no obvious symptoms. Strokes typically cause brain damage without warning. Cancer is often not detected until it has metastasized (spread throughout the body).[11] In fact, most of the degenerative biological processes that result in devastating clinical disease are invisible and silent, and most of these, and the deaths and suffering they cause, can be prevented or significantly delayed.

compromised. The result is ineffectual guidelines and a double compromising of health.

The recommendations for vitamins, for example, continue to be dominated by the RDA (recommended dietary allowance) system. But these address only minimal levels to avoid specific vitamin deficiencies and do not begin to reflect the levels required for optimal health.[12] Dietary recommendations in general are severely watered down. For example, the nutrition guidelines for people with type 2 diabetes fail to recommend sharp reductions in carbohydrates,[13] and the recommendations on fat consumption are the same as for the general public.[14] The guidelines from the American Diabetes Association are completely ineffective, despite the fact that the condition, particularly in its early stages, can be largely controlled through nutrition. The same observations can be made regarding dietary recommendations for avoiding heart disease, the nation's number one killer.[15]

When we discuss the ineffectual nature of public nutrition guidelines with some health professionals, they counter that their patients won't even follow these weak recommendations, let alone stricter ones. Our counter to that is that people don't follow the weak guidelines precisely because they don't work. Actually, following stricter recommendations is *easier* in many ways.

Take, for example, carbohydrate consumption. Eating carbohydrates, particularly those with a high glycemic index (those that convert rapidly into sugar in the bloodstream), causes cravings for more carbohydrates. Attempting to "moderately" reduce consumption of carbohydrates turns out to be very difficult because a moderate reduction does nothing to fend off cravings. It's like suggesting that smokers simply reduce the number of cigarettes they smoke each day. But sharply reducing carbohydrates, particularly high-glycemic-index ones, effectively eliminates cravings, like quitting smoking altogether. It is far more motivating to follow a program that has the potential to make a dramatic difference in your immediate and long-term well-being.

As another example of compromised recommendations, the public health guideline for folic acid supplementation is 400 micrograms (mcg) per day, which may be a reasonable general recommendation. However, for someone with elevated homocysteine levels—a major cause of cardiovascular disease—the recommendation remains 400 mcg per day, which is inadequate to reduce dangerous homocysteine levels. Folic acid supplementation of 2,500 mcg or more per day, however, is safe and effective in reducing homocysteine (as are other recommendations, which we will discuss).[16]

The same situation holds for recommendations on "optimal" blood lipid (fat) levels. Public health guidelines state that total cholesterol should be below 200 milligrams per deciliter and that the ratio of total cholesterol to HDL cholesterol should be under 4.6. But even people who achieve such "optimal" levels suffer heart attacks.

How often does a person who consistently maintains a truly desirable lipid profile suffer a heart attack? The answer is almost never. But are such levels really achievable by most people? The answer is yes, they are. So why not set these as the targets?

Our philosophy is to provide optimal recommendations based on the latest research. A great deal is known about ways to modify the long-term destructive health trends that result in the vast majority of deaths and chronic diseases. We'll offer our best knowledge of effective measures, and you can decide for yourself what changes you are willing to make.

DYNAMIC VERSUS STATIC TESTING

Another unique aspect of our program is the extensive use of dynamic rather than static testing for early detection of abnormalities whenever possible. Dynamic tests measure the body's response to changing, or "stressful," physiologic conditions, while static tests simply provide measurements under baseline (resting) conditions. The exercise stress test, typically done on a treadmill to evaluate cardiac function, is an example of a common dynamic test. In an exercise test, the electrocardiogram (ECG) tracing is monitored both at rest and under conditions of increased workload. Many more patients with early cardiac disease can be detected by an exercise test than by a resting (static) ECG alone.

Another common dynamic test performed by conventional physicians is the glucose tolerance test for the diagnosis of diabetes, which is more sensitive than the static fasting blood sugar test. Yet, as we will see in chapter 9, "The Problem with Sugar (and Insulin)," the standard glucose tolerance test measures only blood sugar levels in response to a dietary sugar challenge (ingestion), so it still misses many early cases of diabetes. By also measuring insulin levels, using the glucose-insulin tolerance test we recommend, many additional cases of diabetes and sugar intolerance can be diagnosed. If we go a step further and add a simple insulin challenge test—a test performed by only a handful of physicians around the country—it is possible to detect numerous cases of insulin resistance, one of the most dangerous risk factors for a host of chronic diseases suffered by a significant segment of the population.

In chapter 13, "Methylation—Critically Important to Your Health," we discuss abnormal homocysteine metabolism, a risk factor for heart disease, stroke, and Alzheimer's disease that is carried by more than one-third of the adult population. Yet many cardiologists still don't perform even the static test on their patients to determine risk levels, and most large U.S. cities don't have a single cardiologist outside of a teaching hospital who performs the dynamic and far more accurate, yet inexpensive, homocysteine stress test that we recommend.

Early detection of risk factors is a hallmark of our program. By performing dynamic stress tests when appropriate, you can substantially increase the effectiveness of your screening processes.

THE PILLARS OF OUR LONGEVITY PROGRAM

We've organized Ray & Terry's Longevity Program around the activities and primary physical and metabolic processes that lead to either disease or sustained health. Our program combines the best of both conventional and alternative medicine. Many people have the view that conventional medicine is scientific, whereas alternative medicine reflects unverified folk traditions. The reality is that there are many conventional medicine practices that have not been scientifically verified, while there are many "alternative" practices supported by impressive research and verification.

Alternative medicine is not a single integrated methodology. Rather, it consists of a broad array of ideas that fall outside of conventional medical practice. Indeed, many of these ideas are not well grounded in science or in practical results. We've drawn our ideas from the best of conventional medicine, alternative medicine practices with convincing research on safety and efficacy, and cutting-edge developments in biotechnology and nanotechnology.

PARTNERING WITH YOUR HEALTH PROFESSIONAL

It would be difficult to follow a program of this comprehensive nature without a personal guide. Our philosophy has been to draw upon the best from both conventional medicine and alternative schools of thought in an unbiased fashion. So to follow the ideas in our program, you will need access to both worlds.

Your personal physician is trained to deal with diagnosing and treating catastrophic illness, but most physicians are not well prepared to provide guidance in the type of aggressive illness prevention that we address in this book. Unfortunately, disease prevention is not a major focus of mainstream medicine. Moreover, the critical issue of nutrition receives almost no attention in our nation's medical schools. An ideal approach is to find a physician who combines the best of multiple traditions.

More and more physicians have seen the limitations of practicing orthodox conventional medicine. They have begun to transcend the deep conditioning from their years of medical training, and they (and even more so their patients) have started to experience the joy that comes from thinking outside the box. Many such physicians have joined professional associations that serve as resources to train physicians in cutting-edge nutritionally based medical therapies, offering formal education and examinations to ensure competency. (For a list of certified practitioners and physicians in your area, see Fantastic-Voyage.net.) Even within the field of nutrition, we are dismayed by how many dietitians—people in the field of nutrition—rigidly follow the highly compromised public health recommendations.

THE MOST IMPORTANT PRINCIPLE: CONTINUAL EXPLORATION

The knowledge represented here is inherently dynamic. This is not a fixed program that one simply adopts. The most important principle of the program is continual active exploration of new knowledge from multiple sources:

- Newly available diagnosis and treatment options resulting from the emerging biotechnology revolution
- New insights into natural therapies
- Your own growing personal knowledge of available health information
- New personal knowledge about your own condition
- We plan to update the information in this book on our Web site (see Fantastic-Voyage.net) and through future editions of this book. A list of resources also appears on the site.

Most health books offer just one or two new ideas. Ours is different in that it provides dozens that are incorporated into a single integrated program. Based on our research, we believe that the recommendations in *Fantastic Voyage* will enable you to dramatically reduce your risk of disease in the future while quickly boosting your well-being in the present. Our core idea is that we now have the knowledge to determine where each of us is located in the progression of these decades-long degenerative processes and reverse them.

The support for this concept is rooted in decades of investigation and years of collaboration. Many of the simpler ideas presented in other contemporary health books are valid, but there is no single silver bullet that can address all of the key issues, given the complexity of our bodies and brains. Typically, other health books present one or two ideas combined with a lot of preaching. Instead, we provide a high density of ideas on how to harness contemporary longevity knowledge to transform your health.

Ideas have immense power to transform reality, but only if they are put into practice. There are two ways to use this book:

- Select ideas you find appealing and add them to your personal health program. We expect this is how many readers will benefit from this book.
- Follow all of the recommendations of Ray & Terry's Longevity Program, which we designed as an integrated and comprehensive approach to nutrition, lifestyle changes, and cutting-edge medical therapeutics.

Health is not simply the absence of diagnosed disease; it's a path toward ever-greater physical, emotional, and spiritual well-being. There is always the potential to improve your personal health.

2

THE BRIDGES TO COME

"Life expectancy will be in the region of 5,000 years . . . by the year 2100."
—Aubrey de Grey

Biological systems are remarkable in their cleverness. In the 15th century, Leonardo da Vinci wrote, "Human ingenuity may make various inventions, but it will never devise any inventions more beautiful, nor more simple, nor more to the purpose than nature does; because in her inventions nothing is wanting and nothing is superfluous." We share da Vinci's sense of awe at the designs of biology, but we do not agree with him on our inability to improve on nature. Da Vinci was not aware of nanotechnology, and it turns out that nature, for all its apparent creativity, is dramatically suboptimal. For example, the neuronal connections in our brains compute at only 200 transactions per second, which is millions of times slower than today's electronic circuits.

Despite the elegant way our red blood cells carry oxygen in our bloodstream and deliver it to our tissues, it is still a slow and cumbersome system, and robotic replacements (respirocytes) already on the drawing board will be thousands of times more efficient than red blood cells. The reality is that biology will never be able to match what we will be capable of engineering once we fully understand biology's principles of operation.

Another major component of the coming revolution is molecular nanotechnology, which will ultimately enable us to redesign and rebuild, molecule by molecule, our bodies and brains.[1] The timing of these two revolutions—biotechnology and nanotechnology—is overlapping, but the biotechnology revolution is leading the full realization of nanotechnology by a decade or two. That's why we describe these as the second and third bridges, respectively, to radical life extension. Most of the material in this book is Bridge One material—ways to take maximum advantage of the most advanced diagnostic testing and preventive strategies currently available so you can get to Bridges Two and Three.

A BRIDGE TO A BRIDGE TO A BRIDGE

This book describes three bridges.

1. The First Bridge—**Ray & Terry's Longevity Program**—consists of present-day therapies and guidance that will enable you to remain healthy long enough to take full advantage of the construction of the Second Bridge.

2. The Second Bridge is the **biotechnology revolution.** As we learn the genetic and protein codes of our biology, we are gaining the means of turning off disease and aging while we turn on our full human potential. This Second Bridge, in turn, will lead to the Third Bridge.

3. The Third Bridge is the **nanotechnology-AI (artificial intelligence) revolution.** This revolution will enable us to rebuild our bodies and brains at the molecular level.[2]

These emerging transformations in technology will usher in powerful new tools to expand your health and human powers. Eventually, the knowledge represented in this book will be automated within you. Today, however, you have to apply that knowledge yourself. We will talk about each of these three bridges as they relate to the topics under discussion. In each chapter, we will begin with Bridge One strategies that you can apply starting today. Where relevant, we will include a tantalizing look at what Bridges Two and Three have to offer in the near future.

BRIDGE TWO: THE BIOTECHNOLOGY REVOLUTION

As we learn how information is transformed in biological processes, many strategies are emerging for overcoming disease and aging processes. We'll review some of the more promising approaches here, and then discuss further examples in the chapters ahead. One powerful approach is to start with biology's information backbone: the genome. With gene technologies, we're now on the verge of being able to control how genes express themselves. Ultimately, we will actually be able to change the genes themselves.

We are already deploying gene technologies in other species. Using a method called recombinant technology, which is being used commercially to provide many new pharmaceutical drugs, the genes of organisms ranging from bacteria to farmyard animals are being modified to produce the proteins we need to combat human diseases.

Another important line of attack is to regrow our cells, tissues, and even whole organs, and introduce them into our bodies without surgery. One major benefit of this therapeutic cloning technique is that we will be able to create these new tissues and organs from versions of our cells that have also been made younger—the emerging field of rejuvenation medicine.

As we are learning about the information processes underlying biology, we are devising ways of mastering them to overcome disease and aging and extend human potential. Drug discovery was once a matter of finding substances that produced some beneficial effect without excessive side effects. This process was similar to early humans' tool discovery, which was limited to simply finding rocks and natural implements that could be used for helpful purposes. Now that we can design drugs to carry out precise missions at the molecular level, we are in a position to overcome age-old afflictions. The scope and scale of these efforts is vast; the examples in this book are only a small sampling of the most promising ideas. We'll provide additional compelling examples in the chapters ahead.

NOT JUST DESIGNER BABIES, BUT DESIGNER BABY BOOMERS

Gene technologies will comprise three stages: (1) influencing the metabolic expression of genes, (2) blocking or modifying gene expression, and (3) somatic gene therapy. Let's discuss how these imminent technologies might impact your personal voyage into the future.

Influencing the metabolic expression of genes. Science does not yet have the ability to change your genes (although this is starting to work), but by knowing what genes you have, you can make appropriate lifestyle choices and engage in preventive strategies to influence their impact. As we'll discuss in chapter 11, you already have the tools to read a portion of your personal genetic makeup and use this information to guide your lifestyle, nutritional, and supplement choices. You can use this information to design an individualized protocol to avoid diseases and progressive degenerative conditions for which you are genetically predisposed.

Blocking or modifying gene expression. Although we do not yet have the means to alter genes themselves, we are beginning to be able to alter their expression. Gene expression is the process by which your genetic blueprint is read and its instructions are implemented. Every cell in your body has a full set of all your genes. But a specific cell, such as a skin cell or a pancreatic islet cell, gets its characteristics from only a small fraction of all the genetic material it carries—the portion of genetic information relevant to that particular type of cell.[3] Since it is possible to control this process outside the cell nucleus, it's easier to implement these genetic blocking strategies than therapies that require access to the inside of the nucleus.

Gene expression is controlled by *peptides*, molecules made up of sequences of amino acids and short RNA strands. Scientists are just beginning to learn how these processes work.[4] Many new therapies now in development and testing are based on manipulating this gene expression process to either turn *off* the expression of disease-causing genes or turn *on* desirable genes that may otherwise not be expressed in a particular type of cell.

Two evolving therapies for blocking or modifying gene expression are antisense therapy and RNA interference. The target of this therapy is the messenger RNA (mRNA), which is transcribed (copied) from DNA and then translated into proteins. For damaged or mutated genes, researchers are exploring ways to block the mRNA created by these genes so that they are unable to make undesired proteins. The repair process uses mirror-image sequences of RNA, called antisense RNA. These sequences stick to the abnormal protein-encoding RNA, preventing it from being expressed.[5]

In the RNAi (RNA interference) approach, researchers construct short double-stranded RNA segments containing both the "sense" and "antisense" strands. These match and lock on to portions of the RNA that are transcribed from mutated genes. This blocks the native RNA segment's ability to create proteins, effectively silencing the defective gene. In recent tests, using both RNA strands in this way has been dramatically more effective than using just the antisense strand. In many genetic diseases, only one copy of a given gene is defective. Because you get two copies of each gene, one from each parent, this approach leaves one healthy gene to make the necessary protein.[6]

Somatic gene therapy. This is the holy grail of bioengineering. This third stage will effectively change the genes inside the nucleus by "infecting" the nucleus with new DNA, essentially creating new genes.[7] The concept of changing the genetic makeup of humans is often associated with the idea of "designer babies." But the real promise of gene therapy is to actually change our *adult* genes.[8] These new genes can be designed to either block undesirable disease-producing genes or introduce new ones that slow down and reverse aging processes.

Animal studies began in the 1970s and 1980s, and now a range of "transgenic" animals, including cattle, chickens, rabbits, and sea urchins, has been successfully produced. The year 1990 marked the first attempts at human gene therapy. The challenge remains to transfer therapeutic DNA into target cells so that the DNA will then be expressed in the right amounts and at the right time.

Let's look first at how transfer of new genetic material occurs. A virus is often the vehicle of choice. Long ago, viruses developed the ability to deliver their genetic material to human cells, often resulting in disease. Researchers now simply remove the virus's harmful genes and insert therapeutic genes instead, so the virus then "infects" human cells with these beneficial genes. This approach is relatively straightforward, but viral genes are often too large to pass into many types of cells, such as brain cells. Other limitations of this process include the length of DNA that can be transferred. The precise location where the new viral DNA is integrated into the target cell's DNA sequence has also been difficult to control. In addition, such "infections" can trigger an immune response, resulting in rejection of the new genetic material.[9]

The deaths of two participants in gene therapy trials a few years ago caused a temporary setback, although research has since resumed. One patient died from an immune response to the virus vector. The second patient, suffering from "bubble boy" disease—essentially, he was born without an immune system—developed leukemia, which was triggered by the improper placement of the gene transferred into his cells.[10] This second death points to two major hurdles that must be crossed for gene therapy to succeed: how to properly position the new genetic material on the patient's DNA strands and how to monitor the gene's expression. One possible solution is to deliver an imaging "reporter" gene along with the therapeutic gene. The reporter gene provides image signals that allow the gene therapy to be closely monitored. The process is permitted to proceed only if the placement of the new gene is verified as correct.[11]

Gene Therapy Starts to Work

A team led by University of Glasgow researcher Dr. Andrew H. Baker has successfully used adenoviruses to bypass the liver and "infect" specific organs or regions within organs. For example, the researchers were able to direct gene therapy at the endothelial cells lining the inside of blood vessels. Baker said that his work could "improve the selectivity, efficiency, and safety of gene delivery to the cardiovascular system."

Another approach is being pioneered by a research team led by Craig Venter, the head of the private effort that successfully transcribed the human genome, which has already demonstrated the ability to create synthetic viruses from genetic information.[12] A primary application is designing viruses to deliver new genetic information for gene therapy.

Physical injection (microinjection) of DNA into cells is possible but prohibitively expensive. Exciting advances have recently been made in other means of transfer. For example, fatty spheres with a watery core, called liposomes, can be used as a molecular Trojan horse to deliver genes to brain cells. This opens the door to treatment of disorders such as Parkinson's disease and epilepsy.[13] Electric pulses can also be used to deliver a range of molecules, including drug proteins, RNA, and DNA, to cells.[14]

One option is to pack DNA into ultratiny (25-nanometer) nanoballs for maximum impact.[15] This approach is already being tested on human patients with cystic fibrosis. Researchers reported a "6,000-fold increase in the expression of a gene packaged this way, compared with unpackaged DNA in liposomes."

Yet another approach uses DNA combined with microscopic bubbles. Ultrasonic waves are used to compress the bubbles, enabling them to pass through cell membranes.

RECOMBINANT TECHNOLOGY: BETTING THE FAMILY PHARM

We are already using gene therapy in other species. By modifying the genes of bacteria, plants, and animals, we can cause them to create the substances we need to combat human diseases. Recombinant proteins made by combining DNA from more than one organism are now being manufactured by bacteria, a novel biotech appropriately referred to as *pharming*. In recombinant technology, the genetic material that codes for a desired protein is spliced into the DNA of certain species of bacteria, which then go to work making this protein. Given how fast bacteria multiply, it's easy to create significant amounts of proteins this way. Insulin was the first molecule to be created synthetically by recombinant technology, so that insulin-dependent diabetics are no longer reliant on injections of beef or pork insulin. Many diabetics developed allergic reactions or high levels of antibodies against the foreign proteins found in the beef- and pork-derived insulin preparations. With recombinant *human* insulin, this is no longer a problem.

Children with growth hormone deficiency (dwarfs) used to rely on injections of hGH (human growth hormone) derived from the pituitary glands of human cadavers. It took a lot of cadavers to provide enough hGH for just one child for a year. There was also the risk of certain infections. Recombinant hGH has solved this problem and substantially lowered the price of this therapy. It has enabled adults with growth hormone deficiency to be treated as well.

Genes from proteins have also been spliced into "immortalized" human kidney cells and are now being pharmed to create proteins found useful in treating patients who have suffered strokes, as well as numerous other illnesses.[16] Patients with chronic kidney disease are deficient in a protein made by the kidneys called erythropoietin. Without erythropoietin, severe anemia results and frequent transfusions are needed. By inserting the genes that code for this protein into hamster cells, drug companies have been able to create enough erythropoietin to avoid the need for transfusion for many dialysis patients.

New methods involving traditional farm animals are also being found. Cows produce large amounts of milk, so splicing DNA into the genes that code for milk is a valuable technique. The DNA that codes for egg protein is now being used so that the eggs of transgenic (containing a gene or genes artificially inserted from a different species) chickens will contain useful proteins. In the near future we will have pharms where the animals have had their genes altered so that their milk, eggs, or even semen will produce recombinant proteins to help treat currently untreatable or only partially treatable conditions, such as multiple sclerosis, Parkinson's disease, Alzheimer's disease, hepatitis C, and AIDS.

Pharmers won't be restricted to using animals. Plants, particularly types with high protein content such as corn or tobacco, can be reprogrammed to produce substances of great value. In Japan, for instance, a strain of genetically modified rice contains a protein that will kill the hepatitis B virus.

THERAPEUTIC CLONING

One of the most powerful methods of applying life's own machinery to improve and extend life involves harnessing biology's reproductive mechanisms in the form of cloning. Cloning is an extremely important technology, not for cloning complete humans but for life extension purposes. Therapeutic cloning creates new tissues to replace defective tissues or organs.

All responsible ethicists, including these authors, consider human cloning at the present time to be unethical, yet our reasons have little to do with the slippery (slope) issues of manipulating human life. Rather, the technology today simply does not yet work reliably. The current technique involves fusing a cell nucleus from a donor to a recipient egg cell using an electric spark and causes a high level of genetic errors.[17]

This is the primary reason most of the fetuses created in this way so far have not made it to term. Even those that do survive have genetic defects. Dolly the Sheep developed an obesity problem in adulthood, and most of the cloned animals produced thus far have had unpredictable health problems. Scientists have a number of ideas for perfecting this process, including using alternative ways of fusing the nucleus and egg cell without the destructive electrical spark. Until the technology is demonstrably safe, however, it would be unethical to create a human life with such a high likelihood of severe health problems.

However, the most valuable applications of cloning technology are *not* for the purpose of cloning entire human beings but to create human organs, such as hearts or kidneys. This uses germ line cells—those in the prefetal stage (before implantation of a fetus). These germ line cells go through differentiation, which can then be developed into specific organs. Because differentiation takes place during the prefetal stage, most ethicists believe that this process does not raise ethical concerns, although this issue has been highly contentious.[18]

A team of researchers led by Woo Suk Hwang and Shin Yong Moon of Seoul National University in South Korea has taken an important step forward toward perfecting this technology. In an article published in *Science*, they announced they had successfully cloned a line of human pluripotent stem cells, the type that has the potential to turn into any type of cell the body needs. Their cell line had already undergone 70 reproductions without incident.[19] This research paves the way for significant gains in the production of healthy human-replacement tissues and organs derived from a cloned stem cell line.

Defeating programmed cell death. Therapeutic cloning relates to telomeres, which are strings of a repeating code at the end of each DNA strand. These repeating codes are like a string of beads, in which one "bead" falls off each time a cell divides. This places a limit on the number of times a cell can replicate—the so-called Hayflick limit. Once these DNA beads run out, a cell is programmed for death. Recently, it was discovered that a single enzyme called telomerase can extend the length of the telomere beads, thereby overcoming the Hayflick limit. Germ line cells create telomerase and are immortal. Cancer cells also produce telomerase, which allows them to replicate indefinitely. The identification of this single enzyme creates important opportunities to manipulate this process to either extend the longevity of healthy cells or terminate the longevity of pathological cells, such as cancer.

It is interesting to reflect on the remarkable stability of the immortal germ line cells, which link all cell-based life on Earth. The germ line cells avoid destruction through the telomerase enzyme, which rebuilds the telomere chain after each cell division. This single enzyme makes the germ line cells immortal, and indeed these cells have survived from the beginning of life on Earth billions of years ago. This insight opens up the possibility of future gene therapies that would return cells to their youthful, telomerase-extended state. Animal experiments have shown telomerase to be relatively benign, although some experiments have resulted in increased cancer rates.

There are also challenges in transferring telomerase into cell nuclei, although the gene therapy technology required is making solid progress. Scientists such as Michael West, president and CEO of Advanced Cell Technology Inc., have expressed confidence that new techniques will provide the ability to transfer telomerase into cell nuclei and overcome the cancer issue. Telomerase gene therapy holds the promise of indefinitely rejuvenating human somatic (non–germ line) cells—that is, all human cells.

Progress in growing new tissues and organs from stem cells is developing rapidly. Robert Langer's team at MIT has grown primitive versions of human organs such as liver, cartilage, and neural tissues. Their technique involves growing cells on specially designed biodegradable polymer scaffolds, which are spongelike structures with the approximate shape of the desired organ. Langer and his team wrote, "Here we show for the first time that polymer scaffolds . . . promoted proliferation, differentiation and organization of human embryonic stem cells into 3D structures."

One of the challenges in growing new human organs in this way is creating a functioning system of new blood vessels. Researchers at MIT and Harvard Medical School have constructed a working synthetic vascular system using two computer-etched biodegradable polymers sandwiched together to create capillaries only 10 microns (millionths of a meter) wide, as well as arteries and veins up to 300 times wider.[20]

One exciting approach that bypasses the ethical controversy of using fetal tissue, while also providing a substantial source of stem cells, which are currently limited in quantity, is parthogenesis, or so-called virgin birth. Adding certain chemicals to unfertilized human egg cells can turn them into embryos, which might then act as a source of new stem cells.[21] These embryos, called parthenotes, can never become babies, so there should not be an ethical issue in destroying tissue that is destined for destruction anyway. Another intriguing idea is for a woman to create parthenotes from her own egg cells to create stem cells with her own DNA, thereby avoiding potential rejection of foreign cells by a patient's immune system.

Human somatic cell engineering. This is an even more promising approach that entirely bypasses the controversy of using fetal stem cells. These emerging technologies, also called transdifferentiation, create new tissues with a patient's own DNA by converting one type of cell (such as a skin cell) directly into another (such as a pancreatic islet cell or a heart cell) without the use of fetal stem cells.[22] There have been recent breakthroughs in this area. Scientists from the United States and Norway have successfully converted human skin cells directly into immune system cells and nerve cells.[23] Hematech, a biotechnology company, has reprogrammed fibroblast cells back into a primordial state where they can be converted into other types of cells.

Consider the question: What is the difference between a skin cell and any other type of cell in the body? After all, they all have the same DNA. As noted above, the differences are found in protein signaling factors. These include short RNA fragments and peptides, which we are now beginning to understand. By manipulating these proteins, we can turn one type of cell into another.[24]

Perfecting this technology would not only defuse a contentious ethical and political issue, it would also offer an ideal solution from a scientific perspective. If you need pancreatic islet cells or kidney tissues—or even a whole new heart—to avoid autoimmune reactions, you would strongly prefer to produce these from your own DNA, not the DNA from someone else's germ line cells. And this approach uses your own plentiful skin cells rather than your rare and precious stem cells.

This process would directly grow an organ with your genetic makeup, and the new organ could have its telomeres fully extended to their original youthful length, effectively making the new organ young again.[25] That means an 80-year-old man could have his heart replaced with the same heart he had when he was, say, 25.

The master gene that enables stem cells to remain youthful and pluripotent (able to differentiate into virtually any type of other cell) has been dis-

Growing Younger

Scientists at the European Molecular Biology Laboratory and the University of Rome have identified a protein called mIGF-1 that induces stem cells from other parts of the body to quickly migrate to muscles damaged from disease or injury. Our ability to produce mIGF-1 declines with age, but this age-related process can be reversed by administering the protein. This may be a key step in stem cell therapy.

Even more exciting is the prospect of replacing one's organs and tissues with their "young" replacements without surgery. Cloned telomere-extended cells introduced into an organ will integrate themselves with the older cells. Through repeated treatments over a period of time, the organ will end up being dominated by the younger cells. We normally replace our own cells on a regular basis anyway, so why not do so with youthful telomere-extended cells rather than older telomere-shortened ones? There's no reason why we couldn't eventually do this with every organ and tissue in our body. We would thereby grow progressively younger.

covered and named *nanog* by a team at the Institute for Stem Cell Research in Edinburgh, Scotland.[26] "Nanog seems to be a master gene that makes embryonic stem cells grow in the laboratory," says Ian Chambers, one of the team's scientists. "In effect this [gene] makes stem cells immortal." The insight is a big step in being able to turn any cell, such as a skin cell, into a pluripotent cell, which can then be transformed into any other type of cell.

REVERSING HUMAN AGING

Our understanding of the principal components of human aging is growing rapidly. Strategies have been identified to halt and reverse each of the aging processes. Perhaps the most energetic and insightful advocate of stopping the aging process is Aubrey de Grey, a scientist with the department of genetics at Cambridge University. De Grey describes his goal as "engineered negligible senescence"—stopping us from becoming more frail and disease-prone as we get older.[27]

According to de Grey, "All the core knowledge needed to develop *engineered negligible senescence* is already in our possession—it mainly just needs to be pieced together."[28] He believes we'll demonstrate "robustly rejuvenated" mice—mice that are functionally younger than before being treated, and with the life extension to prove it—within 10 years, and points out that this demonstration will have a dramatic effect on public opinion. Showing that we can reverse the aging process in an animal that shares 99 percent of our genes will profoundly transform the common wisdom that aging and death are inevitable. Once demonstrated in an animal, robust rejuvenation in humans is likely to take an additional 5 to 10 years, but the advent of rejuvenated mice will create enormous competitive pressure to translate these results into human therapies.

Earlier in the evolution of our species (and precursors to our species), survival was not aided—indeed, it would have been hurt—by individuals living long past their child-rearing years. As a result, genes that supported significant life extension were selected against. In our modern era of abundance, all generations can contribute to the ongoing expansion of human knowledge. "Our life expectancy will be in the region of 5,000 years . . . by the year 2100," says de Grey. By following the three bridges described in this book, you should be able to reach the year 2100, and then, according to de Grey, extend your longevity indefinitely.

De Grey describes seven key aging processes that currently encourage senescence and has identified strategies for reversing each. Here are four of de Grey's key strategies:

Chromosomal (nuclear) mutations and "epimutations."[29] Almost all of our DNA is in our chromosomes, in the nucleus of the cell. (The rest is in the mitochondria, which we'll come to in a moment.) Over time, mutations occur, that is, the DNA sequence becomes damaged. Additionally, cells accumulate changes to "epigenetic" information that determine which genes are expressed in different cells. These changes also matter because they cause cells to behave inappropriately for the tissue they're in. Most such changes (of either sort) are either harmless or just cause the cell to die and be replaced by division of a neighboring cell. The changes that matter are primarily ones that result in cancer. This means that if we can cure cancer, nuclear mutations and epimutations should largely be harmless. De Grey's proposed strategy for curing cancer is pre-emptive: It involves using gene therapy to remove from all our cells the genes that cancers need to turn on in order to maintain their telomeres when they divide. This will not stop cancers from being initiated by mutations, but it will make them wither away before they get anywhere near big enough to kill us. Strategies for deleting genes in this way are already available and are rapidly being improved.

Toxic cells. Occasionally, cells get into a state where they're not cancerous, but still it would be best for the body if they died. Cell senescence is an example, and so is having too many fat cells. In these cases we need to kill those cells (which is usually easier than reverting them to a healthy state). Methods are being developed to target "suicide genes" to such cells, and also to make the immune system kill them.

Blocking the telomerase enzyme is one of many strategies being pursued against cancer. Doing this would prevent cancer cells from replicating more than a certain number of times, effectively destroying the cancer's ability to spread. There are many other strategies being intensely pursued to overcome cancer. Particularly promising are cancer vaccines designed to stimulate the immune system to attack cancer cells. These vaccines could be used to prevent cancer, as a first-line treatment, or to mop up cancer cells after other treatments.[30] We'll discuss Bridge Two strategies against cancer in more detail in chapter 16, "The Prevention and Early Detection of Cancer."

Mitochondrial mutations. Another aging process identified by de Grey is accumulation of mutations in the 13 genes in the mitochondria, the energy factories for the cell.[31] The mitochondrial genes undergo a higher rate of mutations than those in the nucleus and are critical to the efficient functioning of our cells. Once we master somatic gene therapy, we could put multiple copies of these 13 genes within the relative safety of the cell nucleus, thereby providing redundancy (backup copies) for this vital genetic information. The mechanism already exists in cells for nucleus-encoded proteins to be imported into the mitochondria, so it is not necessary for these proteins to be produced in the mitochondria itself. In fact, most of the proteins needed for mitochondrial function are already coded by the nuclear DNA. There has already been successful research in transferring mitochondrial genes into the nucleus in cell cultures.

Cell loss and atrophy. Our body's tissues have the means to replace worn-out cells, but this ability is limited in certain organs, says de Grey. For example, the heart is unable to replace cells as quickly as needed as we get older, so it compensates by enlarging surviving cells using fibrous material. Over time, this causes the heart to become less supple and responsive. A primary strategy here is to deploy therapeutic cloning of our own cells, as described on page 22.

Evidence from the genome project indicates that no more than a few hundred genes are involved in the aging process. By manipulating these genes, radical life extension has already been achieved in simpler animals. For example,

A Panoply of Emerging Therapies

A broad variety of hybrid biotechnology strategies is emerging. Consider this small sample of additional research under way:

- Chemists from the Scripps Research Institute have identified a molecule they call reversine that appears to reprogram aging cells to make them youthful.[33] It may be an alternative to stem cell therapies to regenerate cells and tissues that need replacement.
- Researchers have demonstrated a method of creating dried stem cells that are revived just by adding water, providing indefinite shelf life to cells that can be used to re-create tissues such as bone, blood, and organs.[34]
- One of the major challenges in medicine is getting needed nutrients, supplements, and medications into the bloodstream without having to go through the gastrointestinal tract, which often interferes dramatically with a drug's effectiveness (and may also cause digestive upset). Many medications, such as insulin, human growth hormone, and therapeutic DNA, must be introduced through painful daily injections that adversely affect a patient's quality of life. Researchers at Johns Hopkins University have developed plastic polymer spheres that surround tiny doses of medications and can be inhaled. The polymers dissolve inside the lungs, releasing their therapeutic cargo at the correct rate. All of the materials used are already FDA-approved, so it is likely that the approval process for this application will be simplified. Ultimately, we will have the means to easily send optimal levels of all nutrients directly into the bloodstream without requiring pills or injections.

by modifying genes in the *C. elegans* worm that control insulin and modifying sex hormone levels, the life span of the test animals was expanded sixfold, the equivalent of a 500-year life span for a human.[32] As we gain the ability to understand and reprogram gene expression, reprogramming the aging process in humans will become increasingly feasible. The idea that aging and dying are inevitable is deeply rooted, but this age-old perspective will gradually change as gene therapies are successfully demonstrated over the next two decades.

BRIDGE THREE: NANOTECHNOLOGY AND ARTIFICIAL INTELLIGENCE

As we "reverse engineer" (understand the principles of operation behind) our biology, we will apply our technology to augment and redesign our

- The human genome on a chip is now available. An integrated "gene chip" allows researchers to determine exactly which genes are expressed in a diseased organ, compared with a healthy one, which will greatly accelerate drug discovery and the prediction of drug effects.
- We are in the early stages of creating simulated biology to be able to discover and test drugs and other therapies *in silico* (in chips). There are already detailed systems that can simulate organs such as the heart using cell-by-cell simulations. The Defense Advanced Research Projects Agency (DARPA) is developing a program that will provide a virtual army of simulated test subjects to automatically test drugs, procedures, even weapons. The goal is to simulate every aspect of the human body, incorporating all genetic and proteomic (protein expression) information on a molecular level. One of DARPA's objectives is to place a digital version of each soldier's complete body on a dog tag. If that soldier needed emergency treatment, medics would access the "virtual soldier program" on the dog tag to immediately make lifesaving decisions. Howard Asher, director of global life sciences for Sun Microsystems, states, "We believe in 10 years we can eliminate the need for all animal studies, we can eliminate phase one and phase two clinical studies, so computationally we can model a drug or therapeutic agent in the computer against the genomic data."

bodies and brains to radically extend longevity, enhance our health, and expand our intelligence and experiences. Much of this technological development will be the result of research into nanotechnology, a term originally coined by K. Eric Drexler in the 1970s to describe the study of objects whose smallest features are less than 100 nanometers (billionths of a meter). A nanometer equals roughly the diameter of five carbon atoms.

Robert A. Freitas Jr., a nanotechnology theorist, writes, "The comprehensive knowledge of human molecular structure so painstakingly acquired during the 20th and early 21st centuries will be used in the 21st century to design medically active microscopic machines. These machines, rather than being tasked primarily with voyages of pure discovery, will instead most often be sent on missions of cellular inspection, repair, and reconstruction."[35]

Freitas points out that if "the idea of placing millions of autonomous nanobots (blood cell–sized robots built molecule by molecule) inside one's body might seem odd, even alarming, the fact is that the body already teems with a vast number of mobile nanodevices." Biology itself provides the proof that nanotechnology is feasible. As Rita Colwell, director of the National Science Foundation, has said, "Life is nanotechnology that works." Macrophages (white blood cells) and ribosomes (molecular "machines" that create amino acid strings according to information in RNA strands) are essentially nanobots designed through natural selection. As we engineer our own nanobots to repair and extend biology, we won't be constrained by biology's toolbox. Biology uses a limited set of proteins for all of its creations, whereas we can create structures that are dramatically stronger, faster, and more intricate.

One application we'll discuss further in chapter 7, on digestion, is to disconnect the sensory and pleasurable process of eating from the biological purpose of obtaining optimal nutrition. Billions of tiny nanobots in the digestive tract and bloodstream could intelligently extract the precise nutrients we require, call for needed additional nutrients and supplements through our body's personal wireless local area network (nanobots that communicate with one another), and send the rest of the food we eat on its way to elimination.

BioMEMS. If this seems particularly futuristic, keep in mind that intelligent machines are already being injected into our bloodstreams today. There are dozens of projects under way to create bloodstream-based biological microelectromechanical systems (bioMEMS) with a wide range of diagnostic and therapeutic applications.[36] There are already four major conferences devoted to these projects.[37] BioMEMS devices are being designed to intelligently scout out pathogens and deliver medications in precise ways.

For example, nanoengineered blood-borne devices that deliver hormones such as insulin have been demonstrated in animals.[38] Similar systems could precisely deliver dopamine to the brain for Parkinson's patients, provide blood-clotting factors for patients with hemophilia, and deliver cancer drugs directly to tumor sites. One new design provides up to 20 separate reservoirs that can release the different substances at programmed times and locations in the body.[39]

Kensall Wise, a professor of electrical engineering at the University of Michigan, has developed a tiny neural probe that provides precise monitoring of the electrical activity of patients with neural diseases.[40] Future designs are expected to deliver drugs to precise locations in the brain as well. Kazushi Ishiyama at Tohoku University in Japan has developed micromachines that use microscopic spinning screws to deliver drugs directly into small cancerous tumors.[41]

A particularly innovative micromachine developed by Sandia National Labs has actual microteeth with a jaw that opens and closes to trap individual cells and then implant them with substances such as DNA, proteins, or drugs.[42]

Complex structures at the molecular level have already been constructed. In some cases, building blocks are borrowed from nature. In fact, copying or manipulating naturally occurring molecules to accomplish specific goals is a cornerstone of present-day nanotech research. DNA turns out to be a useful structural tool because the unzipped strands can be organized into structures such as cubes, octahedrons, and more complicated designs. A team at Cornell University used portions of a natural enzyme, ATPase, to build a nanoscale motor. Another team at the CNRS Institute in Strasbourg, France, has successfully used carbon nanotubes to deliver a peptide into the nuclei of fibroblast cells. Many approaches are being developed for micro- and nanosize machines to perform a broad variety of tasks in the body and bloodstream.

Programmable blood. One pervasive system that has already been the subject of a comprehensive conceptual redesign is our blood. In chapter 15, "The Real Cause of Heart Disease and How to Prevent It," we will discuss a series of remarkable conceptual designs by Freitas for robotic replacements of our red blood cells, white blood cells, and platelets. Detailed analyses of these designs demonstrate that these tiny robots would be hundreds or thousands of times more capable than their natural counterparts.

Nanopower. Developing power sources for these tiny devices has already received significant research attention. MEMS (microelectronic mechanical systems) technology is being applied to create microscopic hydrogen fuel cells to power portable electronics and, ultimately, nanobots that will be introduced into the human body. One strategy is to use the same energy sources—glucose and ATP—that power our natural nanobots, such as macrophages, a type of white blood cell that is designed to destroy harmful bacteria and viruses. A Japanese research team has developed a "bio-nano" generator that creates power from glucose in the blood. Another team at the University of Texas at Austin has developed a fuel cell that uses both glucose and oxygen in human blood.[43]

Continual monitoring. Sensors based on silicon nanowires have shown the potential to detect disease almost instantly.[44] Using any bodily fluid, such as urine, saliva, or blood, diseases including cancer can be detected at very early stages. According to the study leader, Charles M. Lieber, professor of chemistry at Harvard University, this technology will enable you to "give a drop of blood from a pinprick on your finger and, within minutes, find out whether you have a particular virus or genetic disease, or your risk for different diseases or drug interactions." This approach can also be used for detection of bioterrorism threats.

Within several years, we will have the means of continually monitoring the status of our bodies to fine-tune our health programs as well as provide early warning of emergencies such as heart attacks. The authors are working on this type of system with biomedical company United Therapeutics, using miniaturized sensors, computers, and wireless communication. Researchers at Edinburgh University are developing spray-on nanocomputers for health monitoring. Their goal: a device the size of a grain of sand that combines a computer, a wireless communication system, and sensors for heat, pressure, light, magnetic fields, and electrical currents. In another development, a research team headed by Garth Ehrlich of the Allegheny Singer Research Institute in Pittsburgh is developing MEMS-based sensor robots that can be implanted inside the body to detect infection, identify the pathogen, and then dispense the appropriate antibiotic from the device's internal containers.[45]

One application they envision is preventing bacterial infections, a major cause of hip joint replacement failure. Ehrlich points out that today, "the only recourse for such patients is the traumatic removal of the implant, which results in additional bone loss, extensive soft tissue destruction, months of forced bed rest with intravenous antibiotics, and significant loss of quality of life due to complete loss of mobility."

Nanosurgery. Nanobots will make great surgeons. Teams of millions of nanobots will be able to restructure bones and muscles, destroy unwanted growths such as tumors on a cell-by-cell basis, and clear arteries while restructuring them out of healthy tissue. Nanobots would be thousands of times more precise than the sharpest surgical tools used today, would leave no scars, and could provide continual follow-up after certain surgical procedures. Nanobot surgeons could even perform surgery on structures within cells, such as repairing DNA within the nucleus. These nanobots will require distributed intelligence. Like ants in an ant colony, their actions will need to be highly coordinated, and the entire "colony" of nanobots will need to display flexible intelligence. Distributed systems that display intelligent coordination is one of the key goals of research in artificial intelligence—developing computers that emulate human intelligence.

One of Freitas's more advanced conceptual designs is a DNA repair robot. Billions or even trillions of such robots could go inside all of your cells and make repairs as well as improvements to the DNA in the genes. Freitas points out that it may be more efficient to just replace all the DNA in a gene with a new corrected copy rather than attempt to make changes to individual nucleotides.

Here's an original idea: replace the genetic machinery altogether (the cell nucleus, ribosomes, and related structures) with a small computerized robot. The computer would store the genetic code, which is only about 800 megabytes of information, or about 30 megabytes using data compression. The computerized system replacing the nucleus would then perform the function of the ribosomes by directly assembling strings of amino acids according to the computerized genetic information. These computers would all be on a wireless local area network, so improvements to the genetic code could be quickly downloaded from the Internet. It would not be necessary for the computer replacing each cell nucleus to have a complete copy of the genetic code, since these computers will be able to share their information. One major advantage of this approach is that undesirable replication

Harnessing the Heat of the Sun

Experiments by a Harvard team led by physicist Eric Mazur have shown the feasibility of using sharply focused laser light to perform surgical procedures from outside the patient, including destroying small structures inside cells without otherwise affecting them.[46] "It's a microscopic James Bond type of scenario," according to project contributor and Harvard cell biologist Donald Ingber. "It generates the heat of the sun, but only for quintillionths of a second, and in a very small space." The team has already demonstrated the ability of performing their laser-based nanosurgery from outside an animal and successfully manipulated the sense of smell of the worm *Caenorhabditis elegans*.

processes—for example, of pathological viruses or cancer cells—could be quickly shut down.

Intelligent cells. A hybrid scenario involving both biotechnology and nanotechnology contemplates turning biological cells into computers. These "enhanced intelligence" cells could then detect and destroy cancer cells and pathogens, or even regrow human body parts such as organs and limbs. Princeton biochemist Ron Weiss has modified cells to incorporate a variety of logic functions that are used for basic computation.[47] Boston University's Timothy Gardner has developed a cellular logic switch, another basic building block for turning cells into computers.[48] And scientists at the MIT Media Lab have developed ways to use wireless communication to send messages, including intricate sequences of instructions, to computers inside modified cells.[49] By attaching gold crystals comprised of less than 100 atoms to DNA, they were able to use the gold as antennae and selectively cause the double-stranded DNA to unzip without affecting nearby molecules. The technique could ultimately be used to control gene expression through remote control. Weiss points out that "once you have the ability to program cells, you don't have to be constrained by what the cells know how to do already. You can program them to do new things, in new patterns."

We are also making exponential progress in understanding the principles of operation of the human brain. Our tools for peering inside the brain are accelerating in their price-performance, and the ability to see small features and fast events. An emerging generation of brain-scanning tools is providing the means for the first time to monitor individual interneuronal connections in real time in clusters of tens of thousands of neurons. We already have detailed models and simulations of several dozen regions of the human brain, and we believe that it is a conservative projection to anticipate the completion of the reverse engineering of the several hundred regions of the brain within the next two decades. This development will provide key insights into how the human brain performs its pattern recognition and cognitive functions. These insights in turn will greatly accelerate the development of artificial intelligence in nonbiological systems such as nanobots. With a measure of intelligence, the nanobots coursing through our bloodstream, bodily organs, and brain will be able to overcome virtually any obstacle to keeping us healthy. Ultimately, we will merge our biological thinking with advanced artificial intelligence to vastly expand our abilities to think, create, and experience.

3

OUR PERSONAL JOURNEYS

"To fight a disease after it has occurred is like trying to dig a well when one is thirsty or forging a weapon once a war has begun."

—The Yellow Emperor's Classic of Internal Medicine

Before we embark on our Fantastic Voyage together, beginning with chapter 4, we would like to reveal a bit of our personal histories. In this chapter, we each explain how we arrived at the point where sharing this health information became a priority for us and how our lives intersected to create this book.

RAY

My story begins on the outskirts of Vienna, Austria, in 1924, with the death of my paternal grandfather from heart disease when my father was 12. My father carried on with his two passions: the Boy Scouts and music. In 1938, my father's musical talent came to the attention of an American patron of the arts, who helped sponsor his escape from Hitler's Europe. This enabled my father to immigrate to America, where he developed a national reputation as a brilliant concert pianist, conductor, and music educator.

I came along in 1948 and had the opportunity to study music with my father from the age of 6. When I was 15 he also developed heart disease. My father was the kind of person who, when he encountered (then novel) health ideas, such as cutting down on salt, adopted them immediately without a second thought. Unfortunately, we had very little insight into heart disease in the 1960s, and he died of a heart attack in 1970 at the age of 58. I was 22 years old.

I remained painfully aware of this family health legacy, which hovered over me like a cloud on my future. At the age of 35, I was diagnosed with type 2 diabetes. I was prescribed conventional treatment with insulin, but this only made things worse by causing substantial weight gain, which in turn created an apparent need for more insulin. As is typical in someone with type 2 diabetes, I already had high insulin levels, so this was a very bad idea indeed.

A digression is in order here. Starting at the age of 8, I became a passionate fan of Tom Swift Jr. and read all of the available books in this popular series. In each volume, Tom Swift and his friends would get into a terrible jam (and usually the rest of the world along with them). Tom would retreat into his lab and think about how this seemingly impossible challenge could be overcome. Invariably, he would come up with a clever and ingenious idea that saved the day. The moral of these tales was simple: there is no problem so great that it cannot be overcome through the application of creative human thought. That simple paradigm has animated all my subsequent endeavors.

So, in the spirit of Tom Swift, I decided to take matters into my own hands, approaching the issue of diabetes from the perspective of the available scientific literature. I tried to engage my doctor in a discussion of the issues, with only limited success. While he talked to me to some extent, he clearly had little interest in doing so, and admittedly, I was unusually demanding. Finally, exasperated with my persistent questions, he said, "Look, I just don't have time for this; I have patients who are dying that I have to attend to."

Not one to be easily put off by attempts to appeal to my sense of guilt, I couldn't help but wonder whether any of these dying patients might have benefited from earlier explorations into ways to prevent disease. I decided to change doctors and, fortunately, found a physician, Steve Flier, M.D., with an open mind and, since he was just setting up a new practice, some time on his hands. My personal exploration, assisted through my dialogue with Steve, led to a set of health ideas that enabled me to get off insulin and control my diabetes simply through nutrition, exercise, and stress management. I lost more than 40 pounds and never felt better. I went on to articulate these ideas in *The 10% Solution for a Healthy Life* (Crown Books) in 1993, which became a best seller.

The ideas in the book kept me in good health and off diabetes medications for the next decade. Then, in 1999, I met a brilliant and open-minded fellow traveler, Terry Grossman, M.D., at a futurism conference organized by the Foresight Institute. Terry and I struck up a conversation and discovered a wide range of common interests, particularly in health and life extension. Our discourse quickly evolved into a close friendship and an intense collaboration on a wide range of health issues, with a sprinkling of other futurist issues thrown in as well, which has lasted and grown to this day. I've learned a great deal from Terry and hope that I've contributed ideas and insights to our partnership in return.

I can say that our relationship has been a uniquely fruitful intellectual journey of exploration and discovery. For one thing, I find the scientific issues underlying human health fascinating, particularly now that we are beginning to understand genetic and metabolic pathways in the language of information science. And for someone who has a keen interest in the 21st century and all of the marvels it promises to bring, I particularly appreciate the potential of this knowledge to enable us to actually live to see (and enjoy!) the remarkable century ahead.

This book represents the results of our collaboration, which in turn has built upon each of our decades of study of health issues. It is necessarily a work in progress and will always remain incomplete. My own work on technology trends indicates that human knowledge is growing exponentially and that the pace of progress is accelerating. Nowhere is this insight more evident than in the field of health. It seems that Terry and I discover at least one exciting new health insight each week (perhaps we are now down to one every six days!). It is fair to say that a number of our ideas have evolved significantly during the two-year period it has taken to produce this book.

I continue to devote a significant portion of my intellectual and physical energies to the pursuit of my personal health and health insights. I am able to use the same scientific method and information science skills in this endeavor, and I find the subject as intellectually satisfying as my other career as a pattern recognition scientist and inventor.

Along the way, I have encountered two unexpected conflicts. If you see someone standing precariously on a ledge, oblivious to the danger of a great fall, you feel a sense of obligation to inform that person of his or her unrealized plight. If the person is someone you care about, the urgency is even greater. I have not had to look very far to find many others who are desperately in need of the knowledge I have gained. Typical are adult male friends with elevated cholesterol, strong family histories of heart disease (or diagnoses of their own heart disease), and perhaps a few extra inches around the middle. Others include adult female friends with family histories (or their own diagnoses) of cancer.

Invariably, I get drawn into extended conversations on the topic of preserving health and well-being through nutrition and lifestyle. Often, these turn out to be longer conversations than either of us expected. To make the case, I feel compelled to go through a lot of the evidence. Then there are more subtle issues. *Why aren't the standard medical recommendations good enough? This is mostly genetics anyway, isn't it? What happened to moderation?*

If I make it through these issues, I'm inevitably asked to address the big question of palatability. *Sure, you'll live a long time, but who wants to live that way? If you eat this way, maybe it just seems like a long time!* I maintain that this can be an enjoyable, even liberating way to eat and live, but it takes a bit of explanation.

The second conflict has to do with proselytizing. Being a scientist and a trained skeptic, I was always turned off by people with singular agendas. People out to save my soul or even just my health and well-being were strongly suspect. I have felt very uncomfortable, therefore, in this role myself, telling other people how they should eat or live. Recognizing my own resistance to these types of messages, I also realize what I am up against in terms of getting people to take these ideas seriously.

Ultimately, I feel a responsibility to share my knowledge on these issues, but I also need to achieve a certain loving detachment when it comes to people choosing their own eating and living styles. This is not an easy balance to achieve. It is hard not to feel some pride if someone accepts our ideas and then shares with me their excitement at 30 lost pounds or 50 lost cholesterol points. If nothing else, such experiences demonstrate that I was successful in communicating my thoughts.

I have come to consider it my responsibility to empower people to set their own priorities and to make their own compromises. That's what I object to in the public health recommendations. They come precompromised, as if the American people were incapable of making their own decisions on these matters. As it has been said, "Lead me not into temptation, I can get there on my own." We can deliver a complete message, and readers can consider it on their own terms and in their own time. Any follow-up is up to you.

Even this limited goal of effective communication is a challenging one. We have all, by necessity, erected formidable barriers to messages on health. We could hardly survive if we allowed all of the thousands of messages that bombard us daily to get through. It's particularly difficult to penetrate the subtle yet common misconceptions, fears, and folklore—not to mention conflicting advice from experts—that underlie the public understanding (and misunderstanding) of nutrition and health. Food and its images are deeply interwoven in our rituals, fantasies, and relationships. While most people profess ignorance of nutrition and health, almost everyone maintains strongly held views on the subject and its relationship to the rest of our lives. Getting people's attention, let alone truly broadening someone's perspective, is not an easy task. But that is the challenge of any writer.

I have now influenced many people to adopt our ideas for improving their health, while Terry influences many patients through his longevity-oriented medical practice in Denver. The physical and medical results that friends, relatives, associates, and many others have achieved have been deeply gratifying.

For myself, I feel that the cloud that I so strongly perceived during my 20s and 30s has dissipated, and I look forward to a long and healthy life, indeed to seeing (and enjoying) the century ahead. It is too bad that I cannot go back and share this knowledge with my father. Unlike many people, he accepted health and nutritional advice readily and easily. Unfortunately, the knowledge was not available in time to help him. If it were, he could be alive today.

TERRY

I began my medical career some 24 years ago as a conventional physician. But after 15 years in practice, I found myself being drawn toward "integrative" medicine, "the field of health care that focuses on how biochemical individuality, metabolic balance, ecological context, genetic predisposition, lifestyle patterns, and other factors have the potential to strongly influence human physiology and the push-pull dynamics of health and disease."[1] As I began to study health from an integrative perspective, I became fascinated with the prospects for correcting imbalances in human physiology on a more individualized level. In 1994, I came to the realization that there were avenues available for me to help my patients in addition to conventional medical care. Focusing primarily on control of a patient's symptoms, which is the fundamental basis of what I had been taught in medical school, was no longer enough.

After completing medical school in Florida, I did my residency in Colorado and then moved to the mountains west of Denver. During the 15 years I practiced there, I worked as a young version of an old-fashioned general practitioner. I delivered babies at the local hospital, was the doctor for the local jail, and gave the annual talk about the "birds and the bees" to all the fifth-grade boys. I practiced medicine like a typical small-town GP and, by and large, felt satisfied with the care I was providing. I realized that most people I "treated" weren't really getting better, but they were receiving high-quality conventional care. Through prescription drugs, I was quite adept at bringing *symptoms* of high blood pressure, diabetes, or heart disease under "control." While this meant my patients' numbers were better—blood pressure or blood sugar was lower, or there was less chest pain—the underlying disease processes continued unchecked. This bothered me.

Life is a continual learning experience and, as a physician, I have come to regard pain as among the sternest but most effective of life's teachers. Thanks to a major knee injury suffered on a local ski slope some years ago, I found myself in the formal role of patient for the first time in my life, and I sought conventional medical care. I went to the best orthopedic surgeon I knew, a colleague I held in enormous respect.

After several modalities of conventional treatment still left me with constant residual pain in my knee, I did what I have since discovered many of my patients have been doing for years: I began to look at alternatives. Along with life's teachers are life's angels, who show up in most unexpected places at most unexpected times. My angel appeared in the form of a patient advocate of alternative medicine. Through his persistence, this individual forced me to open my eyes to an entirely new, to me, parallel world of medical alternatives.

In my family, medical doctors were treated with a certain amount of reverence, and conventional medical care was the *only* alternative. Yet, my patient advocate of alternative medicine—and angel—taught me that there was an entirely different paradigm of medical care available, completely separate from the world of prescription medications and surgery in which I had lived for so many years. I learned that vitamins and herbs could actually be used to *treat* diseases. He convinced me to try to treat my painful knee condition with a specific herbal concoction derived from the inner bark of a certain type of pine tree that grew only in the south of France.

Feeling like something of a traitor—perhaps a bit like Adam and Eve nibbling at the prohibited fruit—I squeamishly began to take pine bark capsules. It took more than three months but, much to my surprise and gratification, the pain in my knee that I had been experiencing for over a year and a half went completely away. Being a scientist, I decided to perform an experiment to see if my improvement was really the result of the herbal concoction, a placebo effect, or simply a coincidence. I quit taking it. My knee pain returned with a vengeance. I restarted the pine bark extract and, within a few weeks, the pain went away. I repeated the sequence once again: I quit taking the nutritional extract and the pain returned. I restarted it and the pain resolved. As a physician, I am well aware of placebo effects, but these generally go away after a limited period of time. I continued taking the extract, and after a few years I noticed that the pain was gone whether I took it or not. I suspected this was probably just the natural course of the healing process; nevertheless, the nutritional extract seemed to have given me pain relief earlier on, and my interest in alternative medicine was piqued.

I undertook a serious study of integrative medicine with an emphasis on nutritional medicine. I began to learn how to treat diseases with vitamins and other nutrients rather than, or in addition to, prescription drugs. And the more I learned, the more I wanted to know. I went to numerous complementary medicine conferences and read everything I could find about nutritional medicine. There was so much to learn, I felt like I was back in medical school again.

As my knowledge and understanding increased, I slowly began to offer my patients the option of continuing with the conventional treatments they had been receiving from me (in most cases, prescription drugs) or the opportunity to try treatments involving dietary changes or nutritional supplements, either in place of or in addition to conventional care. I was surprised to find that the overwhelming majority of my patients wished to take advantage of these options.[2]

My patients did far better on combined care than they had on prescription medications alone. Over the past 10 years, I have treated thousands of patients who had serious chronic illnesses with nutritional protocols that I have learned and modified for my practice. For example, I see a large number of people with coronary heart disease. Most of these patients come to me looking for an alternative to some type of heart surgery that has been recommended to them. For some, I concur with their cardiologists and recommend immediate surgical intervention because the disease is too far advanced. Yet, for the significant majority, I find that the nutritional and lifestyle program I recommend for heart disease, involving diet, exercise, aggressive supplementation, detoxification, and stress management—as well as prescription drugs when needed—staves off heart surgery. At the same time, I am able to document quantifiable improvements in these patients' conditions, such as going from abnormal to normal cardiac stress tests, eliminating angina pain, and improving exercise endurance.

I derive particular satisfaction from successfully treating people with diseases for which conventional medical practice has little to offer. Age-related macular degeneration (AMD) is the leading cause of vision loss in older individuals in this country, yet presently there are no prescription drugs or surgical procedures that can help prevent the inexorable decline toward blindness. To their credit, conventional ophthalmologists have recently begun to recommend multivitamin/mineral supplementation for their AMD patients based on the AREDS (Age-Related Eye Disease Study) sponsored by the National Institutes of Health.[3]

Yet, like Ray, I find the conventional supplement recommendations precompromised or too watered down, particularly since I have a family history of macular degeneration, and I don't feel that slowing down the rate of visual decline is enough. I want my patients' vision to improve. This requires a more aggressive nutritional approach: combining dietary strategies with much larger doses of vitamins and minerals and working to correct digestive disturbances that inhibit absorption of nutrients. Using this approach, it is possible for patients to stabilize and even improve their vision.[4] This is rarely seen with one- or two-pills-a-day supplementation alone.

Among the greatest devastations for young parents is learning that their child suffers from one of the autistic spectrum disorders. Yet I get enormous satisfaction treating children diagnosed with such diseases. I have nothing but admiration for the dedicated pediatricians, allied health personnel, and special education teachers who have devoted their lives to working with these children. I am saddened, however, by the ineffectiveness of their approaches, which rarely alter the progression of these disease processes.

Our program involves a special diet (avoidance of wheat and dairy products), aggressive nutritional supplementation, correction of digestive disturbances, and detoxification strategies.[5] The majority of children we treat under the age of 6 experience some degree of improvement on this regimen.

There is a long list of ailments for which conventional medicine alone provides limited benefit: chronic degenerative neurological diseases such as Parkinson's disease and multiple sclerosis; digestive disturbances, including irritable bowel syndrome, colitis, and Crohn's disease; and multisystem diseases such as fibromyalgia and chronic fatigue syndrome. For these, an integrated approach, using complementary therapies, is of considerable benefit. Tens of millions of American adults suffer from type 2 diabetes, obesity, high blood pressure, and elevated cholesterol. In the majority of cases, it has been my experience that where our program is followed strictly, the prescription drugs used to treat these conditions can be either reduced or eliminated entirely.

As the years passed and I gained more experience with nutritional medicine, I decided to write a book to share what I had learned with people outside of my practice. With the assistance of several physician colleagues and friends, I completed *The Baby Boomers' Guide to Living Forever* in April 2000.

In the course of researching the topic of nanotechnology for this book, I met Ray at the 1999 Foresight Institute Conference in Palo Alto, California. He was there as one of the nation's foremost futurists. Overhearing Ray discuss his interest in nutritional supplementation and other life extension therapies, I struck up a conversation. I asked him to look over the manuscript of my book and write a "testimonial" paragraph for the back cover, which he kindly agreed to do. A few months later, he flew from his home in Boston to my clinic in Denver to undergo one of the comprehensive health assessments and longevity evaluations we offer.

My nutritional medical practice in Denver has its share of celebrity patients, and Ray Kurzweil is one of them. Ray is unique in that I devote more of my time attending to his health concerns than any dozen of my patients, celebrity or otherwise, put together. But, then again, Ray is quick to admit that he is unusually demanding. I am not surprised that Ray became frustrated with a previous physician who preferred to spend his time on patients who "were dying."

For my part, I have no regrets whatsoever about the amount of time I spend working with Ray on his personal health issues, or the fact that I have to defend every single opinion or suggestion I offer to him. Ray is another of the angels who have entered my life to guide me in the right direction. I feel a special sense of mission in helping him remain alive and well for many years into the future, as this unusually creative and gifted individual, who has already brought so many wonderful insights and inventions to the world, has much yet to share. He has also helped me to refine my focus and leave no loose ends in any medical endeavor.

Moreover, a number of other benefits accrued from the process of working with Ray. In the course of refining his personal health program, he and I began to explore numerous health-related topics, including diet, nutritional supplementation, exercise, detoxification therapies, hormone replacement, and even protection from NBC (nuclear, biological, or chemical) terrorism. We've both learned a great deal from our intense collaboration on health issues. Our dozens of e-mails back and forth turned into hundreds and now number in the tens of thousands. So much information was passed between us that we decided to organize it as the basis for a book.

As members of the baby boomer generation, Ray and I have more than a casual interest in our program. Since we are both now in our mid-50s, demographic analysis would ordinarily suggest that we each have perhaps 25 years left, with gradually declining vitality and health. By following the advice presented in this book, and with some help from the accelerating technologies that Ray speaks about, we hope to be not only alive but vital and "young" a quarter century from now—right at the time corridor when the prospects for truly radical life extension are likely to occur. It is our fervent hope (bolstered by extensive research) that by following the suggestions offered in *Fantastic Voyage*, we and our readers will be able to significantly increase our chances of being alive when extreme longevity becomes commonplace. To that lofty goal, we raise a toast—of freshly squeezed organic vegetable juice—in the hopes that we can join together with our readers to celebrate the 22nd century.

4

FOOD AND WATER

"Why does man kill? He kills for food. And not only for food. Frequently, there must be a beverage."

—Woody Allen

What is food? The answer depends on whom you ask. Ask an evolutionary biologist, and he or she will tell you that food is the primary determinant of the survival of virtually any species.[1] Other resources are also needed, but finding the next meal (while not becoming a predator's next meal) represents the dominant and most crucial activity of most organisms. No species can survive if its food source is permanently disrupted. The advantage of being human is that we can use our intelligence and our opposable thumbs to manipulate our environment and create new food sources when necessary.

Speak to an anthropologist, and you will discover that up until recently, humans spent most of their time hunting, foraging, raising, cultivating, and preparing food.[2] Prior to the 20th century, families spent many hours toiling to create each meal. As recently as 1900, a third of the employment in the United States was on farms, growing food, with many more employed to distribute and prepare food.[3]

Speak to a nutritionally aware health practitioner, and you will learn that the three leading causes of death—heart disease, cancer, and stroke—are caused predominantly by poor nutrition.[4] Although genetic factors predispose you to develop these diseases at different rates, you can either accelerate the degenerative disease processes or reverse them, primarily through your food choices. Obesity, hypertension, and type 2 diabetes are also determined mostly by your nutritional strategy.

A chemist will tell you food is rather simple. It consists mainly of just four types of atoms: carbon, nitrogen, oxygen, and hydrogen, with some traces of minerals.[5] Most of it is water. However, a biochemist will also point out the vast complexity of the molecules made from these elements, mirroring the intricacies of our bodies—which, of course, our food becomes.

A chef will describe the intricacies of food from a more, well, human perspective. Indeed, we are all culinary connoisseurs. We attach great emotional and cultural significance to the food we eat. We comfort ourselves and one another with food. We celebrate and share life's special moments with food. Food figures notably in the stories told in our religious texts and plays a prominent role in social and ethnic identity. So selecting the best nutritional program from a health perspective is more complicated than simply picking the optimal blend of fuels for our biological machinery.

In this book, we will address food from the perspective of health, keeping in mind the important and pervasive role that it plays in our lives. Because of the strong cultural and social influences on individual eating patterns, and the vast array of personal likes and dislikes, we avoid providing specific meal plans. Instead, we hope to provide a deeper understanding of the role food can play in creating conditions of disease or health and share principles you can apply to your own situation.

Let's begin by looking at food's most common constituent: water.

WATER AND HEALTH

Many of the ways our diets have evolved from that of our primitive forebears are not healthy. For example, we were not evolved to eat large slabs of meat rich in saturated fat. We were not evolved to eat large quantities of refined grains, which quickly turn into sugar in our blood. We were not evolved to consume extremely acidic drinks such as colas. These and many other modern food choices underlie the modern-day epidemic of degenerative disease.

However, we cannot simply assume that biological evolution is on our side, at least not after child-rearing age. From the perspective of biological evolution, older generations are just in the way, using up limited material resources that could be used by the younger, more vital members of society. Biological evolution took place in an era of scarcity, so limited life spans were favored. But we currently live in an era in which the cutting edge of evolution is technological rather than biological. This is a period of increasing abundance, even if there are regions of the world that do not adequately share in these material resources. Today we have the opportunity to override our evolutionary heritage and allow all generations to contribute to our intellectual resources.

What Is Water?

Water is the most abundant substance found on our planet, in our bodies, and in our food, making up 70 to 90 percent of organic matter. The liquids, especially the types of water, we consume also have a profound impact on a full spectrum of health issues. Water is far more complex than common wisdom suggests.

As every grade-school child knows, water is composed of molecules containing two atoms of hydrogen and one atom of oxygen: H_2O. In a liquid state, the two hydrogen atoms make a 104.5° angle with the oxygen atom, which increases to 109.5° when water freezes. That's why water molecules are more spread out in the form of ice, giving it a lower density than liquid water—this is why ice floats.

Although the overall water molecule is electrically neutral, the location of the electrons makes a difference. The side of the molecule with the hydrogen atoms is slightly positive in electrical charge, whereas the oxygen side is slightly negative. So water molecules combine with one another in small groups to assume, typically, pentagonal or hexagonal shapes.[6] These multimolecule structures can change back and forth between hexagonal and pentagonal configurations 100 billion times a second. At room temperature, only about 3 percent of the clusters are hexagonal, but this increases to 100 percent as the water gets colder. This is why most snowflakes are hexagonal.

These three-dimensional electrical properties of water are quite powerful and can break apart the strong chemical bonds of other compounds. Consider what happens when you put salt into water. Salt is quite stable when dry but is quickly torn apart into ions (atoms with an electric charge)—sodium and chlorine—when placed in water. The negatively charged oxygen side of the water molecules attracts positively charged sodium ions (Na+), while the positively charged hydrogen side attracts the negatively charged chlorine ions (Cl–). In the dry form of salt, the sodium and chlorine atoms are tightly bound together, but these bonds are easily broken by the electrical charge of the water molecules. That's why water is "the universal solvent" and is involved in most of the biochemical pathways in our bodies. The chemistry of life on our planet is mostly concerned with water.

The elaborate structures formed by water molecules as a result of its electrical field create a form of memory that has been demonstrated by magnetic resonance imaging (MRI) machines. On its surface, water assumes a stable configuration that results in the phenomenon of surface tension. One Korean scientist describes three organized layers of water molecules, each with different properties, around each protein molecule.[7]

There are three vital health issues concerning water: acid/alkaline balance, impurities, and infrastructure. Acid/alkaline balance is based on amount of ionization. Ionization means molecules lose or gain electrons, so they acquire an overall electric charge. When a water molecule becomes ionized, it is split into two parts: a positively charged hydrogen ion (H+) and a negatively charged hydroxide ion (OH–). These ions can then combine with other substances dissolved in water, such as minerals, to form chemical reactions. Ionization is crucial to most of the chemical reactions in our body.

If hydrogen and hydroxide ions are equal in number, the water is considered neutral. At room temperature, neutral water has one out of 10 million (10^{-7}) of its molecules in an ionized state—broken apart into hydrogen ions—and the same number as hydroxide ions. Such water has a pH (proportion of hydrogen) of 7 (the negative exponent of 10 for the hydrogen ions). If we add an acidic mineral—for example, sulfur, chlorine, or phosphoric acid—to water, the acid grabs electrons from the hydrogen atoms, creating a larger number of hydrogen ions (H+). If the proportion of hydrogen ions increases 10-fold, this acidic water will have just one out of one million (10^{6}) of its molecules as hydrogen ions, or a pH of 6. So pH is measured on a logarithmic scale: increasing or decreasing the pH by just 1 corresponds to multiplying or dividing the number of hydrogen ions by 10.

THE IMPORTANCE OF BEING ALKALINE[8]

To understand why the alkalinity or acidity of the water-based liquids is important to health, we need to understand how the body controls the ionization levels of its fluids. Different pH levels support different types of chemistry, so it's essential that body fluids be maintained within very narrow acid/alkaline limits.[9] Your health is extremely sensitive to the slightest change in the pH level of your body's vital fluids. Stomach fluid, for example, is extremely acidic, with a pH of 1.5 (pH less than 7 is acid, more than 7 is alkaline). Pancreatic fluid, on the other hand, is quite alkaline, with a pH of 8.8. The pH inside of our cells ranges from 6.8 to 7.1. The most important balance of all is maintained in our blood, where the pH is very tightly controlled between 7.35 and 7.45.

Your body will act to neutralize acidic drinks such as colas and coffee with alkaline blood buffers, which are then unavailable to neutralize other acidic waste products continually produced by the body, including organic by-products of digestion such as acetic acid,[10] lactic acid,[11] carbonic acid,[12] uric acid,[13] and fatty acids.[14] There are also inorganic by-products created or

Soft Drinks Are Hard on the Body

Most soft drinks, particularly colas, are extremely acidic. Colas contain high levels of phosphoric acid,[15] a powerful acid capable of poisoning you if not quickly neutralized. Cola (regular or diet) has an extraordinarily low (that is, acidic) pH, around 2.5. Because pH measurement is logarithmic (a decrease of 1 in pH means multiplying the acidity by 10), a pH of 2.5 means that it would take 3,200 glasses of alkaline water with a pH of 8 (or 32 glasses with a pH of 10) to neutralize the acid in just one glass of cola. If the body did nothing to counteract it, a single glass of cola would change the pH of your blood to 4.6, killing you instantly.

To prevent acidic poisoning from cola (or other acid) consumption, the body uses two strategies. One is to use alkaline blood buffers (for example, sodium bicarbonate[16] and sodium phosphate[17]) to buffer (neutralize) the acid.[18] The other strategy is to convert these volatile liquid acids into less-reactive solid acids. However, there were no colas thousands of years ago, so our bodies did not evolve to deal effectively with the onslaught of acids that many people consume today, and there are problems resulting from the body's detoxification strategies.

found in food, such as sulfuric acid[19] and phosphoric acid.[20] When the body's limited supply of alkaline buffers is defeated, these toxic acidic waste products accumulate in the body, causing significant health damage.[21]

The body uses calcium to convert the poisonous liquid phosphoric acid in colas into the more stable solid phosphates, for example.[22] But these phosphates may form into calcified kidney stones, or calcium deposits (which can also result from a urinary infection, inherited metabolic disorders, and other causes[23]). Many people erroneously think kidney stones are caused by excessive calcium. But the real culprit may be the high level of phosphoric acid, which happens to be a primary ingredient of colas. Anyone with a concern about kidney stones should avoid colas.

Consuming acidic foods such as soft drinks may also create an ideal environment for cancer to form. Animal cells survive best in an alkaline environment with a blood pH of 7.35 to 7.45. Plant cells are the opposite; they prefer an acidic environment. As our bodies become increasingly acidic, some cells adapt through an internal evolutionary process and become more like plant cells. These abnormal plantlike cells have a high tendency to become cancer cells, which thrive in an acidic environment.[24] So an important strategy for preventing or treating cancer is to maintain an alkaline environment in the body.

Routine consumption of soft drinks containing phosphoric acid (that is, colas) is a risk factor for bone loss.[25] Consumption of alkalinizing mineral water helps retain bone health and improve digestive functions.[26] A comprehensive review comparing alkalinizing diets to acidic diets reported in *The American Journal of Clinical Nutrition* concluded that alkalinizing diets improve bone density, nitrogen balance, and serum growth hormone concentrations, whereas the low-grade acidosis resulting from acidic diets contributes to bone loss, osteoporosis, and loss of muscle.[27]

Not all acidic foods increase the acidity inside our bodies. Orange juice, for example, has an extremely acidic pH of 3.5 because of its citric acid content, but the citric acid is burned away during digestion. Orange juice also contains potassium and magnesium, which interact with water to create alkaline ions. Thus, the overall effect of drinking orange juice is to increase al-

Increasing Your Alkalinity

There are two strategies that you can use to restore your body's alkaline reserves, which are needed for detoxification and destroying oxygen free radicals:

1. Avoid indigestible acids. These are found in soft drinks, particularly colas. Coffee also contributes to creating overly acidic conditions, so its consumption should be limited. Our advice: Drink green tea instead.

2. Drink alkaline water. Metabolic processes create acidic waste products, so it is necessary to restore your alkaline reserves. An effective way to do this is by drinking alkaline water,[28] produced with an alkaline water machine (see Fantastic-Voyage.net for specific product recommendations). This device, which looks like a coffee percolator, contains an electrical ionization system to split water into its acidic and alkaline portions. The alkaline water should be used for drinking and food preparation. The water's alkalinity can be adjusted, and you should increase the level gradually to a pH between 9.5 and 10, which will provide a powerful detoxification treatment. We recommend that you drink 8 to 10 glasses per day of this alkaline water. It is one of the simplest and most powerful things you can do to combat a wide range of disease processes. It is interesting to note that in Japan, professional sports teams drink alkaline water to improve their performance.

The acidic water produced by the alkaline water machine from a separate outlet need not go to waste—it's perfect for cleansing your skin. It's also ideal to use for watering plants, which thrive in an acidic environment.

Certain teas have alkalinizing effects on the body too; see Fantastic-Voyage.net for specific recommendations.

kalinity, despite its acidic content.[29] The inability of your body to fully detoxify underlies many disease processes, including heart disease and cancer. Oxygen free radicals (molecules unpaired with electrons) are highly reactive substances that are deficient in electrons and can cause enormous damage to the body if not quickly neutralized. Adequate alkaline reserves offer a first line of defense against these oxygen free radicals by providing free electrons, which neutralize the radicals and prevent them from damaging healthy cells.[30]

OTHER HEALTH ISSUES WITH WATER

Tap water can contain impurities, including inorganic poisons and pathogens such as bacteria, viruses, and fungi. The water alkalinizing devices we recommend also include filters that remove such impurities, along with an ultraviolet light system to destroy pathogens. If alkalinized water is not available, we recommend filtered water and bottled water from a reliable source as better choices than most municipal tap water. (We'll discuss the issue of removing toxins from water and other environmental sources in chapter 14 on detoxification.)

Another issue concerns the infrastructure of water. Magnetic resonance imaging reveals that most tap water is organized into microclusters of about 12 water molecules each. In alkalinized water, the microclusters are reduced in size to only six molecules per cluster. This enhances the permeability, solubility, and absorption of the water, thereby boosting its detoxification effects. Nutrients, including water-soluble vitamins such as vitamins B and C, will be absorbed more fully, and medications taken with alkaline water may be more effective.

Finally, most of us don't consume enough water to help dispose of all the waste products our metabolic and digestive processes create. Simply drinking more water can boost detoxification. We recommend drinking one-half fluid ounce of water per pound of body weight daily; that would be 70 fluid ounces for a 140-pound person, or about nine 8-ounce glasses of water each day. Of course, soft drinks and coffee don't count. It takes 32 glasses of pH 10 alkaline water to neutralize one glass of pH 2.5 cola. We recommend avoiding these acidic drinks and consuming alkaline water with a pH between 9.5 and 10.

Dartmouth researcher Heinz Valtin, M.D., has criticized this liquid quota as an urban legend.[31] Valtin's opinion, however, is based on avoiding dehydration, whereas our recommendation for drinking alkaline water is to assist detoxification.

5

CARBOHYDRATES AND THE GLYCEMIC LOAD

"I was eating bad stuff. Lots of sugar and carbs, junk food all the time. It makes you very irritated."

—singer Avril Lavigne, discussing her angry lyrics

Ray was diagnosed with type 2 diabetes more than 20 years ago but today has no symptoms or complications from this disease. Terry sees many patients with metabolic syndrome (characterized by many adverse effects, including excessive insulin levels, when more than a small amount of carbohydrate foods are eaten) and type 2 diabetes. As a result, we are frequently asked for guidance by friends, associates, and patients who have been diagnosed with either the metabolic syndrome (TMS) or type 2 diabetes, or whom we suspect have these conditions (in which case we encourage them to get a diagnosis). Typically, we are disturbed to discover that these patients either receive no nutritional guidance or are given the wrong advice—for example, a diet that does not include significant reductions in high-glycemic-load carbohydrates (the type that raise blood sugar quickly). Invariably, these patients' doctors have not tested their insulin levels and often prescribe medications such as Glyburide that stimulate higher insulin levels, when their insulin levels may already be too high.

Our advice to people with TMS or type 2 diabetes is to adopt Ray & Terry's Longevity Program at the lower-carbohydrate level (less than one-sixth of your total calories) and avoid virtually all high-glycemic-load carbohydrates (sugar and starchy foods). We encourage them to have their insulin levels tested. With TMS or early stage type 2 diabetes, it is not unusual for insulin levels to be substantially higher than normal because the cells are resistant to utilizing insulin. If these patients are on a medication that stimulates insulin production by the pancreas, we encourage them to explore replacing this with a medication such as metformin, which discourages the liver from producing further glucose, and/or Precose, a starch blocker that prevents digestion of a portion of consumed carbohydrates. We also encourage them to adopt our nutritional guidelines described in chapters 4 through 8 in this book, to attain their optimal weight, to exercise (including both aerobic and strength training), and to adopt the supplement and other recommendations in this chapter.

People report back that they readily lose weight on a low-carbohydrate, low-glycemic-load diet, and that they have attained normal glucose levels without the use of harsh insulin-stimulating medications. They find they have relatively few, if any, side effects from the supplements we've recommended.

THE PRINCIPAL CYCLE OF LIFE

Most human societies revere carbohydrate-rich foods. The Bible repeatedly describes bread as the "staff of life," and the story of God providing manna (bread) from heaven to the fleeing Israelites is well known. Consider the importance of rice to the Japanese, pasta to the Italians, and hot dog buns and pretzels to American sports fans. A visit to a contemporary mall or airport reveals an unending array of sugar- and starch-laden products.

Modest portions of grains in their natural form can be a valuable part of a healthy diet, but the human digestive system did not evolve to consume the vast quantities of sugars and refined starches that make up the modern Western diet. Over time, this dietary imbalance can cause the body's systems for controlling blood glucose (sugar) to break down, resulting in TMS or type 2 diabetes, and can accelerate the progression of both heart disease and cancer. Also, consuming these glucose-producing foods is actually habit-forming. Breaking this habit is the key to successfully attaining and maintaining your optimal weight.

To gain insight into the function—and dysfunction—of our digestion of carbohydrates, we must understand what they are. The name *carbohydrate* suggests "hydrated" or "watered" carbon, and the chemical formula for most carbohydrates—$Cx(H_2O)x$—demonstrates that they consist of combinations of water molecules and carbon atoms. Simple sugar, for example, is $C_6(H_2O)_6$. Glucose (also called dextrose), fructose (found in fruit), and galactose all have this formula, but arrange their atoms differently.[1] Glucose is vital to life on Earth; most life-forms get their energy directly or indirectly from units of glucose.[2]

Table sugar is a disaccharide, meaning a double sugar: each molecule consists of one unit of glucose and one of fructose. Table sugar is quickly broken down in the body into its constituent monosaccharides, or simple sugars, such as glucose.

The key to the power of carbohydrates for energy storage is their ability to form large, complex chains of glucose and fructose units. Polysaccharides, meaning "many sugars," provide much of the energy, as well as the structure, of living organisms. Hundreds of types of polysaccharides have been identified. They consist of up to several thousand linked monosaccharide units. The most common polysaccharide, cellulose, consists of a large number of glucose units. It provides the structure for most of the plant world. In the human body, another polysaccharide, glycogen, stored by the liver and the muscles, consists of branching chains of glucose and stores most of the short-term carbohydrate energy.

The immediate energy source for our cells that circulates in the blood is the simple sugar glucose. This fuel can be quickly produced from either dietary carbohydrates and proteins or by the breakdown of glycogen stores. The principal energy storage in the body is in the form of fat, which is a long-term reserve made available to cells at a much slower rate. Interestingly, however, fat cannot be broken down into glucose (blood sugar). Protein, which is our body's principal building block for creating the structure of our cells, is also made indirectly from carbohydrates. Ruminant animals, such as cows and sheep, convert the carbohydrates in grass into protein that we eat as meat.

Carbohydrates are vital to the primary energy cycle in the biological world. In a process called photosynthesis, plants combine carbon from the carbon dioxide in the air, water from the air and ground, and energy from the sun to produce stored carbohydrate energy as well as the waste product oxygen. In a balanced and opposing process, animals breathe the oxygen, eat and digest the carbohydrates from plants, and give off carbon dioxide.

CARBOHYDRATES IN OUR DIET

Carbohydrates have powerful effects on the body. The proportion of your diet that is carbohydrates and, more important, the type of carbohydrates you consume, have vital effects on your health. Of the three sources of calories—carbohydrates, fat, and protein—carbohydrates are the only one *not* necessary for survival.[3] Without certain essential fats and the right protein building blocks, you could not live. But you don't need carbohydrates to build the structure of your cells; you could get all the energy you need from fat and protein. However, we do recommend a healthy balance of carbohydrates because certain carbohydrate foods, such as vegetables, are rich in vitamins, minerals, and other nutrients. But a principal problem with our modern diet is its dependence on a large quantity of the wrong kind of carbohydrates.

BRIDGE TWO

PERFECTING THE STARCH BLOCKER

There are two limitations to starch blockers (medications that block the action of amylase and thereby prevent digestion of carbohydrates). First, they are unable to prevent the starch breakdown caused by amylase in saliva, so only about half of the conversion of starch into glucose is blocked. Second, the undigested starch ends up in the large intestine, where it functions as fiber and becomes food for colonic bacteria. Up to a point, fiber is beneficial. But excessive fiber and undigested starch can cause digestive upset and excess gas.

What if there were a medication that could block carbohydrate from food that has already been broken down into glucose and absorbed into the bloodstream but before it ends up being stored as fat in the fat cells? Suppose it would also block sugar, in both sucrose (table sugar) and fructose (found in fruit) forms, something that starch blockers are unable to do, and it would not cause digestive problems. That is the goal of a GlaxoSmithKline (GSK) drug now in phase I FDA clinical trials. Called "869682," a name obviously not approved by their marketing department, the drug will benefit people with type 2 diabetes, but it also has obvious dramatic appeal as a weight-loss drug. GSK's R&D chairman, Tadataka Yamada, calls it "the chemical Atkins diet." What this means is that you would get the benefit of a low-carbohydrate diet without necessarily reducing carbohydrate consumption. If successful, it will also reduce the effective calories of what you eat. Understandably, GSK has identified it as a potential blockbuster.

A drug from Eli Lilly called Exenatide, which has already completed phase 3 FDA clinical trials, also reduces blood glucose levels after digestion of carbohydrates. It has the added benefit of suppressing hunger, thereby attacking weight loss from two directions. Other drugs that destroy glucose in the blood before it gets converted to fat in the fat cells are also in development.

To understand the proper role of carbohydrates in your diet, you need to understand a few things about how they are digested. Simple sugars are directly absorbed by the epithelial (lining) cells in the small intestine. Disaccharides such as sucrose, table sugar, or lactose, milk sugar, are also absorbed directly by the epithelial cells, but these double-sugar molecules require certain enzymes so they can be broken down into their constituent monosaccharides. The enzyme sucrase, for example, breaks down table sugar, sucrose, into glucose and fructose. Lactase converts the disaccharide lactose (the sugar in dairy products) into glucose and galactose. More than half the adult human population of the world has a genetic deficiency of lactase, which allows lactose to arrive in the colon undigested, where it ferments and causes gastrointestinal upset.[4]

Starchy foods, or polysaccharides, cannot be directly absorbed by the epithelial (lining) cells of the small intestine, but must first be broken down into monosaccharides and disaccharides. This is accomplished by the enzyme amylase, secreted by the salivary glands and pancreas. The most common polysaccharide is amylose, which consists of a long string of glucose units. As its name suggests, amylase is the enzyme designed to break down amylose. The carbohydrates found in refined grains and starchy vegetables such as potatoes are mostly amylose and are digested very quickly. There is not much difference between eating these quickly digested starches and eating simple sugar in terms of quickly boosting the level of glucose in the blood.[5]

Rather than merely forming long, beadlike strands of simple sugars, polysaccharides can also be formed from more complex arrangements of monosaccharide units that include many cross-links between molecules. Fiber is an example of an extensively cross-linked polysaccharide.

Fiber Facts

Fiber comes in two forms: soluble (dissolves in water) and insoluble (doesn't dissolve in water). Oats, as in oatmeal, are an example of soluble fiber. Celery is mostly insoluble fiber, which is indigestible. Although fiber is classified as a carbohydrate and, by law, is required to be included in the total amount of carbohydrates listed on food labels, it has no digestible calories.[6] So the low-carbohydrate tortilla you see in the supermarket that contains 11 grams of carbohydrates, of which 8 grams are fiber, actually has only 3 grams of digestible carbohydrates, and only digestible carbohydrates can raise blood sugar and serve as fuel (or fat) for the body. Fiber calories just pass through the digestive tract pretty much as they started—undigested.

High-fiber foods are also beneficial because the soluble portions are digested slowly. Legumes such as lentils and beans combine digestible and indigestible polysaccharides in complex formations, causing a slowdown of carbohydrate digestion that is key to preventing harmful surges of insulin, as we will discuss in chapter 9, "The Problem with Sugar (and Insulin)." Insoluble fiber also helps keep food moving through the intestines, which assists with normal bowel functioning.

THE GLYCEMIC INDEX

The term *glycemic index* refers to how fast a food is converted into glucose in the blood. Simple sugars become glucose almost instantly, so they have a very high glycemic index, or G-I. Starches consisting mostly of amylose, such as potatoes, rice, and any foods made from refined flour, such as bread, bagels, pasta, and pastries, are digested almost as fast, so they also have a high G-I. Legumes like beans and lentils, because of their high fiber content and complex arrangement of digestible and indigestible carbohydrates, have a relatively low G-I.

Interestingly, fructose, or fruit sugar, even though it consists of two monosaccharide units just like sucrose or table sugar, has a much lower glycemic index than many other sugars because it is absorbed more slowly by the epithelial cells of the small intestine.[7] So fruit, even though sweet in taste, has a lower G-I than sweets made from sucrose, such as pastries. Fruit also contains fiber and many valuable nutrients, such as vitamins and phytochemicals. This does not apply to high-fructose corn syrup, commonly used to sweeten soft drinks. High-fructose corn syrup has a high G-I.

Proteins can also be converted in the body into glucose, but this process requires many more steps and generally takes much longer, which is reflected in a relatively low glycemic index for most protein foods.

When you eat a meal containing many high-glycemic-index carbohydrates, the level of glucose in your blood rises quickly and the pancreas produces a large spike of insulin, which helps move glucose from the bloodstream into the cells. This action keeps blood glucose levels under control. But a major problem is that these temporary high levels of insulin often overshoot the mark, driving blood glucose levels down too low, which leads to a craving for more high-G-I carbohydrates—a vicious cycle.[8]

Over time, the continual abuse of this cycle—eating large amounts of carbs, leading to high blood glucose levels, leading to quick bursts of insulin, leading to low levels of blood glucose, leading to consuming more carbohydrates, etc.—results in a lower level of sensitivity of the body's cells to insulin. This insulin resistance is a principal cause of the metabolic syndrome (also called syndrome X), a major health problem that accelerates atherosclerosis and other aging processes. Insulin resistance can also lead to type 2 diabetes,[9] in which case it doesn't matter how much insulin the body makes—the blood sugar is still too high. Other problems caused by excess levels of glucose in the blood:

- The rapid conversion into triglycerides or fats accelerates atherosclerosis and other degenerative processes (see chapter 6).

- The immune system is inhibited.

- Adrenaline production increases by up to four times. This chronic activation of the fight-or-flight stress reaction (see chapter 23) worsens the damage to the body, including increased levels of cortisone, which further inhibits the immune system, and cholesterol.
- Sugar promotes growth of a broad variety of pathological cells, including candida (yeast), fungal infections, and cancer.
- Sugar and vitamin C compete for the same transport system, so excess glucose in the blood interferes with vitamin C's vital roles in combating infection and building body tissues.
- Sugar causes protein molecules to become cross-linked to one another, a primary cause of the aging process.
- The lack of insoluble fiber in simple starches causes food to pass too slowly through the intestines, which encourages gas, bloating, and formation of toxins. This effect may contribute to colon cancer.[10]

Kicking Your Carb Cravings

Recent research has shown that temporary high levels of glucose in the blood are addictive. Bartley Hoebel, a researcher at Princeton University, fed high levels of a sugar solution to rats. Over a one-month period, the rats became dependent on their sugary diet.[11] Blocking the opiate receptors in their brains through drugs led to classical withdrawal symptoms typical of addiction to opiates. Gliadinomorphin, a wheat protein, and caseinomorphin, a protein found in dairy products, can also be highly addictive for some people. This is why people have a tendency to crave wheat and dairy products.

If you eliminate high-glycemic-index foods from your diet altogether, your carbohydrate cravings will disappear so it becomes much easier to control your appetite.[12] Because these foods are addictive, you may experience a brief withdrawal period, but this lasts only one to two weeks. Completely cutting out high-glycemic-load foods is by far the most important step you can take to attain and maintain your optimal weight. Simply reducing them doesn't work because the addiction is never broken.

Eliminating high-glycemic-index foods offers other health benefits as well. The 2004 Women's Health Study showed that "dietary glycemic index was statistically significantly associated with an increased risk of colorectal cancer."[13] In this UCLA study, women who ate a diet with the highest glycemic index had almost three times the risk of colorectal cancer as women who ate the least.

HOW TO CHOOSE HEALTHY CARBOHYDRATES

The more quickly sugar enters your bloodstream, the more your insulin rises. By making proper food choices, you can largely control your insulin levels. But choosing the right food isn't always simple. Some foods that don't taste sweet at all, such as white potatoes, raise your blood sugar and insulin levels dramatically, while other foods that taste sweeter, like sweet potatoes, have less effect.

The main way to determine how fast a given food elevates your blood sugar and thus insulin is to know the glycemic index of the food (see Table 5-1).[14] A food with a high G-I will raise blood sugar and insulin levels more quickly than one with a low G-I, such as peanuts. So you need to know the actual G-I—the laboratory measurement of how quickly the food raises blood sugar; you can't estimate by how sweet it tastes or how much sugar and starch it contains.

But even knowing a food's glycemic index isn't enough. If you look at the table below, the conventional wisdom would be to eat fewer high G-I foods such as green peas while emphasizing lower G-I foods such as white rice. Green peas, with a relatively high G-I of 75, appear comparable to white bread (70) and not far from table sugar (100). White rice, on the other hand, has a lower G-I of 64. Looking at these numbers alone would suggest that white rice is a better food choice than peas. But is it?

A better idea is to look beyond the food's glycemic index and examine what is called its *glycemic load* (GL). This is the number of grams of carbohydrate in the food multiplied by its G-I. The GL provides a rough measurement of how much insulin your body is going to need to digest a given food, since the amount of insulin generated is based on both the *amount* of carbs and how *fast* they are converted to glucose. And that's what you really want to know. Your body's insulin responses are affected more by the glycemic loads of the food you eat than by their glycemic indices.

Comparing the glycemic load of green peas and white rice leads to some interesting conclusions. A half-cup serving of cooked green peas contains 7.5 grams of carbohydrates—a GL of 6 (7.5 × 75 percent). But a half-cup serving of white rice contains 36 grams of carbohydrates for a GL of 23 (36 × 64 percent), four times as much as peas. So even though white rice raises the blood sugar more slowly than green peas, based on its lower glycemic index, because it has lots more carbohydrates to begin with, it ends up having a much greater effect on insulin levels, as reflected by its much higher glycemic load.

Table 5-1. Glycemic Load of Common Foods[15]
Glycemic Load = Grams of Carbohydrates × Glycemic Index

The table below lists the glycemic load of some common foods (see Fantastic-Voyage.net for a more complete list):

FOOD	CARBOHYDRATE CONTENT (GRAMS)	GLYCEMIC INDEX (AS %; GLUCOSE = 100%)	GLYCEMIC LOAD PER SERVING
Skittles (2-oz bag)	54	70	38
Potato (1 baked)	30	85	26
Cornflakes (1 cup)	26	92	24
White rice (150 g)	36	64	23
Pasta (1 cup cooked)	48	44	21
White bread (2 slices)	28	70	20
Mixed-grain bread (2 slices)	28	55	15
Brown rice (150 g)	33	55	18
Kidney beans (150 g)	25	28	7
Green peas (150 g)	7.5	75	6
Lentils (½ cup cooked)	20	29	6
Carrots (½ cup cooked)	8	47	4
Peanuts (dry-roasted Valencia, ¼ cup)	7	14	1

Studies have demonstrated that high-GL diets are strongly associated with blood markers (indications) of syndrome X, such as lowered HDL-C (beneficial cholesterol) and elevated triglycerides (blood fats).[16] So try to eat lower-GL foods.

It's obvious from the table above that you are better off eating foods like peanuts, beans, lentils, peas, carrots, brown rice, and whole-grain bread as opposed to white potatoes, white rice, pasta, most ready-to-eat breakfast cereals, and candy. It takes a little time to learn which foods to eat and which to avoid, but you'll get the hang of it—once you begin, it's really not complicated.

YOUR OPTIMAL CARBOHYDRATE INTAKE

It's not unusual for people to consume 60 percent or more of their calories in the form of carbohydrates. We divide people into "low" and "moderate" carbohydrate groups, although our recommended level for even our moderate-carbohydrate folks is far less than many people normally eat. Our low-carbohydrate group consists of five subgroups of people who should cut down carbohydrates to no more than one-sixth of their total calories and eliminate all high-glycemic-load carbohydrates. (We'll describe below how to translate this into carbohydrate grams.)

Low-Carbohydrate Group

The five subgroups of the low-carbohydrate group are:

- *People trying to lose weight.* Eliminating high-GL foods will assure success. By ending carbohydrate cravings, you'll find that you can control your eating and feel full and satisfied on fewer calories. Failure to make this change will likely doom your chances of successfully losing and maintaining optimal weight.

- *People with TMS.* We will discuss the diagnosis of the metabolic syndrome (or syndrome X), which affects about a third of the adult population, in chapter 9, "The Problem with Sugar (and Insulin)." People with TMS often have a fasting blood glucose level of 100 or higher, indicating that the body is not able to adequately process carbohydrates. This syndrome results in high blood pressure and greatly accelerates atherosclerosis, which is the underlying process in most heart disease. The most important change you can make to control TMS and prevent it from becoming full-blown type 2 diabetes is to reduce the total amount and type of high-GL carbohydrates in your diet.[17]

- *People with type 2 diabetes.* One of the main characteristics of type 2 diabetes is that the cells have become highly resistant to insulin, so the body's cells can't efficiently use the glucose in the blood, which causes blood glucose levels to remain high. In an effort to drive glucose levels down, the pancreas produces very high levels of insulin, which is itself a problem because insulin accelerates atherosclerosis and other aging processes.[18] Eventually, the pancreas may burn itself out and stop generating insulin, worsening the diabetic condition.[19]

- *People with elevated risk factors for heart disease.* We will discuss how you can assess these risk factors in chapter 15, "The Real Cause of Heart Disease and How to Prevent It." Cutting down sharply on carbohydrates will significantly improve your cholesterol and other lipid levels and lower your risk of heart disease.[20]

- *People who have cancer, have had cancer, or have an elevated risk of cancer.* Unlike other cells and tissues in our body, cancer cells grow rapidly and have a voracious appetite for glucose. Glucose is, in fact, the only food cancer cells can eat. By reducing the easy availability of glucose, you will help to prevent latent cancer cells from becoming full-blown tumors.[21] Reducing carbohydrates, particularly of the high-GL variety, is one of the more important steps you can take to prevent cancer.

These five subgroups compose a majority of people in the United States and Europe. Note that our recommended intake of carbohydrates for people in this low-carbohydrate group is still higher than that recommended by other popular low-carb diets, such as the Atkins diet.

Specifically, for this low-carbohydrate group, we make these recommendations:

- Eliminate high-GL foods, including pastries, desserts of all kinds containing sugar and refined starch, breads, bagels, pasta, and high-starch vegetables such as potatoes and rice.

- Limit total carbohydrate consumption to less than one-sixth of your total calories (see Table 5-2 on page 60).

- Generally avoid grains and fruit juices.

- Eat very small quantities of low-GL fruits, such as berries and melons.

- Eat limited quantities of acceptable carbohydrates such as legumes (beans, lentils) and nuts.

- Eat larger quantities of acceptable carbohydrates such as low-starch vegetables, particularly fresh and lightly cooked. Good choices:

  Cabbage, brussels sprouts, broccoli
  Kale, mustard greens, Swiss chard, collards, spinach
  All types of lettuce (red and green leaf lettuce, romaine lettuce, endive, etc.)
  Chinese cabbage, bok choy, snow peas, celery
  Cauliflower, zucchini, cucumbers
  Green or "above ground" vegetables in general

  Most vegetables that grow underground ("root crops"), such as potatoes, beets, and turnips, typically have many more total carbohydrates and a higher glycemic load than green (above ground) vegetables.

Table 5-2: Maintenance Calorie Level and Recommended Carbohydrate Level for Low-Carbohydrate Group

WEIGHT	SEDENTARY		MODERATELY ACTIVE		VERY ACTIVE	
	Total calories	Carb grams	Total calories	Carb grams	Total calories	Carb grams
90	1,170	49	1,350	56	1,620	67.5
100	1,300	54	1,500	63	1,800	75
110	1,430	60	1,650	69	1,980	82.5
120	1,560	65	1,800	75	2,160	90
130	1,690	70	1,950	81	2,340	97.5
140	1,820	76	2,100	88	2,520	105
150	1,950	81	2,250	94	2,700	112.5
160	2,080	87	2,400	100	2,880	120
170	2,210	92	2,550	106	3,060	127.5
180	2,340	98	2,700	113	3,240	135
190	2,470	103	2,850	119	3,420	142.5
200	2,600	108	3,000	125	3,600	150
210	2,730	114	3,150	131	3,780	157.5
220	2,860	119	3,300	138	3,960	165
230	2,990	125	3,450	144	4,140	172.5
240	3,120	130	3,600	150	4,320	180

- Use a starch blocker (see Fantastic-Voyage.net for specific recommendations), a supplement or medication that binds with the digestive enzyme amylase. By combining with amylase, the starch blocker deactivates the enzyme, which results in the starch passing through the digestive tract without being broken down, as if it were fiber. Effective starch blockers include the prescription drugs Precose and Glyset. Note that excessive undigested starch, just like fiber, can cause gas, bloating, and digestive upset. So we recommend starch blockers as an adjunct to a reduced-carbohydrate diet, not as a substitute for reducing carbohydrates.

Here's how to figure your calorie limit if you're in the low-carbohydrate group: For your current weight and activity level, look up the number of total calories and number of carbohydrate grams in Table 5-2. The number of calories in this table represents the calorie level that will enable you to *maintain* your current weight. (As we will discuss in chapter 8 on attaining your optimal weight, if you need to *lose* weight, we recommend you instead adopt the calorie level, and corresponding carbohydrate level, for your desired or optimal weight. That way, you will gradually achieve your target weight.)

As an example, the maintenance calorie level for a moderately active person weighing 150 pounds is 2,250 calories. That translates into a carbohydrate limit of 94 grams per day.

If you're not in the five subgroups defined above, you still have risks of developing TMS, heart disease, and cancer, so you also need to be concerned with the quantity and choice of carbohydrates.

Moderate-Carbohydrate Group

For the moderate-carbohydrate group—people who are not in the five subgroups above—we recommend these steps:

- Cut down sharply on high-glycemic-load foods such as pastries, desserts of all kinds containing sugar and refined starch, breads, bagels, pasta, and high-starch vegetables, such as potatoes and rice.

Table 5-3: Maintenance Calorie Level and Recommended Carbohydrate Level for Moderate-Carbohydrate Group

WEIGHT	SEDENTARY		MODERATELY ACTIVE		VERY ACTIVE	
	Total calories	Carb grams	Total calories	Carb grams	Total calories	Carb grams
90	1,170	98	1,350	113	1,620	135
100	1,300	108	1,500	125	1,800	150
110	1,430	119	1,650	138	1,980	165
120	1,560	130	1,800	150	2,160	180
130	1,690	141	1,950	163	2,340	195
140	1,820	152	2,100	175	2,520	210
150	1,950	163	2,250	188	2,700	225
160	2,080	173	2,400	200	2,880	240
170	2,210	184	2,550	213	3,060	255
180	2,340	195	2,700	225	3,240	270
190	2,470	206	2,850	238	3,420	285
200	2,600	217	3,000	250	3,600	300
210	2,730	228	3,150	263	3,780	315
220	2,860	238	3,300	275	3,960	330
230	2,990	249	3,450	288	4,140	345
240	3,120	260	3,600	300	4,320	360

- Limit total carbohydrate consumption to no more than one-third of total calories (see Table 5-3 on page 61).
- Eat limited amounts of whole grains.
- Consume limited quantities of fruit juice and fruits.
- Eat good carbohydrates including legumes (bean, lentils) and nuts.
- Eat plenty of low-starch vegetables (see list on page 59), carbohydrates that can be eaten with relatively no restriction.
- Consider using a starch blocker.

Use Table 5-3 to determine your maintenance calorie level and recommended carbohydrate level if you are in the moderate-carbohydrate group.

ALTERNATIVE SWEETENERS

You should avoid sugar in all of its varied forms, since it's the ultimate high-glycemic-load food. This raises the issue of acceptable alternatives.

Saccharin, the original substitute for sugar, was required in 1977 to include a warning label as a potential carcinogen, because it was linked to bladder cancer in laboratory animals.[22] There has recently been a movement by saccharin manufacturers to remove this requirement, but we continue to find the original studies citing cancer risk to be compelling and do not recommend its use.

Aspartame (Nutrasweet and Equal) is also a sweetener that we do not recommend because of its potential health risk.[23] According to a report in the *New England Journal of Medicine*, aspartame can cause a significant imbalance of amino acids and neurotransmitters in the brain.[24] One example of this is a decreased availability of the amino acid tryptophan. This may reduce serotonin levels in the brain, which can cause mood imbalances and sleep disorders.

In addition, methanol (wood alcohol) is released from aspartame when it is consumed. This toxic substance can provoke headaches, fainting, seizures, memory loss, mood swings, depression, numbness in the extremities, nausea, gastrointestinal distress, and fibromyalgia symptoms. There have been reports that aspartame may cause a rise in insulin levels even though it contains no sugar. Since many of the diseases associated with sugar consumption are the result of elevated insulin levels, this finding is of serious concern.

Acesulfame-K (acesulfame potassium), approved by the FDA in 1988, is marketed as a tabletop sweetener called Sunett. Raw acesulfame-K is 200

Tips for Reducing Carbohydrates in Your Diet

- Eat more fiber. Fiber is an important constituent of many carbohydrate foods and offers an array of health benefits. Soluble fiber such as pectin, arabinose, beta-glucan, and psyllium is found in legumes, fruits, root vegetables, oats, barley, and flax and lowers LDL-C ("bad") cholesterol.[25] Insoluble fiber such as the cellulose in celery improves functioning of the large intestine and may reduce colon cancer.[26] Both forms of fiber add bulk and texture to your diet. Under labeling laws, fiber may be listed under carbohydrates, even though it is not digested and has no digestible calories. So in counting carbohydrates, you should subtract fiber grams from carbohydrate grams to determine actual digestible carbohydrates. You can also reduce the calorie count by four times the number of fiber grams (the calorie count includes four calories for each gram of fiber).

- Be patient. It takes one to two weeks for carbohydrate cravings to go away when carbohydrates are significantly reduced in the diet, particularly when high-glycemic carbohydrates are cut or eliminated. It is almost impossible to reduce your weight and maintain an optimal weight without eliminating carbohydrate cravings in this manner.

- Use substitutes. Replace carbohydrate-rich foods with low-carbohydrate substitutes. There is an entire world of low-carb substitutes for high-carb foods that you enjoy: breads, hot and cold cereals, frozen desserts, puddings, pastas, syrups, jams, and many others. See Fantastic-Voyage.net for specific product suggestions and links. These products make adopting a low-carbohydrate diet relatively easy.

- Take it along. Bring some packets of stevia with you when dining away from home. You can whip up a low-carbohydrate, low-fat salad dressing by combining stevia with lemon juice and/or balsamic vinegar.

- Eat low-starch veggies to your heart's content. We suggest eating a broad variety of vegetables of as many colors as possible.

- Switch to fruits. Eat berries and small portions of other fruits for dessert.

- Eat slowly.

- Avoid highly processed foods, such as french fries and baked goods.

- Use a starch blocker to further reduce the carbohydrates actually digested by your body.

times sweeter than sugar. The marketing benefit of this product is a long shelf life, which has been attractive to the diet soft drink industry. One study suggested that it might stimulate insulin release. A 1997 study in mice concluded "in view of the present significant *in vivo* mammalian genotoxicity data, acesulfame-K should be used with caution."[27]

Sucralose (Splenda), approved by the FDA in 1998, is created by modifying sugar molecules so that they are not digested. The modified molecules, which are created by chlorinating sugar, are 600 times sweeter than table sugar. To the extent that sucralose leaves the body undigested, it has less potential to create complications than saccharin or aspartame. Concerns have been expressed about the potential for sucralose to create harmful compounds as it passes through the digestive system, but trials in mice indicated that the sweetener was excreted mainly unchanged, with only minor metabolites. Many of the low-carbohydrate food substitutes now available use this new sweetener. It's a better choice than saccharin or aspartame, based on our current knowledge, but there is reason for caution until we have considerably more experience with its use.[28]

There is one natural noncaloric sweetener that we *do* recommend.

Stevia is one sweetener that is even good for you. This is a plant indigenous to South America and is a natural food supplement that is 30 to 100 times sweeter than table sugar. It has been valued for its medicinal effects and natural sweetness in Paraguay for 1,500 years. It has been used similarly in Japan for the past two decades and has recently become popular in the United States as an all-natural, healthy sweetener.

We are not aware of any adverse reactions reported from the use of stevia. Numerous studies have been performed in Japan and in the United States on stevia's effects on cell membranes, enzyme systems, and cancer,[29] and no negative effects have ever been discovered. In fact, many significant health benefits have been observed:

- It is highly nutritious.
- It can lower blood sugar in diabetics, but also regulates blood sugar in nondiabetics.
- It can lower elevated blood pressure.
- It kills bacteria that cause tooth decay.
- It can increase energy levels and mental activity.
- It helps reduce cravings for alcohol and tobacco.

6

FAT AND PROTEIN

A half billion years ago, thin layers of fat under the skin and around the organs of early animals provided cushioning for the inevitable bruises of long journeys in search of food. Thicker layers of adipose (fat) tissue provided thermal insulation for cold winters. Most important, animal species were subject to extremes in food availability, and fat provided an efficient means of storing energy from occasional periods of abundance for the inevitable periods of scarcity. Plants never developed this adaptation.

Humans now live in an era of abundance, at least of calories if not of high-quality foods. We no longer need to store dozens of pounds of fat, which accelerate long-term degenerative processes that underlie such diseases as heart disease, diabetes, and degenerative arthritis. It is too bad that we cannot reprogram our biochemical software to account for the radically different circumstances from those in which our bodies evolved. We will gain the means of doing exactly this in the next five to ten years. In the meantime, we are stuck with our outdated metabolic programs.

Excessive calories, regardless of the type of food consumed, end up being stored as body fat. Dietary fat is more than twice as dense in calories as other foods—9 calories per gram versus 4 for protein and carbohydrates—so it can be a clear contributor to excess body weight. The most obvious benefit of reducing fat from the diet is to feel full and satisfied on fewer calories.

There are other ways that our food consumption has strayed significantly from the evolutionary design of our digestive system. Two families of unsaturated fats—the omega-6 fats, found mainly in plant-based oils, and the omega-3 fats, found in fish, walnuts, and flaxseeds—were in relative balance a century ago. The modern diet now emphasizes omega-6 fats by as much as 25 to 1. These fats encourage inflammation, while omega-3 fats are anti-inflammatory.[1] Since inflammatory processes have an important role in degenerative disease (see chapter 12, "Inflammation—The Latest

'Smoking Gun'"), this long-term imbalance is a major contributor to chronic disease.

Modern methods of manufacturing cooking oils, margarine, and shortenings also create harmful by-products and distorted forms of fat molecules that did not exist in human diets when our digestive processes evolved. The effects of these pathological forms of fat are worsened by frying foods at high temperatures. For these reasons, diets that severely restrict fat reduce the contribution of dietary fat to excessive body fat, reduce omega-6 fats, and remove sources of pathological fat. The Ornish diet includes recommendations for consuming healthy omega-3 fats. Some people following traditional low-fat diets such as the Pritikin diet have not included this guideline. Cutting fat is a significant improvement over the standard Western diet, but we can refine this recommendation by including healthy anti-inflammatory fats.

WHAT IS FAT?

When we think of fat, we are inclined to think of the soft, somewhat vaguely shaped tissues in the human body that we associate with excess weight. These body tissues are comprised of fat cells, which, as the name suggests, are specialized cells optimized for storing fat. They are like miniature fuel tanks storing energy for future use.

But fat plays a far more complex role in our metabolic system than simply as an efficient form of energy storage. Beneficial fats also help in the body's creation of hormones, phospholipids (used to create the membranes that surround our cells and the organelles inside cells), and prostaglandins (hormonelike substances that control a wide variety of functions, such as platelet stickiness and blood pressure). By understanding the complex role of dietary fat, we can design optimal nutritional strategies for fat consumption.

Small differences in the structure of fatty acids distinguish various types of fat. Although the structural differences are subtle, they result in dramatic differences in the type of biochemical interactions the fatty acid can participate in. These have important implications for your health.

There are many types of fat molecules, but they all contain one unit of glycerol, which provides a backbone for three fatty-acid chains. Like complex carbohydrates, each fatty acid consists of a chain of 2 to 22 carbon-hydrogen units (CH_2). The "fatty" part of the fatty acid consists of this chain with a methyl group (CH_3–) at the end. As the name suggests, the fatty portion of a fatty acid is soluble in fat and insoluble in water. At the other end is a water-soluble, fat-insoluble carboxylic acid group (–COOH), by which

Don't Skimp on the Omega-3s

An important part of Ray & Terry's Longevity Program is ensuring adequate dietary consumption of anti-inflammatory omega-3 fats. A major study conducted by the Harvard School of Public Health, the Nurses' Health Study, which began in 1976, followed 90,000 nurses over many years to determine the effect of nutrition on patterns of disease. One of the most striking findings from the study was that nurses who ate at least 1 ounce (about a handful) of nuts per day had 75 percent less heart disease than those who did not eat nuts.[2]

There are a number of highly beneficial fatty acids, which are the main components of fats. In fact, two of them are essential to human health and are even known by that name—essential fatty acids (EFAs). Alpha-linolenic acid is an essential omega-3 fatty acid, while linoleic acid is an essential omega-6 fatty acid. But the Western diet already includes excessive omega-6 fats. An effective way to correct the distorted ratio of omega-6 to omega-3 fats is to increase the consumption of omega-3 fats, such as EPA (eicosapentaneoic acid) from fish oil, while reducing omega-6 fats, such as corn oil.

Foods that contain healthful amounts of omega-3s include walnuts, seeds (particularly hemp seed), fish, fish oil, flaxseed, flaxseed oil (linseed oil), canola oil, soy, and dark green leaves such as spinach, broccoli, kale, and seaweed (although leaves have little oil, it primarily contains alpha-linolenic acid).

it attaches to the glycerol "trunk" of the molecule. These fatty acids are very small: a single drop of oil comprises a billion billion of them.

Biochemistry is based on carbon because it's a remarkably flexible building block. That comes from having four positions (electrons) that can form links to other atoms. In the fatty-acid chains in Figure 6-1 on page 69, you can see that every carbon atom is linked to four other atoms. Biological cells are capable of inserting a double electron bond between two of the carbon atoms in a fatty-acid chain by removing two of the hydrogen atoms. Note that all of the carbon atoms still have four bonds, which is a structural requirement for biochemical molecules. Fatty acids with one or more double carbon bonds are called *unsaturated* because the portion of the fatty acid with the double carbon bond is not saturated with hydrogen atoms. This enables this portion of the fatty acid to react with other molecular units, such as oxygen, water, hydroxol (–OH), and sulphydryl (–SH) groups. This ability to interact chemically is key to the health benefits of unsaturated fatty acids.

Fatty acids with no double carbon bonds are said to be *saturated*—fully saturated with hydrogen atoms. As a result, saturated fatty acids are relatively inert (they do not interact with other substances), unlike unsaturated fatty acids, which are more volatile. Saturated fats do not facilitate vital biochemical reactions in the body; they store energy, and in excess—that is, excess body fat—and contribute to a wide range of illnesses, including type 2 diabetes, hypertension, stroke, and heart disease. Even if you maintain a normal weight, consuming excess amounts of saturated fats contributes to high cholesterol levels and increases platelet stickiness, which are primary risk factors for heart disease.

If a fatty acid has one double bond, it is considered *monounsaturated*; with more than one double bond, we call it *polyunsaturated*. Each exact placement of these unsaturated positions (double carbon bonds with missing hydrogen atoms) enables specific types of biochemical reactions.

Naturally occurring unsaturated fatty acids have a "cis" configuration (*cis* is Latin for "same side"). With both remaining hydrogen atoms involved in a double carbon bond on the same side, they repel each other and bend the fatty acid.

In the "trans" configuration, the two remaining hydrogen atoms involved in a double carbon bond are on opposite sides, which keeps the fatty acid rigid. Although a *trans-fatty acid* is unsaturated, it acts more like a saturated fatty acid because it is rigid and biochemically stable. Trans-fatty acids, the primary fatty acids found in margarine and shortening, do not occur naturally; they are artificially created in factories. Their stability is attractive from a commercial perspective in terms of providing a long shelf life, but by converting the natural *cis*-configuration into an unnatural *trans*-configuration, the health benefits of the unsaturated fatty acid are lost. The trans-fatty acids found in many commercial foods contribute to heart disease in the same way that saturated fats do.

UNSATURATED FATS

Unsaturated fats, which are liquid at room (or body) temperature, are distinguished by the position of the first carbon double bond. The omega numbering system refers to the location of the first double bond from the methyl end of the fatty acid. Thus, omega-3 fatty acids have their first carbon double bond three carbon atoms from the methyl end; omega-6 fatty acids start with the sixth carbon atom. Both omega-3 and omega-6 families are *polyunsaturated* fatty acids, meaning that they have multiple double bonds.

In general, omega-6 fatty acids promote an inflammatory response in the body, while omega-3 fatty acids inhibit inflammation.[3] Both families are es-

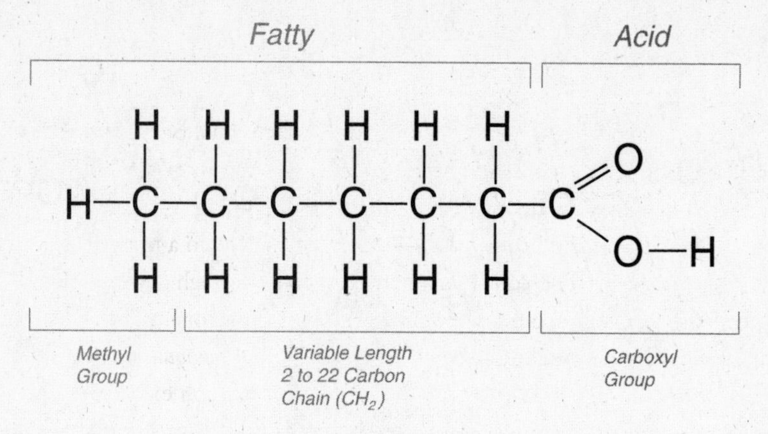

FIGURE 6-1. FATTY ACIDS AND FAT MOLECULES

sential, but the typical Western diet has become highly unbalanced toward the pro-inflammatory omega-6 fats, which are found in vegetable oils such as safflower, sunflower, corn, and sesame. Inflammation underlies a key step in the development of heart disease, stroke, Alzheimer's disease, and other degenerative conditions, and rebalancing these two families of fatty acids is critical to reversing this degenerative process.[4]

The omega-3 fats include:

- Alpha-linolenic acid, sometimes *incorrectly* referred to as linoleic acid. This is one of the two EFAs (essential fatty acids) because it's required for life and cannot be created in the human body from other foods. Alpha-linolenic acid helps:
    - Improve oxygenation of tissues
    - Improve oxidation of food in the mitochondria (the small fuel cells in every cell)
    - Speed muscle recovery during exercise
    - Speed healing
    - Improve sense of calmness
    - Reduce inflammation
    - Reduce platelet stickiness
    - Reduce blood pressure

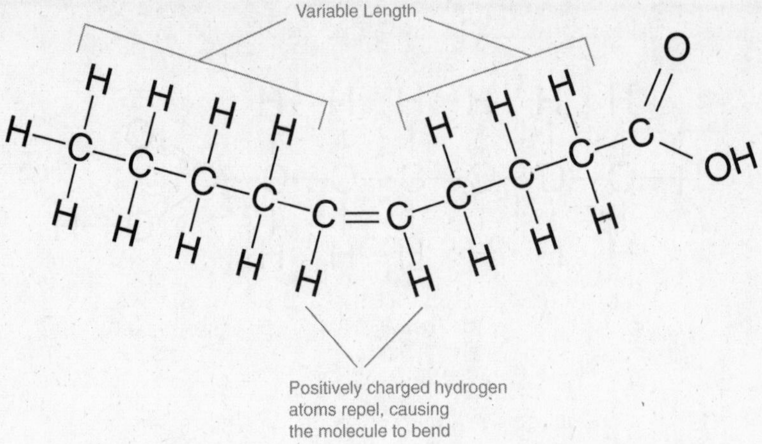
Variable Length
H
H
H
H
H
H
H
H
H—C—C—C—C—C=C—C—C—C—C
OH
O
H
H
H
H
H
H
H
H
Positively charged hydrogen
atoms repel, causing
the molecule to bend

FIGURE 6-2. UNSATURATED FATTY ACID IN CIS-CONFIGURATION (SHOWING BEND)

A cautionary note: Recent research has shown a possible link between a high level of consumption of alpha-linolenic acid and the incidence of prostate cancer.[5]

- EPA (eicosapentaneoic acid) and DHA (docosahexaneoic acid), which are fatty acids derived from alpha-linolenic acid, are two absolutely crucial nutrients. When consumed in adequate amounts, EPA and DHA help:
    - Lower triglyceride levels by up to 65 percent and cholesterol levels by up to 25 percent[6]
    - Encourage dispersal of other fats, counteracting the harmful effects of saturated fat and trans-fatty acids[7]
    - Reduce platelet stickiness, which contributes to atherosclerosis and heart attacks[8]
    - Lower apo(a) and fibrinogen levels, two major risk factors for atherosclerosis[9]
    - Increase production of series 3 prostaglandins (which counteract the series 2 prostaglandins made from omega-6 fats), which lower blood pressure[10]
    - Inhibit the growth and metastasis (spreading) of cancer cells[11]

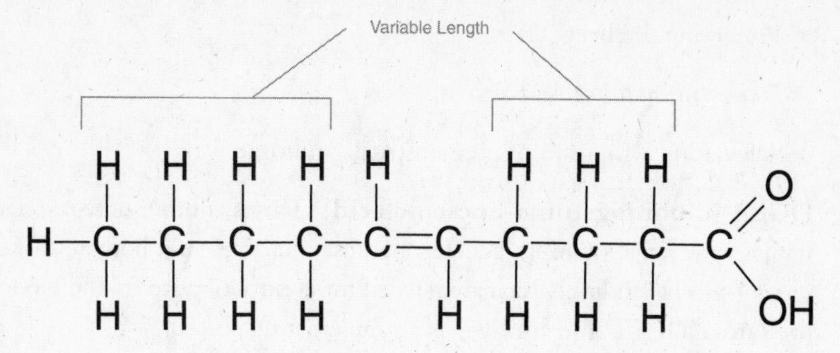

FIGURE 6-3. UNSATURATED FATTY ACID IN TRANS-CONFIGURATION

You should eat foods containing EPA and DHA regularly, but we also recommend supplements to ensure that you obtain optimal levels of these vital anti-inflammatory substances. Even the American Heart Association, usually quite conservative in its supplement recommendations, has recently come out in favor of EPA and DHA supplementation.[12]

The omega-6 family of fats includes:

- **LA (linoleic acid).** This is the other EFA essential to life. LA is a pro-inflammatory EFA. Although "essential" for human life, most Western diets include excessive amounts, so we recommend restricting LA-rich oils, which include safflower, sunflower, soy, pumpkin, and sesame.
- **GLA (gamma-linolenic acid).** Although this is an omega-6 fat, it's a precursor of vital prostaglandins, so it has important health benefits.[13] Human body cells convert LA into GLA, but a variety of abnormal conditions, including enzyme deficiencies, elevated lipid cholesterol levels, diabetes, the metabolic syndrome, and other aging processes, can block this process. If blood levels of GLA are low, it can be taken as a supplement. Evening primrose oil is a recommended source. It has shown potential for:
  - Lowering blood pressure
  - Lowering cholesterol levels
  - Lowering the risk of heart attack and stroke
  - Improving fat metabolism in people with diabetes

- Protecting the liver
- Relieving symptoms of arthritis
- Alleviating symptoms of premenstrual syndrome

- **DGLA (dihomogamma-linolenic acid).** This is another omega-6 fat with a wide range of health benefits, because it enables your body to make series 1 prostaglandins, which are needed for control of platelets and blood pressure. DGLA is also available as a supplement.
- **AA (arachidonic acid).** Found primarily in meats and animal products, it may be thought of as the Darth Vader of the omega-6 fats. It is needed for human survival because it enables the body to make series 2 prostaglandins, which, however, also promote inflammatory processes.[14] It is not considered an EFA because the body can synthesize it. But it has a dark side. The high consumption of meat in Western diets and the resulting elevated levels of AA are key factors in high levels of inflammation, which promote coronary artery disease and other degenerative diseases.

Contemporary Snake Oil

Interestingly, certain snakes such as Chinese water snakes are particularly good sources of EPA and DHA, so those 19th-century snake oil salesmen may have been on to something. Chinese water snakes are in short supply these days, but, luckily, healthful amounts of EPA and DHA can be obtained from a number of cold-water fish and marine animals:

- Salmon. This is a particularly rich source, containing up to 30 percent EPA/DHA; among fish, it's also relatively low in mercury contamination. However, salmon from fish farms are fed cornmeal rather than the usual wild-salmon diet of other small fish and algae. As a result, salmon from fish farms are much lower in EPA and DHA.[15]
- Trout
- Mackerel (although it's been found to be relatively high in mercury contamination)
- Sardines
- Brown and red algae grown commercially as EPA and DHA sources

The monounsaturated fatty-acid family includes:

- **OA (oleic acid),** an omega-9 fat (because its one double carbon bond starts with the ninth carbon atom from the methyl end). OA has anti-inflammatory properties similar to alpha-linolenic acid and is a beneficial fat. OA resists interaction with oxygen, so it's relatively stable. A major benefit is that it helps keep arteries flexible and supple, thereby resisting atherosclerosis.[16] Good sources of OA include olives, (extra virgin) olive oil, avocados, peanuts, pecans, cashews, filberts, and macadamia nuts. Animal products and butter provide small amounts of OA, but this benefit is far outweighed by the high levels of saturated fat and AA in animal-based fats.
- **POA (palmitoleic acid).** An omega-7 fat, POA raises cholesterol levels and is not a healthy fat.[17] It is found in high concentrations in coconut and palm oils, primary ingredients in commercial nondairy creamers.

SATURATED FATS

Saturated fatty acids, which tend be solid at room (or body) temperature, contribute to obesity, type 2 diabetes, and the progression of atherosclerosis, a primary cause of heart disease. Long-chain saturated fats tend to stick together, which causes platelets to stick together and contributes both to atherosclerosis and the formation of blood clots that can initiate heart attacks.[18] Saturated fats also cause red blood cells to stick together, thereby reducing their ability to deliver oxygen to the cells. The resulting hypoxia (lack of tissue oxygen) accelerates atherosclerosis and other degenerative processes.

Saturated fats are a primary family of fat that should be limited. Sources include stearic acid, found in butter and animal meats such as beef and pork; palmitic acid, in coconut and palm oils; butyric acid, in butter; and arachidic acid, in peanuts.

Another problem with meat is that the fat is a major source of pesticides and other chemicals used in farming. Pesticides used to grow animal feed become highly concentrated in animal fat.

PATHOLOGICAL FATS

Although not formally a family of fat, our discussion would not be complete without considering *pathological fats*. Most commercial cooking oils fall into this category. Commercial oil producers start with seeds and nuts that are valuable sources of vitamins, minerals, EFAs, fiber, lecithin, and other es-

Go Extra Virgin

Virgin olive oil refers to oil obtained by pressing whole, ripe, healthy olives with no heat applied—neither external nor from the pressing process—and with no refining or other processing steps such as bleaching or deodorizing. Olive oil that is not labeled "virgin" is refined and will contain the same types of pathological fats found in any other refined vegetable oil.

Extra virgin olive oil is even better. This oil meets a stricter set of guidelines, including the use of only very-high-quality olives. We recommend that you use only organic extra virgin olive oil for salads, cooking, and other culinary uses.

Keep in mind, though, that when oils, even beneficial types such as extra virgin olive oil, are heated to very high temperatures (above 320 degrees Fahrenheit), toxic chemicals are formed, including acrylamides, and toxic cyclic monomers, which have been linked to abnormal liver deposits.[19] Deep-frying creates large numbers of free radicals that interfere with vital metabolic processes.[20]

We recommend that you avoid deep-frying altogether and instead stir-fry. A healthy method is to first put water in the pan (a wok is recommended), then add a small amount of extra virgin olive oil. Keep the temperature at a moderate level (under 180 degrees; if the oil is smoking, the temperature is too high), and cook the food for only a brief period of time. This approach will avoid the creation of most toxic by-products.

sential nutrients, but almost all of these are destroyed during production.[21] Moreover, the high levels of heat and chemicals used during production create toxic molecules, including pathological forms of fat, which are unnatural configurations of fatty acids that the body is unable to use in any metabolic processes other than as energy storage.

Margarine and other solid and semisolid fat products contain high levels of trans-fatty acids, which are not normally found in nature and are a primary contributor to degenerative disease. Margarine made from "100% corn oil," for example, typically contains up to 25 percent trans-fatty acids. Trans-fatty acids in commercial fat products have been shown to:

- Increase lipid cholesterol levels and decrease the "good" HDL cholesterol, according to an extensive, well-designed study published in the *New England Journal of Medicine*[22]
- Raise lipid triglyceride levels[23]
- Interfere with the body's detoxification systems[24]
- Raise Lp(a), another major risk factor for heart disease[25]

- Decrease testosterone levels[26]
- Increase insulin resistance, a primary cause of the metabolic syndrome and type 2 diabetes[27]

Of particular concern is the process of *hydrogenation* of liquid vegetable oils to create spreadable products such as margarine and shortening. Spreadability appeals to the palette and is desirable from a commercial perspective because of the very long shelf life. It is not desirable, however, from a health perspective. Hydrogenated fats consist largely of saturated fat and trans-fatty acids. Both of these cause aggregation of blood cells, thereby contributing to atherosclerosis.[28] The hydrogenation process uses nickel and aluminum, and residues of these harmful metals remain in the finished products.[29] In addition, hydrogenated fats contain pathological fragments of fatty acids and a myriad of other harmful by-products.

CHOLESTEROL

Our discussion of fat would not be complete without considering cholesterol. Cholesterol is not a fat, but its metabolism is closely related to fat in the diet. Cholesterol is a 27-carbon molecule made in the body from two-carbon acetates, which are breakdown products from fats and sugars (as well as protein through a less direct pathway). Cholesterol is a hard, waxy substance that is essential for human health and life. It plays a vital role in maintaining the health of our cellular membranes.[30] It is also the precursor to the male and female sex (steroid) hormones, including estrogen, progesterone, and testosterone, as well as cortisone, the stress hormone.

Cholesterol is, of course, well known as a primary risk factor for atherosclerosis, the process of plaque formation in the arteries that can lead to heart attack or stroke. Elevated levels of cholesterol play an important role in this process, although there are other equally crucial risk factors. Cholesterol is found in your food, but the primary source of cholesterol in the body is manufactured by your cells, especially in the liver. Cholesterol is unique because the body can make it, but cannot break it down. Excess cholesterol can only be removed from the body in the stool, where it is combined with bile acid. This process is facilitated by dietary fiber, which is another benefit of fiber in the diet.[31]

Excess calories contribute to the body manufacturing more cholesterol than is healthy.[32] These come especially from high-glycemic-load carbohydrates and unhealthy fats, specifically saturated fat, trans-fatty acids, and pathological forms of polyunsaturated fatty acids. Stress also contributes to excessive cholesterol levels because the body needs to make cholesterol to produce the stress hormone cortisol.[33]

We'll discuss in detail the role of cholesterol in heart disease, as well as a variety of ways to keep the body's cholesterol levels in a healthy range, in chapter 15, "The Real Cause of Heart Disease and How to Prevent It." In terms of dietary influences on cholesterol levels, the most important thing you can do is eat less saturated fat, trans-fatty acids, and the pathological forms of polyunsaturated fatty acids.[34]

Some observers maintain that dietary cholesterol is not a significant factor in blood cholesterol levels because the body's own regulation of its cholesterol levels overrides what is consumed. This is only true, however, if cholesterol consumption is reasonably low and if the metabolic pathways controlling cholesterol levels are working properly.[35] Neither of these assumptions is warranted for many individuals, particularly those with a susceptibility to heart disease.

The average person has about 150,000 milligrams of cholesterol, most of which is incorporated into cell membranes, with only about 7,000 milligrams circulating in the blood. The daily usage of this circulating cholesterol is typically around 1,000 milligrams. So consuming several hundred milligrams per day of cholesterol can increase blood cholesterol levels, particularly if cholesterol-related pathways are impaired (as indicated by a tendency toward elevated levels). We recommend that dietary cholesterol consumption be kept under 1,400 milligrams per week. If you have elevated risk factors of heart disease, you should reduce this to 700 milligrams per week.[36] The average American diet contains about 800 milligrams of cholesterol each day. Cholesterol is found only in animal products, including egg yolks (one yolk has about 250 milligrams of cholesterol); shellfish, including shrimp and lobster; meats, including beef, pork, and poultry; organ meats (4 ounces of liver has 250 milligrams); and dairy products (each ounce of butter has 60 milligrams).

RECOMMENDATIONS FOR FAT AND CHOLESTEROL IN THE DIET

Dietary fat is a major contributor to obesity because of its high caloric level—9 calories per gram versus 4 calories per gram for protein and carbohydrates. We recommend restricting fat to 25 percent of calories, although virtually all of this should be "good fat." The National Institutes of Health

Sugars and Starches Turn into Fat

Dietary sugars and simple starches are a primary source of body fat. Simple starches, found in high-glycemic-index foods such as breads, pasta, rice, potatoes, and pastries, are quickly converted during digestion into glucose. Glucose that is not immediately used for energy is converted into *triglycerides*—fat molecules consisting of a unit of glycerol with three saturated fatty-acid chains. These triglyceride molecules act like saturated fat: they accelerate atherosclerosis. An elevated triglyceride level is an independent risk factor for heart disease. Many researchers believe this is as significant as elevated cholesterol levels.

The body readily converts excess dietary glucose from sugars and starches into fat, but it has no mechanism for going in the opposite direction. The only way to get rid of excess body fat is to break down fat molecules into chemicals called "ketone bodies," which can be used as an energy source. All of the body's organs, including the muscles, can use ketone bodies derived from fat for fuel, but whether the brain can use ketone bodies or must have glucose is still controversial.[37] For people who are fasting or on ultra-low-carbohydrate diets, the body will also convert dietary protein or its own muscle protein into glucose to feed the brain and other vital organs.

(NIH) recommends that saturated fat be kept to less than 7 percent of total calories.[38] We recommend that you reduce saturated fat to no more than 3 percent of calories. For individuals consuming 2,400 calories a day, the NIH recommendation corresponds to 18 grams of saturated fat daily, whereas our recommendation is no more than 8 grams. By making proper food choices, it is easy to keep saturated fat intake below this guideline. For example, a 3-ounce serving of top round has only 2 grams of saturated fat. A typical fast-food double cheeseburger, on the other hand, with 15 grams, exceeds your daily guideline.

Note that 25 percent of calories means less than 12 percent of food weight because of fat's higher caloric density.

More important than how much fat you eat is what kind. Focus on getting dietary fat from:

- Nuts
- Fish high in EPA and DHA, especially salmon, which is rich in EPA and DHA (wild salmon has even more than farm-raised) and relatively low in mercury

- Extra virgin olive oil
- Flaxseed and naturally pressed flaxseed oil, which can be transformed by the body into EPA and DHA. However, many people lack the critical enzymes needed for this conversion and should consume EPA and DHA directly.
- Vegetables, which contain small but mostly healthful forms of fat
- Tofu

Sources of fat that can be eaten in small quantities:

- Lean meats, especially white meat from chicken and turkey. Free-range poultry grown without hormones and antibiotics is preferable. Lean beef is a good source of protein, but there are several reasons to avoid beef and other red meats:
- Cows, pigs, and other livestock are high in hormones and antibiotics, which are used in the factory farming process.
- The fat has higher concentrations of the chemicals and pesticides used to grow the grain for feed. This is especially true of red meat (versus poultry) because it takes longer to raise cows and pigs.
- Meat is relatively high in cholesterol.
- There is a danger of prion infection with BSE (bovine spongiform encephalopathy), or mad cow disease, which has been epidemic among English cows. As of this writing, one cow with BSE has been detected among American cattle, but only a small sample of American cows are tested. Although the U.S. Department of Agriculture claims that American beef is safe, there can be no assurance that the disease has been successfully isolated. Mad cow disease is caused by prions, which are malformed proteins that self-replicate in the brain of affected animals. Similar prion-based diseases have been widespread in American wildlife, including deer. It is believed that Creutzfeldt-Jakob disease in humans, a fatal disease, is caused by consuming meat from animals infected with mad cow disease.[39] Latent infection with prions is almost impossible to detect, and the incubation period can be 20 to 30 years, so it's possible that there's a widespread prion infection among the American population that has yet to be detected. Our recommendations: never eat beef from England or meat from wild game such as deer from anywhere. The possibility that prion infection has spread beyond these sources is another reason not to eat any beef at all.

Healthy?

"High in Polyunsaturated Fat": Not necessarily.

The polyunsaturated omega-3 fats are healthy because they inhibit the inflammatory processes underlying heart disease and other degenerative diseases. However, the omega-6 polyunsaturated fats found in many commercial cooking oils encourage these dangerous inflammatory processes and are eaten in significant excess quantities in most Western diets. Commercially produced food products such as margarine and shortening are typically high in polyunsaturated trans-fatty acids, which act like saturated fats, accelerating atherosclerosis and other disease processes. The high levels of heat used to manufacture commercial food oils also produce pathological forms of polyunsaturated fats, which also may cause health problems.

"Monounsaturated": Not necessarily.

OA (oleic acid), a monounsaturated omega-9 fat, is anti-inflammatory and healthy. However, POA (palmitoleic acid), a monounsaturated omega-7 fat, raises cholesterol levels and is not a healthy fat.

"No Cholesterol": Not necessarily.

Blood cholesterol levels are affected primarily by consuming excess calories, particularly from high-glycemic-index carbohydrates and unhealthy forms of fat. Corn-oil margarine, for example, has "no cholesterol," but it is high in trans-fatty acids, which increase blood cholesterol levels.

"Cold Pressed": Not necessarily.

"Cold pressed" simply refers to the lack of external heat during pressing. Commercial methods of extracting oil use enormous pressures, which often create intense heat due to friction alone. This will destroy the health benefits of food oils, including olive oil. And the term "cold pressed" refers only to the oil-extraction step; it does not convey any information about other potentially harmful steps used during preparation.

"Extra Virgin": Yes.

"Virgin olive oil" refers to olive oil obtained by pressing whole, ripe, and healthy olives, with no heat applied and with no refining or other processing steps. "Extra virgin" olive oil follows stricter guidelines and is what we recommend for salads and cooking.

Forms of fat that should be avoided include:

- Saturated fat in fatty meats, butter, milk, and other animal products
- Commercial cooking oils (use extra virgin olive oil instead)
- Hydrogenated fats in margarine and shortening, and almost all commercial bakery products, which are high in trans-fatty acids.

Recommendations for food preparation include:

- Avoid deep-frying.
- Stir-frying in healthy oil, such as extra virgin olive oil, is acceptable. An even healthier way to stir fry is to first put water in the pan (a wok is recommended), then a small amount of a healthful oil (extra virgin olive oil), and then cook briefly at a low to moderate temperature.
- Supplementation with EPA and DHA (1,000–3,000 mg per day of EPA and 700–2,000 mg per day of DHA) is recommended.
- Cholesterol consumption in the diet should be limited to 1,400 mg per week. Persons with elevated risk factors for heart disease should limit cholesterol consumption to 700 mg per week.

PROTEIN: THE FOUNDATION OF LIFE

Fat is a relatively recent evolutionary innovation. Protein, on the other hand, is the very foundation of life. The Book of Life—the DNA that compose genes—contains the codes for proteins, which underlie all of the diverse life-forms on Earth.

The mechanism is very clever. The DNA in the nucleus of a cell is copied to create a mirror image in a molecule of RNA. The RNA travels outside the nucleus, where molecular machines called ribosomes read the RNA and create sequences of amino acids. In a remarkable process, these amino acid strings fold themselves into intricate three-dimensional structures. We still do not understand exactly how protein folding takes place, although supercomputers are being developed to model and simulate this process in the next few years. (IBM's Blue Gene/L supercomputer scheduled for 2005, capable of 360 trillion operations per second, is specifically targeted at modeling the protein folding problem.[42])

Proteins, which are elaborate and ungainly three-dimensional structures

BRIDGE TWO

THE FUTURE OF FAT

In the previous chapter on carbohydrates, we mentioned the Bridge Two development of drugs that could block carbohydrates even after they pass through the digestive tract by destroying glucose in the bloodstream. What about destroying fat after it is consumed? A drug being developed by Australian firm Metabolic appears to accomplish this by dramatically increasing the rate of burning fat in the bloodstream before it gets a chance to be stored in the fat cells. In a phase 2 FDA study, obese users of the drug had double the weight loss of a control group using currently available weight-loss drugs. Since the drug merely increases the rate of fat burning, it does not appear to interfere with healthy fats that play vital roles in other metabolic processes.

Omega-3 enriched meat and eggs. We know that increased consumption of omega-3 fats is critical. Presently, the best and easiest way to get these omega-3 fatty acids, EPA and DHA, is by eating fish and taking fish oil supplements. Unfortunately, all the oceans on the planet are now contaminated with mercury, and the form found in fish is highly absorbable. A novel biotechnical solution is in the works. The "fat-1" gene from the *C. elegans* roundworm is able to convert omega-6 into omega-3 fatty acids. Most land-based animals, including humans, lack this ability. Genetic engineers have spliced this gene into laboratory mice, which now convert the omega-6 fats in their diet into omega-3 fats.[40] Since consumers are not likely to find mice a desirable food, scientists are working to apply this technology to more traditional barnyard animals, so soon our eggs, milk, and meat may provide healthful omega-3 fat rather than saturated fat.

Cloned animals for food. Cloning technologies offer a possible solution for world hunger: creating meat and other protein sources without animals by cloning animal muscle tissue. This would allow "growing" beefsteak or chicken breast in a factory without animals. Benefits would include extremely low cost, avoidance of pesticides and hormones that occur in natural meat, greatly reduced environmental impact compared with factory farming, improved nutritional profile, and no animal suffering.

A more immediate application of cloning is improved animal husbandry by directly reproducing an animal with a desirable set of genetic traits. A powerful example is reproducing animals from transgenic embryos (embryos with foreign genes) for pharmaceutical production. Case in point: a promising new anticancer treatment. This antiangiogenesis drug (it inhibits tumors from creating the new capillary networks needed for their growth), called aaATIII, is produced in the milk of transgenic goats.[41]

formed of strings of amino acids, carry out all of the functions of living organisms. Everything we do—breathe, digest, move, think—is performed by structures built up from these proteins.

Eight amino acids are considered "essential" because the body cannot synthesize them and must get them from food. Nutritionists used to advocate eating "complete" proteins that contain all eight essential amino acids. This generally meant animal protein, since animals, being similar to humans, provide proteins with all required amino acids. There are no benefits, however, to eating complete proteins as long as you consume all of the needed amino acids in some form.

It's not necessary to eat each essential amino acid at every meal, although it is desirable to obtain them all each day. Vegetarians should take care to eat a wide variety of foods, because restricting the diet to a narrow range of vegetable products can create amino acid deficiencies. By eating a diverse range of vegetarian products, particularly legumes, it is not difficult to obtain all of the essential amino acids. The daily requirement for most essential amino acids is only about 1 gram per day, so if even small amounts of meat from fish or animals are included in the diet, deficiencies are virtually unheard of.

There are significant advantages to obtaining most of your protein from vegetable sources. Vegetable proteins have much lower concentrations of chemicals, pesticides, hormones, and antibiotics, which are highly concentrated in meat. Meat sources of protein include saturated fat, which raises blood cholesterol levels and can lead to insulin resistance.[43] Meat also contains dietary cholesterol. A better choice is soy protein, which improves blood lipid profiles. There are also obvious ecological benefits: vegetables are about 20 times more efficient in feeding people and avoiding environmental damages of farming than meat protein.

An ideal source of protein is fish (especially salmon), which is also high in omega-3 fats. Wild Alaskan salmon is ideal in terms of reducing levels of mercury.

There are benefits to each amino acid and supplements of specific amino acids have therapeutic value. The essential amino acids include:

- Phenylalanine, a natural antidepressant
- Tryptophan, a natural sedative (although as a supplement, it has had regulatory problems[44])
- Lysine, which combats viruses but is not recommended for people with diabetes

- Threonine, a naturally calming amino acid
- Methionine, involved in a key metabolic pathway, discussed in detail in chapter 13
- Isoleucine, leucine, and valine, branched chain amino acids that enhance protein synthesis in the liver and assist in overall liver function

Two other amino acids are essential for children; one is also of enormous value to adults in therapeutic doses:

- Arginine, which controls the vital nitric-oxide pathway.[45] Supplementing 6 to 9 grams of arginine per day will reduce atherosclerosis and improve blood vessel health[46] (see chapter 15 on heart disease)
- Histidine

Other nonessential amino acids are also beneficial in therapeutic doses:

- Cysteine, which is essential for premature infants and has value for adults to support the body's antioxidant system
- Tyrosine, a natural antidepressant found in thyroid hormones

Protein can be converted into glucose in the body, but the metabolic pathway is less direct than that for starch, so pure protein products, such as soy protein, generally have a lower glycemic index than moderate- or high-glycemic-index carbohydrates.[47] But any source of calories can end up increasing cholesterol and triglyceride levels if intake is excessive.

Studies have shown that diets high in animal protein increase insulin resistance, which is the ultimate source of the metabolic syndrome and type 2 diabetes. It is not clear, however, whether it is the protein or the associated saturated fat also found in animal products that is responsible. We do know that saturated fat increases insulin resistance.[48] Studies have shown that diets high in vegetable protein (as opposed to animal protein) do not appear to increase insulin resistance.[49]

Plenty of low-carbohydrate substitutes for normally high-carbohydrate foods such as bread, pasta, cereals, puddings, and desserts are available in food and nutrition stores and via Web sites (see Fantastic-Voyage.net for recommendations). Many of these products are made with soy protein, which reduces blood cholesterol levels. Be aware that these foods may also contain wheat protein (gluten); some people are sensitive to wheat, as discussed in the following chapter.

Vitamins and Free Radicals

Saturated fats provide more health benefits than unsaturated fats, but even these healthful fats can harm arteries by becoming oxidized if blood levels of antioxidants are insufficient.[50] Vitamins are important to this process for several reasons. Several serve as powerful antioxidants, such as C, E, and beta-carotene. They also act as catalysts in vital metabolic pathways that use the omega-3 fatty acids and other healthful polyunsaturated fatty acids. The body is unable to use these fats without adequate levels of both vitamins and minerals. For example, EFAs in the body will be damaged by free radicals unless protected by vitamin E and beta-carotene. Vitamins and minerals obtained from dietary sources are ideal, but it's generally not feasible to maintain truly optimal levels of these nutrients from your diet alone, so supplementation is also advised (see chapter 21, "Aggressive Supplementation").

RECOMMENDATIONS FOR PROTEIN IN THE DIET

Our carbohydrate recommendation for people in the moderate-carbohydrate group is to restrict carbohydrates to no more than one-third of your calories; for the low-carbohydrate group, to no more than one-sixth of calories. Our fat recommendation is to limit healthy fats to a maximum of 25 percent of calories. This means that for the moderate-carbohydrate group, at least 42 percent of calories should come from protein; for the low-carbohydrate group, at least 59 percent.

Most of this protein should not come from animal sources—meat, whole eggs, dairy products—because it would be impossible to maintain the other recommendations, particularly regarding eating healthy fats. We recommend that most of your protein come from vegetables, which contain healthy proteins as well as very low-glycemic-index carbohydrates and small amounts of healthy fats. It is important to eat the broadest possible variety of vegetables to get all of the amino acids in optimal levels, which are several times greater than the minimum levels needed to avoid deficiencies.

Another source of high-quality vegetable protein is soy protein, which is found in many products, such as tofu burgers, that are intended to substitute for foods normally high in carbohydrates (such as pastries) or high in animal fat (burgers). Other protein options include egg whites (including egg substitutes, which are 99 percent egg white); lean meats, particularly the white meat of chicken and turkey; and fish.

7

YOU ARE WHAT YOU DIGEST

The many nutrients in your food can only help you if they are digested properly. From eating to absorption, your food follows a long, intricate, and hazardous path. The human digestive process, which takes place throughout the alimentary canal—stretching from the mouth to the anus—acts like an elaborate inventory control system. It breaks down foodstuffs into their constituent molecular parts for delivery to their ultimate destination: your trillions of cells, where they are reassembled into the little machines and energy sources that animate your life.

Digestion is a remarkably efficient process, but defects—some genetic, others the result of decades of suboptimal nutrition—underlie many discomforts and the progression of long-term disease. One study revealed that 70 percent of American adults suffer from uncomfortable gastrointestinal symptoms resulting from maldigestion, malabsorption, or unhealthy bacteria in the digestive tract, in addition to specific digestive illnesses.[1] Assessing and fixing your digestive process is critical to both your short- and long-term health.

We have little conscious control over our digestive processes. In the mouth and nose, tastebuds and olfactory sensors provide pleasure and critical feedback to the brain on what we are eating. After food leaves the mouth, however, there are only weak nerve sensors communicating to the brain. For the most part, the process is on autopilot, controlled by reflexes via local nerves and delicate metabolic pathways. We can influence the digestive process through supplements and, when necessary, medications. But adopting a healthy diet is the most important thing you can do to maintain and restore the delicate balance of your digestive processes.

HOW DIGESTION WORKS

The diagram of the human gastrointestinal system on page 87 shows the organs involved in digestion, which takes place in the following steps:

Step 1. Digestion begins in the mouth, where chewing breaks down food into manageably small particles. The salivary glands produce about a quart of saliva per day, which moistens and lubricates dry food and buffers it (changes the acid-alkaline balance) to maintain an optimal (alkaline) pH level. The primary digestive enzyme in saliva is amylase, which begins the breakdown of starch (polysaccharides) into glucose (a monosaccharide) and maltose (a disaccharide). Cleansing of the mouth by saliva is also important for oral hygiene.

Thorough chewing is vital to health. Swallowing solid food before it's been adequately broken down and mixed with saliva places a real strain on the digestive process, and can result in inadequate digestion and malabsorption. Insufficient chewing forces the digestive tract to secrete higher levels of powerful digestive enzymes, which can result immediately in excess gas and bloating. Over time, these enzymes can damage both the stomach and digestive system. So slow down when you eat.

Step 2. Next, food travels through the esophagus, where involuntary peristaltic contractions (wormlike movements) send it to the stomach. The stomach can expand to accommodate different-size meals. It serves as a temporary way station that releases food into the intestines in a gradual and controlled manner, through peristaltic contractions as well as overall contractions of the gastric wall.

Step 3. The gastric mucosal cells lining the stomach wall secrete about a quart of gastric juice each day. This juice is the most acidic bodily fluid, consisting largely of hydrochloric acid, with a pH between 1.0 and 2.0. This acid dissolves food into a nearly liquid form called chyme. Gastric juice also contains digestive enzymes, most notably pepsins, which begin to break down proteins into their constituent amino acids. By the time the chyme leaves the stomach, these digestive processes have broken down about 30 to 50 percent of the starches and 10 to 15 percent of the proteins, but virtually none of the fats.

Gastric juice also contains a special protein called *intrinsic factor*, which is required for the absorption of vitamin B_{12}. Inadequate B_{12} absorption interferes with proper folic acid cycle metabolism, resulting in elevated levels of homocysteine (see chapter 13). To assess this condition, it is not enough to simply test for blood levels of B_{12}, because this will not reveal if the body is able to use this nutrient. A better test is to measure blood levels of methylmalonic acid, which is a metabolic intermediate that requires vitamin B_{12} to break down. If methylmalonic acid levels are elevated in the blood, it indicates the vitamin is not doing its job, regardless of the actual level of B_{12} in the blood. Inadequate B_{12} metabolism can be helped by supplementation

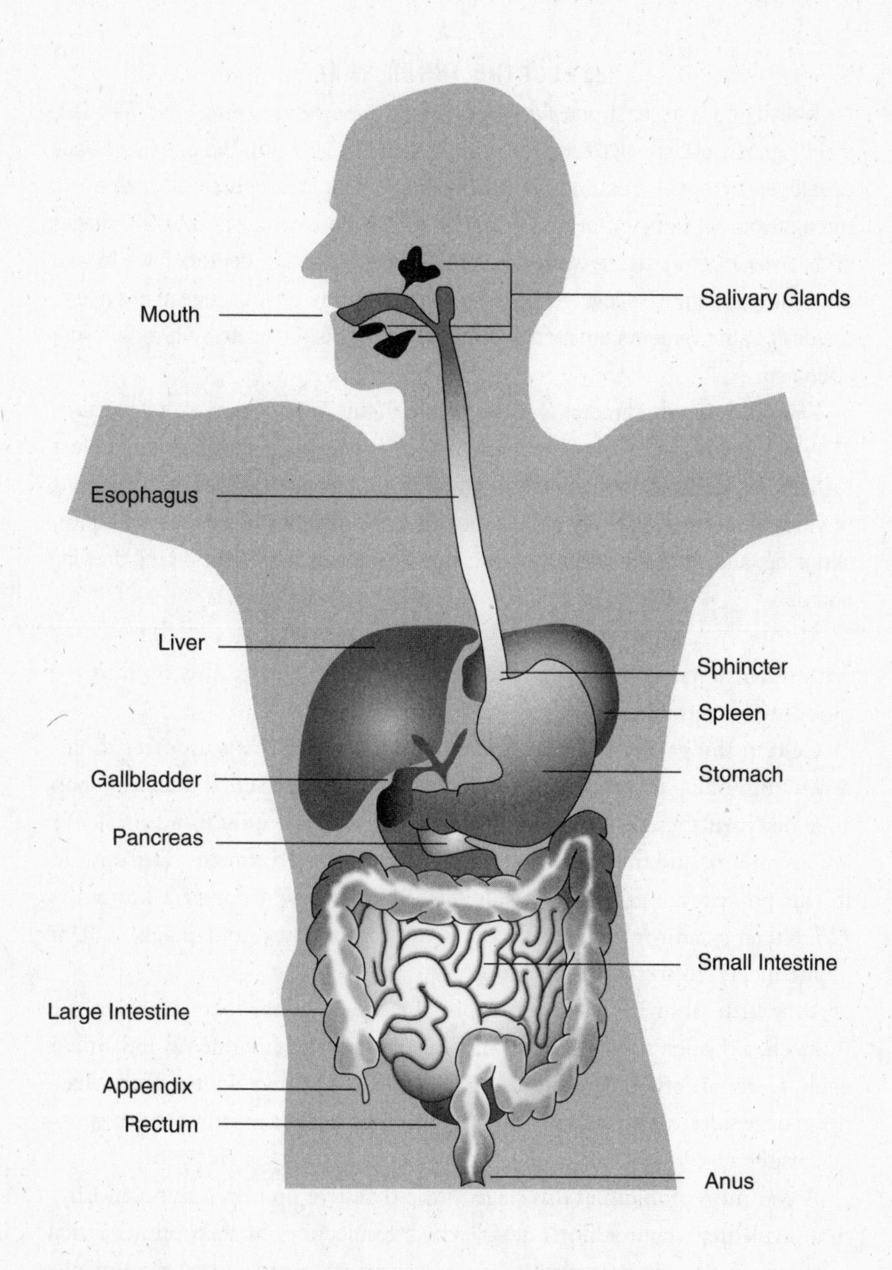

FIGURE 7-1. THE HUMAN GASTROINTESTINAL SYSTEM

BRIDGE TWO

EXPLORING INNER SPACE

We already have at least one *Fantastic Voyage*–type vessel being used for medical diagnosis of digestive function. Called the M2A gut cam, the pill-size device developed by Israeli company Given Imaging Ltd. allows doctors to take a trip through your GI tract. After being swallowed, the capsule takes 57,000 images of its voyage before being expelled and retrieved. Although doctors have to wait until the gut cam is recovered before viewing a movie of your digestive system, the device's movements are recorded in real time by eight sensors placed on your abdomen.

Martin Schmidt, director of MIT's Microsystems Technology Laboratory, says, "The pill is an icon for a whole class of small, potentially disposable, wireless technologies. The technologies for making these types of devices are becoming more pervasive and more accessible." Implantable diagnostic devices are rapidly shrinking, and there are already blood cell–size diagnostic devices on the drawing boards.

with intrinsic factor and introducing higher levels of B_{12} directly into the bloodstream through sublingual tablets or injection.

Cells in the gastric wall also make two kinds of mucus to protect themselves from being digested by their own acidic juices. There is a delicate balance between these acids, pepsins (which are effective only in a very acidic environment), and the protective action of the stomach mucus. Disruptions in this process are often the result of infection with a bacteria known as *Helicobacter pylori* (or *H. pylori*), which can result in severe damage, such as peptic ulcers, to the stomach lining.

Very little absorption of food into the bloodstream takes place in the stomach, although small amounts of simple sugars such as glucose and amino acids do get absorbed through the gastric mucosa. Ethyl alcohol from alcoholic beverages is also readily absorbed directly from the stomach, as is excess water.

A common problem at this stage in the digestive process is hypochlorhydria (inadequate hydrochloric acid), which results in poor absorption of vital nutrients. Indigestion resulting from a lack of stomach acid is often misdiagnosed as caused by excess acid and treated with powerful acid-suppressing medications, which lead to even further digestive maladies.[2] See Fantastic-Voyage.net for recommendations for alleviating this problem.

Step 4. After spending up to about three hours in the stomach (or longer, if the meal has a high fat content or you take antacids), the chyme exits through the pyloric valve and enters the small intestine. About 25 feet in length, the small intestine, or "gut," is the body's longest organ and its principal digestive organ. During this stage we see all three methods of digestion: motor activity, enzyme secretion, and absorption. The primary kind of movement of chyme through the gut is from segment-to-segment contractions, although peristaltic contractions also play a role.

The small intestine has three regions: the duodenum, the jejunum, and the ileum. Slightly less than a foot in length, the duodenum receives digestive enzymes from the pancreas along with bile formed in the liver and concentrated in the gallbladder. Pancreatic enzymes, including trypsin and chymotrypsin (which break down proteins into amino acids), lipase (which breaks down fats), and additional amylase for continued digestion of polysaccharides, are also secreted into the duodenum.

Bile is excreted into the duodenum as a slurry containing many solids, chiefly the bile salts, which assist in fat digestion. Bile emulsifies fats so they mix well with the other digestive juices, and also activates lipase, the pancreatic enzyme that breaks down fats into their constituent fatty acids. The bile and partially digested fat form small colloidal particles called micelles, which are absorbed by special ducts, called lacteals, into the bloodstream. The mucosal cells in the duodenum also secrete digestive enzymes, including a form of pepsin (for protein), amylase (for starch), and lactase (for digesting lactose or milk sugar).

About a third of all American adults, including 80 percent of African-Americans and 50 percent of Hispanics, lack adequate lactase, the enzyme needed to digest lactose (milk sugar). This well-known digestive deficit, known as lactose intolerance, affects a significant portion the world's population.[3]

Hormones control the rate that chyme is released from the stomach, and the duodenum controls that part of the nervous system that regulates the movement of food through the gut.

Step 5. Some absorption of nutrients takes place in the duodenum, but most occurs in the next section, the jejunum. This middle region uses folds in the mucosa called the plicae circulares to increase the surface area of the small intestine for optimal absorption. The most important structures for absorption are the intestinal villi, which are well developed in the jejunum. These fingerlike projections are about a millimeter in height. The epithelial cells on the surface of the villi have their own fingers called microvilli, which increase the surface area even further.

Step 6. Remaining nutrients are digested by less-developed villi (that don't have microvilli) in the ileum, the longest portion of the gut. Vitamin B_{12}, which must be bound to intrinsic factor, can be absorbed only in the ileum.

Step 7. Digestion is essentially complete by the time the chyme passes from the small to the large intestine. The large intestine, also known as the colon or bowel, typically receives about a pint of chyme a day from the ileum. The chyme is moved primarily by segmental contractions similar to those of the small intestine. The mucosal cells in the large intestine mostly secrete potassium, bicarbonate, and mucus to lubricate the chyme and facilitate its movement.

The large intestine contains large, column-shaped epithelial cells that absorb water, sodium, and chloride. The main digestive process that occurs in the colon is the result of intestinal bacteria interacting with nonabsorbable materials, including cellulose and other forms of fiber. These bacteria play an important role in changing the chyme into a form that is suitable for elimination as feces.

The digestive process ends with a bowel movement, which is often triggered by consumption of a meal—but normally, it is not the most recent meal that is being eliminated. The entire trip through the alimentary canal typically takes 24 to 48 hours or longer.

Reduce the Risk of Colon Cancer

Unhealthy bowel bacteria (or flora) create toxins that can result in intestinal dysfunction and, over time, bowel cancer.[4] Your diet plays a powerful role in the type of bacteria that inhabit your colon. Diets high in meat, which typically have highly concentrated toxins, are associated with an elevated risk of colon cancer,[5] as are diets that are low in fiber.[6]

We don't know whether protection comes from the fiber itself or from other healthful nutrients, such as vitamins, found in high-fiber foods; research has not yet clarified that question. But it's possible fiber helps by moving the chyme through the bowel more quickly, thereby reducing the opportunity for toxins to damage the delicate tissues of the large intestine. For this reason we recommend that you consume adequate amounts of fiber-rich foods, including vegetables (such as broccoli, cauliflower, and salad greens) and legumes (such as beans and lentils). Focus particularly on insoluble fiber—at least 10 to 15 grams per day.

EVERYONE IS DIFFERENT . . . TO AN EXTENT

While 99.8 percent of our genome is shared across the human race, we each have our own genetic profile, so the ideal diet will vary from individual to individual. Plus, there are differences in life experience, such as diet and disease, that have an impact on our digestive systems.

There are several ways to fine-tune your dietary program to reflect your individual situation:

- Assess your digestive system function (see the section below).
- The rate of metabolism varies from person to person (see chapter 8, "Change Your Weight for Life in One Day," on weight and caloric restriction).
- Some people do well on a diet that emphasizes animal protein, whereas others may fare better on a more vegetarian diet.[7]

Different people do have different metabolic tendencies, so the best way to assess this is through personal experimentation. If you feel better when eating more animal protein, emphasize fish and lean meats (such as skinless white poultry). In this case, you may also want to increase the overall amount of healthy fats in your diet. If you feel better on a more vegetarian diet, eat less animal fat and protein and get your healthy fats from nuts, avocados, and flaxseed. Persons of all metabolic types do well eating vegetables, so emphasizing vegetables is always a safe bet.

ASSESSING YOUR DIGESTIVE PROCESS

The effectiveness and long-term health of our complex digestive process depends on maintaining many delicate balances. We recommend noninvasive blood, stool, and urine tests. Guidance from a qualified, nutritionally oriented health practitioner to interpret these results and decide on appropriate remedial steps is valuable. Remediation of abnormalities and imbalances may include dietary changes, supplements, and medications. See Fantastic-Voyage.net for specific testing lab recommendations.

Routine Tests

In the absence of specific gastrointestinal symptoms, we recommend the following tests every two to five years.

Comprehensive digestive stool analysis (CDSA). This assesses digestion, absorption, metabolism, yeast levels, and levels of healthy and pathological bacteria.[8] To address inadequate levels of beneficial bacteria or an excess of unhealthful bacteria in the large intestine, we recommend the "seed, feed, and weed" approach.[9] "Multiflora" supplementation, which contain billions of spores of living healthy probiotic (helpful) bacteria, especially lactobacillus and bifidobacteria, effectively seed the colon. Healthful bacteria can be fed through supplements of fructooligosaccharides (FOS).[10] Unhealthful bacteria, fungal infections, and parasites need to be weeded (removed) using appropriate supplements or medications. See the Resource Section on Fantastic-Voyage.net for specific recommendations.

Evaluation of hair minerals for nutrient minerals and toxic heavy metals. Low levels of beneficial minerals such as magnesium and zinc suggest *hypochlorhydria* (low stomach acid), since you need adequate stomach acid to absorb minerals properly. This common condition is easily diagnosed and remedied through supplementation with betaine hydrochloride, a capsule form of hydrochloric acid. Excessive levels of toxic metals, such as mercury and lead, suggest that detoxification with "chelating" agents, chemicals that are able to flush these toxic metals from the body, would be helpful.[11]

Food antibodies blood test. Food sensitivities, revealed by IgG antibodies, can be helped by avoiding the food identified in the test.[12] Depending on the severity of the antibody level, it may be possible to reintroduce a troublesome food in small quantities after a period of avoidance. However, true food allergies, indicated by the presence of IgE antibodies, are usually more problematic.[13] These foods often need to be eliminated completely.

For Chronic Gastrointestinal (GI) Symptoms

For persons with chronic GI symptoms such as indigestion, abdominal pain, diarrhea, irregularity, excessive gas, bloating, other digestive discomforts, the following tests are recommended in addition to those above, in consultation with your health professional.

Evaluation of the stool for parasites. Parasites affect a substantial fraction of the American population severely enough to cause GI symptoms and damage the GI tract. For example, up to 40 percent of U.S. adults have the *Toxoplasma gondii* parasite, and infection with giardiasis, cryptosporidiosis, and many other parasites is widespread.[14]

Evaluation for leaky gut syndrome. See "Leaky Gut Syndrome," opposite.

Breath tests. Lactose intolerance affects more than 50 million Americans, and we recommend a breath test for the condition if you have adverse reactions to dairy products. A breath test can also assess bacterial overgrowth of the small intestine, a common condition that causes a wide range of GI symptoms.[15]

Blood test for *H. pylori*. This reveals IgG antibodies to *H. pylori*, the primary cause of peptic ulcers, gastritis, and a contributing factor to gastric cancer.[16]

LEAKY GUT SYNDROME

A common digestive problem is leaky gut syndrome, which increases with age[17] and affects most people over age 50 to some degree. Leaky gut is caused by chronic inflammation that results in tiny spaces developing between the cells that line the small intestine. This allows toxins, bacteria, and undigested food particles in the intestinal chyme to enter the bloodstream directly, placing significant demands on the detoxification processes of the liver. These foreign substances can trigger an immune system response that, over time, can result in autoimmune reactions to otherwise healthy foods.[18] This in turn leads to chronic inflammatory reactions, which may contribute to the development of arthritis, asthma, and other autoimmune diseases. Leaky gut syndrome also leads to deficiencies of vitamins and minerals, even if adequate amounts are consumed in the diet.

Leaky gut syndrome is caused primarily by years of the wrong diet as well as consuming toxic substances. Over time, the digestive tract becomes irritated, and the integrity of the intestinal lining breaks down from the ingestion of high levels of sugar, processed starches, toxins such as food additives and pesticides, unhealthy bacteria, and harsh prescription drugs. A major contributing factor is the overuse of nonsteroidal anti-inflammatory drugs (NSAIDs) such as aspirin or ibuprofen.[19]

Leaky gut syndrome can be diagnosed through a urine test, which assesses the ability of the digestive system to absorb lactulose and mannitol, two sugars that are not metabolized. In conditions of good health, mannitol is readily absorbed by the gut and appears in the urine, while lactulose is minimally absorbed, so only trace amounts should appear in the urine. If a relatively large amount of lactulose ends up in the urine, it indicates a breakdown of intestinal integrity, suggesting leaky gut.

Treating leaky gut syndrome requires a multifaceted approach.

- Adopt a healthy diet in accordance with our other recommendations. This will reduce the inflammation and other assaults on the delicate tissues of the digestive tract (which caused the problem in the first place) and will allow gradual healing to take place.

- Avoid harsh foods such as caffeine and alcohol as well as medications that can injure the gut, such as NSAIDs.
- Reduce dietary toxins by eating organic foods.
- Take a food allergy and sensitivity test, and consider eliminating foods for which you show significant antibodies.

Be aware that leaky gut is often associated with an overgrowth of yeast or excessive levels of unhealthy bacteria in the intestinal tract.[20] These conditions should be diagnosed and appropriate antifungal and antimicrobial therapies used to kill these organisms.

- Both herbal medications as well as prescription drugs can be used to eradicate ("weed") these undesirable microorganisms. In addition, taking supplemental digestive enzymes can help break down foods, reducing the strain on the digestive system.
- Consume more fiber, which will move the intestinal chyme through more quickly and absorb toxins.
- Take probiotics (supplements containing healthy colonic bacteria, such as acidophilus) to help to seed healthy flora in the gut.
- Consider other helpful supplements, including aloe vera, garlic, and bioflavonoids.

Fructooligosaccharides (FOS) are a special type of fiber supplement that is not digested and provides nutrition for healthy gut bacteria. FOS consists of long chains of fructose units. We recommend 2 to 5 grams of FOS each day.

IRRITABLE BOWEL SYNDROME

Irritable Bowel Syndrome (IBS) is a frequently diagnosed condition with a wide variety of symptoms. IBS is actually not well defined; it has become somewhat of a catchall diagnosis for GI discomforts in the absence of specific indications of disease. IBS is estimated to affect 10 to 20 percent of the population.[21] Symptoms include abdominal cramps and pain, bloating, irregular bowel movements, diarrhea and/or constipation, feelings of incomplete evacuation from bowel movements, excessive passing of gas, nausea, heartburn, and excessive belching. IBS can be the aftermath of a period of actual GI disease, which often leaves the alimentary canal in an overly sensitive state. Another potential cause is a problem with the small intestine's own nervous-system "pacemaker" located in the duodenum.

Most often, IBS results from many years of the wrong diet. Over time, high levels of sugar, high-glycemic-load starchy foods, proinflammatory fats, and other dietary influences result in unhealthy gut bacteria and inflammatory processes. These cause the breakdown of intricate metabolic balances required for healthy gastrointestinal health. Diagnosis of IBS requires ruling out diseases such as Crohn's disease, inflammatory bowel disease, colon cancer, gastroesophageal reflux disease, and other inflammatory conditions.

Adopting the nutritional guidelines of Ray & Terry's Longevity Program is one important step toward resolving IBS and restoring overall GI health. Supplements that have been found to be very helpful in dramatically reducing IBS symptoms include enteric-coated peppermint oil[22] and Seacure, which contains amino acids derived from fish.[23]

EAT ORGANIC WHOLE FOODS

One of the drawbacks of modern factory-produced foods is the excessive use of chemicals and pesticides and the depletion of nutrients in the soil. An understanding of the principles of eating organic whole foods is important for optimal nutrition.

Whole foods are foods in their unprocessed state. Most food is far removed from this. Most grains are "polished," a process that removes the grain's outer coating, which contains selenium, an important antioxidant mineral; fiber; and other valuable nutrients. This processing is done primarily for convenience—longer shelf life and faster cooking time—but the loss in nutritional value is considerable. Avoid quick-cooking products such as instant rice or oatmeal, since these forms of grains are particularly heavily processed.

Another type of non–whole food is fruit juice, which is missing a key ingredient of the original fruit: its fiber. So fruit juice is equivalent to sugar water with some vitamins and minerals. To your body, drinking a glass of orange juice is not a lot different from eating a candy bar, in terms of the sudden increase in the blood levels of glucose and insulin. Eating whole fruit with its fiber content slows down the digestion of the fruit's sugar content.

"Whole" vegetables and grains are grown in "whole" soil that contains all of the minerals and other nutrients required for a nutrient-rich plant. The soil used in factory farms, however, tends to be depleted of important trace minerals. Factory farms also tend to use large amounts of pesticides, insecticides, and other chemicals that end up in the produce.[24] Animals that eat grain produced in this way will concentrate the poisonous chemicals in their fat cells, while those raised on organic produce tend to have significantly lower concentrations of these toxins. Since it takes about 7 pounds of grain to produce 1 pound of meat, toxic chemicals are more concentrated in meat than in vegetables and fruits, so relying on organic food sources is even more important when consuming meat.

Fresh organic vegetables retain their vitamins, minerals, enzymes, phytochemicals, and other nutrients. Frozen vegetables are a close second, because most nutrients (except for vitamin K and some enzymes that are destroyed by freezing) survive in a digestible form. Canned vegetables, on the other hand, look and taste "dead" because they have lost most of their useful nutrients. The basic caloric nutrients—carbohydrates, protein, fat—remain, but most of the vitamins and minerals do not survive. Moreover, most canned vegetables have a lot of added salt. Virtually all of the varied nutritional programs agree on one thing: emphasize fresh organic vegetables in your diet.

Why organic? "Organic foods" refers to produce, grains, and animals produced without artificial fertilizers, pesticides, insecticides, herbicides, hormones, antibiotics, and other chemicals, many of which are approved by the EPA and routinely used in commercial food production. These chemicals are classified by their toxicity, from Category I (highly toxic) to Category V (relatively nontoxic).[25] Consider terbutryn, a Category III herbicide used to destroy weeds in wheat and barley crops.[26] In animal tests, moderate doses of terbutryn caused nervous system defects, convulsions, cancer, and damage to the kidney and liver. Only tiny amounts of terbutryn are allowed on human food, but we simply don't know the impact of long-term continual exposure to small amounts of this chemical. The same is true for the scores of other chemicals, including insecticides and pesticides, routinely used on nonorganic foods.

We recommend that if you do eat nonorganic produce, soak it in dilute hydrogen peroxide before cooking or eating. Add ¼ cup of 3 percent hydrogen peroxide (food-grade hydrogen peroxide can be found in health food stores) to a sink full of water and soak the food in the solution for 20 minutes.

HOW TO EAT FOR HEALTH

The following principles and recommendations should be the basis of your healthy diet.

Eat a variety of foods. Eating the same foods day after day may cause sensitivities and allergies to develop.[27] Also, eating the same foods will cause "taste fatigue" and encourage overeating. You will be more satisfied and enjoy your food more if you vary what you eat. Eating a variety of foods also will promote a balance of nutrients. Each vegetable has specific amino acids, vitamins, minerals, and other nutrients, but no single vegetable provides all of what you need. One way to practice this recommendation is to rotate your foods.

Reduce or eliminate wheat. Wheat, which is a relatively recent agricultural innovation, is eaten in enormous quantities in Western countries. This has led to high levels of wheat sensitivity, particularly to gluten, which is a major protein component of wheat. Many people have discovered that going off wheat has resolved long-standing digestive problems. You can experiment with this yourself by avoiding wheat for two weeks to see what impact this has on otherwise unresolved digestive symptoms. You can also assess your sensitivity or allergy to wheat through the type of food antibody testing discussed above, as well as by specific blood tests for "gluten intolerance." See Fantastic-Voyage.net for specific recommendations on tests.

Eat your vegetables. We cannot emphasize enough the benefits of eating fresh, organic, low-starch vegetables. They contain a myriad of valuable nutrients and fiber, and are low in glycemic index and caloric density. Be careful not to overcook them, though. Overcooking will deplete vegetables of their vitamins, phytochemicals, and other nutrients. We recommend light steaming as the ideal way to cook vegetables. Many can also be eaten raw, although excessive consumption of raw vegetables may cause GI distress.

Eat colorful foods (but not moldy meat!). By eating a variety of naturally colored vegetables, you obtain a broad spectrum of vital nutrients. An entertaining book on this topic is *Eat Your Colors* by Marcia Zimmerman.[28]

Drink freshly squeezed vegetable juice. About the healthiest drink available is made by putting fresh, organic, low-starch vegetables through a juicer. The result will be a low-calorie drink extremely high in vitamins, minerals, and phytochemicals. Ideal vegetables for juicing include celery, cucumber, and fennel. You can also use smaller amounts of red and green leaf lettuce, romaine lettuce, endive, escarole, spinach, parsley, and kale. Carrots and beets are high in sugar, so use these in moderation, if at all.

Drink tea instead of coffee. We are not opposed to moderate consumption of caffeine, which is useful for improving concentration. We strongly recommend tea rather than coffee, however. We talked about the highly acidic nature of coffee in chapter 4. Also, coffee contains very high levels of caffeine; black tea has about one-third the level of caffeine, and green tea has about one-fourth. Moreover, there are many healthful con-

(continued on page 100)

BRIDGE THREE

HUMAN DIGESTIVE SYSTEM VERSION 2.0

If we look a couple of decades into the future, we see that we will be able to fundamentally reengineer the way we provide nutrients to our trillions of cells. These nutrients include caloric (energy-bearing) substances such as glucose (from carbohydrates), proteins, fats, and a myriad of trace molecules, such as vitamins, minerals, and phytochemicals, which provide building blocks and facilitate enzymes for diverse metabolic processes.

Our species has already augmented the "natural" order of our life cycle through our technology: drugs, supplements, replacement parts for virtually all bodily systems, and many other interventions. We already have devices to replace our joints and extremities, teeth, skin, arteries, veins, and heart valves. Systems to replace more complex organs—our hearts, for example—are beginning to work. As we learn the principles of operation of the human body and the brain, we will soon be in a position to design vastly superior systems that will be more enjoyable, last longer, and perform better without being as susceptible to breakdown, disease, and aging.

We have already gone a long way toward disconnecting the relational and sensual aspects of sex from its biological role in reproduction. So why don't we provide the same separation from biological purpose for another activity that also provides both social intimacy and sensual pleasure—namely, eating? Ultimately, the exact nutrients each individual needs will be fully understood and easily and inexpensively available, so we won't have to bother with extracting nutrients from food at all. This technology should be reasonably mature by the late 2020s. Nutrients will be introduced directly into the bloodstream by special metabolic nanobots. Sensors in our bloodstream and body, using wireless communication, will provide dynamic information on the nutrients needed at each moment.

A key question in designing this technology will be how these nanobots make their way in and out of the body. The technologies we have today, such as intravenous catheters, leave much to be desired. Unlike drugs and nutritional supplements, nanobots have a measure of intelligence, so they could keep track of their own inventories and slip in and out of our bodies in clever ways. They will be able to coordinate their activities and pool their collective intelligence in the same way a supercomputer is made up of many smaller computers operating in parallel. One scenario is that you would wear a special "nutrient garment" such as a belt. This garment would be loaded with nutrient-bearing nanobots, which would make their way in and out of your body through the skin or other body cavities.

At that stage of technological development, you will be able to eat whatever you want, whatever gives you pleasure and gastronomic fulfillment, and thereby enjoy the culinary arts for their tastes, textures, and aromas. At the same time, you will be provided with an optimal flow of nutrients to your bloodstream, using a completely separate process. One possibility would be that all the food you eat would pass through a digestive tract that's disconnected from any possible absorption into the bloodstream.

This would place a burden on your colon and bowel functions, so a more refined approach will dispense with the function of elimination, instead using special nanobots that act like tiny garbage compactors. As the nutrient nanobots make their way from the nutrient garment into your body, the elimination nanobots will go the other way. Periodically, you'll replace the nutrition garment with a fresh one.

Ultimately, you won't need to bother with special garments or explicit nutritional resources. Just as computation will eventually be available everywhere, basic metabolic nanobot resources will be embedded everywhere in your environment.

In addition, an important aspect of this system will be maintaining ample reserves of all needed resources inside the body. Our "version 1.0" bodies of today do this to only a very limited extent, for example, storing a few minutes of oxygen in your blood and a few weeks or months of caloric energy in glycogen and fat reserves. "Version 2.0" of the human body will provide substantially greater reserves, enabling you to be separated from metabolic resources such as air and nutrition for greatly extended periods of time.

Our adoption of these technologies will be cautious and incremental. We won't dispense with the old-fashioned digestive process when these technologies are first introduced. Most of us will wait for digestive system version 2.1 or even 2.2. After all, most people didn't throw away their electric typewriters when the first generation of word processors was introduced. People still hold on to their vinyl record collections many years after CDs came out. People still keep their film cameras, although the tide has turned in favor of digital cameras. On the other hand, how many people still have a mechanical typewriter?

The same phenomenon will happen with our reengineered bodies. Once we've worked out the inevitable complications that will arise with a radically reengineered gastrointestinal system, we will begin to rely on it more and more.

stituents of tea. A recent study published in *Circulation*, the journal of the American Heart Association, found that drinking at least two cups of tea a day reduced the risk of dying from a heart attack by a remarkable 44 percent.[29] This finding applies to both black and green teas, but not to herbal teas. Tea also contains l-theonine, which reduces cortisol levels and promotes relaxation. Green tea is particularly beneficial, with additional antioxidants that reduce the risk of both heart disease and cancer.[30]

Go easy on the alcohol. Moderate consumption of alcohol has been linked to reduced rates of heart disease and stroke, apparently by promoting better vessel health.[31] Keep in mind, however, that even though alcohol is technically not a carbohydrate, it is metabolized similarly and has a relatively high glycemic load. The dangers of excessive use and dependence on alcohol are well known.

Eat breakfast, and eat frequently. It is important not to skip the initial meal of the day, as this will create fatigue and potentially low blood sugar levels. It is best to eat a healthful breakfast and have several smaller meals rather than one or two large ones. Eating less at a time, but more often, avoids straining the digestive system and also minimizes insulin spikes that lead to insulin resistance and carbohydrate cravings.

Avoid unhealthful snacks. Typical snack foods are high in high-glycemic-index starches, sugar, salt, and unhealthy fats that encourage cravings for more snack foods. Instead, satisfy hunger with low-starch vegetables and small amounts of fruit.

Plan ahead. If you are going to a restaurant or a party, or traveling, bring healthful foods and condiments with you so you can stick to your nutritional program. For example, lemon juice and a stevia-based sweetener makes a tasty but zero-calorie salad dressing.

Take supplements. It is desirable to obtain as many nutrients as possible directly from food, which contains synergistic nutrients such as phytochemicals that will support the many metabolic pathways in your body. However, it is not possible to obtain all of the nutrients you need for optimal health from food; also, "reprogramming" your biochemistry through supplements becomes increasingly important as you get older. Remember that evolution is not on your side after child-rearing age.

"Everything in moderation, including moderation." It's particularly important not to be discouraged by going off your nutritional plan, even if unplanned. Many nutritional programs are destroyed by the slippery slope of becoming overly discouraged by a temporary lapse of discipline. If you "fall off the horse," get right back on!

Be aware: Sugar is everywhere! Your "health food" supermarket is likely to have an entire section stocked with supposedly healthy milk alternatives: soy milk, rice milk, almond milk. Yet almost every one of these is loaded with sugar. We recommend that you find one of the few alternative milks (see Fantastic-Voyage.net) that are unsweetened, and then add stevia, if desired. Sugar is added to most cereals, even so-called low-sugar cereals. Sugar goes by different names, so be on the lookout for its various forms on ingredient lists, most of which end in "-ose": sucrose, fructose, glucose, and maltose. Foods that are basically sugar include honey, molasses, maple syrup, sucanat, amasake, and high-fructose corn syrup.

A REVIEW OF OTHER NUTRITIONAL PROGRAMS

A number of observers have described our nutritional program as "the best of the popular low-fat and low-carb diets." We do provide the valid ideas from these two opposing poles of nutritional philosophy, but there are key issues that both approaches miss.

The late Dr. Robert C. Atkins (author of *Dr. Atkins' New Diet Revolution* and many other books)[32] was correct in stressing the importance of glycemic index, eliminating high-glycemic-index carbohydrates, and reducing carbohydrate levels overall, for all the reasons we've cited. The plan has traditionally ignored, however, the issue of good versus bad fats and includes excessive

Avoid Acrylamide

A meeting of the World Health Organization was held in 2002 to respond to the recent discovery of unexpectedly high levels of acrylamide, a potent carcinogen, in a wide variety of carbohydrate-rich foods cooked with traditional high-temperature cooking processes, such as baking and frying.[33] The EPA limits acrylamide levels in tap water to 0.12 micrograms per 8-ounce serving. Levels in french fries from popular fast-food chains were found to be as high as 72 micrograms per 6-ounce serving. Potato chips and other baked and fried snack foods contained as much as 25 micrograms in a 1-ounce serving. Potatoes and other vegetables and grains have virtually no acrylamide when raw, so this chemical is created as a result of the cooking process.

There are many reasons not to eat french fries, snack chips, and similar foods, including high levels of unhealthy fat, starch, sugar, and salt. We can now add concerns about acrylamide to the list.

Table 7-1. The Ray & Terry Diet

FOODS TO AVOID	FOODS TO EAT IN LIMITED QUANTITIES	GENERALLY GOOD FOODS TO EAT (THE FOLLOWING SHOULD BE THE MAINSTAY OF YOUR NUTRITIONAL PROGRAM
White-flour products, such as white bread, pastries, pasta	Whole grains	Legumes such as lentils and beans
Sugar, in its raw state as well as in packaged foods	Fruit	Low-starch vegetables, fresh and lightly cooked, particularly spinach, kale, collards, cabbage, broccoli, cauliflower, red and green leaf lettuce, romaine lettuce, endive, other salad greens, bok choy, fennel, Chinese cabbage, celery, cucumbers, brussels sprouts, zucchini, green vegetables in general
Hydrogenated fats	Lean meats, particularly white poultry without the skin	Vegetable proteins
Commercially prepared oils	Wheat	Foods rich in omega-3 fats, such as nuts, avocados, flaxseed, ostrich (a tasty alternative to red meat that contains almost no fat)
Fried foods	Stir-fry foods	—
Soft drinks	Alcohol, caffeine	Filtered water
Unfiltered tap water	—	—
Baked or fried breads and potatoes	—	—
Trans fats, such as margarine and shortening	—	Extra-virgin olive oil
MSG	—	—
Artificial preservatives and chemicals	—	—
Artificial sweeteners such as aspartame and saccharin	—	Stevia
Foods with a high glycemic index, such as cereals and starchy vegetables	—	—
Foods "fortified" with iron (because iron may accelerate the oxidation of LDL cholesterol, an early step in atherosclerosis)	—	—
Larger fish such as tuna, swordfish, sailfish, and shark (they may contain concentrated amounts of mercury)	—	Fish such as anchovies, sardines, and wild salmon (farm-raised salmon is much lower in omega-3 fats)

Tips on Reading Food Labels

- FDA regulations require fiber to be included in both carbohydrate listings and total calories. However, fiber is not digested, so you can delete the number of fiber grams from the carbohydrate count. You can also delete 4 calories per fiber gram from total calories.
- The carbohydrate gram count includes both high- and low-glycemic-index carbohydrates, so you will need to examine the ingredients list.
- Similarly, the fat gram count includes both healthy and unhealthy fats, so study the ingredients list.
- Avoid products with additives, particularly if you are not familiar with the additives listed.

levels of saturated fats and omega-6 fats, although the program has recently taken a modest step in the right direction. Paul D. Wolff, CEO of Atkins Nutritionals, was recently quoted in the *New York Times* as acknowledging, "The way the book was promoted was, here's the program that is counterintuitive. 'You can eat a lot of bacon and steak.' The media saw it as a sexy story." In her seminars on the Atkins program, Colette Heimowitz, director of research and education for Atkins Nutritionals, is now recommending that followers of the Atkins diet restrict saturated fat to 20 percent of calories. This is a change from the 2002 recommendation of 35 percent protein, 60 percent fat, and 5 percent carbohydrates. We believe that 20 percent is still much too high. We recommend a limit of 3 percent calories from saturated fat.

Dr. Dean Ornish (author of *Eat More, Weigh Less, Dr. Dean Ornish's Program for Reversing Heart Disease*, and other books)[34] has historically emphasized cutting down on all fats, although he has for many years also recommended omega-3 fats. Cutting down on all fats does tend to flatten the omega-6 to omega-3 ratio, and Ornish maintains that only small amounts of omega-3 fats are needed. The mainstay of the diet is vegetables, which fully complies with our recommendations. Historically, he has encouraged the consumption of foods high in fiber, and discouraged consumption of processed foods, sugar, white flour, and other high-glycemic-index foods, and recently has been emphasizing this more.

An earlier "low-fat" advocate, Nathan Pritikin, was quite sensitive to glycemic index, although this term did not exist when he wrote during the 1960s and 1970s. Pritikin was as anti-sugar as anti-fat and opposed to refined grains. Ray & Terry's Longevity Program recommends sharply restricting high-glycemic-load carbs and moderate reductions in all carbs. Regarding

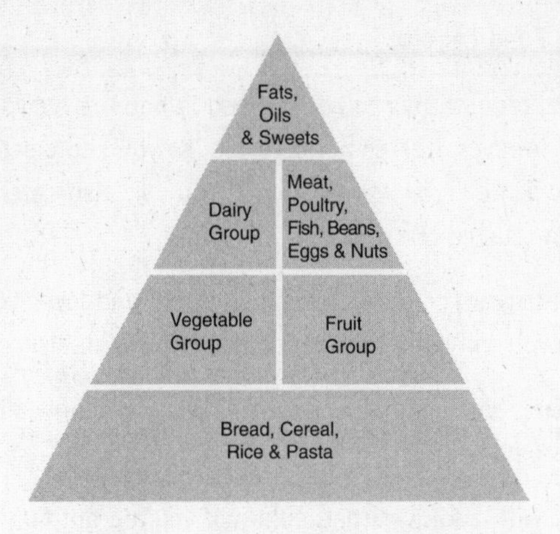

FIGURE 7-2. THE USDA FOOD PYRAMID

fats, we recommend sharp reductions in pro-inflammatory fats (omega-6 fatty acids, saturated fat, trans-fatty acids) and emphasize anti-inflammatory fats (omega-3 fats in fish and nuts, oleic acid in olive oil). We also recommend aggressive supplementation based on age and testing to identify specific issues.

THE FOOD PYRAMID—THEIRS VERSUS OURS

The U.S. Department of Agriculture (USDA) published its now well-known food pyramid in 1992, with small revisions in 1996.[35] Although this influential recommendation may have had a few beneficial results, such as reducing saturated fats (along with all fats) and modestly encouraging vegetables, these recommendations are unhealthy in many ways:

- The mainstays (and largest level) of the USDA food pyramid are high-glycemic-load starches. No distinction is made between low- and high-gylcemic-load carbohydrates.
- Similarly, no distinction is made between healthy and unhealthy fats. All fats and oils are lumped together, as are meat and fish.
- Undue emphasis is made on dairy products, which critics have linked to extensive lobbying from the dairy industry.

In 2002, researchers at Harvard Medical School published their own food pyramid in a challenge to the USDA pyramid.[36]

The Harvard pyramid is a significant improvement because it moves high-glycemic-load starches from the bottom (largest) to the top (smallest).

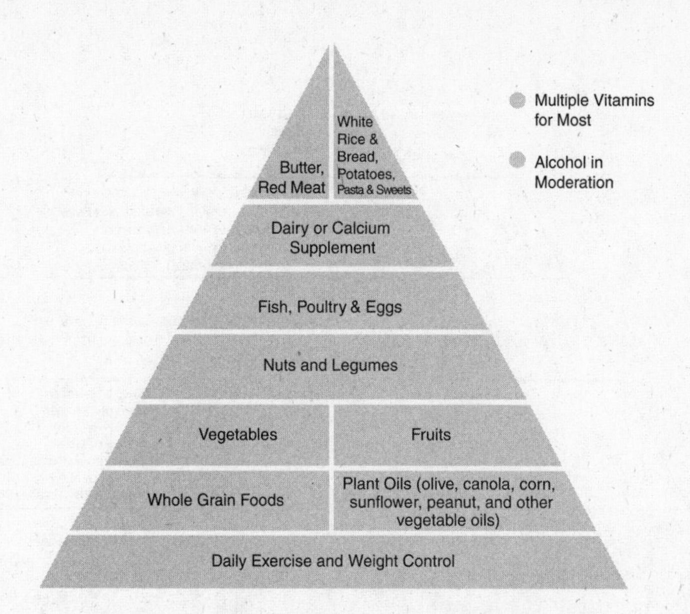

FIGURE 7-3. THE HARVARD MEDICAL SCHOOL FOOD PYRAMID

Although this a big step forward in public health recommendations, these guidelines also fail to make some important distinctions:

- Plant oils make up half of the largest nutritional level. No distinction is made between a healthful oil such as extra virgin olive oil, which is rich in oleic acid, and commercially processed vegetable oils, which are high in pro-inflammatory omega-6 fats, as well as pathological forms of polyunsaturated fat. The booklet accompanying the Harvard food pyramid encourages the consumption of polyunsaturated fat but, as we pointed out in chapter 6, "Fat and Protein," not all polyunsaturated fats are healthy.

- Whole-grain foods are placed at a more prominent level than vegetables. Whole grains, although much healthier than refined grains, are nonetheless relatively high in glycemic load. The proper emphasis should be on vegetables, not grains. Vegetables should also have priority over fruit, which is also relatively high in glycemic index.

- Fish has the same priority as poultry and eggs here, but it should be given a higher priority because of its high omega-3 fat level.

- We do not believe that high-fat dairy products should be on the chart at all.

Ray & Terry's food pyramid emphasizes low-glycemic-load vegetables, which are ideal sources of antioxidant vitamins and other nutrients. It also em-

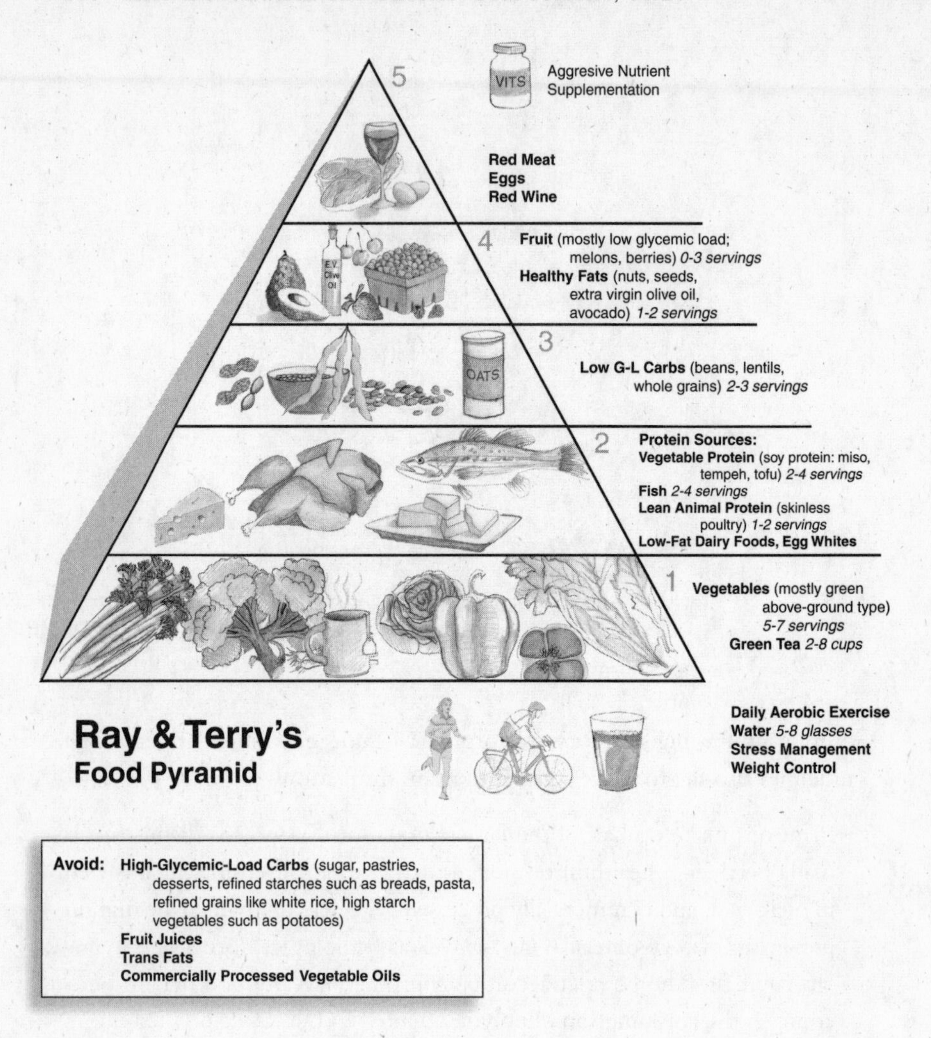

FIGURE 7-4. RAY & TERRY'S FOOD PYRAMID

phasizes healthy fats, including fish, nuts, seeds, extra virgin olive oil, and soy products such as tofu. Unlike the Harvard pyramid, it does not recommend all plant oils and polyunsaturated fats because many of these are high in pro-inflammatory omega-6 fats, trans-fatty acids, and pathological forms of fat.

To conclude our nutritional recommendations, we recognize the many-faceted role that food plays in our cultural, social, and emotional lives, and that it is not always possible to follow a "perfect" set of nutritional guidelines. Our purpose here is to communicate what our current scientific understanding indicates will best facilitate our health and well-being. The point is to strive for a set of guidelines that reflects all of our current knowledge and does not come precompromised.

8

CHANGE YOUR WEIGHT FOR LIFE IN ONE DAY

"I have measured out my life with coffee spoons . . ."

—T. S. Eliot, "The Love Song of J. Alfred Prufrock"

According to a 2003 report in the *Journal of the American Medical Association* (*JAMA*), an obese 20-year-old man has a life expectancy 13 years less than contemporaries of normal weight.[1] Recent research has shown significant health benefits from maintaining the weight you had when you were 20 (assuming that it was an optimal weight) throughout your life.[2] You can significantly reduce your risk of all degenerative diseases, including heart disease, cancer, type 2 diabetes, and hypertension.[3] Being 20 percent overweight triples your risk of high blood pressure and diabetes, doubles your risk of elevated cholesterol (more than 250 mg/dL), and increases your risk of heart disease by 60 percent.[4] At your optimal weight, you'll also have more energy and feel better in general.

Of course, you'll look better too, which is perhaps the main reason losing weight has become a national preoccupation. At any one time, a majority of adult Americans are attempting to lose weight.[5] Americans spend $41 billion a year on diets and diet aids. [6] The sad fact is that 95 percent of dieters will regain all of the weight they have lost—or more. [7] The cycle of taking off weight and then putting it back on, called yo-yo dieting, is actually worse for your health than never having lost it because continual weight changes stress the body. [8]

The key to a successful program is to change your attitude toward losing weight. Rather than thinking of dieting as a temporary period of deprivation, think of it instead as a long-term commitment to a healthy pattern of eating, but one you can make in a single day—like today! While the full benefits, including reaching your optimal weight, won't be achieved in one day, you can start feeling better about yourself and your commitment immediately. And many of the immediate benefits will become evident in only a couple of weeks.

BRIDGE THREE

NANOBOTS FOR DINNER

In the previous chapter, we described nanobots that will provide digestive functions in the future. These nanobots will start out by augmenting your biological digestive system: adding nutrients to the bloodstream that are in short supply and removing unwanted substances such as toxins and excess nutrients. Ultimately, this "version 2.0" digestive system will be so reliable that it will completely replace the "version 1.0" system we rely on today. Even at the earlier augmenting stage, this technology will effectively eliminate excess weight as a problem by destroying unneeded calories. We expect to see technology of this type emerging around 2020, and a refined system that can replace your digestive system is likely by around 2030.

IMPLEMENTING YOUR WEIGHT-LOSS PROGRAM

By following our program, outlined below, you will begin to lose your excess weight slowly and gradually approach your ideal weight. The idea is to place yourself on the right course, one that avoids feelings of deprivation. You must also be patient.

Step One: Determine your body frame size

The first step in our "change your weight for life in one day" program is to determine your body frame size. This will help you pinpoint your ideal weight.

a. First, measure your wrist circumference. You can use a tape measure or a piece of string, and then hold it against a ruler.
b. From Table 8-1, use this wrist measurement to determine your innate build, or frame size. (Unless you're very obese, your wrists don't change size with weight, so their circumference is a good indicator of your natural build. Many overweight people assume they have large builds, whereas what they really have are medium or even small builds with a lot of extra padding.)

Table 8-1: Assessing Frame Size from Wrist Circumference

	SMALL FRAME	MEDIUM FRAME	LARGE FRAME
Adult males	Under 6¼ in.	6¼–7 in.	Over 7 in.
Adult females	Under 5¼ in.	5¼–6 in.	Over 6 in.

Step Two: Determine your optimal weight range

Now that you've determined your frame size, use this information along with your gender and height to determine your optimal weight from Table 8-2 on page 110 (your optimal weight is at the bottom of the range).

Step Three: Determine and adopt your target calorie level

Here is the main principle of our "change your weight for life in one day" program: Begin to consume the number of calories you would need if you were already at your optimal weight. This is your target calorie level.

There are two alternative methods to do this:

Method A: Look up your maintenance calorie level (the number of calories you need to consume to maintain a particular weight) in Table 8-3 on page 111, based on your optimal weight (the bottom of the optimal weight range from step two above) and exercise level. If your optimal weight is in between two rows of Table 8-3, interpolate to get the calorie level. For example, if you weigh 155—half (0.5) of that between 150 and 160—and are moderately active: (2,400–2,250) × 0.5 = 75; so your daily maintenance calorie level would be 2,250 + 75 = 2,325.

We recommend that you maintain an exercise program that is at least at the moderate activity level. If you maintain the maintenance calorie level for your optimal weight, your weight loss (or gain if you are underweight) will gradually taper off as you approach this optimal level.

Method B: Look up the maintenance calorie level in Table 8-3 for your *current* weight and exercise level, then subtract 500 calories per day to lose 1 pound per week, or 1,000 calories per day to lose 2 pounds per week.[9]

Our preference is that you use Method A, the "change your weight for life in one day" program, where you begin by consuming the number of calories needed to maintain your target (optimal) weight. Keep in mind that everyone's metabolism is different, so you will need to experiment to adjust these figures appropriately.

With either method, it is important to get adequate, balanced nutrition while you are losing weight. For long-term success, your goal should be to adjust to a new way of eating—that is, to change your taste preferences and attitudes toward food. You do not want to feel deprived and hungry, which would defeat the process. You should eat at least 10 calories for each pound of your ideal (optimal) weight—no less than 1,000 calories per day for a woman or 1,200 calories per day for a man.

Table 8-2: Determining Estimated Optimal Weight (weight in pounds, in indoor clothing)[10]

HEIGHT AND WEIGHT TABLE FOR WOMEN

Height (feet/inches)	Small frame	Medium frame	Large frame
4'10"	102–111	109–121	118–131
4'11"	103–113	111–123	120–134
5'0"	104–115	113–126	122–137
5'1"	106–118	115–129	125–140
5'2"	108–121	118–132	128–143
5'3"	111–124	121–135	131–147
5'4"	114–127	124–138	134–151
5'5"	117–130	127–141	137–155
5'6"	120–133	130–144	140–159
5'7"	123–136	133–147	143–163
5'8"	126–139	136–150	146–167
5'9"	129–142	139–153	149–170
5'10"	132–145	142–156	152–173
5'11"	135–148	145–159	155–176
6'0"	138–151	148–162	158–179

Women between 18 and 25 should subtract 1 pound for each year under 25.

HEIGHT AND WEIGHT TABLE FOR MEN

Height (feet/inches)	Small frame	Medium frame	Large frame
5'2"	128–134	131–141	138–150
5'3"	130–136	133–143	140–153
5"4"	132–138	135–145	142–156
5'5"	134–140	137–148	144–160
5'6"	136–142	139–151	146–164
5'7"	138–145	142–154	149–168
5'8"	140–148	145–157	152–172
5'9"	142–151	148–160	155–176
5'10"	144–154	151–163	158–180
5'11"	146–157	154–166	161–184
6'0"	149–160	157–170	164–188
6'1"	152–164	160–174	168–192
6'2"	155–168	164–178	172–197
6'3"	158–172	167–182	176–202
6'4"	162–176	171–187	181–207

Table 8-3: Maintenance Calorie Level

This table provides an estimated maintenance calorie level based on your optimal or current weight and activity level.

WEIGHT	SEDENTARY	MODERATELY ACTIVE	VERY ACTIVE
90	1,170	1,350	1,620
100	1,300	1,500	1,800
110	1,430	1,650	1,980
120	1,560	1,800	2,160
130	1,690	1,950	2,340
140	1,820	2,100	2,520
150	1,950	2,250	2,700
160	2,080	2,400	2,880
170	2,210	2,550	3,060
180	2,340	2,700	3,240
190	2,470	2,850	3,420
200	2,600	3,000	3,600
210	2,730	3,150	3,780
220	2,860	3,300	3,960
230	2,990	3,450	4,140
240	3,120	3,600	4,320

Sedentary: You sit most of the day, walking only occasionally, and do not have a regular exercise routine.

Moderately active: Your normal routine involves frequent walking or physical motion. Alternatively, your normal routine is sedentary but you have a regular exercise program equivalent to walking or running 20 or more miles per week.

Very active: Your normal routine involves continual vigorous physical activity (such as construction work, carrying mail, gardening). "Very active" is equivalent to a sedentary lifestyle plus walking or running approximately 50 miles per week.

Note that your maintenance calorie level will change as your weight changes. Since metabolic rates vary from individual to individual, this chart provides only an approximate value, which you will need to adjust based on your own experience.

ACHIEVING YOUR OPTIMAL BODY FAT PERCENTAGE

Lean muscle mass is actually heavier than body fat. Two people can be the same height, build, and weight, yet one may be far less fat than the other. The real objective for weight loss is to lose fat, not muscle, water, or temporary glycogen deposits. In fact, it's more important to determine your ideal percentage of body fat than your ideal weight. You can follow the instructions in body-fat charts (see Fantastic-Voyage.net), but a more accurate method is to use a body-fat measuring device or scale that uses feedback from a small electrical current through your body. These devices are widely available (see the resource section on Fantastic-Voyage.net).

A certain amount of body fat is needed to cushion your vital organs and as your body's primary form of energy storage. Women need a little more to provide support for childbearing and breastfeeding, along with secondary sex characteristics. The ideal percentage of body fat is 12 percent to 20 percent for men, and 18 percent to 26 percent for women. But it's not healthy to be significantly above or below this percentage, and we recommend staying on the lean side of these ranges.[11]

It also makes a difference *where* excess fat resides on your body. Accumulation of visceral fat around your middle—central adiposity, commonly known as a potbelly—is particularly harmful. In a study published in *JAMA*, researchers showed that the risk of endometrial cancer was 15 times

Obesity Triggers Inflammation

Recent research has revealed an unexpected mechanism underlying the health damage from excess weight. In a study published in the *Journal of Clinical Investigation*, researchers reported that once obesity passes a critical threshold, the fat tissues become filled with enormous numbers of macrophages (immune system cells), which secrete inflammatory chemicals.[12] The macrophages also secrete tumor necrosis factor-alpha, which initiates insulin resistance, a cause of the metabolic syndrome and type 2 diabetes. The chronic activation of this type of inflammatory response underlies many degenerative diseases, including heart disease, stroke, and type 2 diabetes. The researchers hypothesized that once the fat cells become overly crowded, some start to break or leak, which triggers the macrophages to begin cleaning up the debris. These macrophages then release chemicals to call for more immune system support, which results in a cascade of inflammatory reaction.

higher in women with a high waist-to-hip ratio, compared with women with a low ratio.[13] Other studies have found strong links between central adiposity and type 2 diabetes, hypertension, and heart disease.[14]

SOME TIPS FOR HEALTHY—AND PERMANENT—WEIGHT LOSS

Your weight reflects your total calorie consumption, how much you exercise, and your metabolic rate, but the composition of the food you eat is also important. Here are some tips.

Reduce carbs. We have found that it's almost impossible to lose weight and keep it off without eating substantially fewer carbohydrates, particularly those with a high glycemic load (GL). As we discussed in chapter 5, "Carbohydrates and the Glycemic Load," consumption of high-GL carbohydrates leads to a desire for more carbohydrates. Eating a low-carbohydrate, low-GL diet will help you control your appetite and decrease cravings. You'll feel full sooner, you'll find it far easier to stop eating once you're satisfied, and you'll find yourself less hungry between meals. If you are trying to lose weight, we recommend you keep total carbohydrates under one-sixth of your calories and eliminate all high-GL carbohydrates such as sugary foods, pastas, and breads.

Reduce fats. Reducing fat in the diet aids weight loss because high-fat foods are more calorically dense—9 calories per gram versus 4 for carbohydrates and protein.

Go for veggies. Emphasize foods that are low in caloric density (that is, low in calories but high in weight). The ideal category: low-starch vegetables, which have a low glycemic index and are rich in valuable nutrients of all kinds, high in fiber, and filling.

Eat fiber. Consume at least 25 grams per day, including at least 10 grams of insoluble fiber.

Don't switch foods radically. While you are losing weight, we strongly recommend against diets that involve eating in a significantly different way from how you intend to eat when not "dieting."[15] People count the days until they are released from this type of gastronomic prison. They do not associate the benefit of weight loss with learning proper eating habits—changing tastes, desires, and attitudes—but rather with the artificial eating patterns that they are anxious to leave.

By adopting our "change your weight for life in one day" advice—to consume the maintenance number of calories for your optimal weight as your starting point—you need to make only a single adjustment once. It may take a couple of weeks to adapt, but you will be on a track that you can maintain indefinitely.

Make health, not weight loss, your goal. If you set a healthy lifestyle as your goal, you are more likely to succeed in both improving your health and attaining permanent weight loss. Don't be too anxious to drop pounds right away. Enjoying the experience is crucial. You want to associate the experience of reaching a healthy weight with that of healthy eating. It may take a few months longer, but it will ensure that you'll never have to lose weight again.

A major reason people get discouraged and drop out of weight-loss programs is *weight plateaus*. Gained muscle mass and blood-vessel expansion due to exercise may temporarily halt weight loss or cause a small gain, but these are actually very desirable phenomena. Since muscle weighs more than fat, you can lose body fat and inches without dropping pounds if you are building muscles at the same time. Changes in medication, menstruation, constipation, water retention, and other factors may also cause weight loss to slow down or even reverse. Remember that your goal is to lose body fat. None of these factors causes an increase in body fat, so do not be discouraged by minor shifts of weight in the wrong direction. Be patient.

Don't rush weight reduction. One of the most important issues in weight loss is recidivism. Most people who lose weight end up gaining it back. Preliminary research on the *ghrelin* hormone, which is secreted in the stomach, may explain part of this troublesome problem. Ghrelin stimulates appetite at the same time that it slows down metabolism. Both of these effects contribute to increased fat storage. Levels of this hormone spike before each meal and drop after you're full. People given injections of ghrelin become extremely hungry, and studies show they eat much more when unlimited food is available, such as at a buffet.

A recent study at the University of Washington showed that ghrelin levels increase substantially after a period of rapid weight loss.[16] Dr. David E. Cummings, the lead scientist on the study, thinks this was an evolutionary adaptation to encourage the body to regain the lost fat as protection from possible future famine. This genetic program no longer applies to our modern situation. Research is currently under way to develop medications that block ghrelin and its stimulation of appetite and storage of body fat.

Slow, gradual weight loss does not appear to cause the same spike in ghrelin levels, however. This is another important reason to approach your ideal weight gradually. Setting your daily caloric level to match your target weight's maintenance level is the best way to lose weight once and to keep it off.

Get exercise. Physical activity is very important for burning calories, lowering your "set point" (the weight your body gravitates toward), and increasing your metabolic level (rate of burning calories), even while you are not exercising.[17] We recommend burning at least 300 calories daily through exercise.

Raise your metabolic rate. A primary factor in determining your metabolic rate—the rate at which you burn calories—is the number of mitochondria in cells. Mitochondria are tiny energy factories that fuel every cell. The more you have, the more energy you will burn, which will keep you leaner.[18] Unfortunately, we cannot simply take a mitochondria supplement. However, fat cells have very few mitochondria because fat cells store energy rather than burn it, whereas muscle cells have many because they need energy to perform their job. So as you build muscle cells from a regular exercise program, you increase your mitochondria, thereby permanently raising your metabolic rate, even when you are not exercising.

CALORIC RESTRICTION

Closely related to maintaining a lean body weight is the practice of caloric restriction, or CR. There is now extensive evidence across a wide range of animal species that restricting calories slows down aging and can extend both life and youthfulness. These experiments have not run long enough to demonstrate actual life extension in humans, but studies of humans practicing caloric restriction show the same reduction in disease and aging markers (changes associated with increasing age) that we see in animal populations.[19] More than 2,000 animal studies show the same dramatic results across many different species.[20]

A highly publicized experiment in 1982 involving rats first introduced the world to CR.[21] The control group was fed a normal diet and lived a normal maximum life span of approximately 1,000 days. Typically, the control rats died from the deterioration of their hearts, kidney disease, or cancer. The diet of the experimental group (the CR rats) contained one-third fewer calories than the control group, but otherwise had adequate nutrients, including vitamins, minerals, protein, and essential fatty acids. The CR rats lived for about 1,500 days, or 50 percent longer.

Equally significant, the researchers noted a slowing of the aging process. Not only did the CR rats live longer, they largely avoided the feebleness, poor health, sluggishness, and grizzled appearance that accompanied the old age of the normal-eating group, even toward the very end of their extended

BRIDGE TWO

SUPPRESSING APPETITE

One approach to weight-reduction drugs is to suppress appetite, the idea behind a nasal spray being developed by Nastech Pharmaceutical. The spray, now in phase I trials, is based on the hormone PYY, which the stomach normally releases when it is full. The drug triggers the feeling of satiety before you have actually filled your stomach.

Extensive research demonstrates that the hormones leptin and ghrelin play a powerful role in appetite control, which research at Harvard led by Barbara Kahn suggests may be due to their effects on the enzyme AMPK (AMP-activated protein kinase).[22] In mouse experiments, inhibiting AMPK caused the animals to eat less and lose weight, whereas increasing AMPK levels had the opposite effect. The authors of the study describe AMPK as "a 'fuel gauge' to monitor cellular energy status." This finding indicates that drugs to control AMPK levels in humans have the potential to have the same impact.

In a study published in the journal *Science*, researchers reported another mechanism by which leptin and ghrelin affect appetite: these hormones actually cause the brain to rewire itself.[23] Previously it was thought that they acted like other hormones in affecting the behavior of brain cells directly. This research showed that leptin strengthened neural connections that inhibited eating and weakened connections that increased appetite. Ghrelin had the opposite effect and could undo the neural changes from earlier administration of leptin. "It is almost as if the brain is developing a memory for the weight it wants the animals to be," commented Dr. Jeffrey Flier of Beth Israel–Deaconess Hospital in Boston. The research underscores the power of these hormones to affect our eating behavior, so drugs that alter their balance have the potential to reprogram our eating in a healthier direction.

lives. For example, the coats of the normal-eating rodents, which are smooth and white early in life, typically turn gray and oily by 24 months of age. The CR rodents, in contrast, kept their fur white and shiny for 40 months or more. The low-calorie rats were also significantly more successful at running mazes than normal-eating rats of the same age. Rates of diabetes and cataracts and the strength of the immune system were all dramatically better in the CR rodents.

Long after the normal-eating rats had died, the calorically restricted rats continued to have shiny coats, very low rates of cancer and other diseases, and the higher levels of energy and responsiveness associated with youth. When these CR rats finally died, they often appeared to do so for no obvious reason—probably just old age. According to Dr. Edward Masoro, a physiologist at the University of Texas Health Science Center in San Antonio, "When we look inside them, they're completely clean."[24]

Other experiments have exposed both calorically restricted and normally fed rats to high levels of carcinogens. CR rats showed significant resistance to the cancer-causing chemicals, whereas normal-eating rodents easily succumbed.[25] Even strains of rats that are specially bred to be prone to cancer and to autoimmune and other diseases gained significant protection from a low-calorie diet. Dr. Richard Weindruch, a gerontologist at the University of Wisconsin at Madison, comments, "Any kind of screwed-up animal seems to benefit from caloric restriction."[26]

Numerous experiments conducted on a wide range of other animals have shown consistent results. The CR animals live about 30 to 50 percent longer, age more slowly, and are generally much freer of disease, even toward the end of their lengthened life spans.

HOW DOES CR WORK?

So why does caloric restriction work? Fortuitously, an important piece of evidence arrived just as we were completing this very chapter. Dr. C. Ronald Kahn, executive director of the Joslin Diabetes Center at Harvard Medical School, and his colleagues created a genetically modified mouse that lacked a single gene that controls insulin's ability to enable fat cells to store fat. [27] These FIRKO (Fat-specific Insulin Receptor Knock Out) mice ate substantially more than normal mice—in fact, as much as they wanted—yet they had 50 to 70 percent less body fat. They were also resistant to diabetes, remained healthier longer than the control animals, and lived 18 percent longer.

Low body fat. Although the diet of the FIRKO mice was just the opposite of caloric restriction, they obtained at least some (although not all) of the benefits of CR anyway. This suggests that at least one mechanism behind CR is simply maintaining a low level of body fat. We discuss the potential of this research to provide human treatments in the Bridge Two section opposite.

Blood glucose level. Both FIRKO mice and low-calorie animals have significantly lower blood glucose levels.[28] That's because these animals burn glucose for fuel at the same rate as the normal-eating animals; but with less caloric intake, there is less unused glucose left over

Level of free radicals. These highly reactive molecules are by-products of improper metabolism of food. Free radicals cause a gradual deterioration of body tissues, particularly fragile cell membranes. Many researchers attribute some aging processes to the effects of free radicals circulating in the bloodstream.[29] The CR animals had substantially lower levels of free radicals (the result of less food metabolism), so they had less free-radical damage to cell membranes. Researchers have also discovered that the levels of a liver enzyme that detoxifies free radicals are about 60 percent higher in low-calorie animals.[30]

DNA repair. Other researchers have discovered that CR animals have more robust DNA-repairing enzymes. Deterioration in the DNA code causes cancer and accelerates other aging processes, so the greater effectiveness of these enzymes would partially account for the slower aging and lower rate of tumors in these animals.[31]

Life calorie limit. It is interesting to note that the total lifetime quantity of food eaten by CR animals and normal-eating animals was roughly the same. The low-calorie animals ate approximately two-thirds as much food per day but lived 50 percent longer, so the total amount of food eaten over their life span was about the same as that of the normal-eating animals. This is consistent with the idea of living cells as heat engines that wear out with the consumption of fuel rather than the passage of time. Hence the wisdom in T. S. Eliot's quotation at the beginning of this chapter: Our lives are measured, if not in coffee spoons, then in the calories of food that our spoons contain.

Each species seems to have a fixed number of calories that it can burn in the course of a lifetime. By eating a little less each day, there will be more days before these calories are used up. Of course, we intend to overcome this and other biological restrictions, but in the meantime we can take advantage of these insights into biology's limitations.

But there is a limit to the ability of caloric restriction to extend life. Because of the need to obtain sufficient nutrients, we cannot, for example, restrict calories to, say, a third of normal levels and expect to live three times as long. Without adequate vitamins, minerals, protein, and other nutrients, a human or other animal will become ill and ultimately die if the deficiency is not reversed. It appears that, at least in the case of animals such as rats, the optimal level of calories for longevity is about two-thirds that of what the animals will consume if they are eating freely. Below that, it is difficult or impossible to obtain adequate nutrition.

APPLYING CR TO HUMANS

There have been a number of human population studies that illustrate the potential of caloric restriction for humans. For example, the people living in the Okinawa region of Japan have 40 times the number of centenarians (people age 100 or older) than the northeastern prefectures, and they have very little serious disease before age 60.[32] Okinawans remain active much longer than their peers in other regions of Japan. The primary difference in their diet appears to be a lower caloric intake.

In applying the animal studies to humans, some researchers have estimated that our maximum life span might be extended from 120 years to 180.[33] Of course, very few of us live to 120 as it is. These estimates refer to a theoretical maximum potential life span—before we apply the Bridge Two and Three technologies to vastly extend it. Perhaps more significant: this implies that by eating a diet low in calories but otherwise healthy, we are more likely to take full advantage of our current biological longevity.

The benefits of caloric restriction also extend to your remaining life expectancy. If you are 40 and thus have a remaining life expectancy of about 40 years, you will be extending only that remaining period. So the earlier you start CR, the greater the benefits. However, regardless of when you start, you'll quickly realize the benefits of maintaining a lower weight.

GUIDELINES FOR CALORIC RESTRICTION

One result of restricting calories is, of course, losing weight. People who follow strict caloric restriction guidelines end up being very thin to the point of looking gaunt, which we don't recommend. From the recent Joslin study that we cited above, it is clear that at least some of the benefits of CR come from the resultant low level of body fat. Our recommendation, therefore, is to practice a moderate form of CR, not as austere as the 35 percent reduction used in the animal experiments.

We suggest the following guidelines:

- Eat a minimum of 12 calories per pound of your optimal weight. For example, a man with an optimal weight of 150 pounds should eat a minimum of about 1,800 calories per day; a woman with optimal weight of 125 pounds should eat at least 1,500 calories per day. Depending on your activity level, these figures are 10 percent to 33 percent lower than recommended in the above tables of maintenance calories.

- Set your minimum weight at 95 percent of your optimal weight based on the charts in this chapter. For example, if your optimal weight is 200, your minimum weight would be 190 (200 times 0.95). If your weight falls below this minimum number, increase your calorie consumption.
- Select foods low in caloric density. The best way to reduce calories is to eat low-starch vegetables such as broccoli and summer squash, which are filling and have relatively few calories, instead of potatoes and rice.
- Focus on fiber. Another choice is foods rich in fiber, which provides bulk and texture with no digestible calories. Fiber also has health benefits by lowering cholesterol levels, improving regularity, and reducing the risks of colon cancer. Most vegetables are, of course, high in fiber. There are also many foods designed to be carbohydrate substitutes that use fiber (as well as vegetable protein) to replace the bulk and texture of starch, such as low-carbohydrate cereals and breads (see Fantastic-Voyage.net for recommendations of specific products).

CALORIE BLOCKERS

Another strategy for weight loss is to block the digestion of food after you eat it. There are some limited but promising approaches to doing this for carbohydrates and fat.

Starch blockers. "Starch blocker" supplements and medications essentially turn starch into the equivalent of fiber. These blockers combine with amylase, the enzyme responsible for breaking down starch into simpler sugars that the body can absorb. With the amylase deactivated, the starch in food passes through the body undigested. It ends up in the large intestine, where it is acted upon by gut bacteria, similar to what happens to fiber and other indigestible portions of food.

Caution: Using a starch blocker when consuming a large amount of carbohydrates may cause excessive gas, bloating, flatulence, and bacterial overgrowth. With a low-carbohydrate diet, excessive gas is not a problem. Consider a starch blocker as an adjunct to a low-carbohydrate diet if you're trying to lose weight or you have the metabolic syndrome or type 2 diabetes. The starch blocker also reduces the glucose load of carbohydrates, which is worthwhile for someone with insulin resistance. If you use a starch blocker, you should follow our "low carbohydrate" recommendation of no more than one-sixth of your calories.

A starch blocker that has been on the market for many years is Bayer's

BRIDGE TWO

CALORIC RESTRICTION WITHOUT THE RESTRICTION™[34]

A birth control pill that worked by suppressing interest in sex would probably find a limited market. People similarly enjoy the sensual pleasure of eating, so they don't want that enjoyment restrained either. The Bridge One solution is to eat a diet that has a very low glycemic load, limits carbohydrates, and is generally low in fats. However, it would still be desirable if we could eat more—perhaps as much as we wanted—and nonetheless enjoy the benefits of caloric restriction and remain slim, like the "lucky" FIRKO mice. So, while medications to control appetite will continue to play an important role, the holy grail of diet drugs is one that lets us eat as much as we want while maintaining an optimal weight.

In the Joslin study, blocking the FIR (fat insulin receptor) gene in the fat cells of mice enabled the mice to eat a lot and remain thin. Why? "Since insulin is needed to help fat cells store fat, these animals had less fat and were protected against the obesity that occurs with aging or overeating. They also were protected against the metabolic abnormalities associated with obesity, including type 2 diabetes," said Dr. C. Ronald Kahn, who headed the study.[35]

Drug developers are currently working on translating these results into human drugs. Such drugs would clearly be blockbusters, and so we can be confident that efforts to develop them will be intense.

In biotechnology research conducted by Roger Unger and his colleagues at the University of Texas Southwestern Medical Center, a virus genetically engineered to deliver the gene for the hormone leptin, which controls appetite, was injected into the livers of rats. These rats then produced high levels of the hormone and lost weight. Leptin is normally produced by fat cells, but fat cells develop a resistance to their own leptin. Because it was coming from another organ (the liver), the fat cells remained sensitive to it.

Surprisingly, not only were the fat cells of the animals carrying less fat, but they also had an unusually large number of mitochondria. Normally, fat cells have very few mitochondria; muscle cells, which need a lot of energy, have many. Increasing the number of mitochondria in fat cells is remarkable and had not been seen before. Once perfected, this approach would have the effect of permanently increasing metabolism, thereby maintaining a low weight.

Precose, a prescription starch blocker taken with meals.[36] A more recent prescription starch blocker is Glyset. There are a number of "natural" starch blockers available over the counter; however, we have had mixed results with these products in our own informal tests, which suggest that Precose is more effective than nonprescription starch blockers.

These starch blockers do not block sugar because it doesn't need to be broken down. We recommend stevia and sucralose (Splenda) as sugar substitutes (see chapter 5, "Carbohydrates and the Glycemic Load," for a discussion of sugar substitutes).

Fat blockers. Xenical is a prescription drug that, like starch blockers, blocks key digestive enzymes[37] called lipases, which break down fat. Xenical blocks about one-third of the fat consumed from being digested.

Another approach is a "natural" polymer called *chitosan*, a derivative of shellfish and available in health food stores, although it is less effective than Xenical.[38] Chitosan binds to the fatty acids directly. It can bind up to about six times its own weight in fat, which then becomes indigestible and passes through the GI tract.

A word of caution: fat blockers will inhibit fat-soluble vitamins, such as vitamin E, so don't use one within three hours of (before or after) taking fat-soluble supplements. Fat blockers also block the fat-soluble vitamins contained in food as well as healthy fats, such as omega-3 fats and oleic acid. If your fat consumption consists primarily of healthy fats, binding a portion of these fats with a fat blocker may still be acceptable.

More effective calorie blockers, as well as body fat inhibitors, are in the pipeline. In the meantime, there are extensive benefits to restricting calories and maintaining a lower body weight. If you combine moderate caloric restriction with a diet that avoids high-glycemic-load foods, restricts carbohydrates, and is generally low in fat (while emphasizing healthy fats), you will find that you can be slim but still eat plenty of food—and never be hungry.

9

THE PROBLEM WITH SUGAR (AND INSULIN)

"The taste of love is sweet . . ."

—Johnny Cash, "Ring of Fire"

Mary, a 52-year-old lawyer, was tired all the time. She was having problems concentrating at work and was often irritable toward her husband. When asked about her sex drive, she responded, "What's that?" Although not obese, she needed to lose 25 pounds. At her most recent physical, blood work revealed her triglycerides (blood fats) were markedly elevated at 426 (normal 30–150), her HDL-C ("good" cholesterol) was low at 32, and her blood sugar was "impaired" but not diabetic at 110 (normal 60–99). Her blood pressure was slightly elevated at 152/94. A two-hour G-ITT (glucose-insulin tolerance test) showed that she had only slightly elevated blood sugar levels, but markedly high insulin levels.

The diagnosis: syndrome X, or the metabolic syndrome. Mary adopted Ray & Terry's Longevity Program and began to eat a low-carbohydrate diet (less than 40 grams per day), including more nonstarchy vegetables. She included "good" fats in the form of olive oil, nuts, and fish, along with adequate low-fat protein foods. She began nutritional supplementation, including EPA/DHA (fish oil), chromium, and a multiple vitamin/mineral formulation. After about a week of sugar withdrawal symptoms such as irritability and headache, she began to lose weight and feel better.

After 12 weeks on our program, Mary's weight had fallen to 132 from 150, and her energy level had improved dramatically. She stated that her memory was back to what it was when she was much younger. Repeat blood testing now showed her triglycerides had fallen to 48 from 426, and her HDL-C had risen to 52 from 32. Her fasting blood sugar was now 72; her blood pressure, 118/76—all normal.

SUGAR: THE SWEET KILLER

Sugarcane was first domesticated in New Guinea about 10,000 years ago. Along with measles, smallpox, and influenza, it was introduced to North America in 1493 by Christopher Columbus. Sugar consumption has increased steadily over the years since then. Annual per capita consumption of white sugar and related sweeteners such as honey, molasses, and high-fructose corn syrup is now estimated at about 152 pounds in the United States.[1] (Since neither of the authors of this book eats any sugar, at least two people are eating over 300 pounds a year to make up for us!) Most of this rise is relatively recent; 100 years ago, annual per capita sugar consumption was less than 5 pounds. Even today, people in less-developed countries consume much less sugar. Per capita consumption in Afghanistan, for example, was only 2 pounds in 1999.

Much of this sugar is consumed in the form of soft drinks. In 2002, the National Soft Drink Association was proud (we think they should be ashamed!) to announce that Americans consumed 53 gallons of soft drinks per person per year, which averages out to 565 12-ounce servings.[2] Soft-drink consumption has doubled since 1972 and jumped six times since 1945. Sugar is finding its way into many manufactured food products too, because it has many features attractive to food manufacturers: it's inexpensive, tastes good, can be used in numerous types of foods, stores well . . . and can be somewhat addicting.

In the 1960s, the Surgeon General's Report alerted Americans to the dangers of cigarette smoking. More recently, major public health efforts have begun to educate people on the health problems of excess dietary fat, particularly saturated and trans fats. Preventive health programs encourage us to get more exercise, eat more fruits and vegetables, lower our blood pressure and cholesterol, even wear seat belts and practice safe sex. Yet to date, public health officials and the majority of the physician community, at least in the United States, have yet to warn the public about the health hazards of sugar, arguably the most dangerous "food" you eat.

There are many reasons sugar has remained safely out of reach for so long. The use of sugar is deeply ingrained in our culture and collective psyche. Just look at some common phrases: "as American as apple pie," "How ya doin', sugar," or even "Honey, I'm home." Part of the confusion surrounding our love of sugar is the frequent association between sugar and love itself. There is even a specific day of the year, Valentine's Day, designated as a celebration of love and sugar *together.*

There are encouraging signs that this has begun to change. In 2003, the World Health Organization issued its "Diet, Nutrition and the Prevention of Chronic Diseases" report, suggesting that people worldwide cut the total calories they get from simple sugars from 25 percent to less than 10 percent.[3] "Official" U.S. policy is already in accord with the WHO report but, in actuality, the information is not being effectively disseminated. According to Dr. Marion Nestle of New York University, "If you do the sums in the Department of Agriculture's Food Guide Pyramid, you'll find it recommends 7 to 12 percent free sugars. But they're afraid to mention actual figures because of the [sugar] industry, which is being very aggressive at the moment."[4] Even worse, the U.S. Department of Health and Human Services has come out as vehemently opposing many critical aspects of the new WHO proposal, including the importance of reduced sugar consumption.[5]

We believe that any healthy diet or weight-loss program must begin with the elimination or reduction of sugar.

Schooling Our Children on the Evils of Sugar

Early in the Paleolithic era, humans didn't have access to fast food or processed foods of any kind. Lacking the ability to grind grains into flour, our ancestors ate no baked goods either. Their diet consisted mostly of proteins, healthy fats, and vegetables. Fruit was the only source of simple sugar and was only seasonally available in small amounts. From a genetic point of view, this is the diet our bodies are programmed to eat.

It's only in the past 40 years or so that fast-food restaurants and 24-hour "convenience" stores have come to dominate our dietary landscape. Most people reading this book can remember back to the days before "nutrition bars" (candy bars in disguise) were eaten for breakfast or "super-sized value meals" for supper. In the past 25 years, our consumption of refined carbohydrates such as baked goods, pasta, and sugar has increased by 30 percent. Today, some public schools have vending machines in the hallways, and students can bring soft drinks into class.

In January 2004, the American Academy of Pediatrics Committee on School Health took a step in the right direction by suggesting that "school officials and parents need to become well informed about the health implications of vended drinks in school."[6] Some school districts around the country have already responded by removing the soda machines, despite the significant loss of revenue to their budgets such decisions frequently entail.

INSULIN: THE FAT GENERATOR

For many years, conventional medical wisdom held the simplistic and since disproven notion that eating fat is solely what makes people fat. Doctors now know that eating sugar and other simple carbohydrates, which quickly turn into sugar in the body, is at least as important a cause of excess weight. Unfortunately, most American physicians continue to act as if they are oblivious to this fact and have not done nearly enough to encourage their patients not to eat so much sugar.

Sugar causes damage not only by becoming rapidly converted into fat, which is stored as excess weight, but also by elevating insulin levels in the bloodstream. Insulin is a hormone secreted by specialized *islet cells* of the pancreas to help lower elevations in blood sugar. This hormone was "discovered" in 1921 by Dr. F. G. Banting, but it's actually a very ancient molecule.[7] It's been tracked back over 400 million years. Insulin was critical to the development of complex life-forms because its primary role is to allow organisms to store glucose for future use. Without insulin, animals—including us humans—would need to eat constantly to ensure a continuous supply of glucose (or sugar) for their cells. Almost all cells of the body have insulin receptors on their surfaces. When insulin binds to these receptors, channels in the cell membrane open so that the sugar molecules can pass from the bloodstream into the cell. Without insulin, sugar can't gain entry into the cell, and the cell will die. But too much insulin causes serious problems of its own.

Before the relatively recent addition of sugar to the human diet, almost everyone's insulin level was very low—typically, less than 5 ng/dL (nanograms per deciliter or 100 milliliters of blood). Nowadays, levels are much higher, sometimes up to 20 or more. Approximately two-thirds of Americans are either overweight or frankly obese.[8] Much of this is the result of the 152 pounds of sugar eaten by Americans every year and clearly shows how efficient insulin is at turning excess sugar into fat.

When you drink a sugary soft drink or eat a sugary snack such as a candy bar or doughnut, here's what happens:

1. Sugar (glucose) in your blood spikes.

2. The insulin level in your bloodstream also rises rapidly to help prevent your blood from becoming too sugary (read: syrupy). But elevated insulin can elevate blood pressure and increase body fat. It causes fluid retention, hormone imbalances, and more.

Sugar Shortens Life Expectancy

Being overweight or obese—defined as weighing 30 percent more than one's maximum healthy body weight—does not merely cause problems with self-esteem and clothes selection. Excessive weight dramatically affects life expectancy. In chapter 1, we discussed how many people live their lives as if walking toward an unseen cliff. With increasing weight, it's like you pick up the pace and head for the cliff at a run. Almost two-thirds of American adults are now overweight, and one out of three is clinically obese. A decade ago, "only" half were overweight and one in five obese.[9]

A recent study published in the *Journal of the American Medical Association* found that a woman who is obese at age 20 can plan on a reduction in life expectancy of 8 years, while an obese 20-year-old man can anticipate living 13 years less than his normal-weight peers.[10] Gaining weight later in life has a lesser but still significant impact on life expectancy. Merely being overweight (not obese) at 40 shortens the average life span by 3.1 years.

3. The excess circulating sugar combines with proteins in your body to form AGEs (advanced glycation end products), which have been directly linked to numerous diseases, including premature aging.

According to the Centers for Disease Control and Prevention (CDC), about one-fourth of the adult U.S. population suffers from a condition that results in a serious inability to process dietary sugar, known as the metabolic syndrome, or TMS.[11] You may already have TMS and not know it; you would certainly not be alone. It is estimated that at least half of patients who suffer heart attacks have some degree of TMS, making it the most under-diagnosed risk factor for heart disease presently known.[12]

TMS: A DISEASE OF CIVILIZATION

At a public gathering, look at all the people who are shaped more like apples than celery sticks (men) or hourglasses (women). These people carry excess weight in their midsections: they have "potbellies" or "love handles," or what is medically known as central obesity. People shaped more like apples very frequently have TMS, and, in almost every case, this is the direct result of eating too much of the wrong foods. Fortunately, TMS is completely avoidable and curable with proper diet and lifestyle choices.

TMS was initially described by Dr. Gerald Reaven in 1988.[13] This condition, also known as syndrome X or insulin resistance syndrome, refers to a cluster of metabolic abnormalities based on resistance of the body's tissues to the effects of insulin. This means tissues don't respond to "normal" levels of insulin whenever sugar or other carbohydrates are consumed, and the pancreas needs to excrete ever-increasing amounts of insulin to bring blood sugar levels down.

TMS can cause you to age more quickly and predisposes you to a host of potentially catastrophic illnesses, including the Big Three—heart disease, cancer, and Alzheimer's disease—as well as diabetes, high blood pressure, numerous neurological disorders, and arthritis. People with TMS also feel tired much of the time, have difficulty losing weight, and suffer problems with memory, concentration, and irritability.

By definition, a diagnosis of TMS requires that three of the following five criteria be present:[14]

- Excessive waist circumference (more than 40 inches in men or more than 35 inches in women)
- Serum triglyceride (blood fat) level greater than 150 mg/dL (milligrams per deciliter or tenth of a liter)
- "Good" cholesterol (HDL-C) less than 40 mg/dL in men or less than 50 mg/dL in women
- Elevated blood pressure higher than 135/85[15]
- Fasting blood glucose over 99 mg/dL

Although not formally included as a diagnostic factor, new research has shown that people with a relatively larger number of "small" LDL cholesterol particles (the type that can more easily burrow into artery walls and lead to plaque formation) are at significantly increased risk for TMS.[16]

TMS is the most common metabolic disorder of Americans, estimated to affect 47 million U.S. residents. The prevalence increases markedly with age. It affects 7 percent of people in their 20s, but increases to 44 percent of people in their 60s and 70s.[17] There's a direct relationship between TMS and type 2 diabetes, but the good news is that only 20 percent of patients with TMS progress to full-blown diabetes.[18]

SUGAR AND AGING

Persistently elevated blood glucose levels have another undesirable consequence: the formation of AGEs, which are created in the body when sugar molecules stick to proteins (remember, sugar is sticky), a chemical reaction known scientifically as the Maillard or "browning" reaction. These sticky conglomerations of sugar and protein gum up your vital enzymes, increasing free-radical damage to tissues, which accelerates the aging process dramatically.[19]

Age spots on the skin indicate AGE formation. Cataracts in the lens of the eye are another example. If you eat a diet containing less sugar and emphasize food with a lower glycemic load—food that doesn't turn into sugar quickly in the body—you can reduce the formation of AGEs and help slow down visible signs of aging. Seen in this light, perhaps that dish of ice cream doesn't look quite as tempting.

How you cook your food also influences AGE formation. Cooking at high temperatures such as baking, barbecuing, frying, roasting, or broiling increases AGE formation. Boiling or steaming is safer because the cooking temperature won't go above 212 degrees Fahrenheit, the boiling point of water. When foods brown during cooking, as with bread crust, basted meats, even coffee beans, it means that their AGE content has increased. Since most fast foods and processed foods are subject to browning, that's one more reason to avoid them.

We like many aspects of the Mediterranean diet, such as the widespread use of heart-healthy olive oil both as a condiment and for cooking. Other types of Mediterranean food, such as Greek or Middle Eastern, are also good. Unfortunately, some people believe the diet includes large amounts of refined-flour products such as pastas and breads. That applies more to American-style Italian cuisine; in Italy, smaller portions of pasta are generally eaten. One way to make Italian food healthier is by replacing the pasta with the same amount of shredded spaghetti squash—the carbohydrates drop from 40 to an acceptable 5 grams per serving, of which 1.1 grams are fiber.[20]

TESTING FOR TOO MUCH SUGAR AND INSULIN

To get an idea if you have (or are on the way toward) insulin resistance (and TMS), try the following at home right now:

1. Use a tape measure to measure the circumference of your abdomen right at the level of your belly button. More than 35 inches for women, or more than 40 inches for men, indicates a good chance of TMS.

(continued on page 132)

BRIDGE TWO

THE RAZOR'S EDGE OF DIABETES RESEARCH

Information about which genes are involved and how their activity changes in people at risk for diabetes is expected to revolutionize the care of TMS and both types 1 and 2 diabetes mellitus. For example, one recent study, spearheaded by Joslin Diabetes Center and Children's Hospital Boston Informatics Program, explored which genes were turned on or off before diabetes developed. They discovered that reduced activity of two genes, PCG1-alpha and PCG1-beta, set off a cascade of events, including decreased activity of other genes that control fat and carbohydrate metabolism.[21] In other research, bone morphogenetic protein-9 (BMP-9) was identified as a potential new drug through large-scale screening and testing protocols not possible just a few years ago. First, 3 million entries were scanned in 1,000 libraries in the Human Genome Sciences database. Then, 8,000 secreted proteins were inserted into human embryonic kidney cells, and assays were used to evaluate the role of the proteins in limiting the expression of a key enzyme in glucose production.[22] In the tests, the effect of BMP-9 resembled that of insulin, so BMP-9 would help people with type 2 diabetes control their blood sugar with diet and other medications.

Another drug with great potential for the treatment of type 2 diabetes is Exenatide, which mimics the action of the body's glucagon-like peptide-1 (GLP-1), a naturally occurring hormone released into the intestinal tract after eating. GLP-1 and Exenatide stimulates the pancreas to release insulin into the bloodstream. In the late stages of clinical trials, Exenatide also reportedly reduces a family of hormones known as glucagons, which act the opposite of insulin, raising blood sugar and slowing absorption of calories.

New for treating TMS are peroxisome proliferator-activated receptor (PPAR) activators, drugs that operate on insulin receptors in the cell membrane. They help insulin in the bloodstream transport glucose into the cell. A version of this type of drug, called Avandia (rosiglitazone), is currently taken by many type 2 diabetics to help lower their blood sugar. The problem with Avandia is that it stimulates only one type of PPAR receptor, the gamma receptors, but has no effect on the alpha receptors, which help control sugar in the cell from being turned into fat. This causes many type 2 diabetics to gain weight, usually the last thing these patients need to do.[23] A drug called Galida (tesaglitazar), under development by AstraZeneca, targets both the alpha and gamma PPAR receptors. It helps reduce insulin resistance but also lowers blood triglycerides (fats) and increases beneficial HDL cholesterol. Galida may prove to be a blockbuster drug for that segment of the 50 million Americans with TMS who are unable to control their symptoms with lifestyle changes alone.

Researchers are also looking hard at the hallmark of type 1 diabetes, which results when the insulin-producing islet cells of pancreas are destroyed by the immune system. Type 1 diabetes affects a smaller percentage of the population than type 2 diabetes, but doesn't respond as well to Bridge One interventions. While proper lifestyle and dietary choices can help people with type 1 diabetes "control" their blood sugar, frequent injections of insulin and blood sugar checks are still needed. It is not uncommon for a patient to endure more than 2,000 needle sticks a year. Even so, the incidence of severe side effects such as heart attacks, kidney disease, and blindness remain elevated.

Three promising biotech solutions to the fundamental problem of pancreatic islet cell failure in type 1 diabetes are islet cell transplants, new insulin delivery systems, and regenerating the damaged pancreas itself (see Bridge Three research).

Researchers in Alberta, Canada, working on a study known as the Edmonton Protocol for the treatment of type 1 diabetes, have begun transplanting insulin-producing pancreatic islet cells from donors into the livers of recipients. Immediately after implantation, these islet cells begin sensing their new host's blood sugar and secreting the precise amount of insulin needed. This protocol uses newly available bioengineered drugs to overcome previously insurmountable hurdles to pancreatic transplantation, such as tissue rejection.[24] Another approach being explored is to create animal-human chimeras, organisms formed by combining human and animal genes. Somatic stem cells taken from a patient are injected into the fetus of an animal such as a sheep. Human cells from the newborn animal are then separated from the animal cells by cell-sorting machines and harvested for that patient.[25] When islet cell transplantation becomes viable, doctors may even use this procedure to boost the amount of insulin produced by type 2 diabetics, per the Joslin Diabetes Center.[26]

Unfortunately, there are far too few available pancreatic islet cells from donors to give to patients who could be helped by these transplants. Researchers at the University of Florida have successfully cloned islet cells and cured diabetes in mice, so cloning technology combined with biotech immunosuppressive drugs may be an interim solution.[27]

Yet another insulin-delivery option, possibly available for human use in as few as five years and for as little as 10 cents apiece, is the so-called Intelligent Pill (iPill). After it's swallowed, it remains in the stomach, where its tiny micropump and sensors use the body's temperature, blood glucose, and pH level to determine when to release insulin. About the size of a penny, it functions for about 24 hours. When the iPill's payload is emptied, the penny-size device is excreted with the rest of the body's wastes. Experiments using dogs as subjects are expected to begin in 2005.[28]

Finally, excess sugar in the bloodstream increases formation of AGEs, the undesirable cross-linking of useful molecules with excess sugar. Since AGEs accelerate aging, it is fortunate that significant progress at effectively combating this process has been made. An experimental drug called ALT-711 (phenacyldimenthylthiazolium chloride), under development by Alteon for several years, can dissolve cross-links without damaging the original tissue.[29] ALT-711 represents a novel approach to the treatment of diseases resulting from age-related stiffening of tissues such as arteries. Research has shown that this compound can soften hardened arteries and thus lower blood pressure.[30] Other molecules with this capability have also been identified and should be available in the near future.

2. Measure the circumference of your hips at their widest point.
3. Divide your waist measurement by your hip measurement.

This waist/hip ratio is even more predictive of TMS than waist circumference alone. In women, the ratio should be less than 0.8 (hourglass); in men, it should be less than 1.0 (celery stick). If your ratio is higher than these benchmarks, have your doctor perform one of the tests described below.

The most accurate screening test for insulin resistance is measuring the insulin level in your bloodstream. Unfortunately, this test is rarely done by most physicians, who rely instead on indirect measurements, such as high blood fats or elevated blood pressure readings—but these are less accurate for diagnosing insulin resistance. Although you will probably need to ask your doctor for it, a determination of your fasting insulin level is inexpensive and can provide you with a great deal of information.

According to Canadian researchers Despres and LaMarche, even modest amounts of insulin resistance substantially increase a patient's chance of suffering a heart attack. In their view, a consistently elevated insulin level in the bloodstream is the second leading risk factor for heart attack (male gender is first). According to their studies, an insulin level over 12 doubles the risk of a heart attack, while levels over 15 triple the risk.[31]

Fasting Glucose Test. The simplest measurement of blood glucose performed by most physicians as part of screening health panels (groups of tests) is the fasting blood sugar (or glucose) test. Fasting glucose levels of 60–99 mg/dL are generally regarded as normal, but we consider optimal levels to be 60–80. We regard fasting blood sugars of 80–99 as high normal, and find that people with numbers in this range are often already on the way to developing TMS or have it already. Anytime your fasting blood sugar exceeds 100, it suggests TMS or overt type 2 diabetes may already be present.

Two-hour glucose tolerance test. When a patient is found to have an elevated fasting blood sugar, conventional physicians will occasionally recommend a two-hour glucose tolerance test.[32] In this test, the patient comes in to the clinic fasting. First, the blood glucose is measured, followed by the "glucose challenge"—typically, 75 grams of sugar in the form of a sweet drink. Blood glucose is measured again one and two hours later.

The problem is that by the time a patient has an elevated fasting blood sugar, TMS may have been present for many years. By this time, destructive changes to 50 percent to 75 percent of the insulin-secreting cells of the pancreas may already have occurred. If physicians wait to screen only patients with elevated fasting blood sugars, they will miss many people who are on the verge of suffering permanent—but preventable—damage.

Glucose-insulin tolerance test. The two-hour glucose tolerance test is more accurate than the fasting test, but we have found that it is still too insensitive for detecting individuals with insulin resistance. Instead, we recommend a two-hour G-ITT (glucose-insulin tolerance test), in which both glucose and insulin levels are measured, as a screening test, *even in the absence of elevated blood sugar.* This can detect individuals who have already begun to develop destructive changes in the pancreas, long before their blood sugars start to rise.

Table 9-1. Optimal Fasting Glucose and Insulin Levels

	OPTIMAL LEVEL	ACCEPTABLE LEVEL	HIGH RISK	USUAL REFERENCE "NORMAL"
Fasting Insulin	2–3	Under 5	Over 10	6–25
Fasting Glucose	60–80	60–99	Over 100	60–99

In a case of type 2 diabetes, insulin levels might be high, low, or normal throughout the test, but all blood sugar levels would be elevated.[33] See Figure 9-1 opposite for a comparison of typical G-ITT curves for normal and insulin-resistant individuals.

The Insulin Challenge Test. The best test for measuring insulin resistance is to inject a patient with a small dose of short-acting insulin and observe what happens to the blood glucose. Because of the expense involved (an intravenous line must be started and the patient observed continuously throughout the test), the insulin challenge test is not done routinely to screen for insulin resistance. But for patients suspected of having insulin resistance yet who have a normal two-hour G-ITT, the insulin challenge test can be diagnostic.

Living your life with your cells bathed in excess insulin does not lead to optimal health or long-term survival. Hyperinsulinemia (excess insulin in the blood) has been implicated as a major risk factor in numerous other age-related diseases, such as Alzheimer's.[34] In another study conducted by Dr. Gerald Reaven, one out of three individuals in the highest third percentile for insulin levels suffered an adverse "clinical event" such as a heart attack or stroke over the six years of the study, compared with a complete absence of such clinical events among people in the lowest third.[35] So the next time you see your doctor, ask him or her to arrange to check your fasting insulin level.

ADDITIONAL LINES OF ATTACK FOR TMS AND TYPE 2 DIABETES

If you have already been diagnosed with either TMS or type 2 diabetes, there are multiple approaches you can take to treat these conditions and prevent their progression.

Weight loss. The importance of keeping your weight down is nowhere as obvious as in cases of TMS and type 2 diabetes. If you follow the dietary suggestions in this book (especially restricting sugar and reducing total carbohydrates), you should lose weight almost automatically. In the previous chapter, we advised that you begin your weight-loss program by consuming the number of calories you will need at your optimal (goal) weight, as opposed to your present weight. Setting that goal 5 percent lower than your "ideal weight" through modest caloric restriction can provide even more benefits.

Resistance exercise. Progressive resistance training (working out with weights) benefits patients with TMS and type 2 diabetes more than aerobic,

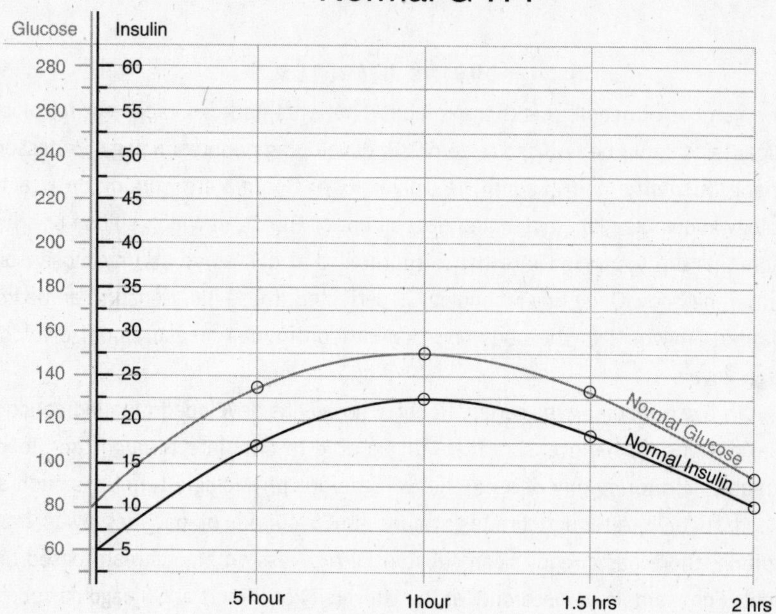

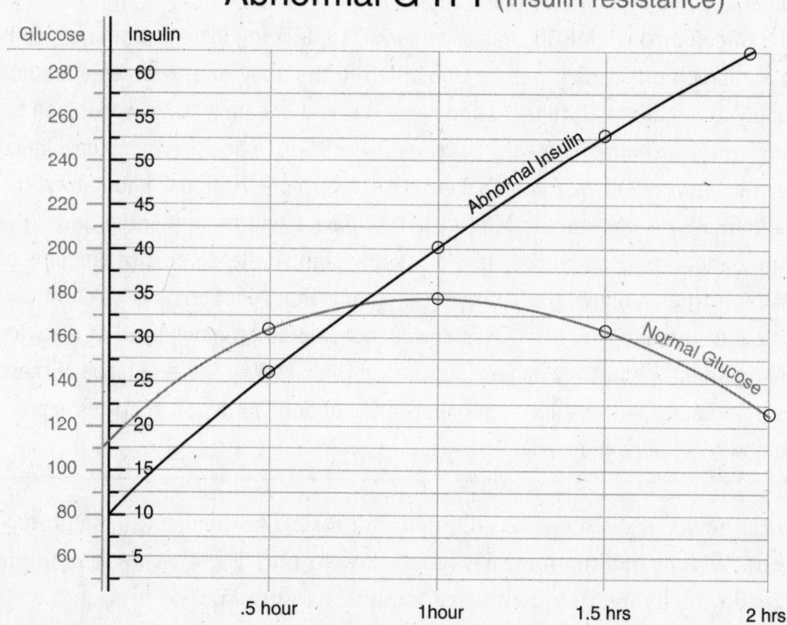

FIGURE 9-1. GLUCOSE-INSULIN TOLERANCE TESTS

BRIDGE THREE

CURING DIABETES

The number of needle pricks that diabetic patients undergo each year for blood sugar tests could be reduced to zero with developing nanotechnology. Professors Zhang, Kisaalita, and Zhao of the University of Georgia are working on a technology known as glancing-angled deposition, or GLAD, in which silicon or other materials are vaporized into nanostructures that can serve as tiny biosensors within the body. Once this technology is perfected, these nanosensors can be implanted anywhere in the body and provide continuous measurement of blood sugar levels.

In his *Nanomedicine* series, Robert Freitas has developed detailed conceptual designs for nanosensors that will be able to circulate through the bloodstream, monitoring any of a number of different physiologic functions such as blood sugar.[36] Although Freitas's plans are a couple of decades away from fruition, there has already been substantial progress on bloodstream-based devices. For example, a researcher at the University of Illinois at Chicago has cured type 1 diabetes in rats with a nanoengineered device that incorporates pancreatic islet cells. The device has 7-nanometer pores that let insulin out but block the antibodies that destroy these cells.

One company, iMEDD, has seen success attaching insulin-producing cells to a microchip that can then be implanted in the body to provide an adequate supply of insulin at all times,[37] and some researchers are already working on totally synthetic hormone organs, such as the artificial pancreas being developed by the Department of Energy's Lawrence Livermore National Laboratory and California-based Medtronic MiniMed. This device will be implanted under the skin to monitor blood glucose levels. Its tiny pumps release precise amounts of insulin using an algorithm (computer program) that functions like our own biological pancreatic islet cells. This device, considered "the holy grail of diabetes management," is already undergoing clinical trials.[38] With the assistance of these novel therapies, a true cure—not merely a treatment—for type 1 diabetes is probable within the next decade.

or cardio, exercise (which is still important for cardiovascular disease protection). Weight training increases blood flow to muscles, so you get multiple benefits, including increased tissue sensitivity to insulin.

Supplements. There are a number of supplements that can help control blood sugar and improve insulin sensitivity. Most of the following supplements are discussed elsewhere in this book, so here we'll mostly offer just dosage and use specifics for TMS or type 2 diabetes.

- Chromium 200 mcg 2 or 3 times daily (with meals) for TMS, 300 mcg 3 times a day for diabetes
- Alpha lipoic acid 100–300 mg twice a day; shown to improve "insulin-stimulated glucose disposal"[39]
- Vanadyl sulfate 7.5 mg once or twice a day; can help lower blood sugar in diabetics, but may cause kidney damage, so close monitoring is needed.[40]
- EPA/DHA (fish oils) 1,000 mg/day; helps increase fluidity of cell membranes and enables insulin to move glucose into the cells more effectively.[41]
- Coenzyme Q_{10} 60–100 mg twice a day[42]
- Carnosine 500 mg 1 or 2 times a day
- Magnesium 200–400 mg/day
- CLA (conjugated linoleic acid) 500–1,500 mg twice a day
- L-carnitine 600 mg 2 or 3 times a day
- Vitamin E 400–800 IU/day
- Vitamin C 2,000 mg/day
- Biotin 3 mg 3 times a day (megadoses); may combat insulin resistance when administered jointly with chromium picolinate[43]
- Arginine 3 grams 3 times a day; helps reduce insulin resistance[44]
- Glutamine 500–1,000 mg; helps eliminate carbohydrate cravings, particularly during the transition period of reducing consumption of sweets and other high G-I foods
- DHEA 15–25 mg 1 or 2 times a day[45]
- N-acetyl-cysteine (NAC) 500 mg twice a day

Medications. Even if you don't have TMS or type 2 diabetes, consider discussing the following medications with your doctor as part of an anti-aging strategy.

- Metformin: a popular prescription drug for type 2 diabetes because it helps make cells less insulin-resistant. It's likely to improve insulin sensitivity in nondiabetics who have TMS. Some researchers believe that metformin may offer powerful anti-aging benefits on a par with those achieved with strict caloric restriction.[46]

- Precose and Glyset: prescription medicines also prescribed for type 2 diabetes to slow down absorption of carbohydrates from the digestive tract. In effect, they lower the glycemic index of the food you eat. Consult your physician regarding the proper dosage.

- Testosterone supplementation may prevent or treat insulin resistance in men.[47]

We have outlined a powerful line of attack of present-day Bridge One recommendations for controlling blood sugar and insulin. However, more powerful strategies for the treatment of diabetes and other blood sugar problems are on the horizon. With Bridge Two biotech therapies, diabetes should become a disease of the past within the next decade. With Bridge Three nanotechnology, you'll probably be able to eat whatever you want. So relax, you don't have to give up ice cream and cake forever—just for a decade or two! If you exert some self-control now, you'll still be around to enjoy eating whatever you want then.

Take Ray, for example. He was diagnosed with type 2 diabetes more than 20 years ago. By following Ray & Terry's Longevity Program, he now has completely normal blood sugar levels and no symptoms, indications, or complications from diabetes. His story follows in the next chapter.

10

RAY'S PERSONAL PROGRAM

When I was 15, my father had a massive heart attack at the age of 51. For the next seven years he was in and out of hospitals with heart failure. His death at the age of 58 cut short a brilliant career as a classical conductor, concert pianist, and music educator. He was a good patient and followed his doctor's recommendations to a T. He lost weight, stopped using salt, and took vitamin E (his cardiologist had heard about cutting-edge research regarding this antioxidant's ability to combat atherosclerosis). However, very little was known in the 1960s about heart disease. At that time, we had almost no knowledge about the role of cholesterol, oxidation, fats, carbohydrates, inflammation, or methylation cycles.

My family's history of heart disease goes back even further than my father; his own father died of the same disease when my father was only 12. This put a cloud over my future outlook, which darkened further when I was diagnosed with type 2 diabetes at the age of 35. Early conventional treatment with insulin caused me to gain weight, which only made the condition worse. Realizing that I would have to take responsibility for addressing this concern, I immersed myself in the research literature and devised a program based on strict restriction of fats (except for fish), eliminating sugar in all its forms, exercising, managing stress, and taking supplements such as chromium. I lost more than 40 pounds and achieved normal blood sugar and cholesterol levels. I reported all of this in my best-selling health book, *The 10% Solution to a Healthy Life.*

As I said earlier, I met Terry Grossman at a Foresight Institute conference in 1999. This started an intense collaboration that has enabled both of us to greatly refine our ideas on health and well-being. This fruitful collaboration occurred just in time for me to confront another serious health challenge: middle age. Whereas some of my contemporaries may be satisfied to embrace aging gracefully as part of the cycle of life, that is not my view. It may be "natural," but I don't see anything positive in losing my mental agility, sensory acuity, physical limberness, sexual desire, or any other human ability. I view disease and death at any age as a calamity, as problems to be overcome. Up until recently, there was relatively little that could be done about our short life span other than to rationalize this tragedy as actually a good thing.

If reversing degenerative disease and aging processes is a war, it's important to have good intelligence on the enemy. With this in mind, I had my genes tested, which only confirmed what I had already figured out: The important Apo E genes were both of the E3 type. This was somewhat of a relief, indicating average risk for heart disease and Alzheimer's disease, at least as far as this one key gene was concerned. However, my CETP gene was heterozygous positive, meaning that a gene from one parent was positive, whereas the gene from the other parent was negative. Having at least one positive CETP gene indicates a disposition to low HDL (good) cholesterol levels. I had several other heterozygously positive genes, indicating an increased risk of atherosclerosis and type 2 diabetes. My MTHFR gene was also heterozygous positive, indicating a slight predisposition to high homocysteine levels. Given that my earlier lipid levels showed exactly this pattern, this was no surprise.

My feeling about my health today is: so far, so good. My blood levels of glucose, insulin, and HgA1c (a measure of glucose levels over the past 90 days) are normal. In years past, this was not the case, until I fully adopted the principles of Ray & Terry's Longevity Program; now I have no indication, symptoms, or complications from diabetes. Over the past year, I had a euglycemic clamp test, an elaborate and sensitive test to measure insulin resistance. My result was "low normal," which is a good result for someone who has been diagnosed with type 2 diabetes. Given my overall program, including a low-carbohydrate diet, my diabetes is fully under control.

My cardiac-related lipid levels are all at ideal levels. I maintain my total cholesterol around 130, LDL around 70, HDL around 55, cholesterol-to-HDL ratio around 2.5, triglycerides around 70, homocysteine around 6.5, and high-sensitivity C-reactive protein about 0.2. All of these values are in the optimum range per our program.

Although I'm 56, a comprehensive test of my biological aging conducted at Terry's longevity clinic measured my biological age at 40.[1] My goal is to be no more than 40, biologically speaking, by the time we have the means to completely arrest and reverse aging in about 20 years. Although we aren't yet able to completely stop aging, my plan is to aggressively apply the means at my disposal to slow down the dozen or so processes that compose aging. Thus, my biological age may slowly creep up from 40, but I then intend to reverse it.

Every few months, I test dozens of levels of nutrients (such as vitamins, minerals, and fats), hormones, and metabolic by-products in my blood. Overall, my levels are where I want them to be, although, in response to these tests, I am continually fine-tuning my program of supplements in consultation with Terry.

I have a personal program to combat each of the degenerative disease and aging processes. Terry and I have a problem with the word *supplement* because it suggests something that is optional and of secondary importance. We prefer to call them "nutritionals" instead. My view is that I am *reprogramming my biochemistry* in the same way that I reprogram the computers in my life. Although I recognize that my body is more complex than my machines, and I still don't have a full copy of my biological "source code," I nonetheless believe this is an apt description.

I take about 250 pills of nutritionals a day. Once a week I go to WholeHealth New England, a complementary medicine health clinic run by Dr. Glenn Rothfeld (I would go to Terry's clinic, except it's 2,000 miles away), where I spend the day. I am provided with an office with high-speed wireless Internet access and a phone, so I work from there. At this clinic, I have a half-dozen intravenous therapies—basically, nutritionals delivered directly into my bloodstream, thereby bypassing my GI tract. I also have acupuncture treatment from Dr. Rothfeld, a master acupuncturist who helped introduce this therapy to this country 30 years ago.[2]

Although my "supplement" program may seem extreme, it is actually optimal. It's fully consistent with Ray & Terry's Longevity Program described in this book, and Terry and I have extensively researched each of the several hundred therapies that I use for safety and efficacy. I stay away from ideas that are unproven or appear to be risky (human growth hormone, for example).

WEIGHT AND DIET

At 5 feet 7 inches tall, I weigh 145 pounds. My body composition is 14 percent fat, which I regard as optimal.

I follow the nutrition guidelines described in this book closely. Because of my concern with diabetes, I keep my carbohydrate consumption below 80 grams per day (which is about one-sixth of my calories), in accordance with our stricter "Group 1" carbohydrate recommendation, and use Precose, a starch blocker.

My typical breakfast starts with a low-carbohydrate cereal sweetened with stevia, unsweetened soy milk, and, often, some berries. I'll also have fish such as salmon, occasionally egg whites or egg substitutes (with no yolks), and green tea. I have recently begun to enjoy a meal replacement shake that Terry and I developed. I don't usually eat a large lunch, so I might have just miso soup and more green tea. I satisfy my desire to nurse a drink all day by drinking about 8 cups of green tea throughout the morning and afternoon.

For dinner, I'll have a protein main course of fish or tofu, sometimes lean chicken or turkey. I eat a lot of low-starch vegetables and salads with olive oil–based dressings. I eat a broad variety of soy-based products. I drink a couple of glasses of red wine a week.

EXERCISE

The mainstay of my exercise program is walking, which is something I can do anywhere. It fits in well with my busy travel schedule. My work shoes are also walking shoes, so I can walk anywhere at any time, for 30 to 60 minutes or more each day. I also use my weight machine three or four times a week; I keep it in my exercise room with a treadmill and small trampoline. I often watch movies and concerts while exercising. In addition, I enjoy bicycling with my family.

STRESS MANAGEMENT

I give a high priority to getting adequate sleep, and I generally sleep well for about eight hours. I report below on my supplement program to enhance sleep. If I'm well rested (which is most of the time), I find that very few problems bother me. However, when I have not had sufficient sleep, then the subtlest problems feel frustrating.

I do a lot of my creative thinking while sleeping. I assign myself a problem before I go to sleep. During a lucid dream period in the morning between sleep and waking, I return to the issue, and invariably I have new insights. I find this lucid dream period a remarkably creative time. By the way, this does not work if I use an alarm clock because waking up suddenly bypasses this in-between stage.

Occasionally, I meditate or have massages. I find exercise relaxing, offering an opportunity to let my mind wander in a meditative fashion. My cortisol levels are in the normal range.

Despite a strong commitment to my ideas and projects since a very young age, I try to maintain balance in my life and seek to keep my relationships with my wife, children, family, friends, and colleagues healthy and vital, with generally positive results. Of course, nobody's perfect.

BRAIN HEALTH

The most important thing I do to keep my brain healthy is to use it. We know from brain-scanning studies that our thoughts literally create our brains, so challenging ourselves intellectually and artistically is a vital anti-aging activity. I stay mentally active with a variety of projects, one of which is an ongoing study of human biology and health. Good sleep hygiene, which I discussed above, is also vital for brain health. I also take an array of nutritionals described below to enhance support for my brain cells.

TOXINS

I do a lot to improve my body's ability to handle and remove toxins. I've taken steps to reduce my exposure to toxins: I've never smoked, and I avoid secondhand smoke. I try to eat organic food whenever possible and drink filtered, alkalinized water. I've had my mercury-containing amalgams removed, I use an ionic air filter in my bedroom and office, and I use an air-tube-based earphone for my cell phone. I drink about 10 glasses of very-high-pH (about 9.5) alkaline water a day (in addition to the green tea). I describe below a number of the nutritional and intravenous therapies that strengthen my body's detoxification abilities.

TESTING

In addition to routinely testing many blood levels, I have had a virtual colonoscopy and a lower-body CAT scan of my organs, which were normal. My thallium stress test, a test of cardiac function, was normal. My blood pressure is in an acceptable range. Extensive cancer screens are all negative. My prostate specific antigen (PSA) is low and stable at 0.4 ng/ml (nanograms per milliliter).

REPROGRAMMING MY BIOCHEMISTRY

A common attitude is that taking substances other than food, such as supplements and medications, should be a last resort, something one takes only to address overt problems. Terry and I believe strongly that this is a bad strategy, particularly as one approaches middle age and beyond. Our philosophy is to embrace the unique opportunity we have at this time and place to expand our longevity and human potential.

In keeping with this health philosophy, I am very active in reprogramming my biochemistry. Overall, I am quite satisfied with the dozens of blood levels I routinely test. My biochemical profile has steadily improved during the years that I have done this.

For boosting antioxidant levels and for general health, I take a comprehensive vitamin-and-mineral combination, alpha lipoic acid, coenzyme Q_{10}, grapeseed extract, resveratrol, bilberry extract, lycopene, silymarin (milk thistle), conjugated linoleic acid, lecithin, evening primrose oil (omega-6 essential fatty acids), n-acetyl-cysteine, ginger, garlic, l-carnitine, pyridoxal-5-phosphate, and echinacea. I also take Chinese herbs prescribed by Dr. Glenn Rothfeld.

For reducing insulin resistance and overcoming my type 2 diabetes, I take chromium, metformin (a powerful anti-aging medication that decreases insulin resistance and which we recommend everyone over 50 consider taking), and gymnema sylvestra.

To improve LDL and HDL cholesterol levels, I take policosanol, gugulipid, plant sterols, niacin, oat bran, grapefruit powder, psyllium, lecithin, and Lipitor.

To improve blood vessel health, I take arginine, trimethylglycine, and choline.

To decrease blood viscosity, I take a daily baby aspirin and lumbrokinase, a natural anti-fibrinolytic agent.

Although my CRP (the screening test for inflammation in the body) is very low, I reduce inflammation by taking EPA/DHA (omega-3 essential fatty acids) and curcumin.

I have dramatically reduced my homocysteine level by taking folic acid, B_6, and trimethylglycine (TMG), and intrinsic factor to improve methylation. I have a B_{12} shot once a week and take a daily B_{12} sublingual.

Several of my intravenous therapies improve my body's detoxification: weekly EDTA (for chelating heavy metals, a major source of aging) and monthly DMPS (to chelate mercury). I also take n-acetyl-carnitine orally.

I take weekly intravenous vitamins and alpha lipoic acid to boost antioxidants. I do a weekly glutathione IV to boost liver health.

Perhaps the most important intravenous therapy I do is a weekly phosphatidylcholine (PtC) IV, which rejuvenates all of the body's tissues by restoring youthful cell membranes. I also take PtC orally each day, and I supplement my hormone levels with DHEA and testosterone. I take I-3-C (indole-3-carbinol), chrysin, nettle, ginger, and herbs to reduce conversion of testosterone into estrogen. I take a saw palmetto complex for prostate health.

For stress management, I take l-theonine (the calming substance in green tea), beta sitosterol, phosphatidylserine, and green tea supplements, in addition to drinking 8 to 10 cups of green tea itself.

At bedtime, to aid with sleep, I take GABA (a gentle, calming neurotransmitter) and sublingual melatonin.

For brain health, I take acetyl-l-carnitine, vinpocetine, phosphatidylserine, ginkgo biloba, glycerylphosphorylcholine, nextrutine, and quercetin.

For eye health, I take lutein and bilberry extract.

For skin health, I use an antioxidant skin cream on my face, neck, and hands each day.

For digestive health, I take betaine HCL, pepsin, gentian root, peppermint, acidophilus bifodobacter, fructooligosaccharides, fish proteins, l-glutamine, and n-acetyl-d-glucosamine.

To inhibit the creation of advanced glycosylated end products (AGEs), a key aging process, I take n-acetyl-carnitine, carnosine, alpha lipoic acid, and quercetin.

MAINTAINING A POSITIVE "HEALTH SLOPE"

Most important, I spend a lot of time researching my own health situation and health issues in general. Terry and I have sent each other well over 10,000 e-mail messages on health over the past five years, and we have had countless discussions. I maintain health dialogues by e-mail and conversations with many other knowledgeable people around the world, including Dr. Rothfeld. I make a minor change in my health procedures about once a week, and a major change 5 to 10 times a year. These revisions stem from newly available knowledge about cutting-edge therapies or from results of new scientific studies. Other knowledge is only new to me. It may be knowledge about myself, or a new awareness of existing information and health wisdom. I am committed to exploring my own health with an open mind and continually seek new perspectives and approaches.

I have also been actively tracking technology trends and developing mathematical models of how technology evolves, especially information technology. Increasingly, health science itself is becoming a form of information technology, subject to its laws of evolution. These models corroborate what I experience on a daily basis: our tools for avoiding disease and aging are advancing at an exponential pace. For this reason, my confidence in my ability—and that of like-minded contemporaries who make the requisite effort—to remain alive and well until the day radical life extension becomes easy has been growing at the same exponential pace.

11

THE PROMISE OF GENOMICS

"We must be trying to learn who we really are rather than trying to tell ourselves who we should be."

—John Powell

"Life consists not in holding good cards but in playing those you hold well."

—Josh Billings, 19th-century humorist

Fantastic Voyage is built on the premise that you can create your own health by taking an aggressive, proactive role. Rather than allowing disease to happen and then seeking treatment, Ray & Terry's Longevity Program provides you with Bridge One tools you can use to create conditions optimal for maintaining your health and, ideally, allows you to avoid disease altogether. Uncovering your personal health risks, largely hidden until now within your genes, will enable you to formulate a preventive health program specific to you.

The current consensus is that there are 35,000 to 40,000 human genes. Your ability to obtain solid information about these has just begun, but it will increase exponentially over the next few years. The tools of modern science can now accomplish in minutes what once took years of trial and error. A new field of medicine called *genomics* now enables you to discover many of the genes you have. As a result, the one-size-fits-all type of medicine that physicians have been practicing soon will be replaced by individualized therapies. This will enable you to create the perfect Bridge One program for yourself: a diet and exercise plan as well as the nutritional supplements and prescription drugs (if needed) personalized for your individual collection of genes: your genome.

THE HUMAN GENOMICS PROJECT

The U.S. Human Genome Project began in 1990 in an effort to create a complete transcription of all the three billion DNA letters found in human genes. The combined efforts of private and governmental agencies, including an international consortium of researchers, led to the successful completion of this project by 2003, 2½ years ahead of schedule and significantly under budget. In the words of James Watson, who along with Francis Crick identified the double-helix structure of DNA 50 years before, "The completion of the Human Genome Project is a truly momentous occasion for every human being around the globe."

The good news is that you will soon be able to know your exact genetic makeup—all 35,000-plus genes—at a very reasonable cost.[1] You could have this done today but, at the present price tag of $100,000, it's still out of most people's reach. But within a few years, and for a few hundred dollars, you'll be able to get a microchip or DVD listing all of your genes, along with an analysis of what much of it means and what you can do to avert some of the potential problems encoded within your genetic heritage.

One of the main problems with this new technology is that data and information are being generated so quickly that scientists and physicians are having trouble making sense of it all. So several new scientific fields have arisen to help make the information accessible to researchers and practitioners alike.

Genomics is the study of the composition of genetic material itself—the DNA in your genes and chromosomes. Genomics testing, the diagnostic part of the genetics revolution, is already in full swing and is an important part of the Bridge One diagnostic tests we recommend.

Proteomics is the study of proteins, both those found naturally in the body and those created in the laboratory. The most important benefit of Bridge Two developments over the next decade or two, in fact, will be proteomics therapies, which will enable patients to receive individualized treatments for diseases based on their underlying genetic structure. Inexpensive tests will make it a simple matter to design the exact compound needed to treat almost any condition or disease process. But figuring out how to create that compound requires massive computing power—more than even our biggest and fastest supercomputers of today contain—so IBM is introducing Blue Gene/L in 2005. This computer, which occupies a room half the size of a tennis court and has a peak performance of 360 teraflops (360 trillion operations per second), has as one of its primary goals solving this problem.[2] This will make proteomic drug design available within the next 10 years and significantly improve our abilities to prevent and treat disease.

Systems biology is the study of how all of the parts of a living organism work together. Systems biology tries "to connect the dots of all the body's RNA, DNA, genes, proteins, cells, and tissues, elucidating how they interact with each other to create a breathing, blood-pumping, disease-fighting, food-processing, problem-solving human."[3] Currently, we are unable to fully explain how a single cell works, so imagine the impact of having an integrated view of the entire system. This daunting task is aided by new software programs for visualizing systems, which allow scientists to work with 100,000 or more parameters rather than the 20 or so that they can juggle in their heads.[4]

Bioinformatics is the new discipline that will help develop the techniques needed to gather and process all of this new information.

Our focus in this chapter will be on *predictive genomics*, a brand-new Bridge One diagnostic tool. There are already a number of genomics tests commercially available to help predict your predisposition to many serious, but preventable or modifiable, diseases, such as heart disease, Alzheimer's, and cancer. The important thing to remember about predictive genomics is that, in almost all cases, your genes merely express *tendencies*. Your lifestyle choices have a much larger role in determining what happens, or how your genes are expressed.

Since genomics only tells you your tendencies, and because proteomics and other therapies that will be able to alter these tendencies are still in their infancy, it's important for you to remain as healthy as possible for the next decade or two, until these new treatments are more fully evolved. The Bridge One therapies and lifestyle choices described in this book will help you avoid or significantly delay *irreversible* physiologic changes (heart attacks, strokes, dementia). Then you'll be able to take fuller advantage of the powerful Bridge Two gene-based proteomic therapies. Within 10 to 20 years, you'll be able to be treated for what are now untreatable or incurable diseases. Soon after that, you'll benefit from the Bridge Three nanotechnology-AI revolution, which will mean that damage to your body that is presently *irreversible* will be completely corrected.

THE LANGUAGE OF THE BOOK OF LIFE

Most of your genetic information is contained within the double-stranded DNA molecules that reside within the nuclei of your cells. DNA molecules

Life as a Game of Cards

One often hears life being compared to a game of cards. Someone born with a serious genetic disease is said to have been dealt a "bad hand," while the 105-year-old we read about who attributes her longevity to eating a jelly doughnut for breakfast and smoking two packs of cigarettes every day clearly started life with exceptionally good cards.

In the 1950s, Roger Williams, M.D., introduced the concept *biochemical individuality*, the idea that every person possesses a specific and unique biochemical blueprint.[5] Until a few years ago, however, uncovering your biochemical individuality had been hit or miss at best, the result of decades of careful trial and error. For example, after many years of observation, you may have noticed that you have more energy after eating protein for breakfast, that strawberries give you a rash, or that you get a headache if you consume artificial sweeteners. But someone else may note just the opposite effects. Truly, "one man's meat is another man's poison," since we are all biochemically unique.

Before Williams came along, Gregor Mendel, the father of the field of genetics, developed the concept of genetic determinism—that the genes you were born with determine your fate. This has given way to the newer idea of genomic relativism—that your genes don't *determine* what diseases you will acquire but rather merely point out your *predisposition* to them. This idea has far-reaching implications for the future of health care and preventive medicine. While there are a few genes, such as those for cystic fibrosis or Huntington's chorea, whose presence means that an individual will definitely get a specific disease at some point, at least based on today's technology, these are only a tiny fraction of the millions of variants possible in the human genome.

The total of your genetic makeup—the entirety of your inherited DNA—is called your *genotype*. But until quite recently, you have been forced to play this card game of life almost completely in the dark, unable to look at the cards you've been dealt. What difference does it make if you're an expert blackjack player, who knows precisely when to take a hit or double down, if you don't know what cards you hold? This has been the scenario for humankind since the beginning of time. You may have a vague idea of your genetic makeup by knowing what diseases "run in the family," but almost no one has ever had access to precise information regarding his or her own specific genetic code.

The new concept of genomic relativism is at once enabling and terrifying. We now see the age-old battle of predestination versus free will being fought on the front lines of our DNA. Fortunately, it is now looking as if very little about your future health is absolutely predestined or predetermined. You can do something about it now!

are so large that if the coiled DNA of a single cell were unraveled into a straight line, it would stretch more than 6 feet; if all the DNA in the human body were put end to end, it would reach to the sun and back more than 600 times. But the basic structure of a DNA molecule is quite simple: just four molecules called nucleotides—adenine (A), guanine (G), thymine (T), and cytosine (C)—cross-linked to one another like the rungs of a ladder.

The unique double-stranded structure of the DNA molecule allows it to unzip itself in places and reproduce perfect single-stranded copies of its complementary strands. In this fashion, genetic information contained within DNA is "transcribed" (copied) into single-stranded RNA "messenger" mol-

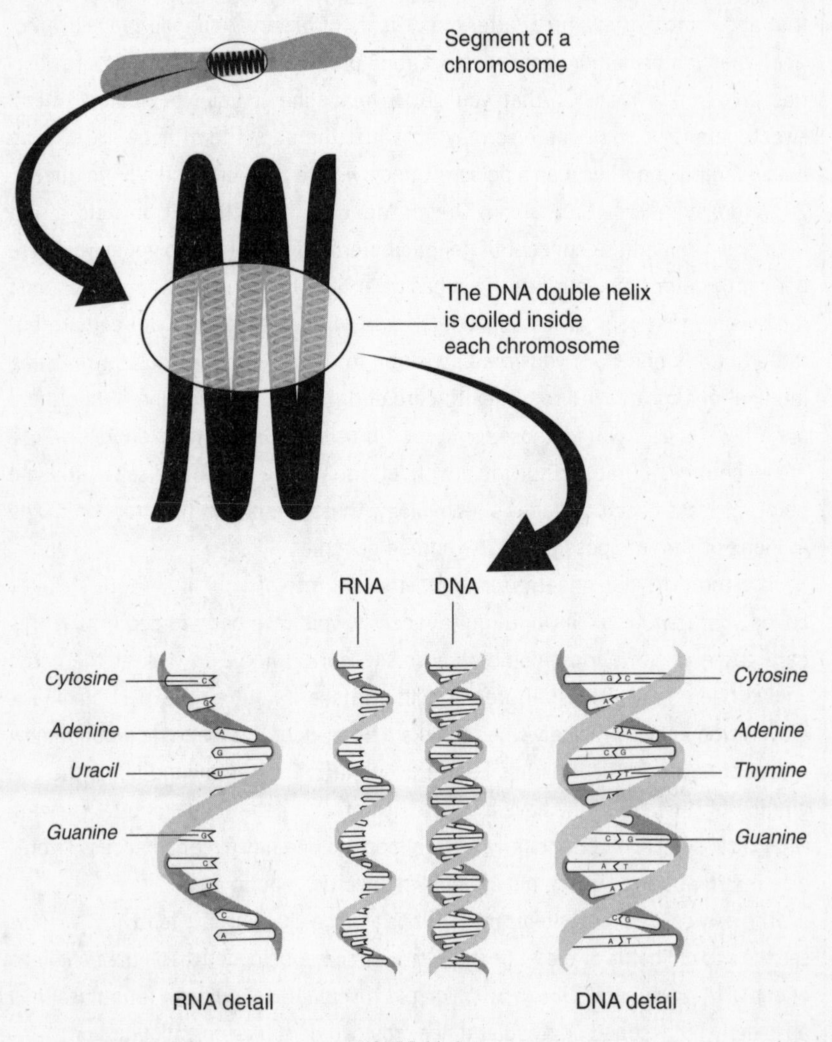

FIGURE 11-1.

ecules. These RNA messenger molecules then "express" (convey) the DNA information to link the 20 amino acids found in the body together to form the proteins that make life possible. These proteins then carry out the daily functions of cellular life.

The actual work of the Human Genome Project involved decoding the entire sequence of the individual letters (A, T, C, U, and G) found in human DNA and RNA. Approximately 3 billion letters were sequenced.[6] The individual letters then form the three-letter "words" (codons) that form the protein sentences, which combine into the 35,000 or so "paragraphs" (genes), which make up the 23 "chapters" (chromosomes) of our genetic Book of Life.[7]

You can find much of this "book" spelled out on the Internet (see www.ncbi.nlm.nih.gov). But subtle differences of a letter or two here and there (misprints) can produce drastic differences. Nearly 99.8 percent of human DNA is identical among all people, and human DNA is even 98 percent identical to chimpanzees. Yet this fraction of 1 percent that is different between us is what creates all the variety of life and ensures that no two humans (other than identical twins, who have the same DNA) will be exactly alike.

THE FUTURE WORLD OF PREDICTIVE GENOMICS

As your DNA molecules replicate themselves trillions of times to create all the cells and tissues of the body, there are numerous opportunities for alterations to occur. These alterations are technically referred to as *polymorphisms* (literally, multiple shapes). It is believed that there are more than 10 million polymorphisms responsible for most of our biochemical individuality. Those that involve only a single nucleotide (A, T, C, or G, as mentioned above) are the most common variety, and these single-nucleotide polymorphisms (SNPs, pronounced "snips") are extremely common. SNPs are important because they can change the way the body functions and, in some cases, predispose you or make you more resistant to specific diseases.

It is estimated that each person may carry as many as a million SNPs.[8] Predictive genomics attempts to identify the most significant SNPs to determine how likely you are to be predisposed to develop a specific disease or health risk, and to evaluate the possibility that this condition might appear under particular environmental circumstances or lifestyle choices.

Furthermore, the same SNP that can be beneficial to a person in one environment can be harmful under different circumstances. For example,

BRIDGE TWO

COMPARING GENES

The time and cost associated with synthesizing and sequencing the billions of bases in human DNA has plummeted in the past 15 years. If current trends continue, "within a decade a single person at the lab bench could sequence or synthesize all the DNA describing all the people on the planet many times over in an eight-hour day. Alternatively, one person could sequence his or her own DNA within seconds."[9] Companies such as U.S. Genomics and the Institute for Genomic Research are building the analytical systems required to maintain that trend, using techniques such as fluorescent tagging of molecules, nanofluid systems, and laser analysis.

These rapid-sequencing speeds facilitate comparisons between different species that will help us better understand the evolution of the human genome. For example, such genetic comparisons are supporting claims that parasites and diseases have caused far more radical mutations in human genes than previously thought.[10] These sequencing speeds also allow genetic comparisons between subpopulations, such as an apparently healthy control group and a group suffering from a particular disease. These kinds of comparative analyses will help identify sets of mutations typically found in certain diseases.

Since the 1990s, microarrays (chips no larger than a dime) have been used to study and compare expression patterns of thousands of genes at a time.[11] The possible applications of the technology are so varied, and the technological barriers have been reduced so far, that huge databases are now devoted to the results from "do-it-yourself gene watching."[12]

one SNP that has historically given individuals a better chance of survival during periods of famine or near starvation is known as the "thrifty gene"—it helps people survive on minimal calories. Nowadays, however, this genetic polymorphism is more of a problem than a benefit, since people who carry it are more prone to obesity when they consume excess or even merely adequate calories. Centuries ago, when famine was much more common, Pima Indians from the southwestern United States who carried the thrifty gene were better able to survive long periods of near starvation, so there was an advantage in possessing this genetic variation, and it became more prevalent. This has led to the majority of modern-day Pima Indians being overweight.[17]

Genetic profiling is being used to revolutionize the processes of drug screening and discovery. Microarrays can "not only confirm the mechanism of action of a compound" but "discriminate between compounds acting at different steps in the same metabolic pathway."[13] As a result, drugs will be brought to market much faster. Their effect will be far more targeted—"designer" drugs will become available, for example, for people with specific genetic mutations. Your doctor will also have a far better idea whether a particular drug will be effective for you. The days of hit-or-miss, one-size-fits-all drug treatment are nearing an end.

We now know that gene expression is controlled by peptides (portions of protein molecules) and short RNA strands. Many new therapies are based on manipulating this process to either turn off potentially harmful genes or turn on desirable genes. Two therapies that do this include "antisense therapy" and "RNA interference." Antisense therapy uses mirror-image sequences of RNA (antisense RNA) to block the expression of harmful genes. One antisense drug, Vitravene, is already on the market, and more of these medications are expected soon.[14]

RNA interference therapy involves placement of segments of double-stranded RNA that bind tightly to the messenger RNA created by a specific gene. This triggers a reaction in which the messenger RNA is cut into small pieces, effectively silencing the gene. By creating an interfering RNA segment, scientists can block a gene's expression and potentially stop a disease from developing.[15]

Also, a new generation of DNA sequencers and synthesizers allows scientists to write, not just read, stretches of DNA. The challenge of writing long sequences with few errors is huge, but a number of machines are in development that can build large segments of DNA automatically. Armed with these technologies, research groups are trying to build synthetic organisms, design new proteins, and create artificial DNA letters to expand the "genetic alphabet."[16]

Different ethnic groups also have distinct SNPs. For example, one of the liver enzymes responsible for detoxification of some environmental toxins is known as CYP2D6 (Cytochrome P456 2D6). This enzyme also metabolizes many common prescription medications as well as the popular herbal remedy kava, which has been safely used for centuries to treat stress disorders among islanders in the South Pacific. Yet when kava is taken by individuals of northern European ancestry, many of them experience liver toxicity. This is because approximately 10 percent of Europeans possess a SNP that makes the CYP2D6 enzyme defective, so it can't metabolize kava. This SNP is rarely found in people indigenous to the South Pacific.

BRIDGE THREE

LIVING WITH BIOBOTS (INSIDE)

Imagine a tiny handheld device with DNA sensors on a microchip that could detect diseases in minutes in your doctor's office, or at home between visits, using nothing more than a drop of saliva or blood. That's a future envisioned by Harvard nanobiotech research chemistry professor Dr. Charles Lieber. He and a research team are developing ultrasensitive nanowire sensors almost as small as molecules, yet 1,000 times more sensitive than the latest DNA tests, such as PCR amplification.[18]

Lieber's initial target is detecting prostate cancer using a microchip with 10 silicon wires, each just 10 nanometers (billionths of a meter) wide. These nanowires are coated with biological molecules that detect PSA, the telltale sign of prostate cancer. When as few as three or four PSA molecules bind to the nanowires, an electrical signal is generated. Expect to see these devices commercially available as early as 2007, says Larry Bock, CEO of Nanosys, which has licensed Lieber's technology. Future models could have thousands of such wires that could detect a wide range of diseases and illnesses.

But once you've detected a genetic fault, how do you fix it? The trick is to deliver the missing genetic components to your cells. Currently, doctors use modified viruses to deliver DNA. But viruses can cause several immune reactions, so they can't be used repeatedly. One idea is to pack DNA molecules into nanoparticles tiny enough to actually enter the nucleus of cells.

This Trojan-horse strategy is exactly what researchers at Case Western Reserve University and Copernicus Therapeutics are developing. They inject DNA into liposomes (fatty globules) tiny enough to pass through the cell's outer membrane. Once inside the cell, the next challenge is to get them into the nucleus. The solution: peptides (a portion of a protein molecule that is only 25 nanome-

GENOMICS TESTING

Almost all of the most common, disabling, and deadly degenerative diseases of our time, including cardiovascular disease, cancer, type 2 diabetes, and Alzheimer's disease, are the result of the *interaction* between genetic and environmental factors. Tests for some of these are now available. Genomics testing allows you and your physicians to gain a deeper understanding of disease processes and develop more specific and effective interventions.

A few genomics testing panels are now commercially available. Each tests for up to a dozen or so SNPs at a cost of less than $30 to $50 per gene,[21]

ters wide—small enough to fit through pores in the membrane of the nucleus). These encase the DNA molecules and release them inside the nucleus, where they can correct the cell's genetic code. The first trials were with 12 patients with cystic fibrosis, who have a faulty gene that causes mucus to accumulate in their lungs. The researchers expect this technique to be available experimentally to doctors within the next few years.

Further in the future, semiconductor nanoparticles called *quantum dots* are likely to play a key role in diagnostics and drug delivery, according to Dr. Shuming Nie, professor of Biomedical Engineering at Emory University and the Georgia Institute of Technology.[19] These molecule-size nanoparticles glow when specific genes and proteins are attached to them, so they can be used to identify tumor cells, for example. They can also monitor the effectiveness of drug therapy and serve as scaffolding in tissue engineering or as "smart bombs" to deliver controlled amounts of drugs into genetically tagged tumor cells.

But instead of waiting until something goes wrong in the body to fix it, Robert A. Freitas Jr. advocates an even more radical approach in his series *Nanomedicine*.[20] "Artificial 'biobots' could be in our bodies within 5 to 10 years," he says. "Advances in genetic engineering are likely to allow us to construct an artificial microbe—a basic cellular chassis—to perform certain functions. These biobots could be designed to produce vitamins, hormones, enzymes, or cytokines in which the host body was deficient, or they could be programmed to selectively absorb and break down poisons and toxins."

Freitas also has developed detailed conceptual designs for a DNA repair robot that goes into the nucleus of each cell and fixes DNA errors. It could also modify the DNA to anything desired. Ultimately, we will be able to replace the cell nucleus altogether with a nanoengineered computer that contains the genetic code with machinery to produce amino acid strings. This will enable us to block unwanted replication and instantly update our genetic code.

---

while only a few years ago it was rare to get a genetic test done for less than $300 each. Within a few more years, in accordance with Ray's Law of Accelerating Returns,[22] for the same few hundred dollars you'll be able to get a panel that tests for *thousands* of genes. By the end of the decade, you will probably have access to DNA chips that will test for most, if not all, of the 10 million SNPs believed to exist.[23]

Problems with Genomic Testing

With our current knowledge and abilities, even if you could know all of the hundreds of thousands of SNPs you possess, all it would do is produce information overload. Neither you nor your doctors would know what to do with most of the information. We will need to wait for the bioinformatics scientists to catch up and provide us with sophisticated computer programs that can make sense of all this information. Today's clinicians, therefore, tend to limit screening to just a few of the more common polymorphisms. Besides, not many patients would want to know that they have a genetic defect that can't be fixed by any presently available therapy. For example, many women whose mothers have been diagnosed with breast cancer frequently decline testing for BRCA1 polymorphisms, since a positive test would suggest a high likelihood of developing breast or ovarian cancer themselves.[24] So most present-day testing is restricted to the hundred or so SNPs that can currently be modified through interventions such as diet, lifestyle, nutritional supplements, and prescription pharmaceuticals.

Another concern associated with genomic testing is patient confidentiality. Health and life insurance companies frequently request copies of a patient's medical records before issuing a policy. It would be unfair if insurers were able to use this type of information, which patients obtain voluntarily to help them better define and reduce the risks from their genetic heritage, as a basis for discrimination. To avoid that possibility, genomics results are typically protected by security codes disclosed only to the patient's attending physician, and most physicians keep the results in a location separate from the patient's regular medical chart.[25] When an insurance company requests medical records for a patient, the genomics results are not sent.

Alexander Pope said in the 1700s, "a little learning is a dangerous thing." With our current state of knowledge, we need to remember that we still have only a "little learning" when it comes to our genomic knowledge base, so genetic testing is not something to be undertaken lightly. Unlike many routine laboratory tests, where results often provide reassurance and decrease worry, the results of a genomics panel invariably reveal some "bad genes"—those that have the potential to increase your risk of serious diseases, like heart attacks, certain types of cancer, or Alzheimer's. Obtaining genomic information can be of great value, and this information may be lifesaving. But we want to emphasize that genomics testing is serious business, and you need to be psychologically prepared for the results.

A Practical Example: Apo E

Genomics testing is now available for several hundred SNPs. Several of these are discussed in the relevant chapters throughout this book; for example, we describe a polymorphism of the MTHFR gene in our discussion of methylation reactions and the GSTM 1 "null-SNP" in our discussion regarding early diagnosis of cancer.

Some Key SNPs

While the interactions between genes may be just as important as the specific SNPs, scientists have started to learn that a few SNPs have very powerful influences.

BRCA1 is a major genetic risk factor for breast cancer. A woman who possesses a defective copy of this gene has a significant chance of developing breast and/or ovarian cancer.

GSTM1, GSTP1, CYP1A1, CYP1B1, and CYP2A6 code for liver enzymes that determine how well you detoxify environmental toxins. Variations in these genes increase or decrease your risk of several types of cancers.

Alpha 1 antitrypsin deficiency predisposes individuals to early emphysema, particularly if they smoke.

Apolipoprotein E has a strong influence on one's potential risk of developing cardiovascular disease and Alzheimer's.

AGT, ACE, and AT1R are associated with blood pressure. Tests of polymorphisms of these genes can suggest if you should avoid salt and what classes of medications would be most helpful for treating your blood pressure if needed.

A very beneficial SNP currently under investigation has been found in the mitochondria of centenarians. In a study of 52 Italians age 100 or older, researchers found a common polymorphism in 17 percent of the centenarians, but which was found in only 3.4 percent of 117 people under the age of 99.[26] Possessing this mutated gene seems to increase fourfold the chances of living past the 100 mark. An even more interesting aspect of this mutated gene is that you can get it either the old-fashioned way, by inheriting it, or as a mutation that arises during the course of your lifetime. This introduces the intriguing idea that there may be specific things you can do to induce this beneficial mutation, increasing your own chances of living past 100. Once we have the ability to change our genes through gene therapy, we will be in a position to turn on genes that promote longevity such as this one and turn off those that promote aging.

As an example of the type of information now available from genomics testing, let's take a look at the Apo E (apolipoprotein E) polymorphisms, which are powerful genetic markers for cardiovascular disease and Alzheimer's. We will first examine the specific risks and benefits associated with the different Apo E polymorphisms, then describe how this information can prompt specific lifestyle recommendations, which can help an individual modify the real-life expression of the more dangerous genetic types.

Apolipoproteins are carrier proteins responsible for transporting lipids such as fat and cholesterol throughout the bloodstream. Fat and cholesterol are oily substances that are not water soluble, so they require specific carrier molecules to help move them from place to place in the body.

Apolipoprotein E comes in three main genetic varieties, called *alleles*: Apo E2, Apo E3, and Apo E4. Because of very minor differences in just one or two of their amino acids, they each differ significantly in their ability to carry fat and cholesterol in the bloodstream. Apo E2, for instance, does a good job of clearing cholesterol from the arteries, while Apo E4 is much less efficient.

Every person possesses two copies of the Apo E gene, one inherited from each parent, so there are six possible combinations: E2/E2, E3/E3, E4/E4, E2/E3, E2/E4, and E3/E4. If you possess one or two copies of the E4 allele, you may have an increased chance of having elevated cholesterol, triglycerides, and coronary heart disease.[27] Even more important, Apo E4 is also correlated with a significantly increased risk of Alzheimer's disease (AD). If you do not have any copies of Apo E4, you have a 9 percent risk of developing AD by age 85. If you have just one copy of this allele—the E3/E4 genotype, which is carried by more than 25 percent of the population, or the much rarer E2/E4 genotype—you have a 27 percent chance of developing AD by the same age; in other words, triple the risk. But if you have two copies—the E4/E4 genotype—the risk rises to 55 percent, a sixfold increase.[28] Furthermore, the average age that AD is diagnosed is much younger, depending on the number of copies of Apo E4 carried: 84 years old if you have no copies of E4, 75 years if one copy, and around 68 with two copies.[29]

The Apo E2 allele, on the other hand, appears to confer some degree of protection against the development of Alzheimer's, and patients with at least one copy of E2 have a 40 percent to 50 percent reduction in their AD risk.[30] Apo E2 is not perfect, however, because some forms of heart disease are more common in patients with this allele. All things considered, Apo E2 is a pretty good hand of cards; the 105-year-old doughnut-eating smoker mentioned above was probably born with one or two copies of Apo E2.

The Apo E3 form is the most common, however—more than 50 percent of the population is E3/E3—and affords some protection against both heart disease and AD.

Although Apo E4 is a potent risk factor for Alzheimer's and may be associated with other forms of dementia, the good news is that most people who carry the Apo E4 allele do not develop dementia, and about one half of people diagnosed with AD do not possess any copies.[31] In some studies, the proportion of patients with dementia that is attributable to the Apo E4 allele is estimated to be about 20 percent.[32] But if Apo E4 were nothing but bad news, it probably would have been selected out of the gene pool long ago. People who carry this variation have a much *lower* incidence of some serious diseases, such as age-related macular degeneration (AMD), the leading cause of blindness in the developed world. Meanwhile, people who carry the more "favorable" E2 are at much higher risk of losing their vision to AMD.[33] Free-radical damage appears to play a key role in developing the specific type of damage seen in AD. So if you discover you carry the Apo E4 allele, special efforts to limit free-radical damage should be implemented.[34] Apo E4 carriers should begin taking aggressive free-radical damage-control measures as early in life as possible.

The following practical recommendations are typical of the type of personalized advice that you can get based on knowing just one aspect of your own DNA—that you carry the Apo E4 genetic variation:

- Take nutrients such as vitamin C, vitamin E, alpha lipoic acid, grapeseed extract, and coenzyme Q_{10} daily to reduce free radicals.
- Consider taking pharmacological agents that may help reduce free-radical production in the brain. These include the monoamine oxidase-B inhibitor selegilene and the hormone melatonin. Also take low-dose aspirin therapy (81 milligrams daily).
- Several nutraceutical agents have been found to be protective of brain neurons. Consider phosphatidylserine (100 to 300 milligrams daily), phosphatidylcholine (900 millligrams twice a day), acetyl-L-carnitine (500 milligrams twice a day), and vinpocetine (10 milligrams twice a day).
- Lifestyle changes, including stress reduction and regular aerobic exercise, can be beneficial.
- Since Apo E4 is also associated with elevated lipid levels, implement a low-fat diet to help keep cholesterol levels down, and maintain a low-glycemic-load, low-carbohydrate diet to lower triglycerides.

Today you have the ability to both know and modify the expression of the genes you were born with through diet, nutrition, and lifestyle choices. These techniques will soon be joined by more powerful biochemical strategies to alter the expression of your genes. Not too long after that, you should be able to change your genes entirely and choose the ones you want.

Predictive genomics testing is available today that can provide previously unknowable genetic information personalized to each individual. This new medical specialty is in its infancy and, as with any new science, there are perils and pitfalls. But today's primitive, incompletely understood tests will lead to ever more sophisticated analyses. Today's magnifying-glass view of the genome will lead to seeing in microscopic detail tomorrow.

12

INFLAMMATION—THE LATEST "SMOKING GUN"

"You can measure [the] inflammatory response, and those of us that have a greater response turn out to be at much higher risk of going on to have either a heart attack or a stroke."

—Dr. Paul Ridker, seminal researcher from Harvard Medical School who helped establish the link between inflammation and heart disease

One of the hottest research topics in medicine right now is the connection between inflammation and a host of disease processes. This is a "hot" topic indeed, as the literal definition of inflammation is to "set on fire." Until quite recently, inflammatory diseases were believed to be confined to conditions with obvious, or acute, inflammation, such as arthritis (inflamed joints), asthma (inflamed airways), and even acne (inflamed skin). Recent studies have shown that there is another less obvious type—chronic, or "silent," inflammation, which plays a significant role in diseases that had not been considered inflammatory disorders at all, including heart disease, Alzheimer's disease, diabetes, and certain types of cancer.

Medical students are taught the cardinal signs of *acute* inflammation by the simple mnemonic: *rubor* (redness), *calor* (heat), *tumor* (swelling), and *dolor* (pain). Although often uncomfortable, acute inflammation plays a critical role in the body's response to injuries such as sprains, strains, and fractures, and to bacterial, viral, and allergenic invaders. The symptoms of *chronic* inflammation are altogether different. In fact, they are barely noticeable until catastrophe strikes, often decades later.

Chronic inflammation is also known as silent inflammation and it can activate potentially harmful genes as well. The aging process, for example, is thought to involve certain "aging" genes being turned on and other "youthful" genes being turned off. So a cornerstone of Ray & Terry's

Longevity Program is minimizing silent inflammation. In this chapter, we will discuss specific lifestyle choices you can make to achieve that goal. We'll also describe a simple test that can tell you how much inflammation you have. First, let's look at some of the serious disease processes where scientists have found silent inflammation lurking just beneath the surface.

ROLE OF INFLAMMATION IN SPECIFIC DISEASES

Chronic low-grade inflammation can smolder silently within your body for decades without causing any obvious or outward problems. But all the while, it is eroding your health and taking years from your life. Taking steps to control silent inflammation gives you a powerful tool to combat several major degenerative diseases, including cardiovascular disease, Alzheimer's disease, diabetes, and cancer.

Cardiovascular Disease

Until recently, cardiologists thought that heart disease was caused simply by the buildup of cholesterol deposits inside the walls of the coronary (heart) arteries. New research suggests that silent inflammation is a fundamental cause for cholesterol being deposited in the arteries in the first place. Silent inflammation has also been shown to be a potent cardiovascular risk factor itself, independent of other well-known risk factors.[1] We know that significant protection against heart disease comes from eating a low-fat, cholesterol-lowering diet, but it is also very important to eat a diet that decreases inflammation as well.

Alzheimer's Disease

Just as inflammation in the heart arteries increases heart attack risk, inflammation of brain tissue increases risk of Alzheimer's disease (AD). Here's how: Silent inflammation in the brain increases production of soluble amyloid protein and increases its conversion into insoluble amyloid fibrils (discussed further in chapter 14, "Cleaning Up the Mess: Toxins and Detoxification"), which are actually toxic waste products that interfere with normal brain functioning and kill brain cells. If the brain cells do not remove these amyloid fibrils immediately, the dead and dying cells stick together to form pleated sheets of crystalline debris called *plaque*. One scenario for development of Alzheimer's includes the following steps:

1. As the fibrils accumulate within brain cells, deterioration of brain function (as seen in AD) begins. Amyloid deposits formed of the dead and dying cells impair the supply of blood to the brain, compounding the problem.[2]

2. Amyloid fibril is identical to one of the amino acid chains that make up immunoglobulins (antibodies), the proteins at the front line in the immune system. So the inflammation started by amyloid fibrils overstimulates the immune system.[3]

3. Overactivity of the immune system then leads to further inflammation.

4. Plaque deposits disrupt normal cellular metabolism, causing further inflammation.

Inflammation in the brain also generates free radicals that destroy neurons in genetically predisposed individuals. This loss of brain cells can result in dementia.[4]

As discussed in the previous chapter, people who carry the Apo E4 allele, a genetic variation seen in over 25 percent of the population, are at increased risk of developing AD.[5] They also develop dementia at a younger age, an average of 10 years earlier than people who do not carry this gene.[6] Apo E proteins are responsible for clearing away soluble amyloid protein quickly, so it doesn't have a chance to form the dangerous crystalline amyloid fibrils or plaque.[7] The Apo E4 variant (as opposed to the E2 and E3 genotypes) clears soluble amyloid protein slowly, so more plaque has a chance to form.

But just because you carry the Apo E4 variant doesn't mean you're destined to develop AD. We need to look for other risk factors. One of these seems to be, of all things, the herpes simplex virus, specifically herpes simplex type 1, which causes cold sores. In combination with Apo E4, the herpes simplex virus can be a trigger for AD.[8] This is an example of how a usually benign environmental factor (herpes simplex) can increase the risk of a serious disease when a certain genetic predisposition (Apo E4) is present.

In any event, if you perform genetic testing and find that you carry the Apo E4 allele, you know that you need to be particularly careful to help control your increased risk of AD and cardiovascular disease. Use your genetic information to individualize your personal health program and minimize risk.

Diabetes

A similar story unfolds in the case of type 2 diabetes mellitus. In Alzheimer's disease, deposits of amyloid form in the brain. In type 2 diabetes, a different

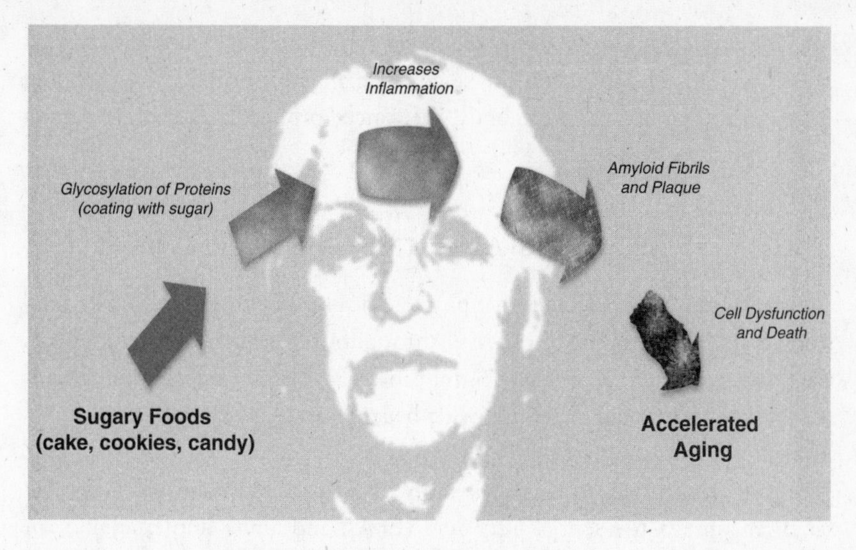

FIGURE 12-1. HOW SUGAR INCREASES AGING

type of amyloid forms in the pancreas.[9] Chronic elevation of blood sugar and insulin levels increases inflammation in the bloodstream, triggering a cascade of events in the pancreas of a type 2 diabetes patient similar to what is seen in the brain of an AD patient. In AD, the trigger may be herpes; in diabetes, dietary sugar is the culprit.

Even if you don't have diabetes, when you eat sugary foods or foods with a high glycemic load, you increase the amount of silent inflammation in your body. The aging process itself is, in a sense, a sugar disease, as seen in Figure 12-1.

By avoiding sugar and other high-glycemic foods, you can break this vicious cycle, decrease the amount of inflammation in your body, and slow aging.

Cancer

Research into the connection between inflammation and cancer has become so intense that Dr. Robert Tepper of Millennium Pharmaceuticals has stated that "virtually our entire R&D effort is [now] focused on inflammation and cancer."[10] Inflammation promotes several different cancers, including colon and lung. NSAIDs (nonsteroidal anti-inflammatory drugs) such as aspirin and ibuprofen decrease inflammation, and people who take these drugs on a regular basis have a decreased risk of several types of cancer, including colon cancer.[11] A study involving more than 14,000 women showed that those who took aspirin regularly had less than half the rate of the most common type of lung cancer.[12] The theory is that silent inflammation alters how certain genes are expressed, setting the stage for malignant growth. By reducing levels of inflammation in the body, the cancer-promoting genes are turned off more of the time, decreasing cancer risk.

PROSTAGLANDINS AND INFLAMMATION

Prostaglandins are molecules that physiologically affect target cells, like hormones do. While hormones travel throughout the body, exerting effects far away from their points of origin, prostaglandins act only in the immediate vicinity of where they are made because of their extremely short lifetimes.

Prostaglandins are made in the body directly or indirectly from linoleic acid, an omega-6 essential fatty acid (EFA), and alpha linolenic acid, an omega-3 EFA. Nuts, grains, seeds, animal products, and most vegetables contain these fatty acids. The diagram below shows the metabolic steps through which EFAs are transformed into prostaglandins.

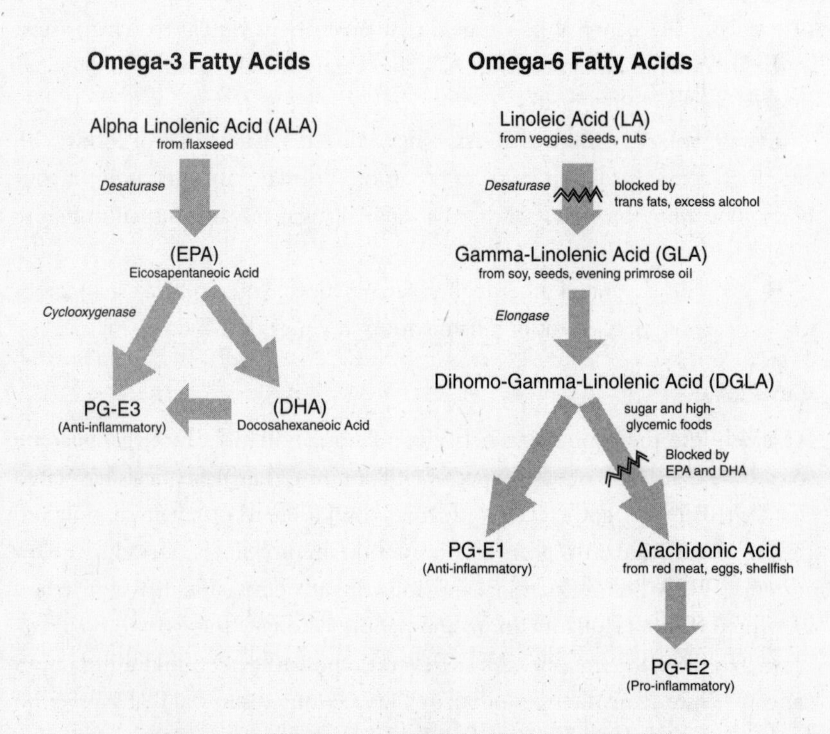

FIGURE 12-2. METABOLIC PATHWAYS OF ESSENTIAL FATTY ACIDS

There are three main types of prostaglandins. Two of them (PG-E1 and PG-E3) are anti-inflammatory, while one (PG-E2) actually increases inflammation in the body. All three are necessary for good health, and excess inflammation results only when there is an imbalance between them. Here's how they are made:

1. As seen in the diagram, the first step in the production of one of these, PG-E1, which is anti-inflammatory, involves the conversion of the omega-6 EFA linoleic acid into gamma-linolenic acid (GLA) under the influence of the enzyme desaturase. You get linoleic acid in your diet from vegetables, nuts, grains, and seeds.

   Note: This step is hindered by intake of trans-fatty acids, such as hydrogenated vegetable oils like margarine, and certain viral infections. Excess alcohol consumption and the aging process also interfere with the function of this enzyme. You can assist your body in this metabolic conversion by eating foods that are rich in preformed GLA, such as soy, sesame seeds, sunflower seeds, and walnuts.

2. Under the influence of the enzyme elongase, GLA is converted into di-homo-gamma-linolenic acid (DGLA).

3. DGLA forms either the strongly anti-inflammatory prostaglandin PG-E1 or pro-inflammatory PG-E2.[13] Which way this goes is largely determined by the level of insulin in your bloodstream. If it's high, as when sugary or high-glycemic foods are eaten, more DGLA turns into arachidonic acid, which then forms pro-inflammatory PG-E2. So, hidden away in this complex biochemical pathway, we find the smoking gun that directly links excess sugar consumption to increased inflammation.

When you consume sugar or high-glycemic carbohydrates, you increase your insulin level, which stimulates production of arachidonic acid and the amount of inflammation within your body.[14] If, on the other hand, you eat a high-fiber, low-calorie, low-glycemic-load diet, as discussed in previous chapters, you will lower your insulin level, reduce production of arachidonic acid, and decrease inflammation. There are other dietary tools to decrease inflammation in your body as well. Red meat, shellfish, and egg yolks are rich dietary sources of preformed arachidonic acid, which is pro-inflammatory. Therefore, consumption of these foods should also be limited.

Another way to block the conversion of DGLA into pro-inflammatory arachidonic acid is by taking more of the omega-3 fatty acids eicosapentaneoic acid (EPA) and docosahexaneoic acid (DHA). EPA is found mainly in fish oils and helps suppress the formation of undesirable PG-E2, while promoting formation of the beneficial third type of prostaglandin, PG-E3. So consuming fish and fish oil is another powerful method of decreasing inflammation in your body.[15] We recommend 1,000–3,000 mg of supplemental EPA and 700–2,000 mg of DHA. Individuals with inflammatory disease may need 5,000–10,000 mg of omega-3 EFAs.

Vegetarians will be pleased to know that EPA can also be made from flaxseed oil, hemp oil, and pumpkin seed oil, although many people lack the ability to make enough EPA through this pathway and ideally should take preformed EPA/DHA (fish oil) supplements. Many people are under the mistaken impression that they can just take flaxseed oil and their bodies will make all the EPA/DHA they need but, in many cases, this will not work. For strict vegetarians, flaxseed oil may be the best alternative, but for everyone else we recommend fish oil supplements, because they contain the exact molecules needed to decrease inflammation, while flaxseed oil must first undergo several metabolic steps.[16]

DIETARY AND LIFESTYLE CHOICES TO REDUCE INFLAMMATION

As you can see, what you eat significantly affects the level of silent inflammation within your body and, in turn, your risk of serious inflammatory-mediated diseases. If you consume a diet rich in high-glycemic foods (such as sugary foods, refined white flour products, and fruit juices) or foods high in preformed arachidonic acid (red meat and eggs), you increase PG-E2 and your amount of silent inflammation. Here's the diet we recommend to reduce inflammation:

AVOID FOODS THAT INCREASE INFLAMMATION, SUCH AS	EMPHASIZE FOODS THAT DECREASE INFLAMMATION, SUCH AS
Red meat, eggs	Cold-water fish
Sugar	Spices and herbs like turmeric (contains curcumin), rosemary, ginger, and hot peppers (contain capsaicin)
Coffee, alcohol	Green tea
High-glycemic carbohydrates (such as pastries and pasta)	Low-glycemic carbohydrates (such as whole grains and green vegetables)

There are other lifestyle choices you can make to cool the flames of inflammation within your body.

- Lose weight. Fat cells are very powerful generators of inflammation. As you lose weight, you automatically reduce your amount of silent inflammation. Maintaining your optimal body weight through diet and regular exercise is critically important, because excess fat tissue increases the level of inflammation in the body.[17]

- Exercise more. A program of regular exercise averaging 30 minutes daily will also significantly reduce inflammation.

- Reduce stress. Stress causes inflammation. A recent study has shown that chronic exposure to stress significantly raises levels of the inflammatory compound IL-6 (interleukin-6).[18] Overproduction of IL-6 has been linked to numerous diseases, including cardiovascular disease, arthritis, type 2 diabetes, certain cancers, and accelerated aging. Taking steps to control stress in your life is an important step toward reducing inflammation in your body.

You can also control silent inflammation with nutrients. We have already mentioned fish oil supplements. Both cold-water fish and fish oil supplements are very rich sources of the important anti-inflammatory omega-3 fatty acids, EPA and DHA, which prevent and help treat heart disease. An analysis of 11 studies published in 2002 showed that omega-3 fatty acid supplementation alone decreased mortality in patients with coronary artery disease.[19] Omega-3 fatty acids can substantially reduce the risk of sudden death from cardiac heartbeat irregularities as well, even in cases of advanced heart disease.[20]

The spice turmeric (curcumin) has powerful anti-inflammatory properties, as do several of the compounds found in green tea. A number of herbs and herbal extracts, such as boswelia (frankincense), licorice, rosemary, and ginger, also have anti-inflammatory properties. Onions and garlic exert a mild anti-inflammatory effect as well.

MEDICATIONS TO REDUCE INFLAMMATION

Anti-inflammatory drugs are among the most commonly used medications in the United States. In cases of acute inflammation, these pharmaceuticals can provide comfort from the severe pain of acute inflammation on a short-term basis. These drugs are commonly used for long-term treatment, but it is important to weigh the risks associated with extended use against the benefits. The most popular anti-inflammatory drugs are NSAIDs, particularly aspirin and ibuprofen. NSAIDs work by blocking a group of enzymes called cyclooxygenases, which begin the process that turns pro-inflammatory

Dental Health and Inflammation

Regular dental visits might reduce your chance of a heart attack. Silent low-grade infections and inflammation of gum tissues (gingivitis) are potential risk factors for heart disease and stroke.[21] Gum disease is one of the most common types of inflammation and affects almost everyone. Left untreated, this chronic infection can lead to bone destruction and tooth loss. This condition, known as osteonecrosis, has been implicated as one of the types of inflammation associated with numerous serious cardiovascular and neurological illnesses.

To reduce the risks associated with gum inflammation, the American Academy of Periodontology suggests that people who are at risk for cardiovascular disease or have signs of gum disease see their dentist regularly. A regimen of dental care, including twice-daily brushing, flossing, and regular visits to dentists and dental hygienists, is not only critical for the health of your mouth but may help prevent cardiovascular disease as well.

arachidonic acid into other inflammatory compounds, such as PG-E2 (see Figure 12-2 on page 164). In some cases, long-term use of anti-inflammatory medications, either prescription or over-the-counter, may be highly beneficial, but often, you are better off with safer nutritional alternatives, such as the foods and spices mentioned above.

Some studies have shown that patients who take NSAIDs regularly have a lower incidence of diseases associated with silent inflammation.[22] A Johns Hopkins study found that individuals who took NSAIDs for at least two years had a 60 percent decrease in Alzheimer's risk. A 26 percent reduction was seen just with the use of aspirin, and other NSAIDs caused a further reduction.[23] For heart attack and stroke prevention, low-dose aspirin therapy (81 milligrams a day) provides the benefits of long-term NSAID use, with lower risk than full-dose aspirin therapy or other NSAIDs. However, the studies also show that the risk/benefit ratio suggests that low-dose aspirin therapy benefits mainly high-risk patients.[24] The reason is that NSAIDs can cause ulcers and gastrointestinal bleeding. This class of drugs has been implicated as one of the leading causes of iatrogenic (literally, "physician caused") hospitalization and death. Therefore, it is our recommendation that, until further research is available, long-term NSAID therapy be confined to high-risk patients and avoided by otherwise healthy people.[25] A group of new anti-inflammatory drugs introduced in 2000, including Celebrex and Vioxx, are "specific COX-2 inhibitors" that are supposed to reduce inflammation while not significantly increasing risk of gastrointestinal bleeding.

Studies as to how much safer these new drugs are compared to older, less expensive NSAIDs, however, have yielded conflicting results.[26]

The popular cholesterol-lowering "statin" drugs are thought to reduce risk of heart disease by lowering cholesterol levels in the bloodstream. Another reason for their effectiveness is that they also decrease inflammation.[27]

Our recommendation is that you take low-dose aspirin regularly only if you are in the high-risk group, as defined in chapter 15, "The Real Cause of Heart Disease and How to Prevent It." This group includes people who, among other risk factors, have a history of cardiovascular disease, have an immediate relative (parent, sibling) with cardiovascular disease, who smoke, or who have elevated cholesterol levels.

TESTS TO ASSESS INFLAMMATION

By following the recommendations contained in Ray & Terry's Longevity Program, you will decrease much of the inflammation present in your body and significantly decrease your risks of the serious diseases discussed above.

The marker (indicator) used to measure the level of inflammation in your body is called the high-sensitivity C-reactive protein (hs-CRP, or CRP). CRP is a protein made in the liver and released into the bloodstream in response to inflammation. We recommend that you obtain a baseline hs-CRP, if you haven't already done so, and then as a regular part of your ongoing health screening evaluations. If you have a history of cardiovascular disease or any chronic inflammatory condition, or a family history of premature cardiovascular disease or Alzheimer's, you should also consider genomics testing and measurement of essential fatty acids.

Role of C-Reactive Protein

With an acute inflammatory process such as a severe bacterial infection, CRP levels can increase from a normal value of less than 5 to as high as 1,000 or more. Chronic or silent inflammation leads to increases that are much more subtle, say, from 0.8 up to 5.2. Still, even this small increase can be greatly significant over time.

Several recent studies have demonstrated that hs-CRP determination can help predict future coronary artery disease.[28] Men in the highest third percentile for CRP have over twice the rate of heart attacks as men in the lowest third.[29] CRP is not only a predictor for heart disease, Alzheimer's, and stroke but it also rises as a result of adverse lifestyle conditions such as physical inactivity, obesity, sleep disorders, and depression.

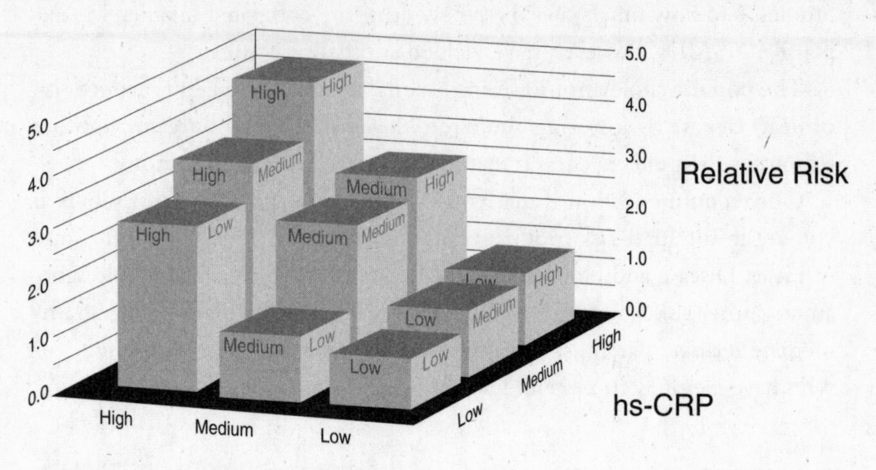

FIGURE 12-3. IMPACT OF INCREASING LEVELS OF HS-CRP ON RISK OF HEART DISEASE[30]

There is more serious disease and disability among seniors who have elevated CRP levels, compared with their peers with lower levels. Healthy people in their 80s and 90s still have low levels of CRP. Some researchers believe that CRP is such a useful tool that it can be used as a marker for biologic aging.[31]

The hs-CRP is an inexpensive blood test that can be easily performed by your regular physician. Blood should be drawn in the morning after a 12-hour fast. The usual reference range of normal hs-CRP is considered to be less than 5, but based on our review of the medical literature, we believe this level is too high for optimal health. You should strive for an hs-CRP less than 1.3, because lower levels are associated with slowing of the aging process and decreased health risks.[32]

TEST CONDITIONS	OPTIMAL hs-CRP	USUAL LABORATORY "NORMAL"
Fasting	less than 1.3	less than 5

Essential Fatty Acids Determination

Another useful test, particularly in cases of inflammatory disease such as asthma, severe allergies like eczema, and rheumatoid arthritis, is the essential fatty acids profile. This blood test is also done under fasting conditions in the morning and provides precise values and ratios between several important fatty acids discussed above, such as the anti-inflammatory fatty acids EPA, DHA, and DGLA as well as the pro-inflammatory arachidonic acid. Knowing your levels of these fatty acids and their ratios provides you with a blueprint for corrective supplementation.

Genomic Markers for Inflammatory Risk

It is now possible to test for several genetic markers that indicate a predisposition toward inflammation. For purposes of illustration, two specific mutations (polymorphisms) will be discussed.

Patients with the IL-1β 31C→T mutation are at risk of developing more inflammation from a given stimulus than people without this genetic variation.[33] If you have genomic testing done and find you have this polymorphism, there are specific dietary measures you will want to follow. Fish oil (EPA/DHA) and milk thistle supplementation will inhibit the inflammation from this genetic defect, as will the herbs curcumin, boswelia, and licorice.

TNF-∝ (Tumor Necrosis Factor—Alpha) is another inflammatory chemical messenger that has far-reaching effects throughout the body. Patients with the TNF-∝-308 G→A mutation are at increased risk for inflammatory conditions such as arthritis and asthma. Consumption of fish oils and green tea can help decrease this excessive inflammation.[34]

Most people don't suddenly get a stroke or a heart attack or develop cancer, Alzheimer's, or diabetes out of the blue. These diseases are often the end result of many years of dietary and lifestyle choices that have increased the amount of silent inflammation circulating throughout their bodies. By making lifestyle changes now, you can begin to reduce silent inflammation and cut your chances of developing serious diseases beginning today.

13

METHYLATION—CRITICALLY IMPORTANT TO YOUR HEALTH

Methylation is a simple chemical process in which a methyl group—one carbon atom and three hydrogen atoms—becomes attached to other molecules. Abnormal methylation can create trouble all throughout life, from womb to tomb. It's the major cause of neural tube defects such as spina bifida and anencephaly, a fatal condition in which the brain is exposed and incompletely developed. This simple biochemical reaction has far-reaching effects on the synthesis of DNA, turning genes within a cell on or off, detoxification, and metabolism.

Because of genetic variations (polymorphisms), abnormal methylation is extremely common. Depending on age and ethnicity, between 10 and 44 percent of the population have a problem with proper methylation, which

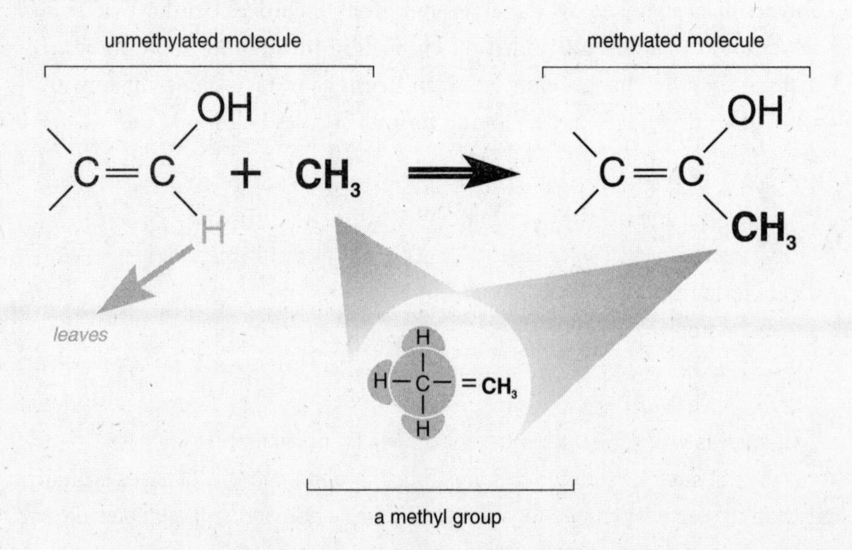

FIGURE 13-1. THE METHYLATION PROCESS

can lead to cervical and colon cancer, coronary heart disease, strokes, Alzheimer's disease, and other conditions.[1] Fortunately, it's simple to detect with the right tests and easy to correct with nutritional supplements. Most physicians, however, do not routinely test for it, and many people discover they have a methylation defect only by suffering its negative effects.

Homocysteine: A Simple Test for Proper Methylation

The easiest way to determine if your body is performing methylation properly is to measure the level of *homocysteine* in your blood. Homocysteine is an example of a "toxic metabolite." These are by-products formed in the course of normal bodily functions that will cause problems unless they are either quickly excreted or rendered nontoxic. Homocysteine forms when you eat methionine, an amino acid normally found in protein foods like red meat and poultry. Your body uses the methylation process to detoxify homocysteine. In a healthy person, this happens quite easily, but when people have a "methylation defect," usually as a result of a genetic problem, homocysteine accumulates to toxic levels.

There are more than 200 scientific papers published every year on the important relationship between homocysteine levels and health, but most physicians still do not test for homocysteine levels properly—if they test for them at all—and do not treat elevated levels aggressively enough.[2]

In conventional medical practice, homocysteine levels are often checked only in high-risk patients. This hasn't been helped by the fact that the American Heart Association still doesn't recognize homocysteine as "a major risk factor for cardiovascular disease."[3] Several people with coronary heart disease have told us that their doctors refused to check their homocysteine levels, telling them that it was "unnecessary" because these patients weren't "high risk." Even when doctors do find elevated levels, they typically write a prescription for folic acid,[4] suggest "some B vitamins," and then feel that they have treated this condition adequately. But often they haven't.

The effectiveness of the methylation process deteriorates and homocysteine levels tend to rise with age, so a critical part of Ray & Terry's Longevity Program involves appropriate testing and a nutritional treatment program aggressive enough to bring homocysteine levels down to the safe range.

If you have elevated homocysteine levels, the proper way to ensure that methylation is working properly in your body is to implement an initial program of nutritional supplementation and then retest in a few months. Changes should be made to your supplement regimen until your homocysteine level plateaus or reaches a good range—ideally, less than 7.5 μmol/L.

METHYLATION PROCESSES

Methylation Can Cause DNA Changes

Methylation is essential in the proper formation of DNA. We previously mentioned SNPs (single nucleotide polymorphisms), where just one nucleotide in the DNA of a gene is changed. For example, if a DNA region that contains the nucleic acid cytosine (C) becomes methylated, it can turn into the nucleic acid thymine (T).[5] This common mutation is referred to as a "C→T (cytosine to thymine) polymorphism" and can lead to dramatic (and sometimes catastrophic) genetic changes. All it takes to produce an often spectacular change in your genes is to put a methyl group somewhere it wasn't before. Such methylation reactions might cause something quite benign, while at the most serious level they can increase the risk of cancer.[6]

Blocked Detoxification

The body also uses methylation to help rid itself of a number of dangerous heavy metal toxins, such as mercury, lead, antimony, and arsenic. If you have defective methylation, these toxic metals may accumulate, which interferes with normal functioning of numerous bodily functions. By performing a hair minerals test, you can find out if you have excessive amounts of toxic heavy metals in your body, which also suggests defective methylation.

The liver uses methylation to assist in excretion of external toxins as well as some of its own chemical wastes, such as hormone by-products. Defective methylation leads to a buildup of external toxins such as pesticides as well as excessive levels of your body's own hormones, such as estrogen. This is a significant problem because excess estrogen has been associated with an increased risk of several types of cancer.[7]

Effect on the Brain

Methylation reactions are also critical to normal brain function. The body uses the amino acid tryptophan to form the calming neurotransmitters (chemicals used by brain cells to communicate with one another) serotonin and melatonin. The amino acids phenylalanine and tyrosine are transformed into the neurotransmitters adrenaline, noradrenaline, and dopamine, which have just the opposite effect—they tend to stimulate the mind and make you excited. Methylation is involved in a number of these reactions. So whether you are excited or calm may be a direct function of methylation reactions in your brain. The brain also uses methylation to form acetylcholine, the neurotransmitter needed for memory. Decreased acetylcholine levels in the brain have been linked to memory loss and Alzheimer's disease.

Methylation defects have been associated with an increased risk of Alzheimer's disease as well as many of the mood disorders commonly seen in the elderly, such as depression and paranoia. Numerous studies have shown a link between defective methylation and abnormal brain function. In one study, an elevated homocysteine level was found in a high percentage of hospitalized psychiatric patients. Patients with mood disorders such as anxiety and depression have a high incidence of elevated homocysteine. Alzheimer's patients often have both elevated homocysteine and low blood levels of folic acid, one of the nutrients that lower homocysteine.[8]

Cardiovascular Disease

Elevated levels of homocysteine can be toxic to the inside linings of your arteries. Homocysteine damages and cracks the inner walls of the arteries directly.[9] In an attempt to repair this damage, the body fills in the cracks with LDL cholesterol particles, thereby initiating the process of atherosclerosis that we will discuss in chapter 15, "The Real Cause of Heart Disease and How to Prevent It." Homocysteine then does even more damage. It increases inflammation, which accelerates the entire plaque formation process and increases the chance that the plaque will rupture. When an arterial plaque ruptures, it can cause a heart attack or stroke.

Both the amount of homocysteine and the amount of inflammation in the body are affected by diet, nutrients, and lifestyle, so here is a common pathway where several key factors discussed in this book—abnormal sugar balance, cholesterol, inflammation, methylation, and genetic risk—intersect to create conditions of health or disease.

Elevated homocysteine levels are an independent risk factor for heart attack and stroke, as confirmed by the noted Framingham Study.[10] Another study that involved almost 15,000 male physicians found that men in the upper percentiles for homocysteine were three times more likely to experience a heart attack.[11] Patients with the highest homocysteine levels also have up to four and a half times the risk of Alzheimer's disease.[12] Elevated homocysteine gives you a risk of cardiovascular disease equivalent to that of cigarette smoking.[13]

Since a significant percentage of patients with cardiovascular disease have abnormal homocysteine, we disagree with the American Heart Association recommendations mentioned above and believe instead that all patients with cardiovascular disease should have their homocysteine level checked and receive appropriate nutritional guidance. In fact, since optimizing methylation processes in the body is so important, we recommend that everyone, not just heart patients, find out about their homocysteine levels.

LOWERING YOUR HOMOCYSTEINE

Fortunately, it is usually easy to lower even dangerously high homocysteine levels with inexpensive and easily available nutritional supplements. Supplementation with vitamins B_6 and B_{12}, folic acid, and some other nutrients will not only lower homocysteine levels but can help prevent the diseases that result from elevated levels.[14]

There is extensive evidence linking elevated homocysteine levels to a lack of vitamin B_6, vitamin B_{12}, and folic acid. Yet what is adequate for one person may not be nearly enough for someone else. Simply measuring the levels of these nutrients in the bloodstream doesn't provide the needed information. Studies have shown that blood levels of these nutrients are often within the so-called normal range, even when more is needed to correct abnormal methylation and to lower homocysteine.[15] So measuring the homocysteine level is a better indicator of the amount of B_6, B_{12}, and folic acid you need than measuring the blood levels of the nutrients themselves.

Because of several common genetic variations, there is often a wide range in the amounts of these nutrients individuals need to adequately lower homocysteine. One person may have an excellent homocysteine level (less than 7.5) by consuming only 2 milligrams of vitamin B_6, 0.6 micrograms of vitamin B_{12}, and 400 micrograms of folic acid (the RDA amounts of these nutrients). Yet someone else may have a dangerously high homocysteine level despite taking 200 milligrams of B_6 (100 times the RDA), 1,000 micrograms of B_{12} (1,600 times the RDA), and 2,000 micrograms of folic acid (five times the RDA). In this case, it may be necessary to administer some of these nutrients (such as B_{12}) by injection or use their activated forms (such as activated folic acid).

What you eat is also important to your homocysteine level. Red meat and poultry contain relatively large amounts of methionine, the amino acid that can turn into homocysteine in the body, so this is another reason that we discourage the consumption of large amounts of animal products. People who have difficulty controlling their homocysteine level should eat fewer methionine-rich foods such as red meat, turkey, and chicken and instead emphasize fish, vegetables, and fruit. One study showed that a vegan diet was able to lower homocysteine levels 13 percent, without supplements.[16]

Smoking and coffee consumption have been associated with increased homocysteine levels, while wine consumption has a "J-shaped" association—a little wine (one to two glasses a day) decreases homocysteine, while larger amounts raise levels.[17] Moderate beer consumption lowers homocysteine levels, perhaps due to its vitamin B_6 content.[18]

The most important thing: don't give up until your homocysteine level is adequately controlled, whatever it takes. A recent paper in *JAMA* claimed to show that "high dose" homocysteine-lowering therapy was of minimal benefit in preventing stroke. The authors of the study compared the effect of low-dose supplementation (0.2 milligrams of B_6, 6 micrograms of B_{12}, and 200 micrograms of folic acid) to what they considered to be high doses (25 milligrams of B_6, 400 micrograms of B_{12}, and 2,500 micrograms of folic acid) on the risk of stroke.[19] They found no significant difference between the two supplement programs and concluded "moderate reduction of total homocysteine . . . had no effect on vascular outcomes." The key phrase here is "moderate reduction." In this study, so-called high-dose supplementation only lowered average homocysteine levels from 13.4 to 11.0, still far above our target range (less than 7.5). But if they had used even more aggressive supplementation instead of stopping at a modest lowering of homocysteine levels to still unsafe levels, it's likely they would have demonstrated more beneficial results.

One meta-analysis (a review of other studies) of 12 studies concluded that homocysteine levels could be realistically lowered 33 percent with a combination of 500 to 5,000 micrograms of folic acid, plus 500 micrograms of vitamin B_{12}, say from a homocysteine level of 15 to 10.[20] The large European Concerted Action Project further showed that each 5-point increase in homocysteine level was associated with an increased cardiovascular risk of 35 percent in men and 42 percent in women.[21] Putting this information together leads to the conclusion that routine supplementation with an inexpensive daily dose of folic acid and vitamin B_{12} could drop the incidence of cardiovascular disease very significantly.

Low-dose supplementation will work for many people, but not everyone. We have mentioned that RDA amounts of these nutrients often don't lower elevated homocysteine levels enough. But sometimes even very large oral doses of these nutrients prove inadequate. The reason has to do with digestion. The same cells (parietal cells) in the stomach that make hydrochloric acid to aid in food digestion also make what is called *intrinsic factor*, the carrier protein used to transport vitamin B_{12} from the intestinal tract into the bloodstream. As people age, their parietal cells become less active, so adequate stomach acid to digest nutrients and the intrinsic factor necessary for B_{12} absorption are often lacking. In this case, people can't assimilate vitamin B_{12} (which requires intrinsic factor) and minerals (which require adequate stomach acid). A low blood level of B_{12} is a major reason homocysteine levels tend to rise with age. To correct the resulting malabsorption of nutrients, it's occasionally necessary to administer these nutrients by injection.

Some studies suggest optimal homocysteine lowering occurs with a folic acid dose of between 500 and 5,000 micrograms a day, while others feel that little more than the RDA of 400 micrograms is needed.[22] Yet, as we will discuss at further length in chapter 21, "Aggressive Supplementation," because of genetic variations, many people have vitamin needs far in excess of RDA amounts. This is particularly true regarding methylation function in the body, since one particular genetic variation (the "MTHFR 677 C→T polymorphism") affects almost half the people in some population groups, and individuals with this genetic variation need more than RDA amounts of folic acid.

NUTRITIONAL SUPPLEMENTS

Here are the nutritional supplements that can lower elevated homocysteine levels. Note that in several instances, many times the RDA amounts are required. Note especially that the amounts of B_{12} typically needed are up to thousands of times higher than the RDA of 0.6 micrograms.

Table 13-2. Nutrients Useful in Lowering Homocysteine

NUTRIENT	TYPICAL DOSE NEEDED TO LOWER HOMOCYSTEINE	RDA DOSE
Vitamin B_{12}	100–2,000 mcg	0.6 mcg
Folic Acid	800–10,000 mcg	400 mcg
Vitamin B_2	25–100 mg	1.7 mg
Vitamin B_6	50–100 mg	2 mg
Betaine	100–300 mg	N/A
Zinc	15–50 mg	15 mg
Magnesium	400–800 mg	400 mg
TMG (trimethylglycine)	500–3,000 mg	N/A

TESTS TO ASSESS METHYLATION

The methionine challenge (homocysteine stress test). Most measurements of homocysteine are done as part of a cardiovascular risk panel, a group of blood tests designed to screen patients for cardiovascular disease. Blood is typically drawn first thing in the morning after an overnight fast. The patient needs to be fasting to get accurate readings for cholesterol, triglycerides, lipoprotein(a) and so on, but the fasting state is less useful for

Your Homocysteine Level—What Should It Be?

There really is no "safe range" for homocysteine. It's like smoking—it seems that any amount is bad. A paper from Norway contends that "total homocysteine is an independent risk factor for CHD (coronary heart disease) with no threshold level," meaning that there is no level that is free of risk, so it remains unclear what constitutes an optimal level.[23] Most large laboratories provide a reference range of what they regard as "normal" for each lab test. One of the main national reference laboratories considers an acceptable range for fasting homocysteine to be less than 15 for men and 12.4 for women.[24] Many nutritional physicians now regard a fasting homocysteine less than 7.0 as low risk, 9.0 as moderate risk, and 15.0 or above as high risk. We have set our optimal goal for fasting homocysteine at 7.5 or lower, which is realistic for most people. In our experience, we have found that with appropriate supplementation, most individuals can drop their levels to this relatively safe range.

assessing homocysteine levels and methylation status. Remember that homocysteine is formed from the breakdown of the dietary amino acid methionine. After an overnight fast, methionine levels are at their lowest level of the day, so measuring homocysteine in this state does not provide the most accurate assessment of the body's ability to break down this amino acid.

An electrocardiogram taken of a patient at rest yields much less information about cardiac risk than an exercise stress test, where the electrocardiogram is monitored while the patient stresses the heart by exercising on a treadmill. Similarly, to adequately assess the body's ability to metabolize homocysteine, instead of the conventional static (fasting) test, we recommend a homocysteine stress test, where blood is taken after the patient consumes an oral methionine challenge.

For a homocysteine stress test, the patient arrives at the clinic fasting. Blood is drawn for a baseline homocysteine level. (Other elements of the cardiovascular risk panel can be drawn at this time as well.) The patient is then given an oral challenge of 25 to 100 milligrams/kilogram of methionine and blood is drawn again 3–4 hours later. Patients with normal fasting homocysteine levels will often have abnormalities after the methionine challenge. By using this method, doctors can identify up to 27 percent more patients with defective methylation than by using the fasting homocysteine determination alone.[25]

Genomic testing. Many people carry genes that predispose them to abnormal methylation. Inexpensive genomics tests can determine if you have a genetic predisposition toward altered methylation and homocysteine metabolism.

The enzyme MTHFR (methylenetetrahydrofolate reductase) helps change homocysteine back into methionine. One of the most common polymorphisms (genetic variations) in the human population is a genetic variant of this enzyme—the so-called MTHFR 677 C→T polymorphism. (This means that the 677th nucleotide in the gene that codes for homocysteine, cytosine, is replaced by thymidine.)

This 677 C→T polymorphism is quite common. Up to 44 percent of the Caucasian and Asian populations have at least one copy of this mutated gene. It is reported to be much less common among individuals of African descent, although one study in sub-Saharan Africa revealed a prevalence of this genetic aberration in over 62 percent of the population in coastal Togo.[26]

Whereas a single copy of the mutated gene rarely causes elevated homocysteine levels by itself, approximately 12 percent of Caucasians and Asians are homozygous, meaning they have two copies of the polymorphism. Without aggressive nutritional treatment, these individuals often have elevated homocysteine levels.[27]

People who have two copies of the MTHFR 677 C→T polymorphism have three times the normal risk of developing premature cardiovascular disease.[28] This polymorphism also increases risk of other diseases. For example, women who have only one copy of the mutated gene still have double the risk of developing CIN (cervical intraepithelial neoplasia), a precancerous lesion of the cervix, while women with two copies have a threefold risk.[29] This polymorphism may also significantly impact life span; one study from Japan showed that while 19 percent of people 14 to 55 years old were homozygous for this mutated gene, it was found in just 7 percent of people older than 80.[30] This suggests that people with two copies of the gene died younger in life.

Luckily, it appears that in most cases the harmful effects of this genetic variant can be corrected by taking adequate amounts of nutritional supplements.

14

CLEANING UP THE MESS: TOXINS AND DETOXIFICATION

"For the first time in the history of the world, every human being is now subjected to contact with dangerous chemicals, from the moment of conception until death."

—Rachel Carson, *Silent Spring*, 1962

"We've got to pause and ask ourselves: How much clean air do we need?"

—Lee Iacocca, former president of Chrysler Corp.

No one can avoid exposure to either external (environmental) or internally generated toxins. During the year 2000, more than seven billion pounds of toxic waste were released directly to the air, land, and water—in the United States alone.[1] The sad truth is that our planet has become so toxic that everyone's health is in danger. Anyone who has lived through a garbage worker's strike knows the consequences of a disruption in waste removal. Even a clogged toilet or drain can wreak havoc on normal household function. Our bodies are no different, and the prompt and effective removal of toxins is critically important to longevity.

You are exposed to toxins in many forms: in polluted air and water, in your workplace, and in your food. You are bombarded with electromagnetic radiation and polluted by heavy metals. You even form some of the most dangerous toxins within your cells yourself. Efficient and effective mechanisms for decontaminating and safely disposing of these toxins are critical to health maintenance and longevity. Let's discuss each of these types of toxins and some steps you can take to prevent accumulation of these contaminants in your body.

Green Cleaning

A study performed by NASA a number of years ago showed that houseplants can reduce airborne toxins such as benzene and formaldehyde from indoor air that air filters can't remove.[2] Spider plants are particularly good at removing formaldehyde, while plants with fuzzy leaves help remove particulate matter. Other useful plants include English ivy, peace lily, bamboo palm, Chinese evergreen, florist's mum, Gerbera daisy, and mother-in-law's tongue.

AIR POLLUTION

Polluted air threatens the health of all air-breathing life-forms on our planet. In addition to its direct toxic effects on the body, air pollution creates smog and acid rain and pokes holes in the protective ozone layer, increasing your risk of skin cancer. It also increases the potential for global warming, also known as the greenhouse effect.[3] The Environmental Protection Agency (EPA) lists 188 hazardous air pollutants, such as benzene, formaldehyde, and dioxins. Some are well-known carcinogens. Others have been associated with respiratory disorders, birth defects, and other serious health problems.

In 1990, the EPA conducted a comprehensive survey, measuring outdoor concentrations of 148 toxic air contaminants. They found concentrations of hydrocarbons greater than the EPA's desired cancer levels in more than 90 percent of the 60,000 regions studied. Two hundred areas had concentrations of pollutants that exceeded their recommended levels by 100-fold or more.[4]

When heavier cold air traps warm air beneath it, an "inversion" results. This can often result in outside air becoming very polluted, and local officials issue air advisories urging individuals with respiratory diseases to remain indoors. Yet indoor levels of many pollutants are several to more than 100 times higher than they are outdoors.[5] Indoor air pollution is a critical issue, because many people spend as much as 90 percent of their time indoors. Common sources of indoor air pollution include cooking and heating fumes, tobacco smoke, commercial household cleaners, pest-control products, outgassing from building materials and carpets, smoke from woodstoves and fireplaces, and radon.

In an effort to improve energy efficiency, engineers have successfully reduced the thermal losses of many buildings by tightening leaks and recirculating the same air again and again. If you work in an energy "efficient" building—you usually can't open the windows—you are exposed to even greater amounts of indoor pollution. This has led to a new group of pollution-related human illnesses, known as sick building syndrome, that are associated with a higher incidence of allergy and respiratory tract irritation as well as cancer and infections.[6]

For some simple steps you can take to avoid or reduce your exposure to airborne pollutants, see Table 14-1 below.

Table 14-1. Reducing Indoor Air Pollutants

TO AVOID TOXICITY FROM	DO THIS
Tobacco smoke	Don't allow smoking in your house and avoid venues where smoking is allowed
Commercial household cleaners	Use vinegar and water
Dry-cleaning chemicals	Wear washable clothes. Let dry-cleaned clothes "outgas" outside or in your garage before bringing into the house
Outgassing from building materials and carpets	Use stand-alone air filters, houseplants
Office machines	Move printers, copiers, and fax machines as far as possible from your workspace; ventilate the area where these machines are used
Smoke from woodstoves and fireplaces	Use gas fireplaces
Radon	Use radon mitigation system

WATER POLLUTION

Drinking water polluted with pathogenic microbial organisms produces short-term effects within hours or days of exposure. Bacteria, viruses, or parasites can cause acute illnesses such as gastroenteritis (stomach "flu" or "turista"), hepatitis, and cholera.

Chronic effects persist after exposure to waterborne contaminants for years or decades. Water pollutants include chemicals used to kill pathogens, such as chlorine; industrial solvents; pesticides; radioactive elements; and toxic minerals such as arsenic. Chronic exposure to such waterborne contaminants has been linked to cancer and liver, kidney, and reproductive-tract problems.[7]

When you increase your water consumption as recommended in this book, it's important that you pay attention to what type of water you drink.

Most municipal water, as it comes out of the tap, is suboptimal for ideal health. Chlorine and fluoride, which are routinely added, are highly reactive chemicals with adverse effects on the human body.[8] Chlorine is necessary to kill the germs that contaminate water as it travels from its original source to your home, so avoid drinking or bathing in chlorinated water. You should filter tap water before use.

Use a water filter that also alkalinizes your drinking water. We discussed the importance of reducing the acidity of your tissues by eating a more alkaline diet and drinking alkalinized water in chapter 4, "Food and Water." Most people are already too acidic from eating excess meat, simple carbohydrates, and sugar. Alkalinizing water is a very effective step to bring this into balance.

You should also filter tap water before bathing or showering. Chlorine from your bath or shower water can be absorbed directly through the skin.

Strengthening Your Detoxification Capacity

There are a number of steps you can take to assist with toxin removal. The following foods, nutritional supplements, and lifestyle modifications are of value.

- A diet rich in garlic, onions, lemon, rosemary, and green tea can help strengthen the liver's enzymatic functions and assist with elimination of heavy metals.
- Cruciferous vegetables, such as broccoli, cauliflower, kale, cabbage, brussels sprouts, and bok choy, contain antioxidants that have detoxification properties. Cilantro is a natural heavy-metal chelator.
- Other helpful nutritional supplements include N-acetylcysteine (NAC), which boosts levels of glutathione, one of the liver's most important Phase II detoxifiers.
- Milk thistle (silymarin) and alpha lipoic acid give the liver a helping hand.
- Adequate levels of vitamin C, many of the B vitamins, magnesium, and selenium are critical for optimal detoxification enzyme function.
- Alkaline water (see chapter 4) helps avoid constipation and enhances detoxification.
- Other useful lifestyle techniques include vigorous aerobic exercise and saunas, since there is some suggestion that heavy metals and fat-soluble toxins may be partially excreted in sweat.[9]

A number of household filtering systems can remove most toxins from tap water for a few hundred dollars.

ENVIRONMENTAL POLLUTION

Your work and play environments can have an impact on your health in many ways. For example, golf-course superintendents have an increased incidence of non-Hodgkin's lymphoma, brain and prostate cancer, and neurological illness.[10] Their work involves prolonged exposure to pesticides, fungicides, herbicides, and fertilizers. Oil refinery workers, another group exposed to numerous environmental toxins, suffer increased rates of mortality from cancers of the lip (384 percent of normal), stomach (142 percent), liver (238 percent), pancreas (151 percent), connective tissues (243 percent), prostate (135 percent), eye (407 percent), brain (181 percent), and leukemia (175 percent).[11] Chimney sweeps are exposed to heavy doses of toxic dust and suffer excess mortality from numerous diseases, including cancers of the lung, bladder, and esophagus, as well as respiratory ailments and ischemic heart disease.[12]

FOOD POLLUTION

Chemicals are used extensively in food production. Fruit trees, for instance, are regularly sprayed with insecticides, fungicides, and herbicides during the growing season. Poison grain is dropped from planes onto apple orchards to kill rodents. The trees are sprayed with chemicals designed to keep the apples from falling off their stems. After being picked, the fruit is coated with wax to improve appearance, then stored for several months in warehouses filled with toxic gases to prolong storage before coming to market. When you eat conventionally grown apples, many of these toxins remain on and even *within* the fruit.

No matter how vigilant you are, you can't avoid accumulating pesticides in your tissues. Pesticide residues are present in all categories of foods, and it is not unusual for a single food item to contain residues of five or more toxic chemicals.[13]

To reduce these toxins, eat organic whenever possible, and soak fruits and vegetables for a few minutes in water mixed with a commercial produce cleanser, which helps remove some of the toxins from the outside of the produce.

Organic Only

According to the Environmental Working Group, certain fruits and vegetables are heavily contaminated with pesticides, while others are much less so. By avoiding the most contaminated fruits and vegetables and eating those least contaminated instead, it is estimated that you can reduce your pesticide exposure 90 percent.[14] Of course, you should still eat organic produce whenever possible, but these recommendations apply to those situations where conventionally grown produce is your only option.

MOST CONTAMINATED	LEAST CONTAMINATED
Bell peppers	Sweet corn
Spinach	Avocado
Celery	Cauliflower
Potatoes	Asparagus
Peaches	Onions
Nectarines	Sweet peas
Strawberries	Broccoli
Apples	Pineapples
Pears	Mangoes
Cherries	Kiwi
Grapes (imported)	Papaya
Raspberries	Bananas

Animal-based foods are higher in toxins than vegetables and fruits, with most toxins being concentrated in fatty tissues. To reduce your exposure to toxins, you should favor plant-based food over animal food, and trim obvious fat from red meat and remove the skin from poultry. Boiling chicken or meat causes a significant amount of fat to rise to the top of the cooking container, where it can be easily skimmed off before eating.

ELECTROMAGNETIC POLLUTION

Your body is constantly bombarded with a wide array of man-made forms of electromagnetic (EM) radiation. Some occurs in your immediate environment, such as from computer displays, cell phones, hair dryers, electric razors, waterbed heaters, and electric blankets. Other exposure comes from remote sources, such as cell phone towers, television and radio stations, satellite transmitters, and radar signals. Radiation from the dozens of radio and television stations and cell phone towers in your area passes through your body every second of the day.

Although still controversial, there is growing evidence that continual exposure to such sources of radiation can produce changes in the function of biological tissues and may lead to a wide variety of adverse health effects.[15] EM radiation can cause DNA damage,[16] and cell phone radiation can alter the proteins in your cells.[17]

Over the course of billions of years, all living organisms on earth have learned to adapt to the underlying 8 cycles per second (8 Hz) "earth frequency"[18] and other natural frequencies. Your DNA and the other cellular machinery of life are well adapted to this radiation but were not programmed to deal with 60 cycles per second (60 Hz) AC (alternating current), 900 MHz cell phone frequencies, or 97.3 FM.

It's impossible to completely avoid the electromagnetic smog you live in, but there are steps you can take to reduce your exposure and reduce dangers. Minimize your use of computer monitors, hair dryers, electric shavers, and other high-power electrical devices. Make sure that you (and particularly your children) sit at least 10 feet away from big-screen TVs. Do not sleep under electric blankets or on a heated water bed.[19]

Cell phones can cause significant EM exposure. No one really knows the long-term health effects of holding a radio transmitter (which is what a cell phone really is) next to your head for hundreds or thousands of hours a year. Therefore, you should try to limit your cell phone use. If you find that impractical, at least take steps to reduce EM exposure while using your cell phone.

Using a hands-free set with a built-in microphone and earbud allows the cellular phone to be located a few feet from the head, so you may think that you are reducing exposure to the radiation coming from the phone. However, the wire traveling to your ear can actually act as an antenna, so your ear and brain may receive an even more concentrated dose of radiation under some conditions.[20] A safer solution is to use a hands-free connection to the ear based on an air tube, with the device's wire portion passing through a ferrite choke[21] to block some of the radiation from reaching your head. You can further reduce radiation by moving its source away from your body, using an external antenna (available for some phones that have car kits).

BRIDGE TWO

BIOENGINEERED MICROORGANISMS TO THE RESCUE

Bioremediation and other genomics techniques are Bridge Two biotechnology developments that offer promising strategies for cleaning up toxic accumulations in both the environment as well as our bodies. Historically, we humans have just dumped our waste products in the air, water, and soil with little regard for where they would end up.

These days, such flagrant disregard for the environment isn't as obvious but is often still as harmful, such as sealing toxic waste in containers and burying it underground in mines or above ground in mounds, or sealing it in concrete and using it in building materials.

Bioremediation. A Bridge Two technology known as bioremediation may be far more effective and less expensive. This refers to using bioengineered microorganisms such as fungi or bacteria to do the work. Researchers have already identified bacteria that can digest and destroy many dangerous toxins, from TNT to dioxin. Breakthroughs in genetic engineering are driving this field forward quickly.

In the shorter term, current bacteria are proving to contain treasure troves of useful capabilities. The common bacteria *Geobacter sulfurreducens*, whose genome was sequenced in 2003, for example, produces electrical charges as it cleans up uranium contamination.[22] This means these bacteria might be able to generate electricity from radioactive waste, power electronic devices in remote locations, or be used in microbial fuel cells. Craig Venter and others are attempting to write genomes for synthetic bacteria specifically designed for particular cleanup tasks.[23]

Genetic engineering is being used to make crops resistant to pests so less insecticide and fungicide is needed.[24] Despite opposition from some groups to the entire idea of genetic engineering, this technology has the potential of not only

HEAVY-METAL POLLUTION

Toxic heavy metals have been shown to increase free-radical activity, a major cause of accelerated aging. Many enzymes critical to good health require a vitamin and mineral cofactor for proper activation. Toxic heavy metals can take the place of the proper mineral cofactors, interfering with normal functioning of these enzymes. In this way, accumulation of toxic heavy metals can contribute to premature aging and age-related diseases. Heavy-metal toxicity can also lead to abnormal immune function, learning disorders, and neurodegen-

helping feed the world but of also reducing the need for synthetic fertilizers, pesticides, and herbicides, rather than figuring out ways to clean up these toxins later.

Understanding how plants utilize trace metals such as zinc and iron will result in crops that "pull minerals from the soil more efficiently" and require less fertilizer.[25] Cover crops can be engineered to self-destruct at planting time so they don't need to be killed with herbicides.[26]

Prion diseases. Researchers are also applying genomics techniques to understanding detoxification pathways in people and how failures in those pathways can lead to disease. Stress, for example, dramatically changes lipid metabolism and detoxification in the liver by altering the expression of the genes that control these processes.[27] Until recently, Parkinson's disease had no known cause, and there was no way to slow the disease. Now the symptoms have been linked to mutations, caused by toxins such as the pesticide rotenone, in two genes that influence how the body handles proteins. These mutations result in accumulation of misfolded proteins (prions). The toxins also kill dopamine cells. "Dopamine is like the oil in the engine of a car," according to Peter Lansbury, an associate professor of biology at Harvard. "If the oil is there, the car runs smoothly. If not, it seizes up."[28] Loss of dopamine cells in the brain is the hallmark of Parkinson's. Understanding this pathway has been a huge leap forward and is expected to accelerate bioengineered treatments.

Misformed proteins form aggregates within cells that can produce harmful effects. Aubrey de Grey has described strategies using somatic gene therapy to introduce new genes that will break down these "intracellular aggregates"—toxins inside cells, such as protofibrils.[29]

A key strategy for combating toxic materials outside the cell, including misformed proteins and the amyloid plaque seen in Alzheimer's disease and other degenerative conditions, is to create vaccines that act against their constituent molecules.[30] Several research groups are pursuing this strategy, which may result in the toxic material being ingested by immune-system cells.

---

erative diseases. Typical symptoms seen in patients with heavy-metal toxicity include fatigue, mood disorders, poor concentration, and hair loss.

Arsenic, beryllium, cadmium, chromium, cobalt, and nickel are naturally occurring heavy metals that are carcinogenic. They have been shown to cause tumor development in experimental animals and to inhibit the body's ability to repair damaged DNA.[31] Other toxic heavy metals include mercury and aluminum, along with dozens of others.

Mercury is a particularly toxic metal that causes numerous problems when given to laboratory animals. Methylmercury, the form of mercury found in seafood, is a known neurotoxin widely distributed throughout the oceans of the Earth. For instance, children who live in traditional Inuit communities in Greenland and eat a diet very high in seafood suffer a higher incidence of neurobehavioral problems from mercury exposure.[32] Mercury has negative effects on immune-system function in laboratory animals.[33]

Since virtually all seafood is now contaminated with mercury,[34] it is important that you limit your consumption of fish and seafood to certain species. Larger fish such as tuna, swordfish, sailfish, and shark have a much higher mercury content and should not be eaten.[35] Smaller fish, like anchovies, sardines, and salmon, are lower on the food chain and have less mercury in their tissues. Wild salmon is preferable to farm grown because it's relatively low in mercury and high in vital omega-3 fats.[36] Our program recommends fish-oil supplementation, but you should be sure your source of EPA/DHA fish oil has been tested to have minimal contamination. The amount of mercury found in some fish and commercial fish products follows.

Table 14-2. Mercury Levels of Common Seafood[37]

LOWEST MERCURY (OK TO EAT REGULARLY)	INTERMEDIATE MERCURY (EAT ON OCCASION)	HIGHEST MERCURY (MOSTLY AVOID)
Salmon (wild Pacific)	Great Lakes Salmon	Tuna, canned and steaks
Haddock	Cod	Halibut
Blue crab (mid-Atlantic)	Blue crab (Gulf Coast)	Shark
Trout (farmed)	Mahi mahi	Swordfish
Flounder (summer)	Pollock	Sea bass
Croaker	Channel catfish (wild)	Walleye
Shrimp	Eastern oysters	Gulf Coast oysters

People also are exposed to mercury from their "silver" amalgam dental fillings, which are over 50 percent mercury. We recommend that you do not have any additional mercury-containing amalgam fillings placed in your mouth or in the mouths of your children. Make sure from now on your dentist uses less toxic composite materials, which are polymers of glass or ceramic,[38] for both new and replacement fillings. If a metal must be used, such as for crowns or bridges, gold is the least bioreactive material available.

In some genetically susceptible individuals, aluminum may play a role in neurodegenerative diseases, particularly those associated with memory deficits.[39] You can reduce your exposure by not using aluminum cookware or aluminum foil; if you use antiperspirant, apply a nonaluminum-containing product, available at natural food stores.

Dietary and lifestyle choices that can assist the body in heavy-metal removal are discussed below. There are a number of chemicals known as "chelating" agents, available in oral and intravenous forms, that are very effective at removing toxic heavy metals from the body as well.

TESTING FOR TOXINS

Heavy-Metal Toxins

It is useful to measure your body burden of toxic metals to determine how aggressively you need to pursue their removal. There are a variety of ways to screen for heavy-metal toxins. Tests of hair, urine, and blood are used most commonly.

Hair mineral analysis. This test offers a cost-effective and painless way of screening for chronic heavy-metal accumulation. A small amount (about 1 gram) of hair is collected from the nape of the neck, where removal is not visible. Hair that has just emerged from the follicle provides a good representation of heavy-metal exposure in the recent past. While not a perfect test, since the results are often skewed by external contamination, there is a reasonably good correlation between heavy-metal content in hair and total body levels. If a hair analysis shows significant abnormalities, additional information can be obtained by performing a urine provocation test for more exact measurement and confirmation of results.

Urine provocation test. Typically, the patient is given an oral or intravenous injection of an agent designed to concentrate heavy metals in the urine, which is then collected for the next 6 to 24 hours for analysis. Based on these results, a course of heavy-metal detoxification, typically using oral or injectible chelating agents, can be determined.

Blood tests. These are less useful for measuring heavy-metal toxins. For toxic metals to show up in blood tests, you need to have significant toxicity or poisoning. Subacute and low-grade chronic accumulations are much more common, yet are rarely detected with blood tests, so in most cases we prefer hair or urine screening tests.

Fat-Soluble Toxins

It is more difficult to measure the levels of fat-soluble environmental toxins in the body. These tests are expensive and generally reserved for patients more seriously affected by environmental illness. Commercial testing is available for only a few hundred of the more than 70,000 known environmental toxins, but it's a simple matter to measure the body's detoxification capacity.

(continued on page 194)

BRIDGE THREE

NANOBIOTIC DETOXIFICATION

Avoiding pollutants and removing toxins with supplements and diet can go only so far. Extreme toxicity from infections, chemotherapy, pollution, and other causes can overload the natural ability of the liver, kidneys, and other detoxification systems in the body. When that happens, inflammation from toxins affects the entire cardiovascular system, causing blood pressure to crash. This state, technically known as "shock," starves the body of oxygen, so the liver, lungs, heart, and other organs begin to fail. Death soon follows, often the result of kidney failure.

Nanotechnology and kidney failure. For kidney failure, the traditional solution has been hemodialysis machines. These washing machine–size devices can take over the important role of cleaning the blood usually done by healthy kidneys. However, they can only filter out certain toxins, the treatment takes up to seven hours, and patients must be hooked up to large, expensive machines several times a week. "Hemodialysis works great for kidney failure but is useless for most other kinds of detoxification," says Axel Rosengart, assistant professor of neurology and surgery at the University of Chicago.[40] While useful for chronic renal failure, more than half of patients with acute kidney failure (such as in cases of "shock") still die despite hemodialysis. Nanotechnology researchers are working on a solution.

Rosengart and codeveloper Michael Kaminski, an engineer at Argonne National Laboratory, have developed a fast and simple solution using magnetized nanoparticles attached to receptors designed to identify and grab target toxin molecules. The nanoparticles are injected into the bloodstream, where they circulate through the body, picking up the target toxins. To remove the particles from the body after treatment, a small shunt inserted into an arm or leg artery quickly routes the blood through a handheld unit with a magnet. Since the nanoparticles are made of polylactic acid, which is biodegradable, any remaining particles will eventually be eliminated from the blood. So far, tests have been limited to rats, but the results have been promising. "Our initial tests have been very successful—I am very confident that we will be able to remove 99 point something of the particles," says Kaminski. Once perfected, this type of approach need not be limited to patients with renal failure but can be used to augment everyone's detoxification capacities.

An even more advanced machine being developed for cleaning the blood is the bioartificial kidney, which uses a plastic cartridge containing a billion human kidney cells inside of 4,000 hollow plastic fibers. It is being developed at Nephros Therapeutics, based on research by University of Michigan internist David Humes.[41] It will deliver the full range of kidney functions, including its immune-system-regulating activities. In a partial clinical trial, 6 out of 10 critically ill patients survived; all but one had been judged to have no more than a 10 percent to 20 percent chance of living. This combination biological/artificial kidney could be available for widespread use by 2006.

For the hundreds of thousands of patients with chronic renal failure, however, it would be better to replace such an external machine with long-term implants that incorporate living kidney cells that can receive nourishment from the body itself. The challenge to this technology is to implant cells that are able to filter some 100 liters of fluid daily while avoiding immune-system rejection. William Fissell, a researcher at Humes's University of Michigan lab, is testing a solution: nanopores stretched into elongated nanoslits.

Nanotechnology and liver failure. For the body's other major detoxification organ, the liver, several labs are developing similar devices, using liver cells to remove the toxins that accumulate in the blood when the liver fails. These bioartificial livers could help patients with chronic liver failure—their only hope today is a rare organ transplant.

A more radical approach to coping with liver failure is to design a "liver chip," a realistic model of a human liver on a mass-produced silicon chip.[42] Currently being developed by MIT tissue engineer Linda Griffith, it would not *replace* the liver but instead allow scientists to test for drugs that combat liver cancer and hepatitis and find out in advance how liver cells would react to various toxic substances.

Nanotechnology and age reversal. The end result of these kinds of nanomedical advances will be a major advance in anti-aging medicine, what nanomedicine expert Robert A. Freitas Jr. calls dechronification, or "rolling back the clock."[43] Dechronification will first stop biological aging, then reduce it by performing three kinds of procedures on each one of the trillions of tissue cells in your body.

Freitas describes three steps toward dechronification. The first step would be to inject nanobots that would enter each cell in the body and clean out accumulated metabolic debris and toxic buildups. Because these toxins would build up again, this would be done by the nanobots either on a continuous basis or, say, once a year, as part of an annual tune-up. The nanobots would also carefully correct any damage that occurred to your genetic DNA. The final aspect of nanobiotic detoxification would be repair of other cellular structures that the cells are unable to fix on their own, such as malfunctioning or disabled mitochondria.

Many of the organic toxins discussed above become concentrated in your fat cells. The primary mechanism the body uses to remove fat-soluble toxins is to convert them into safer, water-soluble metabolites that are more easily excreted in the urine and stool. The liver has the greatest role in decontamination of toxins and has hundreds of specialized detoxification enzymes to assist with this task.

In general, hepatic (liver) detoxification occurs in two steps, known as Phase I and Phase II detoxification reactions. In Phase I, toxins undergo biotransformation via specialized enzymes known as the cytochrome P450 enzymes. In this step, fat-soluble toxins are made water-soluble. The products of the Phase I reactions can still be quite toxic, so it is important that Phase I and Phase II reactions work together, or harmful Phase I toxins can accumulate. In Phase II, a water-soluble molecule is attached to the partially metabolized Phase I toxin, making it much less toxic and easier for the body to excrete.

In a common test, a patient takes three mild toxins: caffeine (NoDoz), acetaminophen (Tylenol), and aspirin. Since caffeine is almost completely metabolized by Phase I reactions, a saliva sample taken a few hours after the caffeine ingestion will indicate how well the Phase I pathway works. Acetaminophen and aspirin require both phases for their elimination. By doing a simple analysis of blood and urine the following morning, the ability of the liver to detoxify environmental toxins can be assessed.[44]

Genomics Detoxification Testing

The main Phase I detoxification system consists of the cytochrome P450 enzymes used by the liver (see above). Genetic variations of these enzymes can lead to impaired (or occasionally enhanced) detoxification capacity. Among the hundreds of cytochrome P450 enzymes, several dozen well-known polymorphisms have been identified. Some of the more common of these (and the types of problems associated with them) are listed below.

Table 14-3. Risks of Some Liver Detoxification Enzyme Genetic Polymorphisms

ENZYME	RISKS OF VARIANT GENES (POLYMORPHISMS)
CYP450 1A1	Higher susceptibility to some cancers, such as lung cancer in smokers[45]
CYP450 2D6	Earlier onset of Parkinson's disease[46]
CYP450 2E1	Increased risk of alcoholism in certain ethnic groups[47]
CYP450 3A5	Changes in doses required for some prescription drugs[48]

THE BIGGEST SOURCE OF TOXINS: MISFORMED PROTEINS

Misformed proteins are perhaps the most dangerous toxin of all and are formed within your own cells. Research suggests that misfolded proteins may be at the heart of numerous disease processes in the body. Such diverse diseases as Alzheimer's disease, Parkinson's disease, the human form of "mad cow" disease, cystic fibrosis, cataracts, and diabetes all result from the inability of the body to adequately eliminate misfolded proteins.

Protein molecules perform the majority of work done in the cell. Proteins are made within each cell according to DNA blueprints. They begin as long strings of amino acids, which must then be folded into precise three-dimensional configurations to function as enzymes, transport proteins, and so on. Specialized "chaperone" molecules protect the amino acid strands while they undergo the folding process.[49] Even so, about one-third of formed protein molecules are folded improperly. These disfigured proteins must be immediately destroyed, or they will rapidly accumulate, disrupting cellular functions on many levels.

Under normal circumstances, as soon as a misfolded protein is formed, it is tagged by a carrier molecule, ubiquitin, and escorted to a specialized region of the cell, where it is broken back down into its component amino acids for recycling into new and, it is hoped, correctly folded proteins. As cells age, however, they produce less of the energy needed for optimal function of this mechanism. Heavy-metal toxins also interfere with normal function of these enzymes, making the problem worse. There are also genetic mutations that predispose individuals to misformed protein buildup.[50]

If the disposal and recycling process is disrupted, improperly folded protein fragments begin to accumulate within the cell. These particles, known as protofibrils, float around in the cytoplasm or fluid portion of the cell and begin to stick together, forming filaments, fibrils, and, ultimately, larger globular structures. Until recently, these accumulations of insoluble material were regarded as the causes of these diseases, but it is now known that the smaller soluble collections of protein fragments, the protofibrils, are the real problem.

The quicker that protofibrils are turned into insoluble deposits known as amyloid, the more slowly the disease progresses. Some people form amyloid quickly, which protects them from protofibril damage. Others turn protofibril into amyloid less rapidly, allowing more extensive damage. These people also have little visible amyloid. This explains the paradoxical finding that some individuals have extensive accumulation of amyloid plaque in their brains but no evidence of Alzheimer's disease, while others have little visible plaque yet considerable manifestation of the disease.

The Apo E4 allele, discussed previously as a genetic risk factor for Alzheimer's disease, slows down the process of amyloid production. If you are a carrier of Apo E4 (as is 25 percent of the population), you are predisposed to protofibrillar damage throughout the brain and should be especially careful to take nutritional supplements specifically targeted to reduce this damage (see chapter 11).

Many other diseases have been linked to protofibril accumulation. Prions, the cause of mad cow disease and its human equivalent, variant Creutzfeldt-Jakob disease, are actually misfolded proteins, that is, protofibrils.[51] Many other serious human maladies, including type 2 diabetes, ALS (Lou Gehrig's disease), and Huntington's disease, are caused by abnormal protein folding.

As cells age, they become less efficient at producing ATP, the energy source for all cellular function, so less energy is available to quickly recycle the misfolded proteins. Also, as heavy-metal toxins accumulate with age, they disrupt normal enzymatic function. Research is currently under way to discover medications that can neutralize protofibril damage. Meanwhile, there are steps you can take.

Strategies to assist the body with protofibril damage control include therapies to improve energy or ATP production. Specific nutrients such as coenzyme Q_{10}, NADH, and carnitine help the body make more ATP. Mineral and vitamin cofactors such as the B vitamins, magnesium, and manganese can assist as well. Chelating agents help remove heavy-metal toxins.

In a couple decades, nanobots will be assisting you in removing toxins and debris from your cells and repairing the resulting damage. In the meantime, it is important to make proper lifestyle choices to reduce your exposure to environmental toxins and optimize your body's detoxification capabilities.

15

THE REAL CAUSE OF HEART DISEASE AND HOW TO PREVENT IT

"A new and emerging understanding of how heart attacks occur indicates that increasingly popular aggressive treatments may be doing little or nothing to prevent them. . . . In 75 to 80 percent of cases, the plaque that erupts was not obstructing an artery and would not be stented or bypassed. The dangerous plaque is soft and fragile, produces no symptoms and would not be seen as an obstruction to blood flow. . . . Some doctors still adhere to the old model. Others say that they know it no longer holds but that they sometimes end up opening blocked arteries anyway, even when patients have no symptoms. Researchers are also finding that plaque, and heart attack risk, can change very quickly—within a month, according to a recent study—by something as simple as intense cholesterol lowering. Even more disquieting . . . is that stenting can actually cause minor heart attacks in about 4 percent of patients. That can add up to a lot of people suffering heart damage from a procedure meant to prevent it."

—*New York Times*, Gina Kolata, March 21, 2004

Heart disease continues to be our leading cause of death among both men and women, killing more than 600,000 Americans each year[1] (although a woman's comparable risk trails a man's by about 10 years; a major risk factor is being 45 years of age or older for a male versus 55 for a woman).[2] Until recently, the conventional understanding underlying the cause of most heart attacks was this: over time, excess cholesterol delivered by LDL-C (low-density lipoprotein, or bad, cholesterol) becomes oxidized, builds up in the coronary arteries, and eventually calcifies into a hard plaque. These calcifications narrow the passageway for blood through these arteries. If this plaque reaches a dangerous level—closing off 80 percent or more of the artery—a blood clot may get stuck in the narrow passageway and obstruct blood flow. This sudden total blockage of the artery—a coronary thrombosis, or heart attack—causes the portion of heart muscle fed by that artery to die, resulting in either the death of the patient or permanent damage to the heart.

This old understanding of heart disease led to the development of two major operations, both of which have become large industries.[3] Bypass surgery "bypasses" the problem of a clogged artery by diverting blood flow through a grafted vein or mammary artery. Coronary artery bypass surgery is one of the most invasive surgeries available and carries a 2 to 6 percent rate of surgical mortality.[4] Among its many complications is a significant decline in mental function and mood in up to 80 percent of coronary bypass survivors.[5]

Even more popular is balloon angioplasty, during which a thin, flexible tube (called a catheter) with a tiny inflatable bulb at the end is inserted into an artery in the patient's groin and guided up through the artery to the blockage. After the "balloon" is positioned, it is inflated to compress the plaque and widen the lumen, or opening, of the blocked artery. A refinement to the technique was introduced around 2000: adding a stent, a rigid tube designed to prevent the compressed plaque from re-expanding and forming future blockages in the same spot. Angioplasty is less invasive than bypass surgery, but it too has significant complications.

As a result of these surgeries, patients may think that they have either a brand-new artery, free of atherosclerosis (in the case of bypass surgery), or one with increased room for blood flow (in the case of angioplasty). However, many studies of these two procedures have demonstrated that they do *not* prevent further coronary events.[6] In fact, many studies have failed to demonstrate any statistically significant increase in survival from either bypass surgery or balloon angioplasty, compared with nonsurgical treatment for most groups of patients[7] such as beta-blocker drugs, aspirin, and statin drugs, with far less risk and cost.

Angina

Angina is experienced as pressure, tightness, or pain in the heart area. The leading theory about its cause is that in an attempt to prevent rupture of the vulnerable plaque and the ensuing catastrophe, the body "walls off" the vulnerable plaques by placing a calcified layer over it. If too much calcified plaque forms, it interferes with blood flow. With physical exertion, not enough blood reaches the heart, resulting in the discomfort or pain called angina pectoris (literally, pain in the chest). It is also possible for a calcified plaque to grow large enough to block blood flow completely, leading to a heart attack or stroke, but this explains less than 15 percent of heart attacks.[8]

Since patients often experience relief of their chest pain (angina) after surgery, they seem to get confirmation that the therapy "worked." We now understand that these surgeries do not show sustained benefit except in a small portion of patients who have severe multivessel disease or who are more than 80 years old.[9] The model of heart disease as a simple plumbing problem is wrong for most heart attacks. The real cause of the vast majority of heart attacks is not addressed—and may even be worsened by these surgical procedures.

THE NEW UNDERSTANDING: MOST HEART ATTACKS ARE CAUSED BY VULNERABLE PLAQUE AND INFLAMMATION

The large, calcified plaque growing on the inside surface of coronary arteries is *not* the cause of most heart attacks. Rather, the primary culprit is the soft, relatively small "vulnerable" plaque that forms within the vessel walls.[10] Large, calcified plaque is actually relatively stable and, because of its hard calcified covering, less commonly cracks. The more dynamic, less stable soft plaque is much more likely to suddenly rupture. As the body forms a clot to try to heal such a rupture, the result may be a total blockage of blood flow; in other words, a heart attack. The soft plaque is hidden inside the walls of the artery and often causes no obvious blockage or loss of blood flow until, of course, the often-fatal rupture.

Yet there is good news hidden in this new understanding because the buildup of soft, vulnerable plaque is much easier to reverse than that of hard, calcified plaque. Levels of the two types of plaque are related, since the same process (see page 203) appears to result in both forms. One prevalent theory on the origin of the hard plaque is that it's the body's attempt to protect the artery from vulnerable plaque by covering it with a hard, calcified layer. However, bypass surgery and balloon angioplasty do not slow down the process of soft or hard plaque formation; they often accelerate them.

An early research study in 1986 by Dr. Greg Brown of the University of Washington at Seattle demonstrated that sudden blockages causing heart attacks were occurring in locations of coronary arteries that had very little plaque, not nearly enough to qualify for bypass or angioplasty surgery.[11] In the late 1980s, Dr. Steven Nissen of the Cleveland Clinic began to examine the coronary arteries of heart patients with an innovative ultrasound camera that he guided into the blood vessels. He found many soft bulges of plaque, often numbering into the hundreds in a single patient, but relatively few areas of calcified plaque. He proposed the idea that it was these widely distributed soft bulges of plaque, not the deposits of hard plaque, that were the primary culprit behind heart attacks.[12] Brown's and Nissen's research, as well as similar studies, were slow to be accepted. Recently, Dr. Nissen has emerged as a leading innovator in fostering new therapies for heart disease, playing a leading role in several new drugs (see Bridge Two sidebar on Pfizer's new Torcetrapib drug and new PPAR drugs, on page 207). Dr. Nissen also conducted an important study (mentioned below) that indicates that lowering LDL cholesterol to levels significantly below the standard recommendations reduces risk.

The pivotal study that began to rapidly change minds on the importance of vulnerable plaque was conducted in 1999 by Dr. David D. Waters of the University of California. In the study, which was called AVERT (atorvastatin versus revascularization treatments), Dr. Waters randomly assigned patients who had been referred for angioplasty surgery to two groups. One received the surgery and standard follow-up care. The other received cholesterol-lowering statin drugs but no surgery. The nonsurgery group actually had fewer heart attacks and fewer visits to the hospital for chest pain than the surgery group.[13] Dr. Waters commented that the research "caused an uproar. We were saying that atherosclerosis is a systemic disease. It occurs throughout all the coronary arteries. If you fix one segment, a year later it will be another segment that pops and gives you a heart attack, so systemic therapy, with statins or antiplatelet drugs, has the potential to do a lot more. There is a tradition in cardiology that doesn't want to hear that. There is a culture that the narrowings are the problem and that if you fix them, the patient does better."[14]

Dr. Eric Topol, a cardiologist at the Cleveland Clinic in Ohio, adds, "There is just this embedded belief that fixing an artery is a good thing." Dr. Topol describes the typical situation in which a patient has symptoms such as vague discomfort in the chest, goes to a cardiologist, gets a heart scan that shows signs of calcified plaque, has an angiogram—itself an invasive procedure—and then quickly receives a recommendation for surgery. "It's this train where you can't get off at any station along the way," Dr. Topol says. "Once you get on the train, you're getting the stents. Once you get in the catherization lab, it's pretty likely that something will get done."[15]

Dr. David Hillis, a cardiologist at the University of Texas Southwestern Medical Center in Dallas, explains some of the motivation. "If you're an invasive cardiologist and Joe Smith, the local internist, is sending you patients, and if you tell them they don't need the procedure, pretty soon Joe Smith doesn't send patients anymore. Sometimes you can talk yourself into doing it, even though in your heart of hearts you don't think it's right." Explaining the

BRIDGE TWO

BLOCKING INFLAMMATION

With the recent recognition that inflammation plays a crucial role in every step of plaque formation, as well as in the final eruption of vulnerable plaque that initiates a heart attack, another major front in the war against heart disease has emerged. A new generation of drugs targets leukotrienes, which are signaling molecules that play an important role in triggering inflammation. In a recent animal study, a new leukotriene-blocking drug from Pfizer called CP-105,696 caused artery plaques in mice to shrink dramatically in a mere 35 days (the control group had no shrinking).[16] Several other new leukotriene-inhibiting drugs are also in development.

There are already leukotriene blockers on the market that treat asthma: Merck's Singulair and AstraZeneca's Accolate. Animal studies indicate that these are also effective at stopping atherosclerosis, and clinical trials are under way to assess the value of the drugs in humans.[17] If their value in heart disease is confirmed, they could be used immediately, since they are already approved and available for another disease.

A related approach is to block the first step of vulnerable plaque formation, the oxidation of low-density lipoprotein cholesterol (LDL-C). This causes inflammation, resulting in the production of VCAM-1 (vascular cell adhesion molecule-1). VCAM-1 in turn causes the LDL-C particles to stick to the surface of the artery. A new drug from AtheroGenics called AGI-1067 blocks LDL-C oxidation and appears to halt VCAM-1 production. A 1999 trial on 300 patients showed significant reversal of atherosclerotic plaque,[18] and a $40 million, two-year, late-stage trial involving 4,000 patients is now under way.

Another promising drug in development that prevents LDL-C from sticking to artery walls is Sankyo Pharmaceuticals' Pactimibe. The drug, now in late-stage trials, blocks the enzyme ACAT (acyl-CoA cholesterol acyl-transferase), which plays a role in this crucial early step in the formation of plaque.[19]

GlaxoSmithKline's "480848" drug also works by inhibiting a critical enzyme, Lp-PLA2 (lipoprotein-associated phospholipase A2). This enzyme is actually a small fragment that hangs off each LDL-C particle; blocking the enzyme appears to prevent the particle from initiating the plaque process. The drug is in late-stage trials.[20]

patients' perspective, Dr. Hillis adds, "I think they have talked to someone along the line who convinced them that this procedure will save their life. They are told, 'If you don't have it done, you are a walking time bomb.' "[21]

The makers of stents acknowledge that the research fails to show a benefit in terms of avoiding heart attacks and death. Paul LaBiolette, senior vice president of Boston Scientific, a leading stent maker, says, "It's really not about preventing heart attacks per se; the obvious purpose of the procedure is palliation and symptom relief."

However, angina pain can be managed without surgery in most cases, often very quickly. "The results are now snowballing," says Dr. Peter Libby of Harvard Medical School. "The disease is more mutable than we had thought."[22]

This new perspective explains why bypass and angioplasty surgeries don't work, and it helps explain why heart attacks typically strike with no warning and often to people who appear to be "perfectly healthy," according to conventional diagnostic methods. Armed with this more accurate model, we can apply noninvasive methods to address each stage of this progressive and degenerative process. By combining targeted therapeutic approaches, we can rapidly and dramatically reduce the risk of a heart attack to very low levels. With few exceptions, no one need suffer a heart attack.

WHAT WE NOW KNOW

Our new understanding of the progression of heart disease offers several important insights.

HDL's role. We continue to credit high-density lipoprotein (HDL), the "good" cholesterol, with removing cholesterol from the arteries before it has a chance to oxidize. Recent research shows that HDL also provides protection by interfering with the inflammatory process itself.[23]

Risk factors. Smoking, diabetes, and high blood pressure are major risk factors for heart disease. Our new understanding explains why. Smoking accelerates the oxidation of LDL, the first step in vulnerable plaque formation. Diabetes, which results in elevated glucose levels in the blood, accelerates glycation of LDL, which is also part of this initial stage. Obesity often leads to type 2 diabetes and also encourages inflammatory processes. High blood pressure puts greater pressure on the artery walls, creating small cracks, thereby encouraging inflammatory processes. Angiotensin II, a hormone associated with hypertension, also increases inflammation.

Aspirin's value. Supplementation with low-dose aspirin (81 milligrams per day) lowers the incidence of heart attacks by reducing inflammation as well as clotting.

Unpredictability. Heart attacks occur suddenly to apparently healthy people. Vulnerable plaque causes no symptoms and is difficult to detect through heart scans or any other procedure, although a new generation of high-sensitivity, ultrafast coaxial tomography (UF CT, also known as electron beam coaxial tomography, or EBCT) and magnetic resonance imaging (MRI) machines may soon be able to detect it.[24] The latest generation of high-speed EBCT scanners can image vulnerable plaque if the heart rate is slowed down to around 55 beats per minute. This usually requires an intravenous beta-blocker drug.

The failure of bypass and angioplasty to prevent subsequent heart attacks. In most cases, bypass surgery replaces a diseased artery, generally one with a high level of calcification, with a vein. Typically, the calcifications in the bypassed arteries are relatively stable, but the vein is subject to rapid growth of new vulnerable plaque. Moreover, since veins have much less capacity to cope with the extremely high pressures of blood flow in a coronary artery, they tend to develop vulnerable plaque even more quickly than the original artery. The attachment sites for veins are also areas that attract inflammatory processes where vulnerable plaque quickly develops. And the surgery does nothing to slow down the underlying soft plaque process.

Angioplasty's problems. Angioplasty treats calcified plaque but, like bypass surgery, it misses the real culprit: the widely distributed vulnerable plaque formations, which lie elsewhere. It does nothing to reduce them. As the balloon compresses obstructions, it actually damages the previously stable calcified plaque, causing it to rupture, opening up many cracks for vulnerable plaque to develop and leak out. The many points of damage from the angioplasty procedure each accelerate the inflammatory process that underlies vulnerable plaque formation.

Arterial-stent problems. Not only do they leave vulnerable plaque untouched, these highly invasive surgical procedures induce a strong inflammatory response that, of course, is highly counterproductive.

HOW SOFT PLAQUE FORMS

Acute inflammation signals that our immune system is fighting off pathogens. But there is also a more insidious form of "silent" inflammation that underlies many diseases, including heart disease, rheumatoid arthritis, Alzheimer's disease, and even several types of cancer.[25] By understanding how silent inflammation leads to formation and eventual rupture of soft, vulnerable plaque, we can apply noninvasive methods to address each stage of this degenerative process and, combined with targeted therapeutic approaches, we can rapidly and dramatically reduce the risk of a heart attack to very low levels.

1. "Bad" cholesterol builds up. In the coronary arteries, the soft-plaque-formation process starts with low-density lipoprotein cholesterol (LDL-C), which contains esters of cholesterol, triglycerides, phospholipids, and other proteins. LDL-C is (correctly) cast as a villain in this story, but it does play an important role in transporting cholesterol from the liver to the body's tissues, which use cholesterol for vital membrane repair tasks. LDL-C is also involved in the creation of our steroid hormones, including sex hormones.[26]

If the level of LDL-C is excessive, it accumulates inside the artery wall in the intima, the layer of the blood vessel closest to the blood, where it undergoes chemical changes, including oxidation. Iron in the blood can act as a catalyst for this process because it facilitates oxidation. This may be one reason that premenopausal women have protection from heart disease: menstruation keeps iron levels low.[27] The LDL-C molecules also undergo glycation, or binding with sugar molecules.

2. Invaders move in. These modified LDL molecules no longer appear to be friendly to the immune system, so they are mistaken for foreign invaders. The endothelial (lining) cells in the blood vessel wall call for help by secreting chemicals that warn of infection, including oxidants, which directly damage the "invaders," along with chemokines and cytokines, which are immune-system signaling molecules.

3. Immune system responds. Immune system cells, including monocytes (white blood cells that ingest dead or damaged cells) and T lymphocytes (white blood cells that attack and destroy foreign substances and germs), respond to the SOS and invade the intimal layer to do battle with the now-pathological LDL molecules. Chemokines and other molecules secreted by both the endothelial cells and the muscle cells in the vessel wall cause the monocytes to multiply and turn into macrophages, which are the fully matured fighters of the immune system.

4. Fatty streak forms. The macrophages—literally, "big eaters"—fight the apparent invaders (the oxidized and glycosylated LDL molecules) by ingesting them. They eventually become filled with fatty LDL. These "stuffed" macrophages, known as "foam cells" because they look like fat bubbles of foam, together with the T cells form a "fatty streak," which is the early form of vulnerable plaque.

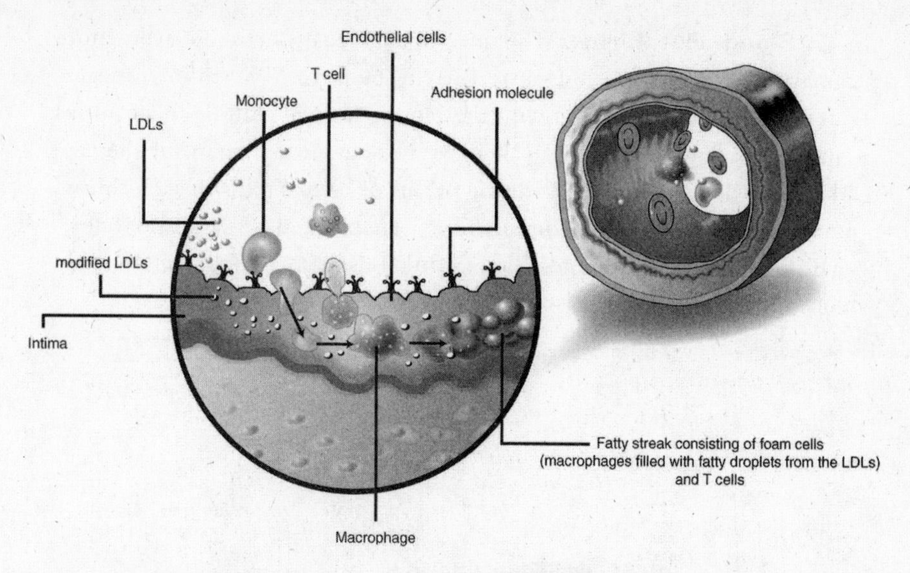

FIGURE 15-1. CREATION OF A FATTY STREAK

5. Fibrous cap forms. Inflammation causes the blood vessel's smooth muscle cells, which normally lie below the intima, actually to travel to the top of the fatty streak, where they form a fibrous cap. This cap adds to the size of the plaque, but also protects it from the bloodstream. Typically, the plaque at this point does not restrict blood flow but, rather, expands the outer diameter of the blood vessel.

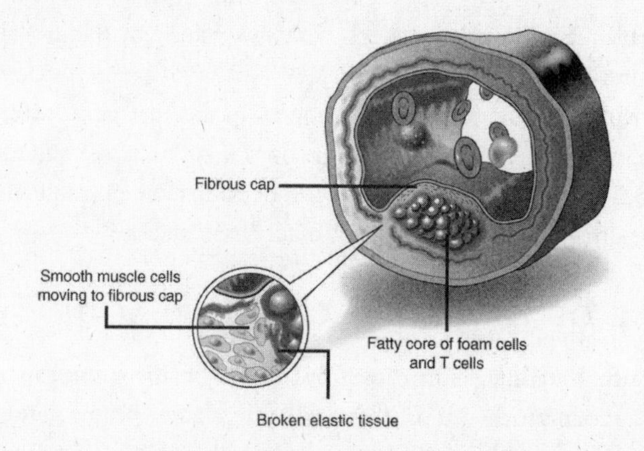

FIGURE 15-2. DEVELOPMENT OF A FIBROUS CAP

6. Cap ruptures. The foam cells undergo further changes, which cause them to secrete additional inflammatory molecules. These molecules damage the muscle cells and other cells that make up the fibrous cap. The result can be a disastrous rupture.

7. Blood clot forms. The macrophage foam cells do even more damage: they secrete a substance called "tissue factor." When the cap ruptures, the tissue factor along with the blood's normal clotting mechanisms can cause a large blood clot, called a thrombus, to emerge over the cap. Although the arteries have chemical means of breaking down such threatening clots, these "coagulation inhibitors" are blocked by the tissue factor.

8. Heart attack. If the resulting thrombus is large enough to completely block the vessel, the result is a coronary thrombosis—a heart attack.

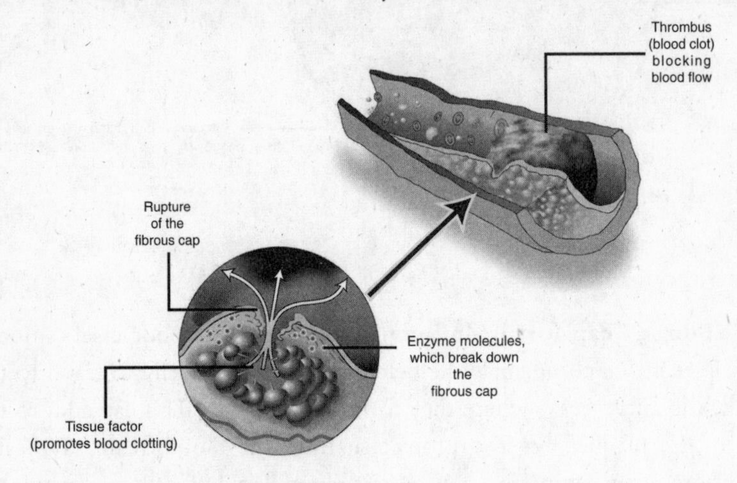

FIGURE 15-3. PLAQUE RUPTURE CAUSING A HEART ATTACK

Vulnerable plaque is difficult to detect, which explains why traditional diagnostic procedures fail to find it. Usually, it does not significantly restrict blood flow before it ruptures. Since vulnerable plaque is located *within* the artery wall rather than on its surface, it merely causes the outside of the vessel wall to bulge slightly, which is easily missed. Yet the process of vulnerable plaque formation outlined above underlies about 85 percent of all heart attacks.[28]

THE ROLE OF HARD, CALCIFIED PLAQUE

Even though most heart attacks are caused by soft plaque, there is still an important reason to be concerned with your level of hard, calcified plaque in the coronary arteries: levels of soft and hard plaque appear to be related to each other.[29]

Unlike soft, vulnerable plaque, it is easy to determine the amount of calcified plaque in the coronary arteries, using the Ultra-Fast Coaxial Tomography (UF CT) Heart Scan. This is a three-dimensional CT scan of the heart that can image calcium in the artery walls. The calcium is presumed to be calcified plaque because there is no other reason for calcium to be there. A computer

BRIDGE TWO

BOOSTING HDL

Now that we are rapidly gaining an understanding of the development of heart disease in terms of the interaction of specific molecules, drugs that precisely target each key step in the process are in the pipeline. Although one can never be certain which drugs will make it through the regulatory process and prevail in the marketplace, the results of trials on a new generation of drugs to reduce heart disease risk are very encouraging.

As this book is being written, some exciting research is being conducted on a synthetic form of HDL cholesterol called recombinant Apo A-I Milano (AAIM). In animal trials, AAIM resulted in rapid and dramatic regression of atherosclerotic plaque.[30] In a phase I FDA trial, which included 47 human subjects, administering AAIM by intravenous infusion resulted in a significant reduction (an average 4.2 percent decrease) in plaque after just five weekly treatments. No other drug has ever shown the ability to reduce atherosclerosis this quickly.[31] Further research is required to confirm these results in larger clinical trials and to determine the impact of this treatment on the incidence of heart attacks and heart disease-related deaths. However, an indication of its potential is reflected by Pfizer's recent acquisition of Esperion, the small company that developed the drug, for $1.3 billion.

Pfizer is also making a major bet on another approach to boosting HDL levels and combating plaque formation. The patent on Lipitor, its $9-billion-a-year blockbuster LDL-cholesterol-lowering drug, is running out, so Pfizer hopes to replace it with a new drug that combines Lipitor with Torcetrapib, which raises HDL. Torcetrapib works by blocking an enzyme that normally breaks down HDL.[32] Lipitor is the best-selling drug of all time, and Pfizer plans to spend a record $1 billion on the phase III trials of Lipitor-Torcetrapib, now under way.

Another method of boosting HDL is to activate certain cellular switches called PPARs (peroxisome proliferator activated receptors), which is how Eli Lilly's "PPAR alpha agonist" works. This drug also lowers triglycerides, another risk factor, and is in mid-stage trials.[33] PPAR is getting a lot of attention: GlaxoSmithKline, Merck, Bristol-Myers Squibb, Pfizer, and AstraZeneca are all working on drugs based on turning on PPARs. These PPAR activators may also lower LDL and help control blood sugar levels, so they may be effective against both type 2 diabetes and heart disease.

calculates a calcium score for each region of calcified plaque, each coronary artery, and a total calcium score for the heart as a whole.

There is some controversy among physicians about the usefulness of this coronary calcium score in predicting the risk of a heart attack. Higher scores are in fact related to a higher risk, but this may be because a high calcium score is associated with higher levels of vulnerable plaque. We believe it is a useful measurement if you understand the different roles of calcified and soft plaque.

Rather than interpret the calcium score as an absolute number, you should compare your score to other people who are your age and gender. If your score is higher than the average shown in the table below, then it is likely that your rate of plaque creation is high and you should give a high priority to lowering your risk factors for heart disease. If your score is higher than 75 percent of the people your age and gender, then you should give this an urgent priority.

Table 15-1. Calcium Scores (average and 75th percentile)[34]

MEN			WOMEN		
Age	Average	75th Percentile	Age	Average	75th Percentile
40–45	2	11	40–45	0.1	1
46–50	3	36	46–50	0.1	2
51–55	15	110	51–55	1	6
56–60	54	229	56–60	1	22
61–65	117	386	61–65	3	68
66–70	166	538	66–70	25	148
70+	350	844	70+	51	231

People whose cholesterol levels and other risk factors are in poor control may increase their score by as much as 40 percent per year,[35] as calcified plaque growth accelerates.[36] (The amount of the increase itself increases, instead of remaining constant. For example, at 40 percent growth rate, starting at, say, 100, your calcium score would rise each subsequent year to 140, 196, 274, 384, and so on, compared with 140, 180, 220, and 260, if it were increasing by a constant 40 points.) People who adopt healthier diets and take certain steps such as taking lipid drugs (discussed below) typically reduce this growth rate to about 10 percent. Note, however, that the usual remedial steps do not stop the growth—the total calcified plaque growth still accelerates, just at a lower rate.

The higher your calcium score in relation to the average of people your age and gender, the more urgent it is to reduce the growth rate to zero or, better yet, reverse it. (Some physicians say it's impossible to reverse calcified

BRIDGE TWO

LATE-STAGE INNOVATIONS

Preventing blood clotting combats heart disease at its final stage by reducing the risk of the fatal blood clot that constitutes a heart attack. Millions of people with elevated risk factors take aspirin, which modestly reduces clotting. But patients with a very high risk often take Coumadin (warfarin), which has been the main prescription blood-thinning drug for the past half century. The problem with Coumadin is that it interacts with many other medicines, foods, and supplements and is extremely difficult to administer safely. A new drug called Exanta, from AstraZeneca, is nearing approval and appears to overcome many of these complications.[37] Some critics of Exanta have pointed out elevated levels of liver enzymes in patients taking the drug, a possible indication of toxic effect, but the liver enzyme increase appears to be temporary, even if the drug therapy is continued.

People suffering from congestive heart failure may soon be able to implant a sensor the size of a grain of rice into their left atrium. The device, developed by Integrated Sensing Systems (ISSYS) in Ypsilanti, Michigan, reads the blood pressure from inside the heart, which is a key indicator of the heart's condition. The sensor's output can be read by a device outside the body that communicates wirelessly with the implant. Currently, to obtain this type of information, late-stage heart-failure patients must undergo repeated invasive procedures in which a catheter is inserted into the heart via an artery that runs through an arm or leg. The ISSYS device needs to be inserted only once. It folds up and is inserted into the heart using a catheter under X-ray control. The Cleveland Clinic is now planning to begin implanting the device in human patients.

Another tiny device intended to diagnose artery plaque from inside arteries using ultrasound is being developed by Stanford University's Butrus T. Khuri-Yakub under a grant from the National Institutes of Health.[38]

plaque, but a number of studies support such reversal. This potential was first established by Dr. Dean Ornish with his heart-disease reversal program, which demonstrated reduction of hard coronary plaque levels.[39])

If you have had bypass surgery or angioplasty, that means that you probably had—and almost certainly continue to have—a very high calcium score. The surgeries did *not* cure you. In fact, they did nothing to reduce either hard or soft plaque. The various inflammatory processes underlying plaque formation are typically accelerated by these surgeries and their aftermath. Perhaps the main benefit from these procedures is in shocking you into taking your cardiac health seriously. It is imperative for someone who has a history of heart attack, angina pain, and/or cardiac surgeries to aggressively implement all of the risk-reduction methods described later in this chapter.

RISK FACTORS FOR HEART DISEASE AND HOW TO REDUCE THEM

Armed with a more accurate model of how heart disease develops, we can now better understand the role that the many risk factors play. Although certain risk factors may be more pertinent for some people than for others, the best strategy is to take reasonable steps to improve your odds in every way possible.

The urgency of reducing cardiovascular risk factors depends on where you are in the progression of the disease. Autopsies of children reveal that coronary fatty streaks, which represent the earliest stage, may already be well developed in childhood. People with more than three of the major risk factors below should have a UF CT Heart Scan every five years, or annually if their calcium score is above the 75th percentile.

For each of the major risk factors, we provide recommendations below for reducing these risks. Certain supplements, such as vitamins C and E, garlic, and policosanol, are included for several risk factors. *You should not exceed the recommended amounts; one dose covers all risk factors to which the supplement applies.* For example, if you take the recommended dosage of vitamin C to handle, say, hypertension, do not take more if you are also concerned about additional risk factors that also call for this supplement.

MAJOR RISK FACTORS FOR HEART DISEASE

Count your major risk factors by adding one point for each category (such as smoking) that is in the risk-factor range. For example, add one point if you smoke now and have smoked for more than 10 years. This self-test is intended only as a rough guide to determine if you should undergo further evaluation.

If you have three or more major risk factors, we recommend:

A full set of blood tests (cholesterol, HDL, triglycerides, hs-CRP, homocysteine)
UF CT Heart Scan with calcium score
Exercise stress test

Add one risk factor for each category that is in the risk-factor range in Table 15-2.

Table 15-2.

RISK FACTOR CATEGORY	OPTIMAL RANGE	RISK FACTOR RANGE
Genetic Inheritance	No parents or siblings with heart disease	Father had heart attack before age 55 and/or mother had heart attack before age 65
Gender and Age	Male 44 or younger; female 54 or younger	Male 45 or older; female 55 or older
Smoking	You never smoked	You now smoke 1 or more packs/day and/or you smoked in the past for 10 years or more
Weight	95% of optimal weight (see Table 8-2 in chapter 8)	20% or more over optimal weight
Cholesterol and Triglycerides	Total cholesterol 160–180; LDL: 80 (if you have less than 3 major risk factors); 70 or less (if you have 3 or more major risk factors)[40]; HDL over 60; ratio of total cholesterol to HDL under 2.5; triglycerides under 100 (see recommendations for optimal cholesterol levels, below)	Any of the following: Total cholesterol over 200; LDL over 130; HDL under 40; ratio of total cholesterol to HDL over 4 (see recommendations below)
Homocysteine	7.5 or less	10.0 or greater
hs-CRP	1.3 or less	over 5.0
Fasting Glucose and Insulin	Glucose 60–80; Insulin 2–3	Glucose over 110; and/or Insulin over 10
Blood Pressure	Systolic 120 or less; diastolic 80 or less	Systolic 140 or higher; and/or diastolic 90 or higher
Stress	Not type A personality with high level of anger (see chapter 23 on stress); not type D	Type A with high level of anger; or Type D
Exercise	Active	Sedentary

Genetic Inheritance

Your genetic profile profoundly affects your predisposition to many of the other risk factors, such as high LDL and low HDL levels, insulin resistance, hypertension, elevated homocysteine, and others. Assess the health history of your parents and, to a lesser extent, siblings. If your father suffered a heart attack before age 55, or your mother suffered a heart attack before age 65, consider this to be a major risk factor.[41] Also examine your own predispositions. If you have a tendency toward hypertension or insulin resistance, it is likely that this results at least partially from your genes.

Take advantage of the recently available genomic (genetic) tests, which we discussed in chapter 11. Some of the more common tests to evaluate your genetic predisposition to cardiovascular risk include the Apo E, eNOS, MTHFR, and PAI-1 genes. For example, the Apo E4 variant of apolipoprotein E results in a disposition to excess LDL levels and increased cardiovascular risk. The eNOS gene directs the production of the NOS (nitric oxide synthase) enzyme, which is responsible for blood vessel dilation. Nitric oxide helps prevent development of plaque progression.[42] Common mutations of this gene lead to increased risk of atherosclerosis. The MTHFR gene is intimately involved in methylation reactions and homocysteine production (homocysteine is one of the most powerful determinants of cardiac disease risk). Mutations in this gene are very common, but easy to correct with simple nutritional supplementation (see chapter 21). The PAI-1 gene produces plasminogen activator inhibitor, which reduces risks of blood clots, a major factor in heart attacks and strokes. People with PAI-1 mutations can take steps to reduce their risk of blood clot formation. The nice thing about genetic testing is that if you discover you have a potentially harmful mutation, you can largely eliminate your risk by following safe and simple lifestyle interventions.

You might consider going back in time and choosing your parents more carefully. Needless to say, the technology to do this is nowhere on the horizon. But as discussed earlier, we are close to being able to block "bad" genes (through RNA interference), and ultimately we will be able to add new healthy genes using gene therapy. In the meantime, it is worthwhile being aware of your own genetic predispositions and aggressively combating those risk factors for which you have a genetic tendency.

Gender and Age

The common wisdom is that only men need be concerned about heart disease. A 2002 survey by the Society for Women's Health Research revealed that 60 percent of women fear cancer the most, compared with only 5 percent for heart disease.[43] But heart disease is the number one killer of both men and women. Of 1.1 million heart attacks each year, almost half are suffered by women.[44] It is true that premenopausal women have a level of protection; statistics for cardiac events as well as levels of coronary calcium score on UF CT heart scans show that women lag men by about 10 years.

If you are a man, upon reaching the age of 45, you will need to consider your male gender and age as a major risk factor. For women, your age becomes a major risk factor at age 55. If you have two additional major risk factors, you should give a high priority to adopting all of the recommendations in this chapter.

Smoking

Smoking significantly increases the risk of dying from a heart attack or stroke.[45] There are 4,000 poisons contained in tobacco and tobacco smoke, and we could devote this entire book to cataloguing the havoc that all of these substances wreak on your metabolic pathways. Cigarette smoke, either direct or secondhand, greatly increases free-radical activity, which accelerates the oxidation of LDL, the critical step that turns the LDL molecules into the pathological variant that invades the coronary vessel walls. Smoking also:

- Increases the overall level of inflammation in the body, which accelerates every stage of heart disease
- Interferes with metabolism of vitamin C, a critically important antioxidant
- Increases heart rate and blood pressure, both of which accelerate damage to the arteries
- Doubles the blood levels of adrenaline, causing vasoconstriction (tightening of the blood vessels) and the aggregation of platelets, which encourages the final eruption of a coronary thrombus
- Interferes with vitamin B_6 metabolism, which is critical to proper homocysteine metabolism

Our recommendations are obvious: Don't smoke, and avoid secondhand smoke.

Weight

Being overweight contributes to a wide range of diseases and to a number of the risk factors listed here. It is a major contributor to developing the metabolic syndrome, type 2 diabetes, and hypertension. The Framingham Study found increased risk of heart disease with increasing levels of obesity for both men and women.[46]

Our recommendation is to maintain your optimal weight, as we discussed in chapter 8.

Cholesterol and Triglycerides

Cholesterol and its LDL and HDL components continue to play major roles in the new inflammation-based understanding of heart disease. The inflammation process starts with excess LDL particles, which enter the coronary artery lining and become oxidized. HDL (the good cholesterol) particles reduce heart disease risk by transporting excess cholesterol back to the liver and also by reducing the oxidation of LDL. Although excess cholesterol can play a key role in the creation of vulnerable plaque, levels that are too low may increase other risks. Very low cholesterol levels can increase the risk of hemorrhagic stroke (stroke caused not by atherosclerosis but by a burst blood vessel).[47] There is also a correlation between very low cholesterol levels and increased incidence of suicide.[48]

Based on the statistics for the general population, total cholesterol of 180 to 200 is optimal. However, people following our recommendations are different from the general population. In our program, the optimal range for total cholesterol is 160 to 180. Ideally, LDL-C should be 80 or less, depending on the number of your risk factors. HDL-C should be 60 or higher. An ideal ratio of total cholesterol to HDL-C is under 2.5. By following our nutritional and supplementation recommendations, you will be reducing inflammation, weight, and blood pressure and will be much less likely to suffer a hemorrhagic stroke.

Recent research has confirmed that reducing LDL cholesterol to much lower levels than the standard recommendation (below 100) substantially reduces the risk of heart disease. The 2004 study by researchers at Harvard Medical School and published in the *New England Journal of Medicine* examined the question of whether reducing LDL-C levels well below 100 (by taking 80 milligrams a day of Lipitor, a statin drug) would substantially reduce heart disease risk. The "experimental" group taking the more aggressive LDL-C lowering therapy had a median LDL-C level of 62, compared with 95 for the control group, who took a more moderate course of statin drug therapy (such as 20 milligrams of Lipitor a day). The experimental group who took the larger dose had substantially fewer heart attacks as well as fewer recommendations for bypass or angioplasty surgery. "This is really a big deal," commented Dr. David Waters, a professor of medicine at the University of California, San Francisco. Waters, who was not involved in the research, added, "We have in our hands the power to reduce the risk of heart disease by a lot."[49] Based on this and other corroborating research, we recommend that you keep your LDL-C levels at approximately 80 (if you have fewer than three major risk factors), or 70 or less (if you have three or more major risk factors).

If your coronary calcium score is higher than 75 percent of the people your age and gender, you probably have a very active plaque creation process, so it's particularly urgent that you reduce your heart disease risk factors as much as possible. At this level, your risk of a heart attack is likely to be much greater than your risk of hemorrhagic stroke or suicide, so you should give a priority to reducing cholesterol and LDL-C levels and bringing total cholesterol levels down to the 150 to 160 range.

Recent research has shown that particular factions of LDL-C and HDL-C are better indications of risk than total LDL-C and HDL-C. Researchers are still studying whether the HDL2 or HDL3 faction of HDL-C is the more protective form.[50] LDL pattern B and lipoprotein(a) are particularly harmful factions of LDL-C.[51] LDL particles in these two factions are smaller than the relatively large, buoyant particles in pattern A, so they more easily invade the vessel wall. Elevated levels of these two forms of LDL-C triple the risk of heart disease as compared with elevated levels of LDL-C alone.[52]

Another independent risk factor for heart disease is the amount of triglycerides (lipids, or free-floating fat) in the blood.[53] Ideally, triglyceride levels should be under 100.

The first step toward improving cholesterol and triglyceride levels is to adopt a healthy diet by following the nutritional recommendations in chapters 4 through 8 of this book. Most important, you should sharply reduce saturated fat, which is the most significant dietary influence. No other major dietary nutrient increases LDL levels more than saturated fat.

There is some controversy regarding dietary cholesterol. Cholesterol levels in the blood are regulated by the liver, so a healthy system is able to maintain healthy levels of cholesterol in the blood despite consumption of dietary cholesterol. However, if you have unhealthy lipid levels, these cholesterol-regulation mechanisms are probably not working optimally. If your blood cholesterol levels are not optimal, we recommend reducing dietary cholesterol to no more than 100 mg per day. One egg yolk has about 220 mg of cholesterol.

There are many effective nonprescription supplements that significantly improve cholesterol, LDL, HDL, and triglyceride levels. We recommend that you try these supplements first and then use prescription statin drugs (discussed below) if these prove insufficient. The supplements described here have mechanisms that are independent from the statins, so they can be used together with the drugs.

The most effective supplements include the following:

Policosanol is a dramatically effective supplement for improving lipid levels, with results comparable to statin drugs.[54] Studies have also demonstrated that combining policosanol with statins provides even greater effects. One study showed that at doses of 10 to 20 milligrams per day, policosanol "lowers total cholesterol by 17 percent to 21 percent and LDL cholesterol by 21 to 29 percent. It also raises high-density lipoprotein cholesterol by 8 to 15 percent."[55] Similar to lipid drugs, policosanol also inhibits the oxidation of LDL, a critical first step in the creation of deadly foam cells.

Table 15-3. Reference and Optimal Blood Levels

BLOOD TEST	STANDARD REFERENCE VALUE	"OPTIMAL" LEVEL
Fibrinogen (mg/dL)	Under 460	Under 300
C-reactive protein (mg/L)	Under 5	Under 1.3
Homocysteine (umol/L)	Under 15	Under 7.5
Ferritin (mg/dL)	Under 180	Under 100
Cholesterol (mg/dL)	100–199	160–180
LDL (mg/dL)	0–129	LDL: 80 (if you have fewer than 3 major risk factors); 70 or less (if you have 3 or more major risk factors)
HDL (mg/dL)	40–59	60 or more
Cholesterol : HDL ratio	2.5–4.0	under 2.5
Triglycerides	0–149	under 100

Policosanol has been used for many years in Europe and has recently become popular in the United States. The main ingredient is octascosanol, a long-chain fatty alcohol found naturally in the covering of leaves and fruits.

Policosanol works synergistically with **gugulipid** (gugulesterones), an ancient natural supplement made from the resin of the *Commiphora mukul* tree in northern India. A study of 125 patients showed that over a four-week period, gugulipid reduced total cholesterol by 11 percent and triglycerides by 16.8 percent. It also increased HDL by 60 percent.[56]

Vitamin E may also be effective in both lowering cholesterol and dramatically reducing overall heart-disease risk. In the 1996 Cambridge Heart Anti-Oxidant Study (CHAOS), 1,000 male heart patients were given 400 or 800 international units of vitamin E, while a control group of another 1,000 men (with the same health profile) were given a placebo. Eighteen months later, the vitamin E groups had 75 percent fewer heart attacks.[57]

Another very effective natural supplement for improving lipid levels is **plant sterols**. These have been marketed in cholesterol-reducing margarines (such as Benecol and Take Control), but these products contain unhealthy fats, so instead we recommend taking plant sterols as a supplement in pill form (see Fantastic-Voyage.net for specific product recommendations).

We recommend you start with policosanol and gugulipid, along with other supporting supplements (see table below) and measure the results two months later. If your levels still need improvement, add plant sterols and test again in another two months. If your levels continue to need improvement, you can consider adding a statin drug in consultation with your physician.

Phosphatidylcholine (PtC) is a major component of your cell membranes. As you age, the level of PtC in the cell wall diminishes, which is an important aging process. By supplementing with PtC, you can stop and even reverse this process. Research indicates that PtC can stimulate reverse cholesterol transport—that is, removal of cholesterol from artery plaque.[58] This is essentially the same process that HDL promotes. PtC (as both oral supplements and intravenous therapy) is widely used in Germany and approved by the German equivalent of the FDA. When taking oral PtC, it is important to use one that is at least 99 percent pure. Many supplements labeled as phosphatidylcholine are actually only about 30 percent PtC. Food-grade lecithin contains PtC but only about 20 to 25 percent of lecithin is PtC. See Fantastic-Voyage.net for recommendations of specific products.

We recommend that you start with the following first round of supplementation to improve lipid levels.

Table 15-4.

SUPPLEMENT	AMOUNT PER DOSE	TIMES PER DAY	TOTAL DOSE PER DAY
Policosanol	10 mg	2	20 mg
Gugulipid	500 mg	2	1,000 mg
Vitamin E	400 IU	2	800 IU
Garlic	900 mg	3	2,700 mg
Curcumin	900 mg	1–2	900–1,800 mg
Niacin*	100–500 mg	2	200–1,000 mg
Phosphatidylcholine	900–1,800 mg	2	1,800–3,600 mg
Soluble fiber**	4–6 g	2–3	8–18 g
Soy protein extract	1½ teaspoons (3 g)	2	3 teaspoons (6 g)
Green tea extract	500–1,000 mg	2	1,000–2,000 mg

* We recommend inositol hexanicotinate, which is a flush-free niacin (it won't turn your face red). Dosages of up to 3,000 milligrams per day are often used, although we recommend starting with closer to 200 milligrams per day. Note that periodic monitoring of liver function is recommended when taking niacin.

** Soluble fiber, such as pectin, guar gum, or psyllium, is recommended, especially before meals high in fat. If you take the prescription drugs nitrofurantoin or digitalis, do not take soluble fiber.

After two months, if levels are still not optimal, we suggest you add 3,600 milligrams (1,800 twice a day) of plant sterols, and then test again after another two months.

If natural supplements fail to get your cholesterol, LDL, HDL, and triglyceride levels to an ideal range, you and your physician may wish to consider enzyme HMG-CoA reductase inhibitors, known as statin drugs. Statin drugs slow down the creation of cholesterol by the liver and increase the rate at which LDL is cleared from the blood. They also appear to inhibit the oxidation of LDL, thereby slowing down the first step of vulnerable plaque formation. There is controversy among researchers on this last point. Some scientists believe that the reduction in vulnerable plaque formation is entirely due to the improvement in lipid levels.

It is important to note that statin drugs may have toxic effects on the liver, so your physician will want to monitor the health of your liver through blood tests that measure key liver enzymes.[59]

Another critically important consideration when taking lipid drugs is that these medications deplete the body of coenzyme Q_{10}, a vital nutrient needed to maintain the health of the mitochondria (the energy furnaces in every cell). It is vital to take this supplement when taking statin drugs. Fifty to 100 milligrams of coenzyme Q_{10} twice a day is recommended if you are on a statin drug.

A particularly effective statin drug is atorvastatin, known as Lipitor. Unlike other lipid drugs, Lipitor is approved as a treatment to reduce triglycerides in addition to improving cholesterol levels. Lipitor can reduce LDL by 40 to 60 percent and triglycerides by 20 to 40 percent. It also boosts HDL by 5 to 10 percent.[60]

Homocysteine

As we discussed in chapter 13, excessive levels of homocysteine, an indication of abnormal methylation processes, have far-ranging negative implications. High levels of homocysteine accelerate the conversion of LDL into macrophage-filled foam cells and block the production of nitric oxide, which is healing to the endothelial cells in the blood vessel walls. We recommend keeping homocysteine levels below 7.5. Our program for achieving this is described in chapter 13.

C-Reactive Protein (hs-CRP)

Recent research shows that C-reactive protein, which measures the level of inflammation in the body, is a critically important risk factor. Lowering hs-CRP should have the same priority as optimizing cholesterol levels. One study in the journal *Circulation* showed that heart attack risk for men increased threefold with levels of hs-CRP over 2.11 versus levels under 0.55, and more than fivefold for women with levels of hs-CRP over 7.3 versus under 1.5.[61] Your level will temporarily rise during periods of infection, so repeat testing is required if your level is elevated, particularly if you are suffering from a cold or the flu. Given the vital role of inflammation in every step of the heart disease process, the association of hs-CRP and heart disease risk is not surprising.

We recommend achieving an hs-CRP under 1.3. We discussed the role of inflammation and our program for reducing it in chapter 12.

Fasting Glucose and Insulin

As we discussed in chapter 9, metabolic syndrome and its more advanced form in type 2 diabetes have far-ranging implications for developing heart disease. Patients with these conditions have insulin resistance, which results in high blood levels of insulin. Insulin is a growth promoter and accelerates coronary plaque formation. It also encourages hypertension (high blood pressure), which is its own risk factor, as we discuss below. The high levels of glucose in the blood encourage the glycation (binding with sugar molecules) of LDL, a key step in turning macrophages and LDL into pathological foam cells. Fat metabolism is also likely to be disrupted by insulin resistance, causing excessive levels of triglycerides, which is another coronary risk factor.

We recommend having your fasting glucose and insulin levels checked and following the guidelines in chapter 9.

Blood Pressure

Even under normal circumstances, blood pressure in the coronary arteries is quite high, which encourages the inflammation that begins the process of atherosclerosis. The level of inflammation in the coronary vessel is worsened by elevated blood pressure. A study of 10,874 men reported in the *Archives of Internal Medicine* showed that people with stage 1 (mild) hypertension (140/90 or higher) had a 50 percent higher risk of dying of coronary heart disease.[62] Even those with high-normal blood pressure (which is now referred to as "prehypertension" and includes blood pressures above 120/80) had a 34 percent higher risk. Many other studies have demonstrated the ability of hypertension to accelerate atherosclerosis and increase the likelihood of a heart attack. Hypertension is also a symptom of the metabolic syndrome (TMS, or syndrome X).

Optimal blood pressure is less than 120/80. If your blood pressure is over this level, we recommend starting with a nutritional and supplement program and using prescription drugs only if that fails. The first step is to adopt the Ray & Terry's Longevity Program nutritional recommendations and attain your optimal weight. Determine if you have TMS or type 2 diabetes and follow our program in chapter 9. These steps, particularly adopting a low-carbohydrate, very-low-glycemic-index diet, are often adequate by themselves to resolve hypertension.

Supplements that are helpful in resolving hypertension include the following:

SUPPLEMENT	AMOUNT PER DOSE	TIMES PER DAY	TOTAL DOSE PER DAY
Vitamin C	1,000 mg	2	2,000 mg
Vitamin E	400 IU	2	800 IU
Coenzyme Q_{10}	100 mg	3	300 mg
Fish-oil (EPA and DHA)	EPA (500–1,500 mg) DHA (350–1,000 mg)	2	EPA (1,000–3,000 mg) DHA (700–2,000 mg)
Alpha lipoic acid (LA)*	250 mg	2	500 mg
L-arginine **	2–3 g	3	6–9 g
Garlic	900 mg	3	2,700 mg
Calcium	500 mg	1–2	500 to 1,000 mg
Magnesium	400 mg	2	800 mg
Potassium	100 mg	1	100 mg
Policosanol ***	10 mg	2	20 mg
Green tea extract	500–1,000 mg	2	1,000–2,000 mg
Hawthorne	250 mg	2–3	500–750 mg

* LA is an important supplement in preventing and treating TMS, as discussed in chapter 9.

** L-arginine has many additional benefits in improving vessel health.

*** Policosanol is very effective in improving cholesterol and related lipid levels.

If these recommendations prove insufficient and prescription drugs are considered, angiotensin II antagonists such as Cozaar or Hyzaar appear to be safer and more effective than other drugs such as calcium channel blockers.[63] Diuretics and beta-blockers appear to increase insulin resistance, which is counterproductive and increases the risk of developing TMS and type 2 diabetes.[64]

Stress

Given the prominent role of inflammation at every step of the long process leading up to a heart attack, it is not hard to understand why stress is a risk factor. Studies have demonstrated that feelings of aggressiveness and rage increase levels of homocysteine. The continual self-imposed stress associated with a type A personality results in higher levels of adrenaline, which worsens inflammation.[65] As we will discuss in chapter 23, "Stress and Balance," not everyone with a type A personality is at risk. People with short tempers who are continually getting angry have the personality type with higher risk. The type D personality, characterized by a lack of social connectedness and inability to express emotion, also has increased heart disease risk.[66]

Exercise

To put this in a positive context, adequate levels of exercise reduce all of the controllable risk factors, including improving insulin sensitivity, which contributes to weight loss and reduces blood pressure, stress, and inflammation. We discuss this key issue in chapter 22.

SECONDARY RISK FACTORS FOR HEART DISEASE

Obstructive Sleep Apnea

Obstructive sleep apnea is a common condition in which the mouth opens widely during sleep, causing a temporary blockage of air and decline in available oxygen. A person undergoing a sleep apnea event will appear to be gagging. People with moderate to severe sleep apnea may have dozens to hundreds of such events each night. This condition, which is found in many patients who snore a lot, can be diagnosed in a sleep clinic. There are also home tests that use a finger-mounted electronic probe to monitor blood-oxygen levels. Most people who suffer from sleep apnea are unaware of the problem.

Sleep apnea results in increased blood pressure and inflammation, so it accelerates the formation of vulnerable plaque.[67] It is a risk factor for heart disease.

BRIDGE TWO

HEALING THE HEART

People who have suffered a heart attack frequently have damaged heart tissue. In general, the heart is unable to regenerate itself, but such regeneration has been demonstrated using a patient's own stem cells. At Beaumont Hospital in Royal Oak, Michigan, and in Pro-Cardiaco Hospital in Rio de Janeiro, patients who were candidates for a heart transplant had stem cells from their bone marrow injected into the left ventricle of their hearts. The damaged tissue of their hearts regenerated, eliminating the need for a transplant. "This is the first approach where you have an opportunity to actually heal a heart," said Dr. Michael Rosen of Columbia University, referring to the benefits of using stem cells instead of a surgical transplant.

Complete human arteries have already been grown in the laboratory. Starting with human muscle cells, a research team from the University of Birmingham in England allowed the cells to multiply, then added a gene called hTERT to increase the longevity of the cells. Next, the researchers used a scaffold structure made of a biodegradable polymer (plastic) to direct the cells to grow into the desired shape. After 2 months, the group was able to harvest functioning arteries that were 8 centimeters long. These arteries have the genetic makeup of the original muscle cells. The method needs to be assessed for safety before the arteries can be used in

Excess body weight contributes to sleep apnea, so achieving optimal weight is one approach to solving the problem. One popular treatment is CPAP (continuous pulmonary airway pressure), in which the patient wears a mask connected to a device that maintains positive airway pressure, reducing sleep-apnea events. This cumbersome device can seem intrusive, but people who suffer from severe sleep apnea find that it's worth it to be able to sleep well.

Before considering CPAP, we recommend that you try a much simpler product, a flexible device called Sleep Angel (see Fantastic-Voyage.net), which appears to be effective for mild to moderate sleep apnea. This elastic garment is worn between the head and the chin and allows the mouth to open partially, but not widely. Because it's less likely to have a sleep-apnea event without the mouth opening fully, this device may reduce apnea events. Another approach is to avoid sleeping on your back as apnea is much more likely in this position. Some people sew a tennis ball in the back of their pajama top to encourage sleeping in other positions.

human patients, but the project represents another encouraging step in building replacement parts for the human body. We will ultimately be able to convert our own cells, such as skin cells, into new, youthful heart cells, and repair and rejuvenate our own hearts without surgery by introducing these cells into the bloodstream.

Finally, computers are now powerful enough to simulate the heart on a cell-by-cell basis. Elaborate simulations of the heart are being developed by several companies[68] and by academic research organizations, including UC San Diego and Oxford University. The overall effort, known collectively as the cardiome project, seeks to accurately simulate every aspect of the cardiovascular system, from single cells up to the entire heart, including the delicate interplay of many complex electrochemical processes. "We can do a good job now of modeling on a computer what happens to cardiac cells in heart failure, and predicting how a heart contraction will respond to a drug or other stimulus," says UC San Diego's Andrew McCulloch. "It's allowing us to answer a lot of experimental and clinical questions."

These tools are already assisting clinical practice. Physicians feed in results from MRI and UF CT scanners and receive diagnoses that otherwise would have required invasive surgery.[69] As heart simulation software and hardware continue to improve, we will ultimately be able to actually discover new drugs, as well as test them and other therapies by simulating them. A drug trial that might have required months will be done in hours, greatly accelerating the drug development process. Drug and therapy development for all diseases is moving in this direction.

Fibrinogen

Fibrinogen is a protein in the blood that contributes to coagulation. Excessive levels will increase the likelihood of a clot forming when the fibrous cap of a vulnerable plaque ruptures.[70] We recommend that you test your fibrinogen levels. The reference range is usually reported as 150 to 460, but we recommend levels under 300.

For individuals with elevated or high-normal levels, we strongly recommend low-dose aspirin therapy (81 milligrams per day), which will reduce the tendency to create blood clots, although aspirin does not appear to lower fibrinogen levels directly.

Reducing blood levels of fibrinogen is difficult, but supplements that may lower fibrinogen, or at least reduce coagulation, include curcumin, EPA and DHA, garlic, ginger, green tea, gugulipid, policosanol, beta carotene, and vitamins C and E. Also note that high levels of homocysteine contribute to higher levels of fibrinogen, so reducing excessive homocysteine levels will also be beneficial here.

Male-Pattern Baldness

An 11-year study at Harvard Medical School of 22,071 male physicians showed increasing risk of heart disease with increasing levels of hair loss.[71] The link appears to be due to the fact that the vasoconstriction of blood vessels that underlies baldness also plays a role in accelerating both atherosclerosis and heart attacks. Attacking the symptom of baldness with such drugs as Propecia is unlikely to improve this risk factor. However, adopting a heart-healthy diet, supplement, and exercise program will reduce both heart disease risk and further hair loss (but it won't reverse prior hair loss).

Iron in the Blood

High levels of iron in the blood, a hereditary condition called hemochromatosis,[72] particularly in combination with elevated levels of LDL, promote the oxidation of LDL, which is the critical first step in creating deadly foam cells. Premenopausal women have lower levels of iron as a result of menstruation. Men can simulate part of this effect through regular blood donation. (It appears that those medieval doctors who practiced bloodletting were not all wrong.) Supplements that reduce iron levels include fiber, calcium, magnesium, garlic, vitamin E, green tea, and red wine. You should not take supplements (particularly mineral supplements) that include iron, and you should avoid iron cookware. You can check the level of iron in your body with a serum "ferritin level," which ideally should be less than 100.

Gum Disease

Periodontal disease, such as gingivitis, is characterized by chronic inflammation and has been linked in studies to increased risk of heart disease.[73] We do not yet know whether the existence of gum disease itself contributes to heart disease, or whether underlying inflammatory and infectious processes are contributing to both gum disease and heart disease. It is also possible that the varied bacteria involved in gum disease may contribute to the process of atherosclerosis. We do recommend proper dental hygiene to reduce the likelihood of gum disease, including daily flossing, regular dental visits, and appropriate treatment if gum disease is diagnosed.

Hypothyroidism

There is a strong link between hypothyroidism (low thyroid function) and other heart disease risk factors.[74] One half of hypothyroidism patients have high levels of homocysteine, compared with 18 percent of the overall population. More than 90 percent of such patients have either excessive levels of cholesterol or homocysteine, compared with only about a third of the general population. The thyroid hormones, particularly triiodothyronine (T3), play important roles in metabolizing fats and cholesterol. This explains the high levels of homocysteine and cholesterol that are associated with hypothyroidism. Checking thyroid function (free T3, free T4, and TSH levels) should be a routine part of your annual examination and impaired thyroid function should be treated.

NONINVASIVE DIAGNOSIS AND TREATMENT

If you have fewer than three major risk factors (see Table 15-2 on page 211), we recommend the following blood tests at least every five years.

- A blood panel including total cholesterol, LDL-C, HDL-C, and triglycerides
- C-reactive protein
- Fasting glucose
- Hemoglobin A1c (if fasting glucose greater than 100)
- Lipoprotein(a)

The results of these tests may add one or more risk factors. If you have three or more risk factors, we also recommend you consider the following.

- A UF CT heart scan with calcium score measures the total amount of calcified plaque. You should ask for the amount of calcium score associated with each lesion, since the distribution of calcified plaque also indicates risk. Plaque concentrated in one place is more likely to block arteries, resulting in angina pain. And if several arteries are severely blocked, this could result in congestive heart failure. The most important finding, however, is the total calcium score (total of the calcium scores for all lesions). If you have a score over the 50th percentile for your age and gender, your risk is likely to be high, and you should follow the recommendations in this chapter on an urgent basis.
- An exercise stress test monitors your ECG (electrocardiogram) during a graduated exercise test on a treadmill, which can reveal if you have a significant degree of blockage in one or more of your coronary arteries and the ability of the coronary arteries to supply adequate blood to the heart. Like the UF CT heart scan, the exercise stress test does not directly assess the key issue of

BRIDGE THREE

NANOBOTS WILL REPLACE YOUR BLOOD CELLS

Nanomedicine author Rob Freitas has created and analyzed detailed conceptual designs to replace our red blood cells, platelets, and white blood cells.[75] Like most of our biological systems, red blood cells perform their oxygenating function very inefficiently, and Freitas has redesigned them for optimal performance. With an ounce or two of Freitas's respirocytes circulating along with your normal blood, you could go hours without oxygen. Although prototypes are still in the future, the physical and chemical requirements have been worked out in impressive detail. These analyses show that Freitas's designs would be hundreds or thousands of times more capable than your biological blood.

It will be interesting to see how this development is dealt with in athletic contests. We may have the specter of teenagers (whose bloodstreams might contain respirocyte-enriched blood) in school gyms routinely outperforming Olympic athletes (who likely will be prevented from using these technologies).

Freitas envisions micron-size artificial platelets that could achieve homeostasis (bleeding control) up to 1,000 times faster than biological platelets. Freitas also describes nanorobotic microbivores (white blood cell replacements) that will download software from the Internet to destroy specific infections hun-

vulnerable plaque. A more sensitive version is a thallium stress test, in which radioactive thallium is injected when you achieve your maximum level of exercise. Immediately after you get off the treadmill, an imaging device called a gamma camera takes pictures of your heart. If part of your heart muscle is not receiving a sufficient supply of blood, this will be reflected in the image. Another set of images is taken after your heart rate has returned to normal. The images combined with the electrocardiogram during exertion are evaluated by a physician to assess both heart and cardiac artery function.

- A new generation of CT and MRI (magnetic resonance imaging) machines is being developed that will be capable of imaging the subtle bulges associated with vulnerable plaque. Once developed, these systems will be of great value in assessing the true source of heart attack risk.

If you have diagnosed coronary artery disease, there are two other intensive forms of ECG that can provide your physician with detailed information about the pattern of artery occlusion and its effect on the heart.

dreds of times faster than antibiotics and will be effective against all bacterial, viral, and fungal infections, even cancer cells, with no limitations of drug resistance. He estimates that his robotic microbivores could destroy pathological organisms like harmful bacteria or a virus in 30 seconds—at least 100 times faster than biological white blood cells. The pathogen would be broken down into harmless amino acids and other nutrients rather than the often toxic result from the action of our biological immune system.

The authors of this book have personally watched through a microscope our own white blood cells surround and devour a pathogen, and we were struck by the remarkable sluggishness of this natural process.

Freitas also has a design for nanobots that could replace the entire circulatory system. A system of trillions of sapphire-based "vasculoid" nanobots would provide all of the functions of our current circulatory system, including circulation itself, replacing the function of the heart. Freitas describes this as a nanobot that would "duplicate all essential thermal and biochemical transport functions of the blood, including circulation of respiratory gases, glucose, hormones, cytokines, waste products, and all necessary cellular components."

Replacing your blood with trillions of nanorobotic devices will require a lengthy process of development, refinement, and regulatory approval, but we already have the conceptual knowledge to envision the engineering of substantial improvements over the remarkable but extremely inefficient methods used in our biological bodies.

- **Holter monitoring** consists of a 24-hour ECG conducted during your normal routine. The patient wears multiple (three to five) leads on his or her chest, connected to an electronics unit no bigger than a cell phone. For 24 hours, the electronic unit records the ECG signals, which are processed by a computer and then analyzed by a physician. This procedure can detect arrhythmias (abnormal heart rhythms) that would go undetected with normal ECGs, including stress tests.

- **Event monitoring** is similar to a Holter monitor procedure, but it takes place over an even longer period of time, generally one month or longer. The unit does not record all of the signals but records ECG tracings whenever the patient presses a button indicating a perceived symptom, such as a heart palpitation or feeling of angina pain. The unit continually stores the past several minutes of the ECG signals, so the stored ECG recording for each triggered event reflects ECG signals a few minutes before as well as a few minutes after the event.[76]

ENHANCED EXTERNAL COUNTERPULSATION

In addition to the noninvasive remedial procedures involving diet and supplements described above, an ingenious method for reducing angina pain and improving cardiac function in patients with heart failure is enhanced external counterpulsation (EECP).[77] This completely noninvasive treatment involves placing air-filled cuffs around the patient's calves, thighs, and buttocks. While the patient lies on a table, the cuffs are compressed with air in a specific rhythm controlled by a computer that receives input from the patient's real-time ECG. The inflation of the cuffs is timed to occur precisely during the resting phase of the heart rhythm, called *diastole*. As the computer inflates the cuffs, blood is propelled from the lower body back into the heart. This treatment, which is approved by the FDA for some cases of angina pectoris and heart failure, rapidly promotes the development of collateral coronary blood vessels (very small coronary arteries that augment the main coronary arteries). In other words, *EECP causes the heart to grow its own natural bypasses.*

EECP greatly accelerates the natural process of growing collateral bypass circulation. It is well known that elderly heart patients, who have had more time to grow collateral circulation, have a lower risk of dying from a heart attack for this reason. With EECP, however, people can grow effective collateral circulation at any age. A typical course of EECP treatment is one hour per day, five days a week for seven weeks. Although this involves a significant commitment of time and inconvenience, it is far preferable to invasive surgery and involves a healthy, healing process, rather than the risks and complications of surgery. EECP is both FDA- and Medicare-approved under certain circumstances.

INVASIVE DIAGNOSIS AND TREATMENT

Invasive diagnostic procedures for heart disease also have many side effects, including the possibility of scratching your arteries, which may accelerate the formation of both vulnerable and calcified plaque. A very popular but highly invasive conventional diagnostic procedure is cardiac catheterization, popularly known as an angiogram. Typically, a cardiologist will recommend an angiogram when a patient "fails" an exercise stress test. The procedure consists of inserting a catheter (a long tube) into a large vein (usually in the leg) and threading it to the heart. A dye is injected and X-ray images are taken. Blockages can be diagnosed by changes in the rate of flow of the dye near occluded portions of the coronary arteries.

BRIDGE THREE

HEARTLESS—BY DESIGN

Once we perfect nanobot-based replacements for our blood (in the 2020s), we will be in a position to replace the heart altogether. It's a remarkable machine, but it has a number of severe problems. It is subject to a myriad of failure modes—as we've discussed at length in this chapter—and it represents a fundamental weakness in our potential longevity. The heart usually breaks down long before the rest of the body, and often very prematurely. Although artificial hearts are beginning to work, a more effective approach will be to get rid of the heart altogether. Among Freitas's designs are nanorobotic blood cell replacements that provide their own mobility. If the blood system moves on its own, the engineering issues of the extreme pressures required for centralized pumping can be eliminated. As we perfect ways to transfer nanobots to and from the blood supply, we can also continuously replace the nanobots that make up our blood supply.

Energy will be provided by microscopic fuel cells, using either hydrogen, other fuels, or the body's own fuel, ATP. Substantial progress has recently been made with both Micro Electro-Mechanical Systems (MEMS) –scale and nanoscale fuel cells.[78]

With the respirocytes providing greatly extended access to oxygenation, we will be in a position to eliminate the lungs too, by using nanobots to provide oxygen and remove carbon dioxide. One might point out that we take pleasure in breathing (even more so than elimination!). As with all of these redesigns, we will go through intermediate stages where these technologies augment our natural systems, so we can have the best of both worlds. Eventually, however, there will be no reason to continue with the complications of actual breathing and the burdensome requirement of having breathable air everywhere we go. If we really find breathing that pleasurable, we will develop virtual ways of having this sensual experience.

The invasive nature of the procedure creates significant risks: it may actually cause heart attacks, heart arrhythmias, and infection.[79] There is also a risk of damaging the sensitive lining of the coronary arteries, thereby encouraging the formation of new vulnerable plaque.

We strongly recommend that patients avail themselves of the growing arsenal of noninvasive diagnostic procedures that can accomplish as much as or more than conventional angiography. Once fully developed, the new noninvasive UF CT heart scans and MRI scans, which can image vulnerable plaque, will be even more informative, particularly since angiograms are unable to detect vulnerable plaque. At the beginning of this chapter, we discussed how the two most popular forms of conventional invasive treatment for heart disease—coronary bypass surgery and balloon angioplasty—fail to address the true cause of heart disease, which is vulnerable plaque.

The number of these procedures used with patients is excessive, even by published medical standards. Many studies show little or no difference in outcomes between groups of patients treated with statin drugs versus surgery,[80] while other studies question the appropriate application of these surgeries.[81] We believe that the vast majority (at least 90 percent) of bypass surgeries could be avoided and that patients would achieve more effective reversal of coronary plaque, both vulnerable and calcified, through the noninvasive means described in this book. In general, bypass surgery is a palliative (pain suppressant) to reduce angina pain, although even this symptom can quickly be reduced through noninvasive means in most cases. There's a small number of cases in which the coronary arteries are so blocked that a heart attack may occur without the eruption of vulnerable plaque. For them, we do recommend bypass surgery or angioplasty. However, only a small percentage of bypass surgeries performed actually fall into this category.[82]

Bypass surgery is extremely invasive and involves actually stopping the patient's heart during the surgery. A heart-lung machine sustains the patient's life functions during this time. Many of the complications arise from the process of stopping the heart, the use of the heart-lung machine, and the difficult and uncertain process of restarting the heart.

Cognitive Decline from Surgery

One of the more disturbing issues in the use of conventional, invasive therapies is the likelihood of a significant decline in mental function and mood, including cognitive decline, depression, and mood swings. Some physicians have dismissed this concern as a temporary phenomenon, but studies have found the decline to be permanent for approximately half of all bypass patients. A study reported in the *New England Journal of Medicine* that followed 261 bypass patients over five years found significant and lasting decline in mental status.[83] Measures of intellectual function declined by an average of 36 percent at 6 weeks after surgery and 24 percent at six months; 41 percent of the patients had significant cognitive impairment five years after surgery. The researchers concluded that cognitive decline immediately after bypass surgery (which is widespread) was significantly associated with continued decline five years later.

There are many risks and complications associated with bypass surgery. We mentioned above the 2 to 6 percent chance of dying from the surgery itself. In addition, there are risks of a nonfatal heart attack, stroke, nerve damage, and prolonged recovery periods.

As we reported at the beginning of this chapter, balloon angioplasty surgery may be effective in temporarily reducing angina pain, but studies have not reported significant reductions in subsequent heart attacks or deaths. Angioplasty compresses calcified plaque but does not address the basic process that creates vulnerable plaque, the true cause of most heart attacks. In fact, this invasive surgery has a high potential to irritate a region of calcified plaque, causing it to become unstable, thereby encouraging inflammation and vulnerable plaque formation. It also has the potential to damage the delicate lining of coronary arteries, which also encourages the formation of soft plaque.

The use of stents, which has become a standard refinement since 2000, has not appreciably changed these outcomes. Another innovation developed by Johnson & Johnson is to coat the stents with a drug called sirolimus, which discourages cell growth and thereby significantly reduces restenosis, the tendency of cell growth in and around the stent, causing it to close up after surgery.[84] We expect that this form of angioplasty will become dominant because of the substantially improved restenosis rate. However, this improved form of angioplasty still addresses only areas of occlusion (blockage) from calcified plaque, so it misses the real danger: the more widely distributed regions of vulnerable plaque, which are far more likely to rupture and trigger a heart attack. All of the other dangers of damaging blood vessels and encouraging inflammation from this invasive procedure remain unaffected by this refinement.

The invasive forms of treatment tend to be crude palliatives with many serious complications and risks and with little if any improvement in outcomes. The great advantage of the noninvasive means of stopping and reversing both vulnerable and calcified plaque is that they truly heal the source of the problem. With sufficient diligence and attention, almost everyone can avoid heart disease, invasive treatments, and the enormous suffering and death toll that this disease causes.

THE END OF HEART DISEASE

We already have the knowledge to dramatically reduce your risk of heart disease. If you adopt all of the methods we have described in this chapter, you can reduce your risk of having a heart attack to a very small level, regardless of your genes. Once the Bridge Two therapies are fully developed, we will have easily available means to reverse the damage already done by atherosclerosis, and even by previous heart attacks.

There are so many different effective strategies being pursued—lowering LDL, boosting HDL, stopping inflammation, preventing blood clotting, and inhibiting other critical steps in the heart disease processes—that you will soon be able to essentially eliminate heart disease. Most of these therapies are already in the approval pipeline, so we believe that heart disease will be easily controllable by the end of this decade.

16

THE PREVENTION AND EARLY DETECTION OF CANCER

"There is no scientific evidence that food or other nutritional essentials are of any specific value in the control of cancer."

—American Medical Association, 1949

"It appears prudent for all adults to take vitamin supplements."

—American Medical Association, 2002

In the year 2004, about 150 people an hour will be diagnosed with cancer. If you are male, the chance that you will develop cancer at some point in your life is about 50-50; for women, about 1 in 3. Cancer remains the second leading cause of death in the United States, with more than 560,000 Americans dying of some type of cancer in the year 2004.[1] Yet you can make some simple lifestyle choices to radically reduce your chances of becoming a cancer statistic.

More than three-quarters of all cancers (77 percent) are diagnosed in people over 55. While many cancers seem to *appear* suddenly later in life, they were often decades in the making. By making better lifestyle choices earlier in life and learning about specific cancers to which you are genetically predisposed, you can reduce your chance of developing cancer to a minimum.

The most common malignancies, and the number of cases diagnosed annually, in the United States are cancers of the:

Lung (1,200,000)
Breast (1,050,000)
Colon and rectum (945,000)
Stomach (876,000)
Liver (564,000)
Cervix (471,000)

Despite billions of dollars spent since Richard Nixon declared war on cancer in 1971, not much of a dent has been made in these statistics. The rates of these common cancers have remained steady or increased slightly over the past 60 years, except for stomach cancer, which has fallen by over 75 percent.[2] Unfortunately, this decrease is offset by a dramatic increase in lung cancer over the same period.

The keys to reducing cancer deaths are *risk reduction* and *early detection*. Of the two, cutting risk is more important, and the best way to accomplish this is by avoiding or quitting smoking. The American Cancer Society estimates that in 2002, more than 170,000 cancer deaths (30.6 percent of the total) were the direct result of tobacco use.[3] One of the major public-health triumphs of the final third of the 20th century was widespread dissemination of information about the dangers of cigarette smoking, resulting in millions of people quitting.

WHY WE GET CANCER AND HOW TO PREVENT IT

Cancer is a disease characterized by uncontrolled cellular proliferation. While normal cells have a fixed life span, as long is there is enough food, cancer cells will continue to grow and multiply indefinitely. This difference in cancer cells has been linked to mutations in their DNA, primarily caused by exposure to highly unstable and reactive chemicals known as free radicals. Free radicals form naturally in the body, but some specific environmental factors increase your exposure to these highly unstable molecules. You are exposed to excess free radicals from:

- Radiation exposure (such as X-rays and bright sunshine)
- Toxic heavy metals (lead, cadmium, mercury)
- Environmental toxins (pesticides, plastics, pollution)
- Cigarette smoke
- Deep-fried foods
- Excessive stress
- Excess dietary iron

Every cell in your body experiences more than 100,000 free-radical attacks each day. To maintain health and prevent your tissues from becoming cancerous, you require powerful mechanisms to counteract and repair this constant barrage. Luckily, each cell comes equipped with a system of built-in enzymes designed

BRIDGE TWO

EARLY DETECTION OF CANCER

Soon, even more genomics tests will be available to assist in early cancer detection and help determine what specific chemotherapeutic agents would most likely destroy it. One test developed by Genomic Health Inc. tests 21 specific genes to determine a tumor's specific "fingerprint."

From these genetic tests, scientists will be able to judge how aggressive a cancer may be, its propensity to spread throughout the body, and the chance of recurrence after treatment. This test will also reveal what drugs will be most beneficial and even which patients are either unlikely to benefit from chemotherapy or don't need it in the first place. One test called the OncotypeDX is under development and should be commercially available in the near future.[4]

A new handheld scanner from Turin, Italy, called the TRIMprob (Tissue Resonance InterferoMeter Probe) can be waved over the body like a wand and is able to detect malignant tumors with a degree of accuracy approaching larger CT scans or MR (magnetic resonance) images. It is still not quite accurate enough for clinical use but, in clinical trials, it did accurately detect 93 percent of prostate cancers and 66 percent of breast cancers. Because of its small size and affordability (it costs about 1 percent of the price of an MRI scanner), with some refinements in the years ahead, patients may simply have a *Star Trek*–like wand waved over their body as part of early detection-screening for cancer.

Another new scanner resembles a photocopier. It uses terahertz radiation, or T-rays, to show both the composition and shape of objects. The differences in how molecules respond to terahertz frequencies allow the scanner to create images—much like X-rays but so far thought to be safer. "Because tumors tend to retain more water, they show up very brightly in terahertz images," according to Don Arnone, a Toshiba researcher. "[T-rays] may fill important gaps between X-ray, MRI, and the naked eye of the physician."[5]

Molecular imaging employs a variety of magnetic, nuclear, and optical techniques to monitor individual molecules rather than sizable tumors. Probes are administered that bind to target molecules. Many of these probes emit light, which creates telltale spots on an image.

Doctors will be able to use this technology to both identify the early changes that lead to cancer and find out within days whether a cancer treatment is working. The result will be "totally different from the way we take care of patients now," according to Dr. Samuel Wickline of the Washington University School of Medicine at St. Louis.[6] "Several companies such as GE Medical are rushing to bring molecular imaging machines to the market."[7]

to neutralize free-radical attacks. Many of these enzymes require a continuous supply of vitamins and minerals to function properly. This is another reason a fundamental aspect of our program involves aggressive nutritional supplementation. We feel that by ensuring that you have an adequate supply of vitamins and minerals in your body at all times, you can decrease your cancer risk.[8] Toxic metals, meanwhile, can bind to your enzymes, taking the place of beneficial minerals and render these "polluted" enzyme molecules useless. So another aspect of our program emphasizes detoxification strategies.

Nutritional antioxidants—such as vitamins A, C, and E, the mineral selenium, and "super-nutrients" like coenzyme Q_{10} and alpha lipoic acid—are very powerful free-radical scavengers. We will discuss these nutritional supplements at more length in chapter 21, "Aggressive Supplementation." Minimizing your exposure to pollution and stress and making other health-promoting lifestyle choices are also critical to optimal enzyme function and cancer prevention.

EARLY DETECTION OF CANCER

There is an enormous difference between cancer *prevention* and *early detection*. Yet in public pronouncements on billboards, magazine ads, and television, this distinction often appears blurred. Many "cancer prevention programs" are really nothing more than programs for early detection.[9] We do not wish to minimize the value of attempting to detect cancer early. Rather, we want to emphasize that prevention and early detection are entirely separate processes.[10]

Early cancer detection is not as important as cancer prevention through risk reduction. Compare the importance of stopping smoking (prevention) versus finding a tiny lung cancer on a chest X-ray or CT scan (early detection). Many studies provide little support for the benefits of early detection,[11] since most of our current tests require that patients already have a very large number of cancer cells growing in the body before the disease will show up. (Exceptions include the newly developed DR-70 test, discussed on page 239, which requires fewer cancer cells to be present for detection.)

Ray & Terry's Longevity Program is based on the fact, derived from extensive research, that the majority of cancers can be prevented by appropriate environmental and lifestyle choices. In this case, early detection would be unnecessary. We feel that rather than trying to discover what type of cancer you may already have, it's more effective to screen for your genetic *predisposition* to specific types of cancer, then make lifestyle choices that will avoid these malignancies altogether.

BRIDGE TWO

CANCER PREVENTION AND TREATMENT

Researchers are looking to develop two types of cancer vaccines: those that prevent cancer and others that treat it. An example from the first group is hepatitis B vaccine, which protects against a dangerous liver infection that often leads to liver cancer. The drug manufacturer Merck is also developing a vaccine against HPV (human papilloma virus), thought to be the cause of roughly three-quarters of all cases of cervical cancer.[12] Both of these vaccines make use of bioengineered recombinant yeast strains. Other vaccines under development may help "immunize" people against leukemia, lymphoma, and cancers of the breast, colon, pancreas, ovary, and brain.

A promising biotechnology strategy uses "therapeutic" cancer vaccines designed to treat existing tumors.[13] A major reason cancer is able to grow in the body is the unique ability of cancer cells to hide from the immune system. It is the job of "dendritic cells" to serve as the sentinels of the immune system. These cells act like advance scouts on a hunting expedition and circulate throughout the bloodstream, looking for foreign invaders. Dendritic cells sound the alarm whenever an invader like a cancer cell is identified, triggering a powerful attack from the immune system to destroy the cancer cells. Considerable attention and research is now being directed toward the development of dendritic cell vaccines.

By taking a small sample of a tumor and incubating it with a patient's dendritic cells, it is possible to create a large number of dendritic cells targeted against this specific cancer. These cells can then be concentrated into a vaccine specific for this particular type of cancer. The dendritic cells from the vaccine will then circulate throughout the bloodstream, identify the targeted cancer cells wherever they are hiding, and alert the immune system so the cancer cells can be quickly destroyed.

Researchers are also using a wide range of other molecules, including purified proteins, synthetic DNA sequences from bacteria,[14] a surface molecule found in embryonic stem cells,[15] and a gene that codes for interleukin-12,[16] to help the body's dendritic cells find cancer cells more easily. More than 50 vaccines designed to stimulate the immune system to attack cancers are currently under development. Some, such as one from Dendreon for prostate cancer, are in Phase 3 clinical trials and may receive FDA approval in late 2004 or 2005.

OLDER SCREENING TESTS

Bridge One cancer-screening tests include regular examinations for breast, cervix, colorectal, and prostate cancer, among others. To determine which tests have the greatest value, the National Cancer Institute has sponsored the Prostate, Lung, Colorectal, and Ovarian (PLCO) Cancer Screening Trial.[17] These four cancers account for about 53 percent of all cancer deaths in men and 41 percent of cancer deaths in women in the United States each year. This trial has enrolled 154,000 men and women ages 55 to 74 at 10 screening centers nationwide and is scheduled to run for at least 13 years. As the results become available, further refinements to the National Cancer Institute guidelines for routine screening will be made. Current recommendations for low-risk individuals include the following.[18] (Your doctor can help determine if you are at high risk and need additional screening.)

Breast cancer. Women should begin self-breast examination beginning in their 20s, with clinical breast exams every three years until age 40, after which time clinical breast exam and mammography are recommended every one to two years.

Cervical cancer. Clinical exams with a Pap smear should begin within three years of sexual intercourse or no later than 21 years of age. Thereafter, it should be done every one to two years, depending on the type of testing performed. After three consecutive normal Pap smears, it can be decreased to every two to three years after age 30. If three smears within 10 years have been negative, screening may be stopped for women older than 70.

Colorectal cancer. Fecal occult blood testing should be done every year after age 50, with flexible sigmoidoscopy every five years and colonoscopy every 10 years.

Prostate cancer. Digital rectal examination and PSA (prostate-specific antigen) blood testing have been recommended annually after age 50, although the evidence to date does not indicate that such screening reduces mortality from prostate cancer.

NEWER SCREENING TESTS

Newer screening tests have value both for early detection of cancer and for determining your genetic predisposition. These tests include blood tests, diagnostic imaging scans, and genomics testing.

Blood Tests (DR-70)

The DR-70 is a tumor marker found in the bloodstream that is able to detect 13 different types of cancer, including lung, colon and rectum, breast, stomach, liver, ovary, esophagus, cervix, thyroid, and pancreas, at an early stage with a high degree of specificity and sensitivity. This test is currently available in some areas outside of the United States and has been submitted to the FDA as a screening test for colorectal cancer.[19]

This test actually measures FDPs—fibrinogen degradation products—which are increased in malignant tissues. Cancer cells release enzymes known as proteases, which break down adjacent tissues and are responsible for the invasiveness of many cancers. Proteases break down a protein known as fibrinogen into FDPs, which are detected by the DR-70 test. The amount of DR-70 found in the blood is directly correlated with tumor activity or degree of malignancy. This test may prove to be very useful as a screening tool in asymptomatic patients and may someday be included as part of routine testing during a regular physical examination.

Diagnostic Imaging Scans

Total body scans incorporating multiple CT scans of different regions of the body are becoming widely available in major metropolitan areas and are sometimes heavily promoted for routine cancer screening. But three major problems have been identified with using scans in this way.

- A typical full-body helical CT scan (which normally includes a measurement of coronary artery calcification, chest CT, and abdominal CT) exposes the patient to the equivalent of 250 chest X-rays.[20] Excessive exposure to radiation is a cancer risk factor itself.
- There are numerous false positives and false negatives.
- They are expensive and typically not covered by insurance.

Dr. Stephen Swensen has been studying the predictive value of screening CT scans for several years. In a study of 1,520 smokers and former smokers, the scans identified abnormalities in more than 90 percent of subjects tested; 37 malignant tumors were found, but so were 2,800 "suspicious" lung nodules. These required further testing, and many subjects needed chest surgery (which itself carries a 4 percent risk of death) to biopsy (examine) these lesions, later found to be benign. Dr. Swensen concluded that even though some patients were saved as a result of screening, many needless operations and some loss of life occurred as well.[21]

BRIDGE TWO

ANGIOGENESIS INHIBITORS AND OTHER NOVEL THERAPIES

Another exciting new avenue of attack is based on a group of drugs called angiogenesis inhibitors. "Angiogenesis" refers to the creation of new blood vessels within the body and is a critical process for a malignant tumor's ability to grow. Without growth of new blood vessels, cancerous tumors are unable to grow beyond a certain small size. Once this size—about that of a pea—is reached, the tumor gives off a hormone called VEGF (vascular endothelial growth factor), which stimulates the growth of new blood vessels so that the cancer has access to additional nutrients from the bloodstream and can continue growing.[22] Not only does VEGF stimulate growth of blood vessels directly adjacent to the tumor itself, it also allows cells from the tumor to escape into the bloodstream and set up residence in distant tissues, a process called metastasis and the main reason cancer is such a lethal disease.

Interest in angiogenesis really began in 1997, when trials with endostatin, an early angiogenesis inhibitor, resulted in complete regression of some tumors.[23] There are now at least 60 antiangiogenic drugs in various phases of clinical trials.[24] In 2003, Genentech announced that its experimental antiangiogenesis drug Avastin resulted in colon-cancer patients living 30 percent longer.[25] Angiogenesis inhibitors appear to be far safer than standard cancer therapies, with relatively few side effects. There is even the possibility that someday soon, healthy people will be able to be treated with antiangiogenesis drugs prophylactically, preventing cancers even before they form.

Disruptions in the normal path of cell death, or *apoptosis*, also hold important clues. As we discussed earlier, DNA strands are capped by telomeres, which shorten as human cells divide, driving cells toward genetic instability and death. Cell death usually occurs after 50 divisions. Some cells, however, avoid this preprogrammed ending by turning on telomerase, an enzyme that synthesizes telomeres and is inactive in most cells. Reactivating telomerase, which makes cells "immortal," does

Rather than total-body scans, we recommend more focused examinations of specific regions of the body—for example, a chest CT for smokers or former heavy smokers, or more frequent virtual colonoscopies for individuals with strong family history of colon cancer. Virtual diagnostic imaging with CT scan and MRI devices is less invasive, less expensive, and more convenient for most patients than direct visualization of their tissues through scopes, biopsies, and surgery.

not cause cancer by itself, but that event combined with other mutations can. "As cells walk this telomere plank into cellular crisis, where there is massive cell death and genomic instability," says Harvard Medical School professor Ron De Pinho, "only a few would-be cancer cells rise from the ashes."[26] Blocking the telomerase enzyme is one promising strategy in stopping cancer progression.

Researchers are exploiting their new knowledge of tumorigenesis (tumor-forming) pathways to add other novel tools to their cancer-fighting arsenal. Some are developing techniques to induce apoptosis—the process of encouraging cancer cells to commit suicide.[27] Gendicine, whose January 2004 commercial launch in China ranks as the first for any cancer-gene therapy, inserts a gene that triggers apoptosis.[28] Another characteristic of interest is the clumping of cancer cells. By modifying a human protein called galectin-3, a team at the University of California in San Francisco has stopped breast-cancer cells from sticking together in mouse experiments.[29]

Many cancer cells produce a molecule called P-glycoprotein, which acts as a "guard dog" for the cancer, removing threatening materials such as anticancer drugs. This molecule may be the key to the growing resistance of tumors to chemotherapy. "The core of a tumor is an extremely hostile environment for anticancer drugs to work in, with a variety of barriers put up to stop drugs from taking effect," says Dr. Richard Callaghan of Cancer Research UK. Blocking P-glycoprotein may increase the efficacy of a range of anticancer drugs.[30]

Other groups are focusing on the hunters (a type of white blood cell called a T-lymphocyte) rather than the hunted (cancer cells). Researchers at Cancer Research UK modified T-lymphocytes to recognize a surface molecule on bowel-cancer cells and bind to the cells. Once attached, the killer white cells execute a two-pronged attack: they release a molecule that perforates the walls of the cancer cells, and they call in more white blood cells. The researchers also stack the deck against the cancer cells. "We would take maybe 10 million cells, expand them to 10 billion cells, and then return them to the patient," said team member Robert Hawkins. In human trials so far, this approach successfully destroyed the cancer cells.[31]

The definitive test for colon cancer has been a colonoscopy: a 6-foot-long, flexible, fiberoptic tube is passed via the rectum throughout the colon, and the physician can view the lining of the colon directly and biopsy suspicious regions. CT virtual colonoscopy, in which the colon is visualized via CT scan images rather than directly through the scope, is now available in many locations. A recent study compared the results of a group of 1,233 asymptomatic patients (average age 58) who underwent CT virtual colonoscopy, immediately followed by standard colonoscopy. Not only was the virtual exam able to detect polyps (both precancerous and cancerous growths in the colon) with the accuracy of conventional colonoscopy, but it even detected some polyps that the conventional test missed.[32] The small percentage of patients who are found to have suspicious lesions with the virtual test then undergo standard colonoscopy and biopsy. Another study reporting on results from multiple study sites found that virtual colonoscopy missed 25 percent of cancerous tumors that were detected by real colonoscopies.[33] Therefore, the decision as to whether to undergo conventional or virtual colonoscopy will remain difficult until a majority of studies suggests a clear benefit for one versus the other. An advantage of virtual colonoscopy is that one can simultaneously obtain a scan of all the abdomen organs for possible early detection of cancerous tumors or other abnormalities.

Genomic Testing

The evidence is clear that your genetics plays a significant role in the type of cancers you are most likely to develop. The Swedish Twin Study from the Karolinska Institute in Stockholm has followed all the twins born in Sweden since 1886. This largest twin study in the world has tracked 140,000 people and revealed a clear association between genetic risk of certain cancers such as prostate, pancreatic, and colorectal cancers, but not others like cervical and uterine cancers.[34]

You can now determine your genetic predisposition to several types of cancer through genomic testing. Only a few genomic tests are available today, but with "DNA chip" technology, the availability of this information will dramatically increase within the next few years. By knowing to which cancers you are genetically predisposed, you can make appropriate lifestyle modifications to reduce your risk.

For example, some of the most important enzymes involved in protecting your cells from becoming cancerous are called the GST (glutathione-S-transferase) enzymes. One of these enzymes, GSTM1, is chiefly involved in protecting your cells from the constant assault of free radicals coming from environmental toxins such as air pollution or pesticide residues in foods. Yet almost half of the Asian and Caucasian populations and one-third of the Black are born without this protective gene. This is known as GSTM1 null polymorphism. If you couple this genetic predisposition with a lifestyle that increases cancer risk, such as smoking cigarettes, breathing polluted city air, or eating pesticide-contaminated food, the risk of several cancers increases dramatically.

GSTM1 serves many other useful functions, including detoxification of the aromatic hydrocarbons found in cigarette smoke. If you're a smoker who doesn't have the GSTM1 gene, you're at increased risk of developing lung and bladder cancer, so you should be especially cautious about chronic exposure to tobacco smoke. Cigarette smoking has been compared to Russian roulette. If you have the protective GSTM1 gene and smoke, you play with one bullet. If you don't have this gene and smoke, you play with two bullets. Defects in other GST genes have also been associated with cancers of the colon, breast, ovary, nasopharynx, and others.[35] Common genomics profiles include tests for GSTM1 and several other genetic mutations.[36]

A number of mutations protect against or increase risk of cancer development.[37] For example, an important genetic marker that carries powerful prognostic information for risk of breast and ovarian cancer is the BRCA1 gene. A woman who has a defective copy of this gene has a 92 percent chance of developing breast cancer sometime in her life.[38]

Another genetic variant associated with increased cancer risk is defective TP53, a tumor-suppressor gene that causes malignant cells to self-destruct before they have a chance to spread. More than 50 percent of cancers are found in people with defective TP53, making it the most common genetic defect associated with cancer.[39] You can now test your TP53 status easily and at a reasonable cost.

PREVENTION OF CANCER

Let's move now from early detection and determination of genetic risk to cancer prevention. An effective program for avoiding cancer in the first place entails diet and nutrition, lifestyle modification, and chemoprevention.

Diet and Nutrition

Thinking has changed since 1949, when the American Medical Association stated, "There is no scientific evidence that food or other nutritional essentials are of any specific value in the control of cancer." Diet, lifestyle, and nutrition have actually been shown to play an important role in determining cancer risk.[40] For instance, research indicates that populations that consume large quantities of plant-derived foods have a lower incidence of several types of cancer. In 1991, the National Cancer Institute incorporated these findings into the 5 a Day for Better Health Program. It recommended five daily servings[41] of fruits and vegetables as part of a low-fat, high-fiber diet. Despite widespread promotion of this program over the past decade, fewer than one in five American children and fewer than one in four adults eat five portions of produce a day, a statistic that hasn't changed in 10 years.[42] Ray & Terry's Longevity Program regards the 5 a Day program as a good start, but we recommend our 5-to-10-a-Day program, encouraging *five to seven servings of vegetables and zero to three servings of fruit daily*. See Ray & Terry's Food Pyramid on page 106. Emphasis should be on low-glycemic-load (low-starch) vegetables—typically, green vegetables as opposed to higher-carbohydrate root vegetables. Fruit is beneficial, but caution is needed—while it's almost impossible to eat too many low-starch vegetables, you *can* eat too much fruit and consume excessive sugar.

Some people feel that by taking nutritional supplements, they can compensate for a diet insufficient in plant-based foods. While supplements are clearly of proven value, taken alone they do not offer sufficient protection against cancer. A diet rich in naturally occurring nutrients, as found in fruits and vegetables, is needed for optimal cancer prevention.[43] Our dietary recommendations include:

Drink vegetable juice. Start your day right with an 8-to-12-ounce glass of freshly squeezed vegetable juice as part of, or instead of, breakfast: juice some cucumber, broccoli, kale, cabbage, a carrot (for flavor, but not more than one, to avoid excess sugar), and other green vegetables you find in your refrigerator. This can provide almost half of your 5-to-10-a-Day requirements even before you leave your house in the morning. We also reemphasize the importance of eating *organic* produce whenever possible to minimize exposure to pesticides and other carcinogenic chemical residues.

Eat a Mediterranean diet. The Mediterranean diet, which is low in red meat and emphasizes whole grains, fish, and fresh fruits and vegetables, has been associated with reduced cancer risk.[44] Digestive-tract cancers (mouth, esophagus, stomach, and colon) and cancers of the lung and prostate are lower. In a recent study of more than 22,000 Greeks, those who followed the Mediterranean diet had a 24 percent decrease in total incidence of cancer, compared with individuals who did not eat this way.[45] The

Secrets of Soy and the Japanese Diet

Besides possessing a lower incidence of heart disease and menopausal symptoms, the Japanese also appear to have less cancer. This is due in part to their greater consumption of soybean-based foods such as soy milk, tofu, and soybeans.[46] The soy isoflavones genistein and daidzein have cancer-protective properties, particularly against hormonally sensitive cancers, such as prostate cancer in men and breast cancer in women. The typical Japanese diet is also low in meat and high in seafood. Fish contains significant concentrations of the important cancer-fighting fatty acids EPA and DHA. Another cancer preventive typically consumed at almost every Japanese meal is green tea,[47] which contains a powerful anticancer agent known as EGCG (epigallocatechin-3-gallate). Drinking several cups of green tea every day is highly recommended.

BRIDGE TWO

HELPING PATIENTS RECOVER FROM CANCER THERAPY

Cancer treatments of the future will be far less invasive than chemotherapy and radiation of today. If cancer vaccines work, for example, malignancies will be destroyed by the body's normal immune response without repeated chemotherapy or surgery. In the meantime, genomics research is pointing to ways to help cancer patients recover. Scientists from the University Health Network in Toronto have identified a new class of human stem cells that may help patients rebuild their blood systems injured by cancer treatment. "This is an exciting discovery because for the first time, we have found human stem cells that rapidly rebuild a blood system," says Dr. John Dick of the University of Toronto's department of molecular and medical genetics.[48]

Mediterranean diet includes generous amounts of extra virgin olive oil, which protects against several types of cancer—colon, breast, and skin—as well as coronary heart disease.[49] The Mediterranean diet also calls for large portions of fresh tomatoes and tomato sauces. Cooked tomatoes, along with most other red fruits and vegetables, are rich in the bioflavonoid lycopene, which has been associated with a lower risk of prostate cancer.[50]

Avoid the white Satan—sugar. Because cancer cells consume sugar so avidly, the PET scan used by doctors to locate cancer in the body involves giving patients radioactive glucose (or sugar), which is concentrated in areas harboring malignancies and shows up as hot spots on the scan. The 1931 Nobel laureate Otto Warburg demonstrated that cancer cells have a fundamentally different metabolism than normal cells and utilize sugar as their predominant food for growth.[51] You can inhibit cancer formation by avoiding dietary sources of simple sugar as well as foods with a high glycemic load, which are rapidly converted to sugar in the body.

A direct relationship between sugar consumption and pancreatic cancer was seen in women who participated in the Nurses' Health Study. The Women's Health Study, published by researchers at UCLA in 2004, found that a high-glycemic-load diet significantly increased risk of colorectal cancer.[52] When coupled with excess weight and a sedentary lifestyle, women in this study who consumed excess sugar had more than three times the average risk of developing cancer of the pancreas.[53] Avoid "the white Satan" whenever and wherever possible.

Lifestyle Modification

Exercise. Exercise has been associated with a lower incidence of cancer, while a sedentary lifestyle increases cancer risk. We are in favor of the following American Cancer Society recommendations:

- Adults should engage in moderate (or even more vigorous) activity for a minimum of 150 minutes a week. This can be done as three 50-minute sessions, multiple 10-minute sessions, or any combination to total two and a half hours a week.
- Children and adolescents should engage in at least 60 minutes of moderate-to-vigorous physical activity almost every day.[54]

It is often good to perform your exercise in the great outdoors. Sunlight exposure is itself protective against many types of cancer. UVB (ultraviolet B) radiation found in sunlight is associated with reduced risk of cancer of the breast, colon, ovary, prostate, and lymphoma. Lower mortality rates are seen with higher amounts of UVB exposure for cancers of the bladder, esophagus, kidney, lung, pancreas, rectum, and stomach.[55] Sunscreen interferes with absorption of UVB radiation, so we disagree with conventional recommendations that people should use sunscreen whenever they're outside. Unless you're someone who sunburns easily, such as people with very fair complexions and redheads, we recommend you use don't use sunscreen all the time. Instead, apply it primarily when risk of sun damage is high: during midday in summer, at high altitudes, or during any prolonged exposure to intense sunlight, such as when boating or skiing on a bright day.

Better yet, cover up exposed skin with clothing or avoid midday direct sun exposure if possible. Regular exposure of skin to nonburning sunlight is itself cancer-protective. Increased consumption of fish and the omega-3 fatty acids EPA and DHA (along with decreased consumption of omega-6 fatty acids such as from corn oil and safflower oil) can be very protective against melanoma, the most dangerous form of skin cancer, which has been associated with *excessive* sun exposure.[56]

Avoid pesticides. Exposure to agricultural chemicals has been linked to numerous cancers. Agricultural workers are at higher risk of cancers of the stomach (40 percent increased risk), rectum (50 percent), larynx (40 percent), and prostate (40 percent). The increased risk of prostate cancer was specifically related to application of pesticides (70 percent increased risk).[57] Again, we stress the importance of eating organically grown foods whenever possible.

Lose excess body weight. Being overweight or obese is an independent risk factor for several types of cancer, a fact that is not widely known. A survey conducted by the American Cancer Society in 2002 revealed that only 1 percent of the American public realizes that maintaining a healthy weight reduces cancer risk. Yet according to the Centers for Disease Control and Prevention, as of the year 2000, 64 percent of American adults were overweight and about 30 percent were obese.[58] A recent study published in the *New England Journal of Medicine* prospectively followed more than 900,000 American adults to assess the relationship between weight and cancer risk. This study showed that being overweight or obese accounted for 20 percent of cancer deaths in women and 14 percent in men.[59] Obesity was specifically linked to cancers of the liver, pancreas, prostate, and cervix, non-Hodgkin lymphoma, and multiple myeloma.

Avoid tobacco. It has been more than 40 years since the first report of the Surgeon General's Advisory Committee on Smoking and Health was released on January 11, 1964. Thanks to widespread dissemination of information linking smoking to multiple health risks, including cancer, emphysema, and heart disease, the percentage of Americans who smoke has decreased significantly. This downward trend is most prominent among American men: 52 percent smoked in 1965, but only 28 percent smoke currently. Thirty-four percent of American women smoked in 1965, while 22 percent do today. Unfortunately, smoking rates in the United States have remained flat for several years, with little decrease since 1990.

The list of illnesses linked to cigarette smoking reads like the little black book of the Angel of Death. Cigarette smoke increases risk of cancer of all the tissues tobacco smoke touches on its way into the body (lung, mouth, throat, and larynx), on its way out of the body (kidney and bladder), and some places in between (cervix and pancreas). Cardiovascular diseases, including heart attack, sudden cardiac death, and stroke, are increased dramatically in individuals who smoke. Lung problems such as emphysema, asthma, chronic bronchitis, and COPD (chronic obstructive pulmonary disease) are all much higher among smokers. And this is only a partial list!

Smoking cessation is a fundamental part of any cancer-prevention program. There are a number of medications and therapies now available to help smokers kick the habit. If you still smoke, we strongly advise that you implement a smoking-cessation program immediately.

BRIDGE THREE

KNIFELESS BIOPSIES

Nobel Prize–winning chemist Richard E. Smalley, a lymphoma-cancer patient himself, told a congressional subcommittee on June 22, 1999, "Twenty years ago, without even this crude chemotherapy, I would already be dead. But 20 years from now, I am confident we will no longer have to use this blunt tool. By then, nanotechnology will have given us specially engineered drugs . . . that specifically [target] just the mutant cancer cells in the human body, and [leave] everything else blissfully alone . . . I may not live to see it. But, with your help, I am confident it will happen. Cancer—at least the type that I have—will be a thing of the past."[60]

Smalley may not have to wait 20 years. Researchers are already developing a wide array of promising Bridge Three nanomedicine tools for diagnosis and treatment of cancer. The diagnostic tools focus on detecting the earliest stages of cancer instead of waiting for them to form into visible tumors. They range from extremely small (atom-size) quantum dots that emit light when they detect molecules associated with cancer cells[61] on up to the Raman Bioanalyzer developed by the Fred Hutchinson Research Center in Seattle. This room-size machine reads molecular structures by beaming lasers into tissue samples, aided by Intel's famed expertise in detecting microscopic imperfections in chips.[62]

For example, the definitive diagnosis of cancer today usually requires biopsy—surgically removing a tissue sample for examination in the laboratory. This is painful, expensive, and dangerous to the patient, both because of risks inherent in the surgery itself and because anytime a tumor is cut, there is a chance of spreading its cells into the bloodstream. This increases the risk of metastasis, a

Chemoprevention

"Chemoprevention" refers to the use of natural or synthetic substances to reduce the risk of cancer. A number of naturally occurring nutrients are chemoprotective, including vitamins, minerals, herbs, antioxidants, and hormones. While insufficient to prevent cancer by themselves, the following natural chemoprotective agents are a valuable part of a comprehensive cancer-prevention program. Other chemoprotective agents are discussed in chapter 21, "Aggressive Supplementation."

Vitamin C. Linus Pauling, the only scientist ever to receive two unshared Nobel prizes, was so impressed with the ability of vitamin C to both prevent and treat cancer that he coauthored a book on the subject.[67] Vita-

disaster for a patient with a localized tumor. So biophysicists at Cornell and Harvard universities are developing knifeless "optical biopsies" using "multiphoton microscopy" to diagnose cancerous tissues in precise detail.[63] They shine an ultraviolet laser through the intact organ, causing certain compounds (such as amino acids) in the tissues to fluoresce (emit light). The result: high-resolution, three-dimensional pictures of tissues with minimal damage to living cells.

Researchers at Triton Biosciences have taken this a step further by combining both diagnosis and treatment.[64] They are targeting micrometastases—clusters of cancer cells far too small for surgeons to find and remove. Unlike chemotherapy, which kills normal cells as well as malignant ones, their technique focuses only on tumor cells. The scientists attach iron nanoparticles and antibodies into "bioprobes" about 40 nanometers long. These are injected into the body, where antibodies sniff out tumor cells and bind to them. Then the scientists use a powerful magnetic field (similar to an MRI machine) to heat up the iron particles, which immediately kills the cancerous cells.

Nanospectra Biosciences uses a similar approach, injecting tiny gold-silica "nanoshells" into the patient.[65] These circulate through the body and accumulate near tumor cells. Doctors then use an infrared laser to heat the shells and kill the tumor tissue. Both companies hope to have these systems available in doctors' offices by 2006.

Another nanotechnology tactic involves antiangiogenesis. A Scripps Research Institute team[66] attaches molecules that bind to "v3" proteins (these are always present on growing blood vessels and also are good at propelling small particles into cells) to nanoparticles, along with a mutant gene called Raf-1. Once inside, this gene interferes with blood vessel cell growth and destroys the cell. A single treatment in mice erased a large tumor in just six days.

---

min C, particularly when combined with the mineral selenium, can induce cells that are "on the way" to becoming cancerous to turn back from "the dark side" and remain benign.[68] Estimates of optimal doses of vitamin C vary between 1 and 10 grams per day. Our program recommends that most adults take 2 grams (2,000 milligrams) of vitamin C daily for chemoprevention.

Selenium. There are four well-known antioxidant "ACES": three are vitamins (A, C, and E), one is a mineral (selenium). Selenium is the mineral cofactor that activates the powerful antioxidant enzyme GSH-Px (glutathione peroxidase). In the Nutritional Prevention of Cancer trial, selenium supplementation reduced the total incidence of cancer, particularly cancer of the prostate.[69] We recommend a chemopreventive dose of 400 to 600 micrograms of selenium daily.

Coenzyme Q_{10}. Coenzyme Q_{10} is critically involved in energy generation within the mitochondria of the cell. Malignant tissues in the body create increased levels of free radicals. Antioxidant enzymes are under increased stress when attempting to control the free-radical damage found in cancerous tumors. The metabolic needs of these protective enzymes increase dramatically, and coenzyme Q_{10} is vital in helping to provide them with the energy needed to fight cancer.

Breast tumors have dramatically decreased levels of coenzyme Q_{10} as a result of free-radical stress,[70] and breast-cancer patients are typically given large doses of supplemental coenzyme Q_{10} by nutritional physicians. Coenzyme Q_{10} has numerous other protective effects in the body, including lowering blood pressure and protecting the heart. We recommend that healthy adults take from 60 to 200 milligrams of coenzyme Q_{10} a day.

Curcumin. This herb, derived from turmeric (a common spice), has been used in Ayurvedic and Chinese medicine for centuries. Curcumin has powerful anti-inflammatory properties and arrests the growth of cancer cells at the G2 stage of their cell division. Combining curcumin with ECGC (epigallocatechin-3-gallate) from green tea provides synergistic cancer prevention.

Curcumin fights growth of cancer cells in at least a dozen separate ways. It blocks estrogen-mimicking chemicals like pesticides from causing excessive stimulation of hormonally sensitive tissues such as those in the breast and prostate. In this way, it works in harmony with other phytonutrients that have similar actions, such as soy isoflavones and cruciferous vegetables.[71]

Curcumin is used as a natural anti-inflammatory to treat patients with inflammatory conditions such as arthritis. It blocks the COX (cyclooxygenase) enzyme, which creates inflammation in the body. It is well known that colon cancer has a significant inflammatory component and that patients who take COX inhibitors such as aspirin have a reduced incidence of colon cancer. Studies have shown that taking curcumin can also help prevent colon cancer.[72]

We encourage the regular use of the spice turmeric, which contains curcumin, in food preparation, as well as taking 900 milligrams of supplemental curcumin a day for cancer prevention.

Melatonin. Many people know that melatonin can help with sleep. A few people also know that it is a powerful anti-aging hormone. Fewer yet are aware of the fact that melatonin has an important role as a cancer-protective agent. One paper reviewed 27 studies on the use of melatonin as

BRIDGE THREE

STEALTH DELIVERY

A big problem with delivering anticancer treatments is dealing with the patient's immune system, which sometimes mistakes helpful molecules as foreign invaders and attempts to destroy them. One innovative solution: "stealth" packaging to deliver drugs directly to tumors. The Center for Biologic Nanotechnology at the University of Michigan has created spherical molecules called "dendrimers" for this purpose.[73] One type of dendrimer serves to find and tag cancer cells, another diagnoses the type of cancer, while a third can deliver drugs directly to the cancer and destroy it and can show doctors the location of the destroyed tissue by acting as markers for X-rays or MRI images. In animal tests, by using dendrimers, researchers were able to destroy 30 times more cancer cells with the anticancer drug methotrexate, but without the toxic side effects such as nausea and hair loss.

Researchers from the University of Alberta have used a similar tactic: "nanoparticle cluster bombs" that carry designer drugs targeted at lung cancer.[74] The nanoparticles, delivered by inhaler, are programmed to escape immune-system surveillance and leave healthy cells alone.

ALZA Corp. has developed another stealth-delivery scheme, using tiny spheres called "liposomes." To prevent the immune system from attacking them, they are coated with polyethylene glycol.[75]

But perhaps the ultimate stealth-delivery system employs "fullerenes" or "buckyballs." These ultra-tiny molecules (only 1 nanometer in diameter and named after inventor-futurist R. Buckminster Fuller) consist of 60 carbon atoms arranged in a spherical shape.[76] They have a convenient hollow interior where various drugs or a radioactive atom can be "hidden" for stealth delivery to a tumor. Buckyballs could also be utilized to safely deliver hazardous radionucleotides (radioactive metal atoms) into cells for use as contrast agents for MRI scans and X-rays.

Progress using buckyballs has been slow, but a slightly different arrangement of carbon atoms called "nanotubes" looks promising. Nanotubes are long, needlelike tubes that can easily penetrate cells. They are also hollow, so they can be easily used to attach and deliver drugs.[77]

Researchers at Memorial Sloan-Kettering Cancer Center have taken an approach similar to buckyballs in delivering radioactive materials. They've developed a molecular "nanogenerator"—a single radioactive atom contained inside a molecular cage and attached to an antibody that homes in on cancer cells, carrying the nanogenerator to the interior of those cells and destroying them.[78]

BRIDGE THREE

NANOSURGERY

A research team at the University of California, Irvine (UCI), has received a five-year, $2.9 million NIH grant to develop a microscopic probe for detecting and treating precancerous and malignant tumors in humans.[79] Another example of a *Fantastic Voyage* vessel, this nanosize probe would be inserted into a patient and, remotely controlled by a surgeon, guided through the esophagus, stomach, and colon to determine if tumors are growing on the walls of the intestine. If successful, the probe could assist in the early diagnosis of cancers and precancers of the gastrointestinal system. Researchers will test the probe in pigs and human volunteers to determine its effectiveness and safety.

"Currently, gastrointestinal cancers and other diseases are diagnosed only by visual inspection of the intestine's surface," says Dr. Kenneth Chang, director of the H. H. Chao Comprehensive Digestive Disease Center at UCI. "Early-stage cancer screening is difficult because you're looking for microscopic changes. An optical nanoprobe could help pinpoint those changes before they turn into advanced cancer. It also may allow physicians to circumvent traditional biopsies that require removing tissues by providing an optical, or virtual, biopsy sampling of much larger areas."

In the future, surgeons will be able to zap cancerous cells inside tissues untouched by human hands, using "laser nanosurgery" being developed by

a cancer preventive or treatment. The authors concluded that "melatonin could indeed be considered a physiological anticancer substance."[84]

Many studies have centered on the use of melatonin in the prevention and treatment of breast cancer. Melatonin can directly inhibit the growth of breast-cancer cells.[85] It also has important antioxidant and immunostimulatory effects. We recommend taking 0.1 to 3 milligrams of this naturally occurring chemopreventive agent daily, at bedtime.

Folic acid. As discussed in chapter 13, folic acid is intimately involved in numerous methylation reactions. These include synthesis of DNA, turning genes within the cell on or off, and detoxification of chemical toxins. Abnormalities in all of these reactions have been linked to the risk of malignancy. Recently, folic acid deficiency has been implicated as a risk factor for developing cancer. In a review article of 34 studies on the connection between folic acid and cancer, a direct link was found between low folic acid levels and cancers of the colon and breast.[86]

physicist Eric Mazur of Harvard University and his colleagues.[80] In a microscopic version of a James Bond scenario, the laser light is focused extremely tightly, using a microscope, into a space just a few hundred nanometers across. Researchers have even destroyed a single mitochondrion *within* a cell without killing the cell.

But a team of scientists led by Dr. Kevin Prise of Gray Cancer Institute in England has discovered that instead of using deadly radiation to hit every cell in the tumor, targeting just a few cells with a "microbeam" can cause massive destruction to other diseased cells.[81] Cancer cells zapped by the microbeam—a stream of helium ions just 1 micron (millionth of a meter) wide—send out suicide signals to other abnormal cells when they die, telling them, Jim Jones–like, to self-destruct as well. Yet another future surgical tool, developed by scientists at Johns Hopkins University's URobotics Lab, is a nonmetallic robot that can work in conjunction with MRI imaging systems.[82] Using a tiny needle, it can achieve surgical accuracy to within one-tenth of a millimeter—far better than is possible with the human hand. It could be used, for example, to perform precise biopsies of precancerous spots in the lungs or for pinpoint delivery of chemotherapy.

We have previously mentioned Robert Freitas's conceptual designs for tiny nanorobots, the microbivores, which would patrol the bloodstream, seeking out and destroying undesirable bacteria, viruses, and other pathogens. This type of nanobot will be able to download software from the Internet for particular problems, and could be programmed to recognize and destroy cancer cells before they would have a chance to grow and spread.[83]

Folic acid, which is important for both heart health and cancer protection, is one of the few nutrients that works better when taken as a separate supplement than as part of food. A minimum of 800 micrograms per day is recommended, but depending on other factors (such as homocysteine level), this can be raised to 5,000 to 10,000 micrograms or more.

EPA/DHA. The cardiac benefits of the essential fatty acid derivatives EPA (eicosapentaneoic acid) and DHA (docosahexaneoic acid) are well known, but these "fish oils" also play important roles as naturally occurring chemoprotective agents. Like curcumin, fish oils possess an anti-inflammatory action that is the basis of the cancer-protective effect. As discussed in chapter 12, EPA and DHA are naturally occurring COX-2 inhibitors. COX-2 is an enzyme that increases levels of inflammatory chemicals in the body such as PG E2 (prostaglandin E2) and is found in high levels in precancerous and cancerous tissues. Increased levels of both COX-2 and PG E2 have been found in cancers associated with inflammation, such as breast and colon cancer.[87] Consumption of cold-water fish, which is rich in EPA and DHA, as well as EPA/DHA supplementation, is anti-inflammatory, cardioprotective, and chemoprotective.

We recommend a minimum of 1,000–3,000 milligrams of EPA and 700–2,000 milligrams of DHA daily.

Beta-carotene (a special case). Not all vitamins are cancer-protective—at least, not for all people. In particular, the Finnish Alpha-Tocopherol, Beta Carotene (ATBC) Cancer Prevention Study showed that supplementation with beta-carotene actually increased the incidence of lung cancer when taken in supplement form by cigarette smokers.[88] Several other studies have confirmed this association, so we recommend that people at increased risk of lung cancer (such as smokers or workers exposed to asbestos) not take supplemental beta-carotene.[89] Here's the solution for anyone seriously concerned about cancer prevention: if you smoke, stop. Stop today, right now. But if for whatever reason you are unable to quit, don't take supplemental beta-carotene.

Between early-detection tests and preventive and treatment strategies, it is likely that the death rate from cancer will soon begin to plummet and, in the near future, cancer will no longer be the gruesome killer that it is today.

17

TERRY'S PERSONAL PROGRAM

"What is the secret to your longevity?" we asked my grandfather at his 100th birthday party. Dropping his voice, so we all had to gather around to hear him, he replied, "Well, as soon as I was born, I took in a good breath, and then I let it out. And I just kept repeating this . . . again and again."

—Jacob Light, September 21, 1986

If you want to live a long time in excellent health, it doesn't hurt to have good genes. I feel quite fortunate in that at least one of my grandparents, my mother's father, quoted above, enjoyed remarkably good health until, at almost 105, he died of a stroke suddenly during lunch. He was hospitalized only briefly twice in his life, for pneumonia at age 96 and appendicitis at 97. Most of his brothers and sisters lived well into their 90s. Knowing that I have at least some of his genes is a comfort to me, because I know that I also have a number of potentially harmful genes as well. For instance, his wife, my maternal grandmother, died of colon cancer at 57 years old, and I have plenty of her genes too.

I have performed a full panel of genomics tests on myself, and this information has played an important role in the fine-tuning of my health-maintenance program. After I recovered from the initial depression of finding out about my "bad genes" (perhaps feeling a bit like Neo after taking the red pill in the initial *Matrix* film and having my eyes opened to "the real world"), I became even more motivated to follow the principles outlined in this book.

Statistically, I should expect to live another 20 to 30 years. The figure of 20 years is based on actuarial tables from the Social Security Administration, and 30 years is based on questionnaires ("How Long Will You Live?") that ask specific questions about one's lifestyle.[1] But this projected life span doesn't take into account the accelerating progression of scientific discoveries. Today's actuarial tables are based on the past.

In my actuarially projected life span, many Bridge Two therapies should be enormously beneficial to me. As mentioned in chapter 16, "The Prevention and Early Detection of Cancer," sophisticated scanning devices and new therapies should soon be able to both detect and destroy any cancer cells in my body before they have a chance to get out of control. If my heart begins to fail me, as it almost undoubtedly will eventually, I expect to be able to receive new heart tissue cloned from my own cells, thereby avoiding the ethical debate involved with using embryonic tissue. I had a sample of my cellular DNA collected and placed in cryonic (frozen) storage a few years ago, so that I will have the most youthful cells available for this type of contingency. Other options include a heart transplant from a transgenic animal (an animal that has had human genes inserted) or even a shiny new bionic heart.

Now, as Ray did in chapter 10, I'd like to share with you some specifics on what I am doing to optimize my chances of living long enough to live forever.

WEIGHT AND DIET

At 6 feet tall, I weigh 178 pounds. My body composition is 17.9 percent fat, within the acceptable range for men of 16 to 20 percent (although some researchers feel the optimal percentage of body fat for men should be as low as 10 percent).[2] I find strict caloric restriction difficult, but with the use of some new low-calorie, low-carbohydrate foods that Ray and I have developed, I have started to practice Caloric Restriction Without the Restriction™ and hope to drop my percent body fat to below 14 percent, which translates to losing 8 pounds.[3]

I follow the dietary concepts outlined in this book fairly strictly. Although my fasting blood sugar is normal, it is "high normal," and before I went on a lower carbohydrate diet, it was often in the 90s. So I regard myself in the low-carbohydrate group and keep my daily carbohydrate consumption to less than one-sixth of my calories.

I enjoy Asian cuisine and lean toward the modified Japanese diet we recommend. I often eat a breakfast of miso soup, salmon, steamed vegetables, nori seaweed, and green tea. Other mornings I drink a protein shake that Ray and I developed as part of our program.[4] I try to drink vegetable juice several mornings a week, and I have several cups of green tea throughout the morning and at least 10 glasses of alkalinized water per day.

My typical lunch consists of steamed vegetables, tofu or skinless chicken, a small amount of brown rice, and green tea. For supper I have wild ocean salmon two or three nights a week with vegetables. I eat organic turkey and chicken. On occasion I will have a grilled salmon, turkey, or buffalo burger without the bun. I eat no sweets or products containing refined sugar, honey, molasses, fructose, and so on. When I have a desire for something sweet, I eat some wild organic blueberries or another low-glycemic-load fruit. I have a glass of red wine a few evenings a week but avoid beer because of its high glycemic load.

While I travel and eat out frequently, I have found a wide variety of restaurants to be very accommodating to my dietary program. Meals consisting of protein and vegetables are easy to find. I never eat at conventional fast-food restaurants. As you can see, I follow the Ray & Terry nutritional guidelines rather strictly.

GENOMICS TESTING

I have undergone a full panel of genomics tests and have taken measures to tailor my diet and supplement program to counter and minimize the risks presented by my specific polymorphisms. For example, high blood pressure is very common in my family. My genomic testing revealed I possess copies of specific ACE, AGT, and AT1R polymorphisms, which predispose me to high blood pressure. So I am careful to limit my sodium consumption, try to exercise regularly, and keep my weight down. So far, my blood pressure remains in an acceptable range.

INFLAMMATION AND METHYLATION

I have tested my hs-CRP (the screening test for silent inflammation in the body), and it is acceptably low at 1.1. To keep it that way, I take two teaspoons (10 grams) of fish oil and two capsules of curcumin daily. My homocysteine level is 7.0, within our optimal range of less than 7.5, but I do carry the common MTHFR mutation, which predisposes me to abnormal methylation. Therefore, to keep my homocysteine in this optimal range, I take folic acid, B_6, B_{12}, TMG, and other nutrients targeted to enhance methylation.

DETOXIFICATION

My detoxification testing was one bright spot in my otherwise sobering genomics profile. My detoxification capacity seems at least average for survival in a polluted world. However, I try to limit my exposure to environmental toxins as much as I can. I eat organic food whenever possible. I drink double-filtered, alkalinized water at home. I bathe in single-filtered water. I have had my mercury-containing dental fillings removed. I undergo two types of intravenous therapies on a regular basis to assist in detoxification: an intravenous amino acid, vitamin, and mineral formula to remove accumulated heavy metal toxins; and a phospholipid exchange to rejuvenate and detoxify my cell membranes. I have an ionic air filter in my bedroom and many ferns and other houseplants throughout my home. I try to limit my cell-phone use and my exposure to electromagnetic radiation. I use a rebounder (mini-trampoline) to enhance lymphatic detoxificiation.

CORONARY HEART DISEASE AND CANCER

I have had a total-body ultrafast CT scan, including a cardiac scan, and perform periodic blood screening. I get a treadmill test and undergo screening virtual colonoscopy on a regular basis. To maintain my cholesterol within the optimal range, I take policosanol with gugulipid.

HORMONES

I check my hormone levels regularly but don't yet use any hormonal supplementation. I also take an herbal formulation designed to increase levels of free testosterone. I take I-3-C (indole-3-carbinol) to reduce conversion of testosterone into estrogen, as well as a saw-palmetto complex for prostate health and to reduce excess formation of DHT (dihydrotestosterone).

BRAIN

I try to engage myself in both intellectually challenging left-brain as well as artistic right-brain activities. I find that writing provides an excellent outlet for both. I take a number of "smart nutrients" to enhance memory, including vinpocetine, phosphatidylserine, phosphatidylcholine, ginkgo biloba, and acetyl-L-carnitine.

SUPPLEMENTS

I supplement quite aggressively. I take many of my supplements in powdered or liquid form, but it's the equivalent of taking about 64 pills and capsules daily. Of course, that's not counting the 24 small "pillules" my wife, Karen, a licensed acupunturist/traditional Chinese medicine practitioner, has me on as well. I consume 4 of the 10 glasses of water I drink every day just taking my supplements.

Essential nutrients. I take a multiple vitamin/mineral/antioxidant formulation. To provide for essential fatty acids, I take a fish-oil EPA/DHA formula (omega-3) and evening primrose oil (omega-6).

Super-nutrients. For their powerful antioxidant properties and other benefits, I take alpha lipoic acid, coenzyme Q_{10}, grapeseed extract, arginine, and resveratrol. To maintain mental clarity and protect brain function, I take the "smart nutrients" listed above. For detoxification purposes I take N-acetyl-L-cysteine, and to inhibit age-related cross-linking of tissues, I take carnosine.

Specific supplements. Because of my family history of macular degeneration, I take supplemental lutein, zeaxanthin, and bilberry. To protect against arthritic complaints, I take glucosamine and chondroitin. To assist with digestive function, I take a digestive-enzyme formula. To help control stress and aid with sleep, I take inositol and melatonin before bed. I am considering taking a low dose of a statin drug, because so much research suggests a benefit.

My program might seem daunting, but I find it very simple to take a few handfuls of pills each day to ensure that my cells are bathed in these powerful antioxidants and nutrients at all times.

EXERCISE

I try to walk 30 or more minutes outside every day. I enjoy more vigorous activities such as cross-country skiing in winter and in-line skating and bicycling in summer. I engage in weight training at home.

STRESS

I have a number of close friends and try to maintain strong relationships with family, which I feel is the most important aspect of my stress-reduction program. I try to get regular massages to assist with lymphatic detoxification as well as stress reduction. I use an alpha-wave stimulator to increase calming alpha waves in my brain, and I attend a Korean yoga/meditation class twice a week.

THE FUTURE

As I watch many people my age try to figure out ways to use the time they have "on their hands" now that they are retiring, a bigger problem for me is trying to maintain balance in my life by not working on so many projects at once. I still have many goals that I want to accomplish, and so I try to incorporate the advice in this book into my daily life. I feel that as a physician and health educator, I must walk the walk as well as talk the talk. By eating well, exercising regularly, controlling stress, and following Ray & Terry's Longevity Program rather strictly, I feel great almost all of the time. And although we can never be absolutely certain of the future, I am confident that my lifestyle choices will maximize my prospects of living long enough to take full advantage of the radical life-extending therapies that lie just ahead.

18

YOUR BRAIN: THE POWER OF THINKING . . . AND OF IDEAS

"Intelligence is that faculty of mind by which order is perceived in a situation previously considered disordered."

—R. W. Young

The 3-pound mass of nerve cells and supporting tissues known as the brain is arguably the most complex and magnificent object that we know about. Although it is easy to wax poetic about the wonders of the human brain, this organ is also replete with limitations and difficulties. Most of our thinking takes place in the interneuronal connections, which use an electrochemical signaling method that is about a million times slower than contemporary electronic circuits. Of greater relevance to the thesis of this book are the myriad problems to which the brain is subject, from fostering addictive behavior to the potential for gradual or sudden decline with age.

However, we are making dramatic and accelerating gains in our understanding of how the brain works, from the biochemistry of neural components such as synapses (interconnections between brain cells) to the principles of operation of large regions of the brain, such as the cerebellum (seat of muscle-skill formation) and hippocampus (involved in memory formation). The Bridge One ideas in this chapter can optimize your mental functioning today while you dramatically slow down brain aging and other disease processes. Bridge Two and early Bridge Three therapies promise to ultimately stop and reverse most causes of mental aging and decline. Ultimately, Bridge Three technologies will enable us to vastly expand our mental capabilities through an intimate merger with powerful forms of nonbiological intelligence.

All thoughts and emotions, as well as the master control of most bodily functions, from the rate of your heartbeat to the dilation of your pupils, are controlled by electrochemical signals from the brain. The brain has about

100 billion neurons—active brain cells—held in place and supported by over a trillion "glial" cells. A typical neuron is connected to other neurons by an average of 1,000 interconnections, or synapses, so the average human brain has some 100 trillion connections. The glial cells also appear to play a role in influencing the actions of the synapses.

In recent years there has been rapid progress in modeling and even simulating significant regions of the brain. American scientist Lloyd Watts and his colleagues have developed a computer simulation of 15 regions of the auditory cortex, which performs similarly to human auditory perception.[1]

Gathering data from multiple studies, Javier F. Medina, Michael D. Mauk, and their colleagues at the University of Texas Medical School devised a detailed computer simulation of the cerebellum, the region of the brain at the back of the head responsible for controlling movement. It includes all of the principal types of cerebellar cells, with over 10,000 simulated neurons and 300,000 synapses.[2]

Thought processes associated with rational decision making, planning, and the ability to use language are concentrated in a thin layer of neurons on the outer surface of the brain called the cerebral cortex, which is only 1.5 to 4.5 millimeters thick.[3] Constituting 2 percent of total body weight, the brain receives 20 percent of the blood coming out of the heart and consumes 20 percent of all the oxygen in the body. By weight, the brain is 60 percent fat—which helps explain the critical importance of adequate consumption of healthful fats in the diet.

There are 50,000 scientists and engineers working on some aspect of understanding the human brain. The power of our tools for looking inside the brain and our knowledge of the human brain are accelerating. We are, for example, doubling the resolution, price performance, and bandwidth of brain scanning (both invasive and noninvasive) each year. Many of these gains will help us emulate human intelligence in our machines. In this chapter, we will concentrate on practical insights that can help you maintain optimal mental function, as well as take a look at some of the Bridge Two and Bridge Three technologies that will greatly assist in this effort in the years ahead.

MAINTAINING THE BRAIN (BRIDGE ONE)

For almost a century, conventional wisdom held that the brain was relatively immutable and our precious brain cells irreplaceable. It was once thought that you were born with a given number of neurons, then lost a certain number every day as part of the wear and tear of life. Unlike other organs of the body,

BRIDGE TWO

BRAIN-COMPUTER INTERFACES, BIONIC LIMBS, AND THOUGHT-CONTROLLED ROBOTS

Two-way, real-time, noninvasive communication between nerve cells and machines is beginning to work.[4] In a recent experiment, a monkey controlled a robot arm as if it were its own limb, and a human stroke patient was able to operate a computer.[5] The key to progress in this area is interpreting the complex pattern of signals both in and out of the brain. It could also allow quadriplegics, for example, to operate a computer or other devices.

A system called BrainBrowser is being developed at Georgia State University to link to people's brains and allow them to surf the Internet by the power of thought alone.[6]

A small implantable system developed by Advanced Control Research (ACR) in Plymouth, England, can determine a person's intentions and then move a prosthetic hand accordingly.[7] According to Professor Roland Burns, a director at ACR, the system will look at the information contained in that neural signal and through a pattern-recognition system tell what the amputee is thinking, then use that information to command the wrist to rotate or the index finger to move. "Our ultimate goal is to have a fully multifunctional hand," he says.

A research team at Stanford University has developed an implantable chip that provides artificial synapses. The device pumps biological neurotransmitters initiated by electronic commands, providing another way to directly connect electronic computing devices to neurons.[8]

We expect that applications of these technologies will be available to patients with paralysis or sensory impairment within several years. Within 10 years, it should be routine for a physically challenged person to move paralyzed limbs or direct a robot to perform daily chores, or for someone with an artificial hand to regain the dexterity needed to cut paper with scissors and enough sensation to feel the hand of a loved one.

Cutting out the middlemen—nerves and muscles in the arms and legs—and directly connecting the brain to external machines could enhance, not just replace, human performance. This goal is being aggressively pursued by a major research program at the Defense Advanced Research Projects Agency, which wants to slash soldier reaction times, improve decision making, and direct robots and weapon systems with the mind.

which have powerful regenerative abilities—consider how a bone will mend after being broken into pieces, or how the liver recovers from the destructive effects of a night of excessive alcohol consumption—the brain was believed to possess very limited ability to heal itself after injury. But one of the most profound understandings of contemporary neuroscience research is that the brain actually has an enormous capacity for self-healing and regeneration.

For more than a century, scientists believed specific regions of the brain were hardwired for specific tasks. In 1857, French neurosurgeon Paul Broca related injured or surgically affected regions of the brain to certain lost skills, such as fine motor skills and language ability. Although specific areas tend to control particular types of skills, we now understand that these assignments can change after a brain or spinal cord injury that paralyzes part of the body. In a classic study in 1965, Hubel and Wiesel showed that extensive reorganization of the brain could take place after serious damage, such as from a stroke,[9] as other regions of the brain took on the tasks of the damaged sections.

According to this new concept of brain plasticity, different regions of the brain can undertake functions usually performed by other areas. Scientists are already using this approach to treat patients who have suffered damage to a specific region of their brain. For example, a part of the temporal lobe called Broca's area, located just above the top of the ear, is strongly associated with interpretation of language and ability to speak. A patient who has suffered a stroke that damaged Broca's area may have difficulty verbalizing and be unable to speak intelligibly. But other regions of the brain can be taught to take over the speech function so the patient regains the ability to talk.

Moreover, the detailed arrangement of connections and synapses result directly from how much you use a particular region. As brain scanning has developed sufficiently high resolution to see dendritic spine growth (the growth of small protrusions from dendrites that can form new interneuronal connections) and the formation of new synapses, we can watch our brain grow and literally "follow our thoughts." One experiment with monkeys conducted by Michael Merzenich and his colleagues at the University of California, San Francisco, placed the animals' food in a position where they had to manipulate one finger in a very dexterous fashion to get it. Brain scans before and after showed dramatic growth in the interneuronal connections and synapses in a region of the brain responsible for controlling that finger.

Edward Taub at the University of Alabama confirmed this with a study of the region of the cortex responsible for evaluating the tactile input from the fingers. Comparing the brains of right-handed nonmusicians to experienced players of stringed instruments, he found no difference in the brain regions devoted to the fingers of the right hand—but a huge difference in the regions for the left-hand fingers involved in the complex positioning on the strings.

What's more, scientists have learned that even in adulthood, the brain is replete with stem cells, the remnants of embryonic development. Intense research is currently aimed at developing mechanisms to stimulate these brain stem cells, which are able to transform themselves into new neurons and can help repair or, ideally, reverse the damage of neurodegenerative diseases, including that of aging itself. A key to this process appears to be related to nitric oxide, one of the body's major signaling molecules. In one animal experiment, by suppressing nitric oxide production in the brain, scientists were able to increase the number of stem cells that matured into new neurons by 70 percent.[10]

Neural stem cells can become neural precursor cells, which, in turn, mature into neurons themselves and into supporting glial cells (called astrocytes and oligodendrocytes). These prototype neurons further evolve into specific types of neuron cells, from tiny Golgi type II cells to corticospinal neurons several feet in length. However, this differentiation cannot take place unless the neural stem cells move away from their original home in the brain's ventricles or hippocampus—and only about half of them successfully complete this journey. Scientists hope to bypass this faulty neural migration process by injecting neural stem cells directly into target regions and by creating drugs to promote neurogenesis (the process of creating new neurons) to repair brain damage from injury or disease.[11]

An experiment by genetics researchers Fred Gage and Henriette van Praag at the Salk Institute for Biological Studies in San Diego showed that neurogenesis is actually stimulated by your experiences. Moving mice from a sterile, uninteresting cage to a stimulating one (with an exercise wheel) approximately doubled the number of new dividing cells in the hippocampus. You can reap the same benefit by providing your brain with interesting experiences on an ongoing basis. There are numerous ways you can grow new brains cells—take continuing education courses, travel to new places, meet interesting people, and engage in stimulating conversation. The possibilities are endless.

A "COOL" WAY TO SAVE BRAIN TISSUE

Each year about 700,000 Americans suffer strokes, the third leading cause of death after heart disease and cancer. Most strokes occur when a blood clot travels to the brain, usually from the arteries supplying blood to the brain, and blocks blood flow to a specific area, causing it to die from lack of oxygen. But the death of brain cells after a stroke doesn't occur instantly; it takes a number of hours. Scientists already know that cooling brain tissue slows down its metabolic rate and enables it to endure longer periods of time without oxygen. That's why some individuals who have "drowned" in cold water, even for over an hour, then were resuscitated, have almost no loss of brain function.

BRIDGE TWO

NEW TREATMENTS FOR STROKE

A drug currently in human trials, Desmoteplase, has been developed from a protein in vampire bat saliva; it breaks up the brain blood clots associated with stroke but does not prevent a patient's blood from clotting elsewhere in the body, such as at a scratch on the arm.[12] An Israeli team has also built a mesh device like a tea strainer that blocks clots larger than 300 micrometers from traveling up the carotid artery in the neck into the brain.[13]

Though not yet in human trials, a number of approaches to repairing or regenerating nerve tissue are showing promise too. University of Rochester Medical Center researchers are working on developing bundles of human nerve cells that could be injected into the spinal cord to produce new cells. They have genetically manipulated spinal progenitor cells to divide indefinitely, thus solving the problem of generating a sufficient number of these cells for clinical use. Progenitor cells are already committed to becoming a particular type of cell, so one type of progenitor cell might be injected into a Parkinson's patient who needs dopamine-producing neurons and another into a multiple sclerosis patient who needs myelin-producing neurons.[14] Another possible source of replacement nerve cells is the patient's own body. Researchers have had preliminary success in reducing Parkinson's symptoms by extracting a patient's neural stem cells, growing them in the lab, and then reimplanting them.[15]

There are at least eight growth factors that come into play as nerve cells are created and mature. Several biotech companies, including Amgen and ViaCell, are targeting these factors for drug development. By stimulating or inhibiting these factors, it may be possible for neurogenesis in different parts of the brain to be turned on or off. The ability to control neurogenesis may help patients with diseases in which nerve cells die off and even enhance the performance of people with healthy brains.[16] Biogen and Yale University neurobiologist Stephen Strittmatter are using rodent models to determine how to block a protein that inhibits regrowth, thus opening the door for the body to regrow its own nerves in the spinal column after trauma.[17] Using gene therapy, a team at Children's Hospital in Boston stimulated what they called "dramatic regeneration" of nerve cells in rat eyes.[18]

At the International Stroke Conference in early 2004, a team of Japanese researchers presented its findings related to applying a cooling helmet to the heads of stroke patients, which dropped brain temperatures about 7 degrees Fahrenheit, from 98.6 to about 92. By slowing brain metabolism, survival rates were much improved.

A group from UCLA has developed the MERCI Retriever, an apparatus inserted through an artery in the groin that can expand into a corkscrewlike device when it reaches the affected area of the brain. It can then latch on to the clot that caused the stroke and remove it, effectively ending the stroke.[19] Within a couple of decades, these crude devices will be replaced by nanobots that roam the bloodstream and destroy clots as soon as they form. Strokes will then be effectively conquered and become just another story in the annals of medical history, like syphilis and leprosy in most parts of the world today.

ASSESSING BRAIN FUNCTION

Until recently, scanning techniques such as magnetoencephalography (MEG), which measures the brain's magnetic fields, could show only approximately where brain activity was taking place. New tools now allow doctors to get a closer look.

Magnetic resonance imaging

An MRI (magnetic resonance imaging) scan lets physicians easily visualize changes and problems of brain structure such as brain tumors, atrophy (such as found in Alzheimer's disease), hemorrhages, and strokes. The standard MRI, however, provides minimal diagnostic information about brain function like memory, emotion, or thought. Doctors can tell from MRI scans, for example, that Alzheimer's patients have enlarged ventricles (fluid-filled areas in the brain), but this is a nonspecific finding also seen in people with schizophrenia and chronic alcoholism. The hippocampus (short-term memory processing region) in patients with post-traumatic stress disorder appears smaller than in normal patients, but this is also commonly seen in Alzheimer's. And the standard MRI provides no information about brain function.

Functional magnetic resonance imaging

An fMRI (functional magnetic resonance imaging) machine can look inside the brain with far greater spatial and temporal resolution to watch the brain perform tasks in real time. The fMRI works by measuring the magnetic effects of iron in blood hemoglobin, which indicates the ratio of oxygenated to deoxygenated blood in various areas of brain. More active regions show increased delivery of oxygenated blood (from vasodilation).[20]

The fMRI scanner was one of neuroscience's most significant diagnostic advances during the 1990s. It enables scientists to see what regions of the brain are stimulated (light up) when the brain is involved in different activities—for example, in deep thought versus engaged in so-called mindless chatter, during moments of deep religious fervor, or while viewing an inspirational work of art or falling in love.

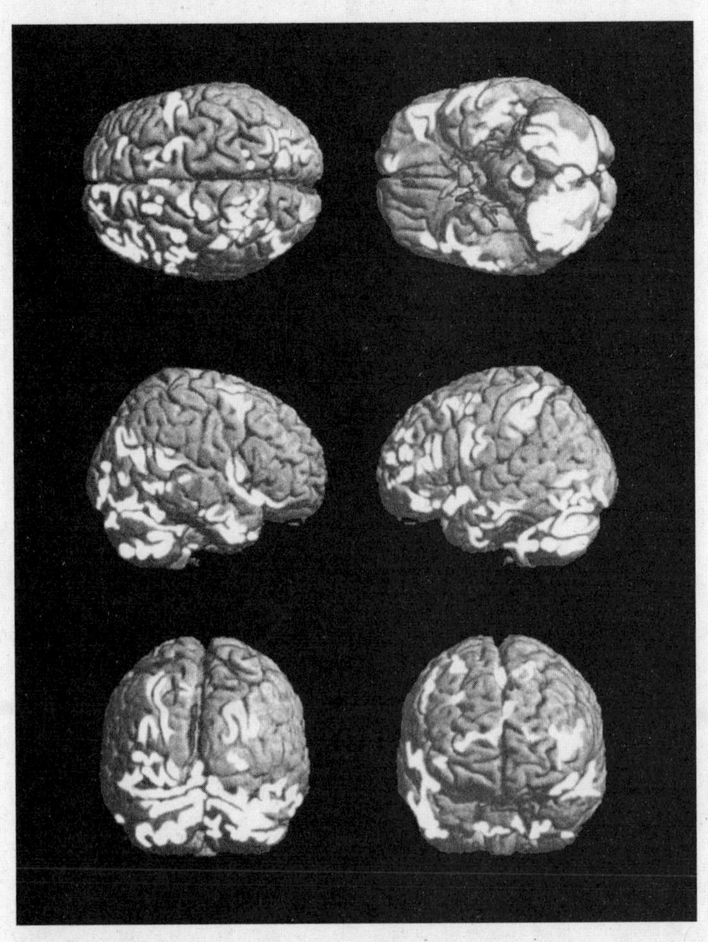

FIGURE 18-1. FMRI IMAGE OF RAY'S BRAIN[21] (COURTESY OF INC. MAGAZINE)

One of the authors, Ray, underwent an fMRI as part of a feature story for *Inc.* magazine in 2002. It showed how different regions of his brain became active as he was asked to change from thinking about routine matters to performing a creative mental task.[22]

Although fMRI is still unable to view the activity of individual neurons and synapses, several new scanning technologies, now in experimental use, are capable of doing this.

Emerging Scanning Technologies

One dramatic new approach is called two-photon laser scanning microscopy (TPLSM).[23] Two-photon laser scanning creates a point of focus in three-dimensional space that allows very high resolution scanning in multiple slices. It uses laser pulses lasting only a millionth of a billionth (10^{-15}) of a second to detect the excitation of single synapses by measuring the intracellular calcium accumulation associated with activation of synaptic receptors.[24] This method can provide extremely high resolution images of individual dendritic spines and synapses in action.

Physicist Eric Mazur and his colleagues at Harvard University have demonstrated the ability to use high-power lasers to perform precise modifications of cells, such as severing an interneuronal connection or destroying a single mitochondrion without affecting other cellular components. "It generates the heat of the sun," says Mazur's colleague Donald Ingber, "but only for quintillionths of a second, and in a very small space."

Another test of cortical functions is called the P300, a large positive wave occurring about 300 milliseconds, or 0.3 seconds, following presentation of a stimulus. In this test, both the size and latency (speed) of the electrical waves in the brain are measured in response to a series of random auditory or visual stimuli. Normal response time is 300 milliseconds plus the patient's age.

Weak signals and slower response time (than expected for the subject's age) are found in mental disorders such as depression, schizophrenia, or addictive personalities; memory and thinking disturbances; Alzheimer's and Parkinson's diseases; and in older people who don't exercise.[25] These changes can be seen many years before symptoms become overt, so this test can be useful in predicting the future risk of serious cognitive problems.

Another approach to identifying brain function is transcranial magnetic stimulation, which involves applying a strong-pulsed magnetic field from outside the skull using magnetic coils precisely positioned over the head. Although not, strictly speaking, a scanning technology, by either stimulating or inducing a virtual lesion (by temporarily disabling small regions of the

BRIDGE TWO

NEURAL IMPLANTS

The age of neural implants is well under way. We have brain implants based on neuromorphic modeling (that is, reverse engineering of the human brain and nervous system) for a rapidly growing list of brain regions.[26] A friend of ours who became deaf as an adult can now engage in telephone conversations again because of his cochlear implant, a device that interfaces directly with the auditory nervous system. He plans to replace it with a new model with a thousand levels of frequency discrimination, which will enable him to hear music once again. (He laments that he's had the same melodies playing in his head for the past 15 years, and he's looking forward to hearing some new tunes.) A future generation of cochlear implants now on the drawing board will provide levels of frequency discrimination that go significantly beyond that of "normal" hearing.[27]

"Rather than treat the brain like soup, adding chemicals that enhance or suppress certain neurotransmitters," says Rick Trosch, an American physician helping to pioneer these therapies, "we're now treating it like circuitry." MIT and Harvard researchers are developing neural implants to replace damaged retinas.[28] There are brain implants for Parkinson's patients that communicate directly with the specific regions of the brain—namely, the ventral posterior nucleus and the subthalamic nucleus—to reverse the most devastating symptoms of this disease.[29] An implant for people with cerebral palsy and multiple sclerosis communicates with another brain region (the ventral lateral thalamus) and has been effective in controlling tremors.[30] A silicon chip that serves as an artificial hippocampus (short-term memory center) is being tested in rats.[31]

A variety of techniques are being developed to connect the wet analog world of biological information processing and digital electronics. Researchers at Germany's Max Planck Institute have developed noninvasive devices that can transmit to and receive from neurons.[32] They demonstrated their neuron transistor by controlling the movements of a living leech from a personal computer. Similar technology has been used to reconnect leech neurons and coax them to perform simple logical and arithmetic problems.

Scientists are now experimenting with quantum dots, ultratiny chips comprised of crystals of photoconductive (reacting to light) semiconductor material. These are coated with peptides that bind to specific locations on neuron cell surfaces. They could allow researchers to use precise wavelengths of light to remotely activate specific neurons for drug delivery, for example, replacing the invasive external electrodes normally used to activate neurons.[33]

brain), skills can be diminished or enhanced.[34] Transcranial magnetic stimulation can also be used to study the relationship between different areas of the brain to specific tasks and has even induced sensations of mystical experiences. It holds promise as a treatment of psychiatric disorders such as depression.[35]

Genomics Testing

Genomics testing can also provide useful information. A number of genetic markers indicate increased risk of neurological and mental disorders, such as Alzheimer's, autism, schizophrenia, and depression. As we mentioned, the E4 type of the Apo E gene can increase risk of Alzheimer's disease as much as 18 times.

An identical twin of an autistic child has a 60 percent chance of having autism—600 times the normal risk. Variants of three genes on chromosome seven have been associated with increased risk of autism: a polymorphism of *WNT2*, a developmental gene associated with language development; abnormal Reelin genes, which code for a protein that helps direct developing neurons to their proper location in the nervous system; and Hoxa 1, which is critical to hindbrain development. Mice lacking this gene develop symptoms that resemble autism.[36]

An identical twin of a schizophrenic has a 45 percent chance of developing the disorder, significantly higher than the 1 percent average in the general population. Numerous genes are associated with increased risk of schizophrenia, but recent interest has focused on the gene that codes for the brain enzyme calcineurin. Genetically modified mice lacking this gene are more likely to develop schizophrenic symptoms.[37]

Clearly, genetics plays a strong role in determining predisposition to these brain diseases, although its contribution is not absolute. The DNA of identical twins is 100 percent the same, but not all identical twins develop the same diseases—genetic markers represent only tendencies. Someday in the near future, physicians may begin to evaluate patients for neurological and mental disorders by using genetic testing to determine to which diseases they are genetically predisposed. They would then follow up with functional brain imaging and P300 tests to see if there are any changes suggesting specific neurological or mental conditions.

By diagnosing a disease before its full expression, it will be possible to outline programs specifically designed to reduce or eliminate the expression of the diseases to which an individual patient is predisposed. We will then have a realistic mechanism for preventing serious neurological or psychiatric diseases. For an aging population, this will be critically important.

Averting an Alzheimer's "Epidemic"

As Bridge Two and Bridge Three therapies become widespread over the next few decades, there will be drastic extensions of the human life span. Currently, 4 million Americans have Alzheimer's disease and 12 million more have a less severe condition known as mild cognitive impairment, while a high percentage of the 80 million or so Americans over age 50 have some degree of memory loss known as AAMI (age-associated memory impairment). At present, more than 50 percent of the population over 85 suffers some degree of senility, with even higher percentages for older age groups. Since life spans of over 120 are likely within the next couple of decades, this could present a major problem. But these Bridge Two and Bridge Three therapies that allow for life extension will also prevent the age-related loss of mental faculties.

The worst thing would be for people to end up like Tithonus, a character out of Greek mythology. Eos, the immortal goddess of dawn, made the mistake of falling in love with Tithonus, a mortal. Not wanting to spend eternity without him, she begged Zeus to grant him immortality. Her wish was granted and Tithonus became immortal but, unlike the other immortals on Mount Olympus, he didn't remain eternally young. His biological clock kept ticking, and with each passing year he grew older and more decrepit. No one seeks extreme longevity if that is how their additional years are to be spent.

HOW TO KEEP YOUR BRAIN HEALTHY

"Use it or Lose it"

New brain cells are formed primarily in two areas: the fluid-filled ventricles of the forebrain and the hippocampus, the region of the brain where new memories are created. As mentioned above, mouse experiments have shown that regular exposure to new experiences results in a dramatic increase in new neuron formation.[38] You can even "bulk up" your brain by doing exercise, just like for your muscles below the neck. A German team recently taught a group of young volunteers how to juggle. These individuals had increases in the size of the gray matter in specific areas over the course of their training, but the gains disappeared when they quit practicing.[39]

Some studies in humans also suggest that maintaining intellectual activity throughout life can protect against cognitive decline in later years. The Victoria Longitudinal Study currently under way in western Canada has shown that middle-aged and older individuals who engage in intellectually stimulating and challenging projects, including everyday activities such as reading the daily newspaper, are less likely to suffer declines in cognitive functioning.[40]

In addition to exercising your left brain, you should also keep your right brain in top form by expressing your creative or artistic urges: study a musical instrument; learn to paint, sculpt, or sing; take up a new hobby. Stay connected to others. Make new friends and continue to maintain long-standing personal relationships. With these simple techniques, you can help avoid the deterioration of intellectual function as well as many of the mood disorders so common among older individuals.

Thinking about Your Health

Another way in which you can keep your brain intellectually engaged is by thinking about your health and continually researching new ways to improve it. Such activity serves a dual purpose, somewhat like Thoreau's adage about wood heat being the one type of fuel that warms you twice—once while you cut it and again as it is burned. Thinking about your health benefits your brain twice: first, as you engage in the intellectual activity of trying to determine your optimal path to wellness (reading this book is a step in that direction); second, as you derive the benefits that such lifestyle choices produce in improving brain health.

Avoid Substance Abuse

Addictive drugs and habits provide powerful stimulation of dopamine receptors in the brain, since dopamine is the neurotransmitter of pleasure. Our brains come hardwired to seek pleasure, and people who can't get it through socially accepted outlets sometimes seek out other routes to get the dopamine they crave. Under conditions of health, dopamine stimulation results from such things as the feeling of accomplishment that accompanies the successful completion of a difficult task, a loving and nurturing relationship, or appreciation of a work of creative genius. Dopamine receptors also fire as a result of victory, such as winning a football game or a battle in a war. An even greater release of dopamine occurs when the positive outcome comes as a surprise, such as winning the lottery or having all 7's show up on a slot machine. This helps explain why the variable reinforcement provided by winning at gambling is more addictive than the predictable reinforcement of earnings at work—more dopamine is released by positive outcomes that are unexpected or come as a surprise.

Genomics research has revealed that certain mutations of dopamine-receptor genes in the brain predispose some people to experience unusually

BRIDGE THREE

BREAKING THROUGH TO THE BRAIN

By the late 2020s, nanobots should be reaching the brain by traveling through the capillaries. One issue that these nanobots will need to contend with is the blood brain barrier (BBB), a biological system that protects the brain from foreign substances. Recent studies have shown that the BBB is a complex system, with biological gateways complete with keys and passwords that allow entry into the brain.

A number of strategies have been proposed for nanobots to defeat the BBB. One possibility: projecting a robotic arm through the BBB and into the extracellular fluid that lines the neural cells. The nanobot itself would remain in the capillary, so it can be large enough to have sufficient computational and navigational resources. Since almost all neurons lie within two or three cell widths of a capillary, the arm would need to reach up to about 50 micrometers. Analyses conducted by Rob Freitas and others show that it is feasible to restrict the width of such a manipulator to under 20 nanometers, the size of the BBB.

Another strategy suggested by Freitas would be for the nanobot to literally barge through the BBB by breaking a hole in it, exiting the blood vessel, and then repairing the damage. Since the nanobot could be constructed from diamondoid—a flexible but very strong material built from carbon—it would be far stronger than biological tissues.

As we have seen, there are already a variety of means for electronic devices

large amounts of pleasure from addictive behavior. Mutations of the dopamine-receptor D2 gene (DRD2) have been associated with alcoholism and other addictions, including cocaine, heroin, smoking, and even overeating. People with mutated DRD2 experience less dopamine release in their brains from a given stimulus, and they apparently turn to substances of abuse to help raise dopamine levels to normal.[41]

Yet giving in to these addictions by taking drugs of abuse or engaging in pathological gambling or sexual addiction is doomed to failure. The reasons are biochemical, not moral or ethical (although individuals who engage in addictive behaviors to excess often find themselves in moral or ethical dilemmas). Here's why: Say someone drinks a small amount of alcohol. This increases the release of pleasurable neurotransmitters such as dopamine in the brain. Excessive consumption, however, depletes the brain's supply of dopamine as well as several other neurotransmitters associated with feelings

to interact with biological neurons. Contemporary neural implants, such as the implant for Parkinson's disease described elsewhere in this chapter, provide for such direct communication. Scientists have already demonstrated that biological neurons receiving signals from electronic devices act just as if they had received these signals from other biological neurons.

These nanobots effectively provide a neural implant, but circa-2029 implants will have dramatic advantages over today's relatively crude devices. Nanobot-based implants could be introduced simply by swallowing or inhaling them. Also, today's implants can affect only one or a very small number of locations in the brain, while with nanobot-based technology you could have billions of them distributed throughout the brain. Today's implants can already download upgraded software from outside the patient. The circa-2029 nanobots will be able to communicate with one another on a wireless local area network, with the Internet, and with your biological neurons.

With this technology, your brain will evolve into a hybrid of both biological and nonbiological intelligence. Today you are limited to a mere 100 trillion interneuronal biological connections that are also extremely slow, computing at only about 200 transactions per second. With nanobot-based neural implants, you'll amplify your thinking by adding trillions of new connections that operate millions of times faster than your biological connections. This will enable you to profoundly expand your pattern recognition, cognitive, and emotional capacities as well as provide intimate connection to powerful new forms of nonbiological thinking. You will also have the means for direct high-bandwidth communication to the Internet and to other people directly from your brain.

---

of well-being and pleasure. In an attempt to feel better, alcoholics drink more and more, further depleting levels of these neurotransmitters and creating an uncontrolled downward spiral. The same general mechanism also applies to other addictive behaviors and drugs.

LIFESTYLE CHOICES FOR BRAIN HEALTH

Brain health is also linked to proper diet, a physically active lifestyle, stress management, adequate sleep, and targeted nutritional supplementation.[42] Let's take a closer look at nutrients that maintain or improve mental function.

In 2003, the journal *Nutrition* examined a number of nonprescription nutritional compounds used as brain nutrients. It reviewed only double-blind placebo-controlled studies and found positive effects for most of the nutrients listed below.[43]

Vinpocetine

Vinpocetine, a nutrient derived from the periwinkle plant, increases blood flow to the brain and ATP production (for energy). Vinpocetine has memory-enhancing effects for people with normal memory as well as those with memory impairment. When combined with ginkgo biloba (discussed below), vinpocetine improves the speed of how quickly short-term memories can be stored in the brains of people with normal memory.[44] In patients with age-associated memory impairment, memory-enhancing effects were significant.[45] Suggested dose: 10 milligrams twice a day.

Phosphatidylserine[46]

Phosphatidylserine is a naturally occurring phospholipid found in the cell membranes of all tissues in the body, but in particularly high concentrations in the brain. Supplemental phosphatidylserine slows down and, in some cases, reverses memory loss in patients diagnosed with age-associated memory decline. It also helps lower levels of the stress hormone cortisol, the main hormone of aging. Suggested dose: 100 milligrams twice a day for one month, decreasing to 100 milligrams daily thereafter.

Acetyl-L-carnitine

ALC (acetyl-L-carnitine) is a naturally occurring substance that helps transport fatty acids to the inside of the mitochondrion, where they can be burned for fuel. A similar molecule, carnitine, had long been used to treat certain cardiac and muscle diseases, and scientists observed that some patients also experienced improved mood and concentration. Further research demonstrated that ALC worked even better than carnitine on the brain. ALC also protects normal brains from the effects of aging by slowing down oxidative damage and inflammation of brain tissues.[47] Suggested dose: 500 to 1,000 milligrams twice a day.

Ginkgo biloba

The leaves of the ginkgo biloba tree—the most ancient tree known, dating back 300 million years—have been used by practitioners of traditional Chinese medicine for thousands of years. Ginkgo biloba increases cerebral circulation and improves general brain function, and numerous studies have shown it reduces short-term memory loss in the elderly.[48] Sales of ginkgo biloba are over $1 billion a year in the United States. In Europe, where it's a prescription drug, physicians prescribe ginkgo more commonly than any other pharmaceutical agent for memory loss. Suggested dose: 80 to 120 milligrams twice a day.

Pregneneolone

Recently, pregneneolone has aroused research interest as a "nerve hormone." Animal studies have shown it can improve communication between neurons. Mice with learning and memory deficits had lower levels of pregneneolone than control animals without these problems, and their ability to retain what they had learned improved after taking pregneneolone.

Pregneneolone is made in the supporting cells of the human brain, the glia, where it seems able to calm brain activity by stimulating GABA (gamma aminobutyric acid) receptors. Many antianxiety drugs bind to these same receptors to reduce anxiety. Suggested dose: 5 to 25 milligrams once a day. Men must monitor PSA (prostate-specific antigen) levels when taking pregneneolone, just as when taking testosterone or DHEA.

EPA/DHA

The brain is 60 percent fat; one-third of that is polyunsaturated fatty acids.[49] The polyunsaturated fatty acid DHA, one of the most flexible of the fatty acids, is found in high levels in brain tissue. Supplemental DHA or its dietary precursor, EPA, found largely in fish, helps keep cell membranes in the brain flexible. When consumption of omega-3 fats is inadequate, the body will substitute cholesterol, omega-6 fatty acids, or even dietary trans fats. When that happens, cell membranes become more rigid, which leads to deterioration in the transmission of nerve signals between cells in the brain and altered brain function.[50]

The ratio of omega-6 to omega-3 fats consumed has increased drastically over the past century. Currently, consumption of omega-6 fats, such as found in corn oil and safflower oil, is 20 times as high as that of omega-3 fats.[51] This is a far cry from the one-to-one ratio recommended by many experts.[52]

Numerous studies have shown that supplementation with EPA/DHA can help treat neuropsychiatric symptoms such as depression, aggression, and anxiety,[53] so a key aspect of our program is daily consumption of EPA/DHA. For preventive health purposes and protection of brain tissues in particular, we recommend 1,000–3,000 mg of EPA and 700–2,000 mg of DHA daily, but for treatments of neuropsychiatric syndromes, larger amounts may be necessary.

Phosphatidylcholine

Another key component of the cell membrane is phosphatidylcholine (PtC), another flexible fat. It's found in high concentrations in the cell membranes of the body, particularly the brain, in young people; its concentration decreases with age. PtC improves memory of mice with dementia[54] and can aid memory and learning in normal human subjects.[55]

Phosphatidylcholine also helps the liver with detoxification and has general benefits as an anti-aging supplement. Most PtC is sold in the form of lecithin, which is only 10 to 20 percent PtC, and some PtC supplements are only 25 to 35 percent pure. Since oral PtC is only 20 percent absorbed, it is best to take a 100 percent pure form. Suggested dose: 900 milligrams two to four times a day.

THE POWER OF IDEAS

The authors are convinced of this basic philosophy. No matter what quandaries we face— business problems, health issues, and relationship difficulties, as well as the great scientific, social, and cultural challenges of our time—there is an idea that can enable us to prevail. Furthermore, we can find that idea. And when we find it, we should implement it. As we relayed in chapters 10 and 17, both of us have had health issues in our lives. Rather than simply accept compromised public-health recommendations or limited medical guidance, we have each sought to understand the true nature of our bodies and to apply all of the ideas that we could gather to overcome these problems. We are both satisfied that our aggressive pursuit of the right ideas has enabled us to overcome these challenges.

"Smart" Drugs

Many research and development companies are attempting to develop a Viagra for the brain.[56] As with Viagra, people are using a number of medications approved for specific diseases of the brain or mental conditions, even when they don't have these diseases.

Modafinil (Provigil) is FDA-approved for the treatment of narcolepsy, but many college students have found it can help them focus while studying for exams. Others use it as a cognitive enhancer; they find they have more energy, need less sleep, and can work and play harder when they take it.

Donazepril (Aricept) is FDA-approved for slowing down memory loss in Alzheimer's patients. Studies have shown this medication helps normal individuals score better on memory tests and maintain better focus.[57]

Often the "good idea" that overcomes a problem is not a single idea at all but rather a set of ideas, each of which chips away at the challenge until there is no problem left. For example, there is no single silver bullet to avoid heart disease (at least not yet). We have half a dozen ways of improving cholesterol and related lipid levels, other means of reducing homocysteine, yet other ways of diminishing harmful inflammation, and so on. We don't go to battle with only a single weapon. We need to harness all the tools available, in war and in health.

Also note that the good ideas you apply to solve a problem don't have to be your own ideas. We have hundreds of thousands of scientists around the world who are advancing our knowledge in the area of health, and the real challenge is finding which of their ideas apply to your individual issues.

The many suggestions in this book will be a good start for anyone seeking to overcome predispositions to health problems, and to achieve and maintain optimal health. But this book is not a simple cookbook of recipes that you can follow without thinking. If there is a single message that we would underscore as most important, it is that you make the commitment to apply your own mental powers to improving your own well-being. As stated above, this has the dual benefit of exercising your brain, which in and of itself will help to keep your brain healthy, and applies your thinking to a worthwhile and achievable goal.

19

HORMONES OF AGING, HORMONES OF YOUTH

"You're never too old to become younger."

—Mae West

It has been well known since antiquity that hormone levels decline with age. The ancient Greeks, Egyptians, and Indians (from India) attempted to restore their declining sexual performance and increase energy by ingesting extracts made from animal testes. Today we understand that declining hormones also account for an increase in many age-related degenerative conditions, such as cardiovascular disease, osteoporosis, and cancer. Other age-associated symptoms, such as loss of muscle mass, increased body fat, and cognitive defects, have also been linked to hormonal fluctuations. Many of these undesirable changes now seem to be caused not only by the declines in the absolute amounts of these hormones in the body but also by shifts in balance between the levels of the various hormones.

Hormones found naturally in the body can be divided into two main types: *anabolic* and *catabolic.*

Anabolic hormones cause tissues to grow or build—for example, they're behind larger muscles and stronger bones. You may have heard of anabolic steroids, which are synthetic chemicals used by bodybuilders to develop exaggerated musculature (and that get the occasional amateur banned from competing in the Olympics). But sex hormones such as testosterone, growth hormones, and DHEA (dehydroepiandrosterone) are naturally occurring anabolic steroid hormones, and their levels almost always decline after the reproductive years.

Catabolic hormones, on the other hand, stimulate tissues to be broken down. The main catabolic hormone is the stress hormone cortisol, secreted by the adrenals. To some extent, insulin (produced by the pancreas) and estrogen (in men) also behave like catabolic hormones. Unlike anabolic hormones, cortisol and insulin levels in both sexes, and estrogen levels in men, often do not decrease with age; sometimes they decrease a little, but they can also remain the same or, in some cases (like estrogen in men), increase instead.[1] This can lead to an imbalance between hormone ratios, which appears to play an important role in the aging process.

Insulin is the hormone secreted by the pancreas in response to elevations in blood sugar. Insulin isn't always a catabolic hormone. In lower amounts it is anabolic and stimulates tissues to grow. In larger amounts, though, as when a lot of sugary or high-glycemic carbohydrates are eaten, it mainly causes growth of only one particular tissue type—adipose tissue, or fat. With age, the cells of the body become less sensitive to the effects of insulin, and insulin levels rise. That's a major reason so many people gain excess weight or increase body fat as they age. With age, the balance of hormones shifts from anabolic to catabolic.

The anabolic hormones—testosterone, estrogen (in women), progesterone, growth hormone, melatonin, and DHEA—build tissues and can keep you youthful, so we will refer to them as the Hormones of Youth. Conversely, we call cortisol, insulin, and estrogen (in men) the Hormones of Aging.

HORMONES OF AGING

What steps can you take to maintain a more youthful balance between the two groups of hormones? Let's begin by discussing ways to slow down or even reverse the gradual increase in dominance of catabolic hormones.

Cortisol

The body responds to acute stress by calling for a quick burst of cortisol from the adrenal glands, which increases cardiovascular and lung function while suppressing the immune, digestive, and reproductive systems.[2] A quick spurt of cortisol quickens your heart rate so you can run faster, dilates your pupils so you can see better, and increases your blood sugar so you can think more clearly. But chronic exposure of tissues to elevated cortisol levels accelerates aging and disease processes, breaks down muscle (sarcopenia) and bone (osteoporosis), causes sodium retention and high blood pressure, increases blood sugar, and depresses the immune system.

Patients with Cushing's disease (linked to excessive cortisol levels) or who need to take synthetic forms of cortisol for prolonged periods develop substantial muscle wasting and loss of bone. In the novel *Dune* by Frank Herbert, we read, "Fear is the mind destroyer." Indeed, fear increases cortisol levels, which, as research has shown, can destroy the mind. For example, in his work with Alzheimer's patients in Tucson, Dr. D. S. Khalsa has shown that chronic stress can damage memory.[3]

As seen in Figure 19-1, all steroid hormones (including cortisol) are synthesized from cholesterol. Cholesterol is first converted into pregnenolone, which can then form either progesterone or DHEA, the "mother hormone" of testosterone and estrogen. When stress is chronic or excessive, more cortisol is produced at the expense of DHEA, testosterone, and estrogen. Normal aging is associated with this same shift toward more cortisol production, with a simultaneous decrease in production of the other hormones.[4]

One of the easiest ways to determine how well your Hormones of Youth are faring against your Hormones of Aging is to calculate the ratio of your levels of DHEA (an anabolic Hormone of Youth) to cortisol (a catabolic Hormone of Aging). You can find out this information, as well as the health

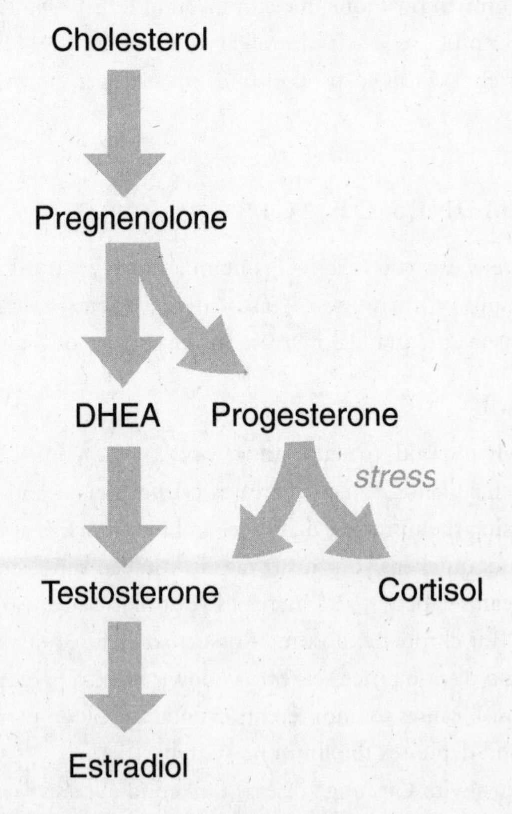

FIGURE 19-1. STEROID HORMONES

of your adrenal glands, by performing an adrenal stress index test. You can get a test kit from a complementary physician or health practitioner, and you don't even need to get your blood drawn. You perform this test at home by collecting saliva samples four times during the day—upon awakening and at lunchtime, suppertime, and bedtime. A normal response shows higher levels of cortisol in the morning, with a gradual decline throughout the day. Under conditions of chronic stress, this daily variation is often diminished, so that the curve appears flat. See Figure 19-2.

In an adrenal stress index test, the ratio of DHEA to cortisol is also measured.[5] In younger people, this ratio is usually high, while in elderly individuals it is often lower.

Specific suggestions to improve your ratio include supplementing with DHEA (dosages discussed below);[6] taking herbs such as natural licorice,[7] used in Chinese herbal medicine, or the Ayurvedic herb ashwaganda;[8] and implementing lifestyle choices to lower cortisol, including a low-glycemic-load diet, stress reduction, regular exercise, and adequate sleep.

Insulin

If a race were held between insulin and cortisol to see which could tear up the body faster, we would place our money on insulin. Barry Sears in *The Anti-Aging Zone* calls elevated insulin "your passport to accelerated aging."[9]

Excess insulin increases body fat, cortisol levels, and insulin resistance; ac-

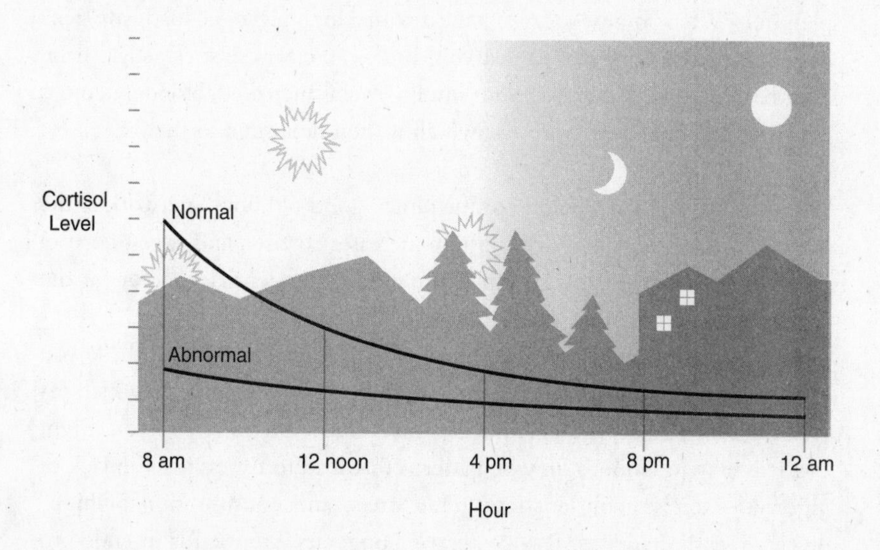

FIGURE 19-2. THE ADRENAL STRESS TEST

BRIDGE TWO

KNOWLEDGE OF THE NEMATODES

C. elegans is a roundworm commonly used by research scientists. This nematode gained fame by becoming the first multicellular organism to have its genome mapped in its entirety in 1999. In 2003, *C. elegans* entered the spotlight again by living an astounding 188 days, which in worm years is equivalent to a human life span of 500 years.

In earlier experiments, life spans of almost 150 days were achieved by manipulating roundworm genes that coded for IGF-1, a protein closely related to human growth hormone. Just one problem with this therapy: while the worms enjoyed dramatically increased longevity, they appeared very lethargic throughout the course of their long lives. Further study by Cynthia Kenton of the University of California, San Francisco, involved manipulating insulin levels and removal of some gonadal tissue. This resulted in even further longevity for the worms—without the lethargy. Since *C. elegans* and human beings share the majority of their genes, this research may lead to therapies for extending human life spans through manipulation of insulin and other hormone levels, but without the removal of any organ tissues.[10]

celerates atherosclerosis and raises the risk of heart disease; and interferes with functions of other Hormones of Youth. Insulin is secreted when calories are plentiful. When you eat sugar or a meal with a high glycemic load, you generate high levels of insulin so that your body can clear the sticky sugar from your bloodstream. Whenever your insulin level is increased, blood glucose is converted more quickly into fat, which is then deposited in the fat cells of the body.[11]

Cortisol and insulin belong to the same "good old boys" network at the hormonal country club. Excess insulin increases cortisol, and excess cortisol increases insulin.[12] Excess insulin is also an independent risk factor for the development of coronary heart disease.[13]

Insulin slows down the effectiveness of your Hormones of Youth, so you age more quickly. For this reason, sugary or high-glycemic foods, which increase insulin, make you age quicker than almost anything else you might eat. It is easy to fall into lifestyle patterns that lead to excess insulin: lack of vigorous exercise, constant low-grade stress, and consumption of high-glycemic carbohydrates. Ray & Terry's Longevity Program is intended to sharply decrease elevated insulin levels.

HORMONES OF YOUTH

Lowering the levels of catabolic hormones will improve their ratio to the anabolic Hormones of Youth. But historically, the most common method of favorably affecting the ratio has been by direct hormone replacement. Usually, the term "hormone replacement therapy" (HRT) refers to use of the sex hormones estrogen, progesterone, and testosterone, which we will discuss in the next chapter. Here, we'll discuss the less commonly used but equally important Hormones of Youth: DHEA, growth hormone, and melatonin.

DHEA

DHEA, or dehydroepiandrosterone, is the most abundant steroid hormone produced in the human body. In the past, DHEA was thought to be simply a precursor to other hormones, with no particular physiological function of its own. But the late William Regelson, M.D., a well-known anti-aging physician-researcher, has referred to DHEA as the "superstar of the super hormones."[14] DHEA levels peak around 25 years of age, then steadily decline by 50 percent in your 40s and fall further to around 5 percent of youthful levels by age 85. Does this mean DHEA could play a role in longevity?[15] Some animal experiments have indeed confirmed that taking DHEA can slow down aging and increase longevity.[16]

There is evidence that men with higher DHEA levels are less likely to die of cardiovascular disease.[17] DHEA has anti-inflammatory properties and can lower the levels of IL-6 (interleukin-6) and TNF-α (tumor necrosis factor alpha), which are powerful and often dangerous causes of inflammation in the body.[18] In Dr. Regelson's cancer research, DHEA appeared to block the ability of cells to divide uncontrollably, a hallmark of cancer cells.[19]

Benefits of DHEA

Decreases risk of heart disease	Prevents bone loss
Fights stress	Enhances libido
Boosts immune function	Improves insulin sensitivity and glucose tolerance
Reduces depression	Increases lean body mass
Improves memory	
Relieves symptoms of menopause	

DHEA can "tame" cortisol. When your body is under stress, the extra cortisol produced depresses your immune system, which increases the likelihood of illness and hastens the pace of aging. Studies have also linked a suppressed immune system with imbalances of cortisol and DHEA. By taking supplemental DHEA, people have improvement in the immune-system depression that comes with taking cortisone or other steroids.[20] Since DHEA is a precursor to testosterone, it can have a positive effect on libido, particularly in women.[21] DHEA also improves the body's ability to transform food into energy and burn excess fat.[22]

Before implementing DHEA supplementation, have your DHEA-S (DHEA-sulfate) level checked, and then recheck it six or eight weeks after you begin supplementation to make sure the desired level has been achieved. Try to achieve a serum DHEA-S level of about 300 for men and 250 for women. Men should start with 15 to 25 milligrams of DHEA per day, and women with 5 to10 milligrams per day, and increase as needed to achieve these levels.

Caution: Because DHEA is an androgenic hormone with predominantly male effects, it can be easily converted into testosterone. DHEA supplementation, then, can raise the PSA (prostate-specific antigen) level in men, an important marker for prostate cancer. Men must check their PSA levels before beginning to take DHEA and on a regular basis (every 6 to 12 months) while taking it. If PSA levels rise, discontinue taking DHEA and consult a physician.

Human Growth Hormone

Enthusiasm over growth hormone (GH) as an anti-aging therapy began in 1990 with publication of an article by Daniel Rudman, a researcher at the Medical College of Wisconsin. He reported the results of a placebo-controlled study of 21 healthy men aged 61 to 81, comparing the effects of GH on several physical parameters.[23] Among the benefits of growth-hormone therapy he found: increased muscle mass, decreased body fat, stronger bones, improved cholesterol profile, and improved insulin sensitivity.

Numerous other researchers have reproduced similar results. A recent online search of the National Library of Medicine revealed more than 48,000 articles related to growth hormone.[24]

There seems little doubt that even in the absence of dietary or exercise changes, growth-hormone therapy can reduce body fat and increase muscle mass.[25] GH can provide profoundly beneficial effects on cardiovascular function, lipid profiles,[26] and blood pressure.[27] The decline in insulin sensitivity that typically occurs with age is reversed by GH therapy in patients who have been treated continuously for up to seven years.[28]

Injectible GH has been the treatment of choice for children with growth hormone deficiency for many years. But GH deficiency in adults, currently called adult growth hormone deficiency (AGHD), was recently recognized as a separate syndrome and its treatment with GH has received FDA approval.[29]

While there are clearly benefits to patients with growth hormone deficiency, there are some downsides to GH replacement therapy. It is relatively expensive, costing from $2,000 to $8,000 a year, depending on dosage needed, and is only sporadically covered by health insurance. It also requires daily injections, and its use in healthy adults remains highly controversial. In a 2002 study funded by the National Institutes of Health, 121 men and women were followed after using injectible GH therapy with or without synthetic HRT between 1992 and 1998. The increase in muscle mass and loss of fat mass originally reported by Rudman was confirmed, but a large number of subjects also experienced significant side effects: 24 percent of the men developed glucose intolerance or diabetes, 32 percent experienced carpal tunnel syndrome, and 41 percent had joint aches. Thirty-nine percent of the women in the study developed edema. The authors of this study concluded that "because adverse effects were frequent (importantly, diabetes and glucose intolerance), GH interventions in the elderly should be confined to controlled studies."[30] There is currently debate about whether GH therapy can increase a patient's chance of developing cancer. But Shim and Cohen at UCLA state, "Cancer risk on GH therapy probably does not increase substantially above that of the normal population,"[31] and other studies do not show increased rates of cancer of the colon or prostate.[32]

Still, because injectible GH has been used as an anti-aging strategy in healthy adults for only a few years, one should keep these potential side effects in mind. It is important to realize that long-term studies on the safety of GH injections in healthy adults have yet to be performed. Luckily, there are a number of lifestyle choices you can make to obtain the benefits of increased GH levels without resorting to injections, however.

- Sugar and high-glycemic carbohydrates decrease GH release from the pituitary, while a protein-dominant diet increases it. Following our recommended low-sugar, low-glycemic-load diet will raise GH levels.

- Deep sleep and anaerobic exercise such as weight lifting are the two major ways to stimulate secretion of GH in healthy individuals. Adults who continue to exercise throughout life maintain their lean body mass and have higher levels of GH.[33]

- Consuming certain amino acids such as arginine, ornithine, glycine, and glutamine causes the pituitary to release more GH. Supplements containing varying amounts of these amino acids, called *secretagogues* because they stimulate the pituitary to secrete the growth hormone it has in reserve, are widely available.
- DHEA supplementation is also an inexpensive way to raise GH levels.[34]

For most people who would like to experience the anti-aging benefits of higher growth hormone levels, we advise making these lifestyle changes. Until more research is available, we recommend reserving GH injections only for adults who have documented AGHD based on careful evaluation and testing by an experienced physician.

Before implementing any of these measures to raise GH levels, have the level of IGF-1 (insulin-like growth factor-1) in your blood checked. IGF-1 is more useful than the GH level itself, because it gives you an average value for your GH level, which fluctuates constantly in the bloodstream. Your physician can tell what an optimal level for you would be, based on your age and sex.

BRIDGE THREE

OUR OBSOLETE ORGANS

In chapter 7, "You Are What You Digest," we discussed future technologies that will maintain optimal levels of all nutrients and other substances in our bloodstream at all times. Once these are developed, we won't need the organs that produce chemicals, hormones, and enzymes. In the human body version 2.0, hormones and related substances (to the extent we still need them) will be delivered via nanobots, and controlled by intelligent biofeedback systems to maintain and balance required levels. Ultimately, it will be feasible to eliminate most of our biological organs. This redesign process will not be accomplished in a single design cycle. Each organ and each idea will have its own progression, intermediate designs, and stages of implementation. Nonetheless, we are clearly headed toward a fundamental and radical redesign of the extremely inefficient, unreliable, and limited functionality of human body version 1.0.

Melatonin

Although sleep is essential to good health, it is estimated that at least 50 percent of Americans over the age of 65 suffer from some form of sleep disturbance. Sleep deprivation for a prolonged period of time can lead to stress and a depression of immune function.

Melatonin is a light-sensitive hormone secreted rhythmically by the pineal gland, located deep within the brain. Humans have an inner clock that keeps our bodies on a 24-hour cycle. The level of melatonin in the bloodstream is low through the daylight hours and begins to rise in the early evening before the onset of sleep. It reaches its peak about midnight or soon after and then declines, whether or not sleep ensues. Melatonin secretion is affected by the night-day cycle. The duration of secretion depends on the length of darkness, so the total amount of melatonin secreted is greater during winter than in summer.

Melatonin secretion peaks at age 7, then declines precipitously during adolescence. At about age 45, the pineal gland begins to shrink and loses the cells that produce melatonin. Hormone production becomes erratic. By age 60, you produce only 50 percent of what you made during your 20s, which explains why so many older people suffer sleep problems. Recent double-blind placebo-controlled studies have shown that melatonin supplementation can improve sleep patterns in people over 55.[35]

Melatonin is a powerful antioxidant and may have value in the treatment of several cancers, most notably breast cancer.[36] Without an ample level of melatonin, a vicious cycle develops:

1. You lose the ability to produce more melatonin, and you begin to age more rapidly.

2. With aging comes an even further decrease in melatonin output.

3. The drop in melatonin alerts the other glands and organ systems that the time has come to wind down. In women, the ovaries stop functioning, the estrogen level drops, and menopause comes on. In men, although sperm are still produced, testosterone declines.

4. In both sexes, the immune system begins to decline, leaving you ever more vulnerable to diseases ranging from infections to cancer and autoimmune ailments (conditions in which your body's immune system turns on and attacks your own tissues).

5. Damage to organs and body systems follows, and the speed of the slide accelerates.

You may be able to slow this downward spiral by taking small amounts of melatonin daily. Because some people who take melatonin every night develop a tolerance to its effects, you should start by taking melatonin only four or five nights a week (although many people take it every night without problem).

Melatonin is inexpensive and widely available without a prescription, but it can exert powerful effects on the body. We suggest you try low doses if you are using it strictly as part of your anti-aging regimen. For most healthy people who have no sleep problems, we suggest beginning with 0.1 milligram about a half hour before bed. You can increase the dose to 0.5 to 1.0 milligram, but more is usually not needed by people without sleep disorders.

If you have persistent trouble falling asleep, a sublingual version of melatonin is recommended because it is rapidly absorbed. Start with doses of 3 to 5 milligrams, and increase to 10 milligrams if needed. It has been our experience that increasing the dosage beyond this will not produce additional benefit. If melatonin doesn't work, you should consult your physician.

If you wake up frequently throughout the night, try timed-release melatonin. Note, however, that you may be tired in the morning. If you have trouble falling asleep and also wake up frequently, try combination products that have both fast-acting and slow-release formulations. To help prevent jet lag, you can take 1 to 3 milligrams at bedtime at your new location the first day you travel and for the following three days. You will have to experiment to find the dose that works best for you.

20

OTHER HORMONES OF YOUTH: SEX HORMONES

"People ask me what I'd most appreciate getting for my 87th birthday. I tell them, a paternity suit."

—George Burns, comedian, 1896–1996

Modern sex hormone replacement therapy began during the late 19th century, when French physiologist Dr. Charles Edouard Brown-Sequard ground up a vial of guinea pig and dog testicles and injected himself with the resulting extract. He believed this extract might be able to restore some of his youthful vigor.

He presented his findings before a meeting of the French Societe de Biologie in 1899. He stated that the injections "have taken 30 years off my age . . . and I recovered at least all the strength I had possessed many years ago." Much to the relief of the guinea pigs of the world, most hormone products for human use are now created synthetically in the laboratory.

There is a reason sex hormones are called "sex" hormones. At normal bloodstream concentrations, these hormones keep both men and women deeply interested in sexual matters. Sex has historically been crucial to the survival of our species. Sex hormones evolved to create powerful urges that ensured that sexual activity would occur often enough to guarantee there would always be a next generation. With the loss of children common in earlier times due to infection, starvation, malnutrition, accident, and other hazards, each human couple needed to have more than two offspring just to keep the population stable. Frequent sexual activity was most urgent during a couple's reproductive years, which were frequently shortened by the same maladies that affected their children.

Changes in availability of food and modern medical care now ensure that the majority of pregnancies go to term and that most children survive to adulthood, particularly in the developed world but also largely true in the developing world. Today, of the many occasions people have sex in the course of a lifetime, only a small fraction are directly related to procreation. Most sexual activity in modern times is a pleasurable activity aimed at improving communication between couples and enhancing quality of life.

There is no such thing as a "male" or "female" hormone. Although commonly regarded as "female hormones," estrogen and progesterone are naturally present in men, where they exert profound physiological effects. Similarly, "male hormones" such as testosterone and DHEA are present in women and necessary to health.[1] Many people are surprised to learn that the average 50-year-old man has as much, if not more, estrogen circulating in his bloodstream as the average 50-year-old woman. But in this particular case, what is good for the goose is not good for the gander, so we will discuss the specific benefits (and detriments) of each of the sex hormones for both men and women separately.

Confusing matters even further, there is perhaps no other field in medicine where medical opinion changes so often and so drastically. It makes the frequently changing popular dietary advice seem stable by comparison. One year it seems a particular hormone is recommended, only to be derided and replaced by something else the next, with the original treatment again advised the following year. This has been particularly true about estrogen replacement for women.

ESTROGEN FOR WOMEN

For decades now, numerous benefits have been attributed to HRT (hormone replacement therapy) for women, and ERT (estrogen replacement therapy) in particular.

- Reduction of menopausal symptoms ("hot flashes" and vaginal dryness)
- Prevention of osteoporosis
- Improved cholesterol profile and protection against heart disease
- Improved mood and reduced incidence of depression
- Improved concentration and sleep
- Prevention of Alzheimer's disease
- Increased libido

There are two types of HRT available to women today. The first and most commonly prescribed type relies on synthetic drugs, which are really patented chemicals foreign to the human body. These include chemically altered estrogen, the most common form of which is Premarin, and chemically altered progesterone, which isn't really progesterone at all but is made of chemicals called progestins such as Provera, which act like progesterone in the body.

We'll refer to chemically altered estrogen replacement therapy as CERT. The other type of HRT is true hormone replacement therapy, since hormones are replaced with hormones, not chemically altered drugs; in fact, the replacement hormones are identical to those found naturally in the body. We will refer to this type of bio-identical estrogen replacement therapy as BERT. As we'll soon see, there is a big difference between CERT, which has been receiving a tremendous amount of bad press lately, and BERT, which has not. In fact, BERT has been gaining favor and is now being discovered by millions of women.[2]

Premarin, the most popular CERT drug, contains forms of estrogen completely foreign to the human body. Premarin is actually made from *pre*gnant *mar*e's ur*in*e, and in the medical literature is often referred to by its generic name, conjugated equine estrogen (CEE). CEE has a totally different composition of estrogens than those found naturally in the human body.[3] The three estrogens found normally in women are estriol (90 percent), estradiol (7 percent), and estrone (3 percent). CEE contains almost no estriol, but lots of estrone (75 percent) and estradiol (5 to 15 percent). In addition, horses have a number of estrogens unique to their species, most notably equilin (6 to 15 percent).

Menopausal Symptoms

As women approach menopause, circulating estrogen levels decrease dramatically. The average age of menopause is 51 years, and about 4,200 American women enter menopause every day.[4] Fluctuations in hormone levels and loss of ovulation may precede complete cessation of menstruation by several years. During this interval of a woman's life, called perimenopause, she often has a number of uncomfortable sensations and symptoms. Most noticeable are the characteristic hot flashes and mood swings. There is no question that conventional estrogen replacement therapy can help relieve these symptoms[5] as well as other uncomfortable menopausal symptoms, such as depression and sleep disorders.

The Movement Away from Chemically Altered HRT

The results of several large studies involving tens of thousands of women on CERT have become available in the past few years and have led to disturbing results. Based on these studies, it is now the consensus of many mainstream physicians that estrogen—that is, chemically altered estrogen replacement therapy (CERT)—is not as beneficial as once thought.[6]

More important, these studies show that CERT is not without risk, so the decision to use it for menopausal women has changed drastically.[7] As recently as mid-2002, about 38 percent of postmenopausal American women were using some form of estrogen replacement, with the overwhelming majority—15 million—on CERT. With so many women utilizing this therapy, you would have expected solid data in the medical literature to support its safety and efficacy. Yet, as the results of study after study were scrutinized, much of the rationale for the routine administration of CERT seemed to disintegrate.

It is remarkable that there is virtually no discussion in the mainstream health press about the fact that so-called estrogen replacement therapy did not use human estrogen at all. The common wisdom now is that long-term use of estrogen increases the risk of diseases such as heart disease and cancer, but the studies purporting to show this were done using CERT, not BERT. Yet we know that very slight changes in the chemical structure of substances can have dramatic effects. For instance, altering only a few carbon-hydrogen bonds can change a healthy fat such as an omega-3 into an unhealthy one such as a trans-fatty acid.

The health problems from CERT should not be surprising. We had exactly the same experience with testosterone replacement therapy. Initially, synthetic forms of testosterone were used, and these chemically altered hormones caused serious health problems, such as liver abnormalities. The medical profession then changed to bio-identical testosterone, which, as we report below, has resulted in numerous health benefits, including reduced heart disease risk.

Osteoporosis

CERT has the reputation of preserving bone mass in postmenopausal women and helping avoid hip fractures and collapse of the vertebral bones in the spine. However, among the 2,763 postmenopausal women participating in the HERS trial (the Heart and Estrogen/Progestin Replacement Study), "there was no evidence of a reduction in the incidence of fractures or rate of height loss in older women."[8] In another study involving ballet dancers who became amenorrheic (stopped having periods) due to low body fat and who were placed on supplemental Premarin and Provera, the chemically altered HRT did not help reduce bone loss.[9] But the large Women's Health Initiative (WHI) study did show that Premarin and Provera helped prevent osteoporosis, and bone density increased 3.7 percent in drug users, compared with only 0.14 percent in the placebo group.[10]

Alzheimer's Disease

CERT has been thought to be protective against Alzheimer's dementia. Yet in a meta-analysis (a study of several other studies) that looked at the relationship of postmenopausal CERT to dementia, some studies showed that women who took CERT developed Alzheimer's disease at a lesser rate, while others suggested that CERT had no effect. The authors of the meta-analysis concluded, "We do not recommend estrogen for the prevention or treatment of Alzheimer's disease or other dementias until adequate trials have been completed."[11]

Heart Disease

After menopause, levels of HDL cholesterol (the "good," or cardioprotective, type) fall. Since estrogen replacement causes a significant increase in HDL levels, estrogen has long been regarded as protective against heart attack and stroke for postmenopausal women. But several new large studies, including HERS and WHI, showed just the opposite. Women on CERT have an *increased* risk of heart disease.[12]

Breast and Ovarian Cancer

The famous Nurses' Health Study from Harvard showed that women who took estrogen alone had 32 percent more breast cancer, while those who took estrogen plus a progestin (an artificial form of progesterone) had 41 percent increased risk. For women over 60 on ERT for 5 years or more, the risk increased 71 percent.[13] The mortality rate from ovarian cancer for women who used CERT was shown to increase by 40 percent after 6 to 10 years of use, and by 71 percent for 11 or more years of CERT.[14]

The Women's Health Initiative

Findings from the Women's Health Initiative drove the final nail into CERT's coffin. This study began in 1997 and was headed by Dr. Bernadine Healy and the National Institutes of Health (NIH). WHI involved 16,608 postmenopausal women and was designed to assess the major health benefits and risks of Prempro, which contains Premarin and Provera and was the most commonly used combined hormone preparation in the United States. In 2002, the NIH halted this trial because the risks of taking the medication appeared to exceed the test guidelines. In the words of the safety monitoring board, "The risk-benefit profile found in this trial is not consistent with the requirements for a viable intervention for primary prevention of chronic diseases."[15] As shown in the graph below, Prempro users suffered 29 percent more heart attacks, 41 percent more strokes, a 26 percent increase in breast cancer, and almost double the risk of blood clots (indicated are the number of patients who suffered from one of these events).[16] Some doctors feel that the WHI results were skewed because the mean age of the women in the study was 63.2 years. They feel that CERT still remains a healthy option (that is, more benefit than risk) for younger "perimenopausal" women closer to age 50. Further study will help clarify this issue.

The monitoring board's decision to terminate a study of this magnitude prematurely because of the excess risks it uncovered shot an arrow through the heart of routine administration of synthetic, chemically altered HRT. Therefore, as of mid-2002, healthy menopausal women could no longer turn to Premarin and Provera as viable anti-aging strategies. Women and their doctors took the widely publicized WHI results seriously, and prescriptions for Prempro fell 66 percent and declined 33 percent for Premarin within one year.[17]

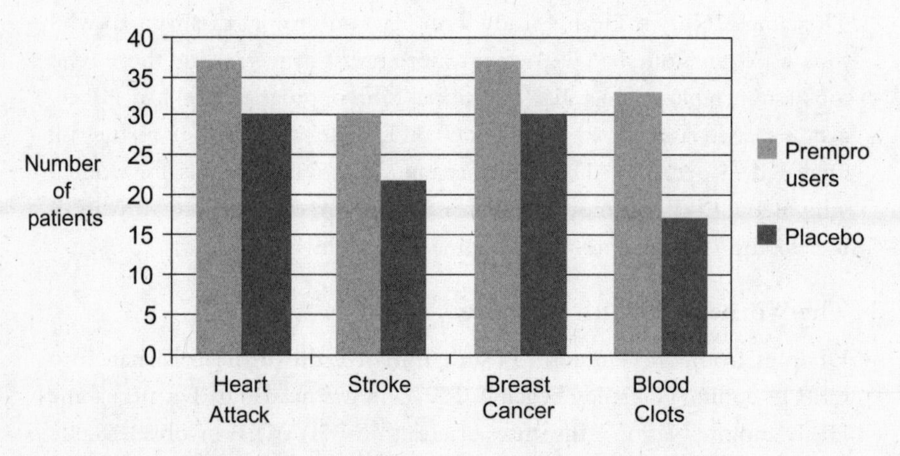

FIGURE 20-1. WOMEN'S HEALTH INITIATIVE TRIAL, 1997–2002

Bio-Identical Hormones

In the meantime, many patients and their physicians turned to bio-identical estrogen replacement therapy, or BERT, as an option during menopause. But can a woman really receive the benefits without the risks found with artificial chemically altered hormones by taking bio-identical estrogen and progesterone? Unfortunately, we don't have large, drug company–sponsored studies to prove this beyond a shadow of a doubt, but a number of small studies suggest this may be so. In one, bio-identical hormones appeared to confer the benefits without the risks of conventional artificial HRT.[18] Another small study in 1991 showed that bio-identical estradiol was able to confer the same level of circulating estrogen and health benefits as Premarin, but without some of the side effects of the latter.[19] In a study in 2000 from the University of Connecticut, the benefits of osteoporosis prevention were seen, with minimal side effects, when low-dose BERT was given to women over 65.[20] BERT can provide protection against Alzheimer's disease and even help maintain skin youthfulness. Several small studies also provide evidence that BERT does not increase cardiac risk and has beneficial effects on blood lipids.[21] A recent angiographic (heart catheterization) study demonstrated that BERT did not lead to progression of coronary artery disease in postmenopausal women already known to have heart disease.[22]

BERT is now widely available from compounding pharmacies (pharmacists who create or compound their patients' prescriptions from the original active drugs as opposed to simply counting premade pills out of bottles). The ratio of estrone (E1) to estradiol (E2) to estriol (E3) in BERT is compounded to closely mimic the ratio found naturally in a woman's body. Typical doses of 2.5 to 5 milligrams per day of "Tri-Est" (E1, E2, and E3) or "Bi-Est" (E2 and E3) are given in combination with bio-identical progesterone and, if needed, bio-identical testosterone. Side effects associated with bio-identical HRT are typically much less than those seen with the synthetic drugs.

Estrogen Fractions and Metabolites

As mentioned, human estrogen is really three estrogens in one: estrone, E1 (3 percent of the total), estradiol, E2 (7 percent), and estriol, E3 (90 percent). These types of estrogen have different effects. Estradiol is the most powerful of the naturally occurring estrogens and helps prevent osteoporosis, but has also been implicated as a *cause* of cancer. Estriol, on the other hand, seems to provide some protection against a number of female cancers (breast and uterus in particular). One major advantage of BERT formulations is that they usually include cancer-protective estriol, which is not found in conjugated equine estrogens.

Estradiol is chemically converted into 2OHE1 (2-hydroxyestrone) and 16∝OHE1 (16-∝-hydroxyestrone). 2OHE1 is similar to estriol—it is a weak estrogen and has anticancer properties—while 16∝OHE1 is more like estradiol, helping prevent osteoporosis but also increasing cancer risk.

It is better to have a higher ratio of 2OHE1 to 16∝OHE1 (which we will refer to as the 2:16 ratio) because this may reduce risk of breast cancer.[23] As part of a comprehensive hormone evaluation, women should have their 2:16 ratio checked and strive to achieve a ratio of greater than 1.0 (more of the beneficial "2" than the dangerous "16" type). This is easy to do with appropriate dietary changes and nutritional supplements, as explained on page 300.

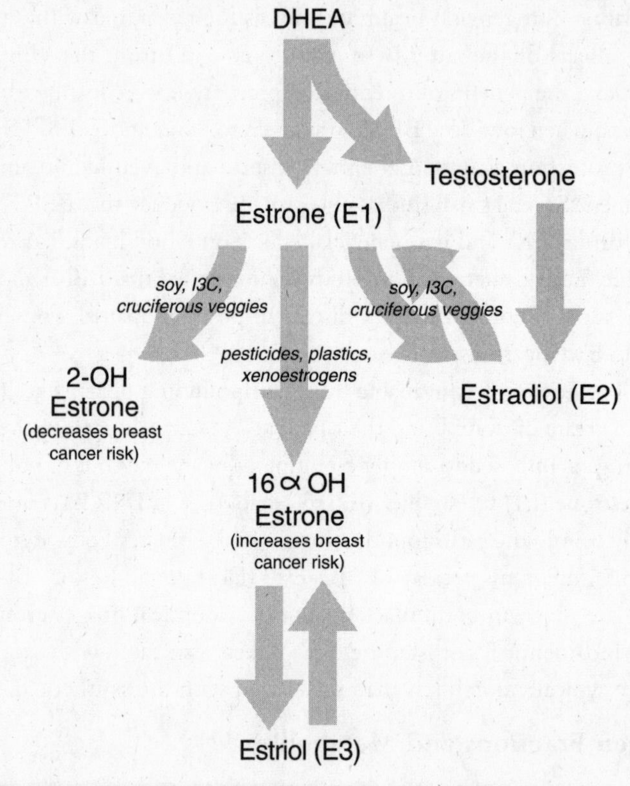

FIGURE 20-2. ESTROGEN METABOLISM

ESTROGEN FOR MEN

Although estrogen is predominantly a feminizing hormone, men still need *some*. Small amounts help to balance testosterone, and estrogen in males protects brain function.[24] The problem for most men older than 50, however, is having too much estrogen, not too little. A man converts testosterone into

BRIDGE TWO

SELECTIVE HORMONE RECEPTOR MODULATORS

The holy grail for sex hormone replacement therapy is drugs that produce the benefits of youthful sex hormones such as estrogen and testosterone, but without side effects or dangers.

One promising avenue for new drug research centers on a group of drugs known as selective hormone receptor modulators, or SERMs. To produce their effects, hormones must first bind to specialized receptors on the membranes of a cell nucleus; but a particular type of hormone receptor is not specific for a given hormone. For instance, besides estrogen, the estrogen receptor is also affected by many other molecules, which are said to have estrogenic activity. These molecules can either act just like estrogen itself or completely oppose the hormone's action.

The first SERM to undergo extensive evaluation was tamoxifen, which is still widely used as a treatment for certain types of breast cancer. But because of an increased incidence of hot flashes and a slightly increased risk of uterine cancer and blood clots, tamoxifen is unsuitable as a routine anti-aging therapy.

A newer SERM that comes closer to the goals of helping to control menopausal symptoms, preventing osteoporosis and breast cancer, and protecting the heart and brain is raloxifene (brand name Evista). Raloxifene has many beneficial effects, working like estrogen to prevent osteoporosis while also lowering harmful LDL-cholesterol levels. But even raloxifene is not ideal; preliminary results have shown that it does not relieve hot flashes—in fact, it seems to cause them—and, like tamoxifen, it is associated with an increased risk of blood clots. Researchers around the world are working to develop "the perfect SERM," and numerous drugs of this class are currently under investigation.[25]

On the other side of the genetic fence, research is also intense to discover the perfect SARM (selective androgen receptor modulator). SARMs have potential application to treat osteoporosis (which men develop as well), sarcopenia (age-related loss of muscle mass), and prostate enlargement. Although there are still no SARMs available for commercial use, GTx, a biopharmaceutical company that specializes in developing drugs related to men's health, has developed more than 250 potentially useful SARMs to date. Just like SERMs, SARMs can either mimic or oppose the action of the natural hormone—in this case, testosterone. The goal is to develop a drug that can produce the beneficial effects of testosterone, such as maintaining libido and bone and muscle mass, while avoiding the side effects or toxicities of testosterone replacement, such as prostate enlargement and male-pattern baldness.[26]

estrogen by a chemical process called aromatization under the influence of the enzyme aromatase. In young men, aromatase activity is relatively minor, and estrogen levels are low. With age, however, aromatase activity increases and estrogen levels go up.

As we will discuss below, testosterone replacement may be a beneficial anti-aging strategy for men. But one of the biggest problems with testosterone replacement therapy is that in many of the men who need it most—namely, older men—aromatase activity is often quite high. In these men, some of the

Natural Approaches to Controlling Menopausal Symptoms

As an alternative to estrogen, there are several nonhormonal approaches to menopause. **Soy products** contain isoflavones, predominantly *genistein* and *daidzein*, which are phytoestrogens (plant-like estrogens). Consumption of soy products has been found to reduce hot flashes and protect against osteoporosis, heart disease, and cancer.[27] Breast cancer is relatively rare in Japan, where average daily consumption of soy isoflavones averages 50 milligrams daily, compared with 1 to 5 milligrams a day in the United States. Soy consumption has also been shown to prevent osteoporosis and Alzheimer's and lowers potentially harmful LDL-cholesterol levels.[28] In addition, soy moves a woman's estrogen metabolism toward a more favorable 2:16 ratio.[29]

To achieve these beneficial hormonal effects, you need to consume about 50 milligrams of soy isoflavones. You can obtain these amounts by eating about 6 ounces of tofu or 4 ounces of a fermented soy product (tempeh) or drinking 2 cups of soy milk. One caution: Be certain to look for unsweetened soy milk; many popular brands contain significant amounts of sugar. You can also obtain soy isoflavones in the form of soy isolate powder or as nutritional supplements.[30]

The herb **black cohosh** has been shown to help control hot flashes and other menopausal symptoms. Sold around the world under the brand name Remifemin, black cohosh helps with uncomfortable menopausal symptoms but does not protect against osteoporosis, so it must be combined with other therapies.

Several other supplements, including dong quai, evening primrose oil, and vitamin E, are frequently included in menopause formulations, but studies have not shown consistent benefit with these products.[31]

Cruciferous vegetables such as broccoli and brussels sprouts can help improve your 2:16 ratio,[32] thanks to their active ingredient I3C (indole-3-carbinol).[33] I3C is also available as a nutritional supplement. Women who find they have adverse 2:16 ratios should eat more soy products and cruciferous vegetables as well as take supplemental I3C (200 milligrams twice a day).

Table 20-1: Estrogen Testing and Treatment Recommendations

WOMEN	MEN
When checking levels of female hormones, particularly in menopause, all three types of estrogen—estrone, estriol, and estradiol—should be checked individually.	Estrogen levels should be checked and excess levels treated with aromatase inhibitors such as I3C, chrysin, and Arimidex (see below).
If the decision is made to proceed with ERT, use bio-identical estrogen in the form of Bi-Est (E2 and E3) or Tri-Est (E1, E2, and E3) from a compounding pharmacy. If your doctor is unfamiliar or uncomfortable with using compounded products, he or she can write a prescription for Estrace, which is bio-identical estradiol (E2), but this is suboptimal as it doesn't have any of the protective E3 (estriol).	Supplementation with estrogen is not recommended for men.
Bio-identical estrogen has a short half-life, so ideally it should be taken twice a day, about every 12 hours.	In the rare case where estrogen levels are too low, some increase is possible with DHEA.
Follow up by rechecking blood levels of estrogen until a proper level is achieved.	—
To assess for breast cancer risk, fractionated estrogens (2OHE1 and 16∝OHE1) should also be measured with a goal of a 2:16 ratio greater than 1.	—
For menopausal symptoms, increased consumption of soy products and cruciferous vegetables is recommended. Black cohosh supplementation can be tried as well.	—

supplemental testosterone becomes aromatized into estrogen, so that while they still get the benefit of higher testosterone levels, it is often at the expense of too much circulating estrogen. So men who use testosterone replacement therapy should also check their estrogen levels and take steps to lessen the amount of testosterone that gets converted to estrogen (discussed on page 309).

XENOESTROGENS—FOR NOBODY!

Xenos is the Greek root for "foreigner." Xenoestrogens are foreign chemicals that are not normally found in the human body but mimic the action of estrogen. Many man-made substances are xenoestrogens, including pesticides, plastics, birth control pills, and the artificial estrogens found in CERT formulations. As you go about your activities of daily life, your tissues are in constant contact with xenoestrogens, which creates a condition of relative estrogen excess and upsets the important balance between all the hormone types. This imbalance is a possible cause for many cases of hormonally sensitive cancers, especially breast cancer in women and prostate cancer in men.

We recommend the following simple changes to reduce your exposure to these common foreign estrogens.

Pesticides have estrogenlike effects on the human body. Eating conventionally grown foods exposes your body to significant amounts of pesticides, so you should eat organically grown produce and animal products whenever possible. Avoid the use of pesticides both inside your house and in your yard. Use natural methods of pest control instead.

Plastics exert estrogenlike effects on the body as well. Try to limit your exposure to plastics, particularly as they come in contact with your food. For example, store food in the refrigerator in glass containers rather than plastic wraps. Never microwave food in a plastic bowl. Softer plastics tend to leach more than harder types, so, for example, try to avoid drinking water that comes in soft plastic bottles. Drink water from glass bottles instead.

We also advise women to avoid taking birth control pills or other synthetic forms of estrogen unless absolutely necessary. Spermicidal creams should be avoided as well. We recommend natural methods of contraception instead.

PROGESTERONE—FOR WOMEN

Until recently, progesterone was thought of as a less important hormone than estrogen. Recent work done by the late Dr. John Lee suggests that, far from

Benefits of progesterone

- Balances estrogenic symptoms and estrogen effects
- Serves as a precursor to estrogen and testosterone
- Enables pregnancy to continue to term
- Protects against breast and endometrial cancer
- Helps prevent fibrocystic breast disease
- Normalizes blood sugar and promotes fat-burning

being of lesser importance, progesterone may be the more dominant hormone in premenopausal women.[34] Rather than being merely estrogen's ugly stepchild, useful primarily to balance estrogen's more powerful effects, some doctors now regard progesterone as the dominant female hormone, while estrogen serves to balance *it*.

Clinically, physicians have found that progesterone balances estrogen-dominant symptoms, such as decreased sex drive, depression, abnormal blood sugar levels, fatigue, fuzzy thinking, irritability, thyroid dysfunction, water retention, bone loss, fat gain, and low adrenal function.

Menopausal Symptoms

Researchers from St. Luke's Hospital in Bethlehem, Pennsylvania, have found that bio-identical progesterone skin cream can serve an effective treatment for menopausal symptoms by itself. In their study, 83 percent of women who used 20 milligrams of progesterone cream daily experienced improvement or complete resolution of their hot flashes.[35]

Osteoporosis

Despite recommendations in the lay press by numerous health writers regarding the benefits of bio-identical progesterone in the prevention of osteoporosis, the medical literature does not confirm this. In fact, most studies suggest progesterone does not assist in osteoporosis prevention.[36]

Cancer Prevention

Cancer researchers have discovered that mutations that cause overexpression of the p53 gene increase the risk of cancer. Studies show that estrogen increases the expression of the p53 gene while progesterone slows this process.[37] Another study showed that whereas application of topical estradiol cream increased breast-cell growth by 230 percent, topical progesterone caused a 400 percent decrease.[38] In this way, progesterone may help prevent breast cancer.

Dangers of Synthetic Progesterone

Most practicing American physicians have never actually prescribed progesterone itself, but they have written millions of prescriptions for artificial progestins such as Provera. A number of serious side effects have been associated with artificial progestins, including blood clots, depression, breast tenderness, abnormal vaginal bleeding, abnormal glucose tolerance, edema (fluid retention), acne and skin rashes, facial hair, and weight gain.

Bio-identical progesterone causes fewer side effects.[39] The liver problems

Table 20-2: Progesterone Testing and Treatment Recommendations

WOMEN	MEN
Progesterone levels should be measured as part of routine hormone screening. Postmenopausal women can have their level checked at any time. Premenopausal women should be screened on days 17–20 of their cycle.	Men should not use progesterone unless directed to do so by their physicians.
If levels are suboptimal, bio-identical hormone replacement with either bio-identical micronized oral progesterone or progesterone cream should be used. Dosage should be under a physician's recommendations.	—

sometimes resulting from chemically altered progestin therapy are not seen when bio-identical progesterone is used.[40] Again, it is important to emphasize the profound difference that even small differences in chemical structure can make, and the differences between bio-identical progesterone and drug progestins are significant.

Bio-identical progesterone has, historically, not been recommended by most physicians because it does not survive the trip through the digestive tract in normal pill or capsule form. But physicians who wanted to prescribe it found a way around this conundrum by having their patients apply the hormones in topical (skin) formulations as creams or gels.

In the past few years, micronized oral preparations have become available by prescription from either compounding pharmacists as generic progesterone or from standard pharmacies under the brand name Prometrium. In these preparations, the hormones have been encapsulated into microclusters that are able to pass through the GI tract and enter the circulation intact. So, today, oral formulations of bio-identical progesterone are widely available, and the main advantages of the artificial hormones are gone.[41] Nonetheless, artificial progestins, rather than bio-identical progesterone, are still prescribed by most doctors.

PROGESTERONE—FOR MEN

A number of articles in the lay press have begun to advocate the use of progesterone as an anti-aging strategy for men, suggesting that progesterone can reduce the incidence of prostate disease, including cancer. Since there are no

BRIDGE THREE

VIRTUAL HORMONES

One of the important implications of the Bridge Three nanobot technology is that we will be able to have experiences in virtual reality that are just as realistic and compelling as real reality. There will be many advantages to virtual experiences in terms of variety, intensity, and safety.

You will be able to have any type of experience with anyone—from business negotiations to sensual and sexual encounters—in virtual environments, which you can leave as easily as hanging up on a phone call. It works like this: When you wish to enter virtual reality, the nanobots in your brain suppress all of the inputs coming from your real senses and replace them with the signals that would be appropriate for the virtual environment. You—that is, your brain—can decide in the "normal" manner to move your muscles and limbs, but the nanobots again intercept these interneuronal signals, suppress your real limbs from moving, appropriately adjust your vestibular system, and provide the appropriate movement and reorientation in the virtual environment. So it seems like you are in the virtual environment.

The Web will provide a panoply of virtual environments to explore. Some will be re-creations of real places; others will be fanciful environments with no "real" counterpart. You will be able to go to these virtual places and interact with other real, as well as simulated, people. Ultimately, there won't be a clear distinction between the two.

In addition to encompassing all of the senses, these shared environments can include emotional overlays, since the nanobots will be capable of triggering the neurological correlates of emotions, sexual pleasure, and other derivatives of our sensory experience and mental reactions. We won't need sex hormones to create sexual pleasure, which may eliminate a possible source of medical problems. We will be able to have experiences in virtual environments while having influence over our emotional and sexual response.

Sex hormones also play a key role in your overall health and aging. By the 2020s, we will have completed the reverse engineering of all metabolic processes in the body, and then we'll be in a position to determine optimal levels of your hormones and all other substances in your cells. One of the primary responsibilities of the nanobots in your bloodstream will be to continually adjust and fine-tune these levels to maintain well-being and youthfulness. The nanobots will ultimately be capable of making repairs inside cells to restore their ability to responsively utilize hormones.

studies in the medical literature to support this hypothesis, we cannot recommend it at the present time.

TESTOSTERONE FOR WOMEN

At about age 20, a woman experiences her peak levels of estrogen, progesterone, and testosterone. By the time she reaches 40, her body is making only half of these amounts. By menopause, many women undergo hormone replacement therapy with estrogen or a combination of estrogen and progesterone to treat menopausal symptoms. Even so, many still do not feel quite as well as they would like. Often they are suffering from the effects of declining levels of testosterone.

In women, the ovaries and adrenal glands produce about 5 percent of the testosterone present in males, but this amount is crucial to good health and well-being. In fact, puberty and the onset of menstruation are triggered by the "male hormones," testosterone and DHEA. Just prior to ovulation, a surge in testosterone creates an increase in the sex drive of most women.

For women, two of the more common sexual complaints are difficulty achieving orgasm and lack of libido. Both of these problems can be helped by applying small amounts of supplemental bio-identical testosterone cream to the skin.[42]

Testosterone is a natural anabolic steroid, meaning it helps build up body tissue. By administering safe amounts of this hormone, the age-related loss of lean body mass may be slowed. Testosterone undergoes conversion in the body into estradiol, which is associated with preventing bone loss.

Women have about ½0th the circulating levels of testosterone as men, and it is typically supplemented at about ½0th the male dose. Most women receive dosages of bio-identical testosterone of 1 to 5 milligrams per day.

TESTOSTERONE FOR MEN

Testosterone levels in men remain relatively constant until about age 50, when they begin to fall slowly. This counterpart to the female menopause is called the *andropause*, or the male menopause. Unlike women, however, where the decline in hormone production is often sudden and abrupt, in men the process is much more gradual. They have decreased sexual desire and erectile function, decreased intellectual activity, increased fatigue and depression, decreased muscle mass, increased visceral fat and obesity, thinning skin, decreased body hair, and decreased bone density.[43]

Libido

The two main sexual complaints of aging males are erectile dysfunction and loss of libido. While testosterone supplementation is only occasionally beneficial at correcting erectile dysfunction, the effect of testosterone supplementation on a man's libido is often dramatic.

Insulin Resistance and Cholesterol

Testosterone decreases insulin resistance and lowers total cholesterol and unfavorable LDL cholesterol. As men age, the balance between testosterone and estradiol tilts in favor of estradiol production. Excess estradiol decreases the level of LH (luteinizing hormone), the pituitary hormone responsible for stimulating testicular production of testosterone.

Osteoporosis

Although osteoporosis is commonly viewed as a female disease, by their late 70s men are just as likely to run the risk of osteoporosis-related hip fractures as women. The lower incidence of fractures in younger middle-aged men compared with women may simply be due to the fact that male bones are initially bigger and denser.

Testosterone and Dihydrotestosterone

In the past few years, there has been a dramatic reversal in our thinking about testosterone. In the old days—a dozen years ago—testosterone was regarded as the enemy. It was believed to be beneficial only for younger men, for whom it provided increased muscle mass, stronger bones, powerful sexual urges, and the "three Vs" of vim, vigor, and vitality. For older men, testosterone was something to be feared and usually avoided. It was thought to be the cause of such age-related complaints as male pattern baldness, urinary difficulties, and prostate cancer.

Today's conventional wisdom has completely changed: testosterone replacement therapy is often *recommended* as a treatment for andropausal symptoms.[44] Many authorities now recommend that men over age 50 have their testosterone levels checked and consider testosterone replacement therapy under the guidance of a physician. Many anti-aging benefits such as improved energy, stronger muscles and bones, and increased libido are possible by supplementation with bio-identical testosterone.

At about the same time testosterone returned to grace, accusatory fingers began to point at one of its by-products, DHT (dihydrotestosterone), a powerful form of testosterone associated with prostate enlargement and male pattern baldness. The thinking was that levels of testosterone in young men are quite high, while levels of DHT are low. In older men, this situation is reversed. The reason for this reversal seems to center on the enzyme 5-∝-reductase, which converts testosterone into DHT. This enzyme becomes more active with age, so levels of DHT increase.

Scientists began to focus on ways to block conversion of testosterone into DHT. The prescription drug finasteride, or Proscar, was developed as a specific 5-∝-reductase inhibitor. It received FDA approval in 1994 for the treatment of benign prostatic hypertrophy, the malady that causes many of the urinary difficulties experienced by older men due to prostate enlargement. Shortly thereafter, a lower dosage formulation of finasteride (Propecia) was approved as treatment for male pattern baldness.[45] Finasteride is not free of side effects, however, including decreased sex drive and erectile difficulties.

These problems are rarely seen with a natural plant-derived product known as *serenoa repens*, which comes from saw palmetto berries. Saw palmetto is a small shrub native to the southeastern United States, and Native Americans have used its berries for centuries to treat urinary problems. Recent studies have confirmed that saw palmetto extract is an effective 5-∝-reductase inhibitor.[46] There is also some evidence that saw palmetto works as an aromatase inhibitor, blocking the conversion of testosterone into estrogen. (See Figure 20-3.)

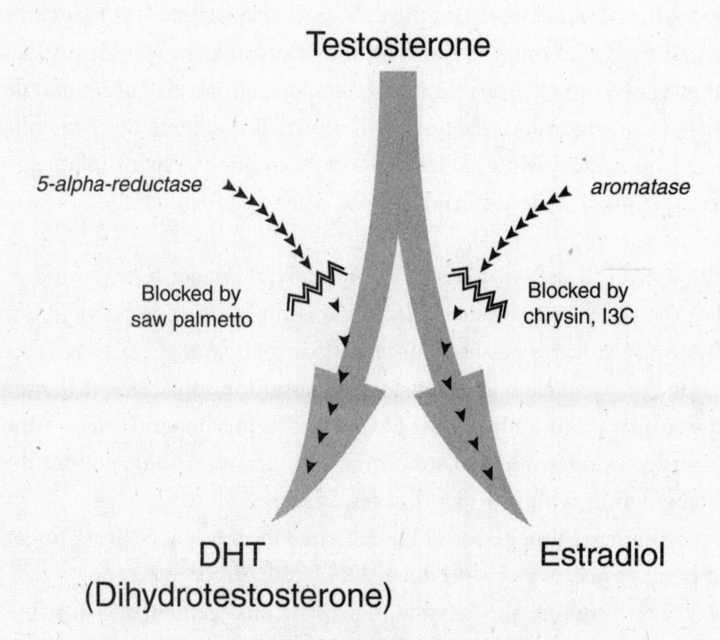

FIGURE 20-3. TESTOSTERONE METABOLISM

None of the studies to date have shown any significant side effects or toxicity to saw palmetto extract, although larger studies need to be done.[47] Saw palmetto use does not interfere with measuring PSA (prostate-specific antigen), a marker for prostate cancer, a problem occasionally seen with finasteride.

The recommended dosage of saw palmetto is 320 milligrams a day. We recommend that most men who take androgenic hormones such as testosterone or DHEA add saw palmetto to their regimens.

Lowering Estrogen

Until recently, it was thought that testosterone and DHT were bad guys. Current thinking centers on increased estrogen levels instead.[48] The conversion of testosterone to estradiol takes place under the influence of the enzyme *aromatase*, another enzyme that becomes more active with age.

Very often, when a man begins testosterone supplementation, his body turns some of it into estrogen. While men need a little estrogen, they can experience numerous adverse effects, including prostate problems and increased abdominal fat, if their estrogen levels increase too much.[49] So whenever testosterone supplementation is used, men also want to block any increased estrogen activity. I3C (indole-3-carbinol), the cruciferous vegetable concentrate mentioned above as useful for breast-cancer prevention in women, reduces the level of estradiol in men as well.[50] The combination of I3C with soy can be even better at blocking estrogen production.[51]

There are other ways to block the conversion of testosterone into estrogen, including chrysin, a natural plant-derived aromatase inhibitor.[52] By taking chrysin—usually, 1,000 to 3,000 milligrams a day—a man is able to get many of the benefits of supplemental testosterone therapy without simultaneously experiencing increased estrogen production.

A number of over-the-counter combination products, which may contain herbs such as stinging nettle and maca, are also available for this purpose. The spices turmeric, ginger, and boswelia are useful as well. If these products are unable to lower estrogen levels to a satisfactory range, then you can consider the use of prescription drugs known as aromatase inhibitors. These drugs lower estrogen levels in women and are FDA-approved for treatment of breast cancer, but also block conversion of testosterone into estrogen in men. If chrysin or a combination OTC product proves insufficient, we recommend you confer with your doctor about using the prescription aromatase inhibitor Arimidex. In many cases, this will help bring estrogen levels down.[53]

In addition, our dietary recommendations of a modified Japanese-style diet—green tea, fish, seaweed, and soy products, but limiting high-glycemic-index white rice—can help reduce the level of aromatase and lessen the conversion of testosterone into estrogen. Diets high in saturated fat, sugar, and alcohol do just the opposite and have been found to increase the risk of prostate cancer.[54]

Another way to raise testosterone levels is with the use of clomiphene, or Clomid. Clomiphene is most commonly used as a fertility drug because it

Table 20-3: Testosterone Testing and Treatment Recommendations

WOMEN	MEN
Levels of free and total testosterone should be measured as part of a full hormone evaluation.	Before beginning testosterone. replacement, we recommend hormonal evaluation, which includes both free and total testosterone levels, DHT, estradiol, and a PSA test for prostate cancer.
If levels are suboptimal, testosterone replacement therapy can be done with topical testosterone cream 1–5 mg daily.	If the decision is made to implement testosterone therapy, a topical (skin) bio-identical testosterone formulation is recommended. Oral testosterone formulations can cause liver problems. Typical doses prescribed are 25–50 mg of bio-identical transdermal testosterone once or twice daily. Follow the directions for application.
—	Saw palmetto 160 mg twice daily will help prevent conversion of testosterone into DHT and should be taken.
—	I3C 200 mg twice daily and chrysin 1,000–3,000 mg daily can help prevent conversion of testosterone into estrogen and should be taken as indicated. The prescription drug Arimidex should be used if estrogen levels are still high despite doing the above.
—	Men on androgen replacement therapy (both testosterone and DHEA) must undergo regular prostate cancer screening with digital rectal examination and blood tests for PSA.
—	Blood levels of testosterone, estradiol, and hematocrit should be checked periodically to ensure adequacy of dosage and that too much testosterone is not being converted into estradiol. Too much testosterone can cause the blood to become too thick, so your doctor will monitor this by checking your "hematocrit" as well.

stimulates ovulation in women. Clomiphene has a direct stimulatory effect on certain cells in the brain, which cause the Leydig cells of the testes to release more testosterone. While administration of testosterone is useful in improving the symptoms of andropause, it is often of limited benefit in treating erectile dysfunction. Clomiphene helps both andropausal symptoms and erectile dysfunction.[55]

Testosterone replacement can serve an integral role as a Bridge One therapy in our program, but it should not be used in cases of prostate disease. However, testosterone replacement does not seem to increase the risk of prostate cancer in healthy men.[56] We suggest that most men not start testosterone replacement until they begin to develop signs of testosterone deficiency (such as decreased libido or erectile dysfunction).

Just as with DHEA, when a man begins testosterone replacement therapy, he must check his PSA level before beginning therapy and regularly afterward, at least every six months. Testosterone replacement therapy is not recommended for men with either known prostate cancer or elevated PSA levels.

Widespread use of supplemental testosterone by healthy middle-aged men is relatively recent. Prescriptions written for topical testosterone for men increased from about 650,000 in 1999 to more than 2 million in 2003. Many short-term studies suggest testosterone replacement therapy is safe for most men and show health benefits such as improved cholesterol profiles. However, large-scale trials have not yet been completed, and we cannot assume that long-term therapy is entirely benign. It will be beneficial to see the results of these trials. In the interim, men should approach testosterone replacement with caution, getting regular physical exams and blood monitoring.[57]

21

AGGRESSIVE SUPPLEMENTATION

"Most people do not consume an optimal amount of all vitamins by diet alone."

—*Journal of the American Medical Association*, June 2002

Many people believe that if you eat properly, vitamin and mineral supplementation is unnecessary. But there are compelling reasons why this is untrue. New research has demonstrated that much of the population is born with genetic defects that can be corrected *only* by taking megadoses of the appropriate supplements. A key example that we have discussed is people with elevated homocysteine levels, who have a dramatically increased risk of cardiovascular disease and Alzheimer's. Such individuals need substantially more vitamin B_6, vitamin B_{12}, and folic acid to lower homocysteine and reduce these risks. Recommended Dietary Allowance (RDA) amounts won't do the trick. With new genetics testing now available, it appears many people have a need for one or more nutrients far in excess of RDA amounts, and for optimal health, these needs cannot be ignored.

Many other factors support the need for supplementation. It's often difficult to consume sufficient quantities of even nutrient-rich food to attain optimal levels of vitamins and minerals. Food storage, preparation, and cooking reduce nutrient levels further. In addition, years of intensive farming has depleted once-fertile farmland, so most available foods are deficient in essential nutrients. The ability of the intestinal tract to absorb nutrients also varies among individuals and decreases with age.

It is estimated that even in developed countries, much of the population consumes less than the RDA amount of one or more vitamins. In the underdeveloped world, the situation is often critical, involving multiple deficiencies. But the consequences of lacking even one nutrient can be quite serious. Deficiencies of vitamins C, B_6, B_{12}, folic acid, and the minerals iron and zinc, for example, lead to DNA damage and can cause cancer.[1]

Taking nutritional supplements makes economic as well as scientific sense. In a study conducted by the Lewin Group, commissioned by Wyeth Consumer Healthcare, if all Americans over age 65 took a multiple vitamin daily, Medicare would save an estimated $1.6 billion over five years because of the decreased risk of coronary artery disease and improved immune function.[2]

In proper dosages, nutritional supplements can not only prevent deficiency diseases but also have therapeutic effects, much like traditional medications. Supplements can reduce the risk of cataracts, prevent memory loss, support the immune system, lower cholesterol, prevent prostate problems, and ease the effects of menopause, to name just a few benefits. Also, consider the important role that silent inflammation plays in many serious chronic diseases, as we have discussed. The results of a double-blind, placebo-controlled study published recently in the *American Journal of Medicine* showed that such inflammation was reduced by 32 percent if patients simply took a multiple vitamin daily.[3]

RUST-PROOFING YOUR BODY

Leave your bike out in the rain and it rusts. Leave a cut apple on the counter and it turns brown. Forget to put the butter back in the refrigerator and it becomes rancid. These changes are due to oxygen combining with the exposed molecules in a process known as oxidation. We cannot live without oxygen, yet this same element that is so vital to our existence is also instrumental in the aging process. So does that mean as we age, we get rusty? Well, sort of. To better explain what happens, we have to talk about free radicals.

Very simply, free radicals are molecules that are missing an electron in their outer shell. Molecules are composed of atoms that are bound together by shared pairs of electrons. All stable molecules are composed of a set number of paired electrons. When a bond between an electron pair is broken, free radicals are formed.

These free radicals are highly unstable and must restore their paired electron status by getting hold of another electron immediately. They do so by stealing electrons from whatever molecules happen to be close by, damaging these other molecules and turning *them* into free radicals. If that molecule is DNA, the genetic blueprint for replication of all cells in the body, the damage will persist as the DNA replicates. This will cause cellular damage and increase the risk of cancer.

Free radicals are generated constantly within cells as oxygen is burned to create energy from nutrients. Other conditions that can increase the load of free radicals in the body include inflammation, infection, not eating enough

BRIDGE TWO

ANTIOXIDANT ADVANCES

There is currently intense research activity to create more powerful synthetic antioxidants. As one example, an international team of scientists has developed *pyridinols*, synthetic antioxidants derived from vitamin E but with 100 times the antioxidant strength.[4] As other powerful compounds are developed, the number of pills you'll need to take will decrease.

With more effective and efficient delivery systems on the horizon, it will also be easier and more convenient to take nutritional supplements than it is today. One option is inhalation: drugs inhaled into the lungs are immediately absorbed into the bloodstream. At least three such products are currently in development.[5]

Polymerix Corporation, a spin-off from Rutgers University research, is taking a different approach to drug delivery, forming the drugs themselves into special polymer backbones and coatings that the body breaks down for gradual medication release. Polymerix pills contain 70 to 90 percent active ingredients rather than the 30 percent in current systems that use nondrug polymer coatings and matrices.[6]

Yet another way to deliver supplements is through your food. By genetically modifying crops, researchers can not only reduce the need for herbicides, pesticides, and fertilizer, but also increase the nutritional value of the food. A strain of genetically modified (GM) rice has been developed containing higher levels of vitamin A, for example, and the mapping of the rice genome is expected to lead to yet other beneficial variants.[7]

antioxidant-rich foods, and exposure to high levels of free radicals in the environment.

Antioxidants, also known as free-radical scavengers, function by offering easy electron targets for free radicals. In absorbing a free radical, antioxidants "trap" (de-energize) the lone free-radical electron and make it stable enough to be transported to an enzyme, which combines two free radicals together, neutralizing both. Without these antioxidant buffers, free-radical damage would quickly spiral out of control and destroy the entire organism.

OUR FREE-RADICAL DEFENSES: ENZYMES AND ANTIOXIDANTS

Fortunately, your body is equipped with two sets of defense systems that are able to neutralize these free radicals before they lead to tissue

About 60 percent of the vitamin E obtained in the American diet comes from vegetable oil, primarily soybean oil. Seventy percent of the vitamin E in most vegetable oils is gamma-tocopherol, while only 7 percent is alpha-tocopherol, the type with more powerful antioxidant activity. Dean DellaPenna is using a genomics-based approach to identify the gene that codes for the more powerful type of vitamin E. With the help of computer models, his team was able to insert this gene into a strain of bacteria, clone it, and then insert multiple copies of the gene back into the oil-producing plants. The result: a new, genetically modified plant that produced up to 10 times as much alpha-tocopherol.[8]

Genetic modification has produced numerous other potentially beneficial effects, including a GM "protato," a potato that contains higher amounts of protein,[9] as well as a genetically modified strain of soybean with a gene that causes soybean allergy in children silenced.[10] Efforts are also under way to increase the beneficial omega-3 fatty-acid content of eggs, milk, and steak, to make these healthier food choices.[11]

Efforts to develop GM crops with higher levels of antioxidants will benefit from understanding the pathways involved in the plants' responses to stresses such as bright sunlight. The same chemicals that enable plants to endure the constant barrage of free radicals that comes from sitting out in bright summer sunshine all day every day also help us fight free radicals in our own bodies. This is why eating vegetables helps prevent cancer and heart disease, protects your vision, and supports your immune system.[12] Helping plants resist stress may also make them better food sources.[13]

---

damage—one that is built into your cells and one that you take in through foods or supplements. Your built-in free-radical scavengers are called *antioxidant enzymes* such as SOD (superoxide dismutase). These beneficial antioxidant enzymes stabilize free radicals so they can't damage tissues in the body.[14]

Your dietary antioxidant system includes *vitamins, minerals, and other antioxidant nutrients*, the best known of which are vitamins A, C, E, and the mineral selenium (our "ACES"). Other vitamins, such as vitamins B_2, B_3, and B_6, as well as some nonvitamins such as coenzyme Q_{10}, alpha lipoic acid, and the proanthocyanidins (grapeseed extract), also serve as powerful antioxidants.

The genes found in your DNA are the blueprints that direct your body to build the proteins it needs. Among the most important of those proteins are your enzymes, which help your cells make the other critical compounds they need for healthy functioning. Of the 3,870 enzymes in the human body identified to date, 860 (22 percent) require a vitamin-derived cofactor.[15] Many enzymes require mineral cofactors as well. Without these cofactors in place, the enzymes are useless. So vitamins and minerals have two critical roles in providing antioxidant protection and maintaining health: they themselves serve as antioxidants and they are needed for built-in antioxidant enzymes to work. By taking supplemental vitamins and minerals, you ensure that your body will have the necessary building blocks to create an adequate supply of functioning antioxidant enzymes at all times.

You could take these antioxidant enzymes as supplements, but they tend to be poorly absorbed when taken by mouth, so it is better simply to take the vitamin and mineral cofactors and let your body build its own enzymes. But genomic research has revealed that many people have genetic defects that cause these cofactors to attach improperly to the enzymes that need them, a problem that can be corrected only by taking larger amounts of these nutrients. This has led us to the important discovery that most people need one or more vitamins or antioxidants far in excess of the RDA amount, and to the even more important conclusion that optimal health is not possible without supplementation.

WHAT GENOMIC TESTING CAN TELL YOU ABOUT YOUR VITAMIN NEEDS

Amazingly, as many as one-third of all the genetic polymorphisms (variations) that have been identified result in enzymes that don't bind their needed vitamin cofactors properly.[16] This means these enzymes come off the DNA/RNA assembly lines with a variety of geometric abnormalities on the usual "docking sites" where their mineral and vitamin-derived cofactors should attach. Since the cofactors have a much harder time attaching to (and thus activating) defective enzymes, the only way to get these enzymes to work is by providing much larger amounts of these nutrients. The most practical way to correct this is to take nutritional supplements.

If enough vitamins and minerals are circulating in the bloodstream, the problem with defective enzymes can be overcome. It's like throwing partially cooked spaghetti at a wall. Most of the strands will fall to the floor, but an occasional one will stick. Even though these defective enzymes (the wall) don't bind their vitamin-derived cofactors (spaghetti strands) as well as they

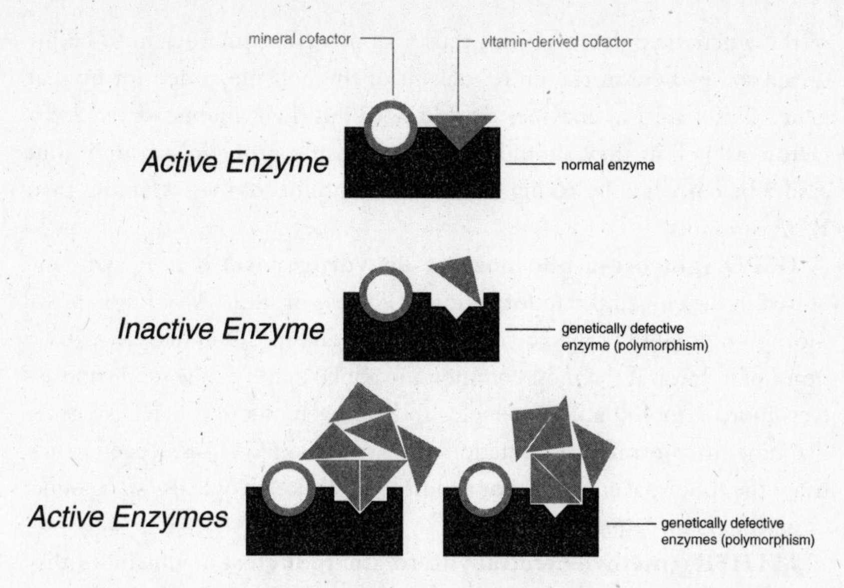

FIGURE 21-1. HOW GENETIC POLYMORPHISMS INCREASE THE NEED FOR VITAMINS

should, if there are enough vitamin molecules around, the job will get done. Sometimes hundreds of times the usual RDA amounts of vitamins are needed.

Valuable research on this topic has been done by noted nutrition researcher Bruce Ames and his group at the University of California, Berkeley. More than 50 genetic diseases have already been identified involving defective vitamin-cofactor binding sites that can be corrected by aggressive nutritional supplementation.[17] So instead of the old notion that taking supplements is wasteful, we now have solid scientific evidence supporting the use of high-dose ("mega") vitamin therapy for health maintenance and the treatment of many diseases.

Since each person has about one million genetic variants and up to one-third of them lead to enzymes that bind vitamins improperly,[18] it's hard to imagine anyone who doesn't need some nutrients far in excess of published RDA amounts. By performing the genomic tests that are available today, you can gain some precise information about which of your enzymes might "need help" through supplementation and which nutrients would be of most value to you.

As some practical examples, let's take a look at three very common genetic polymorphisms. Between them they affect *billions* of people. Each of these genetic polymorphisms causes an incorrect amino acid to be inserted somewhere in the protein chain of the enzyme coded for by that gene. This results in enzymes that do not bind their vitamin-derived cofactors as well as they should. In each case, the problem can only (but easily) be corrected by taking much larger amounts of these vitamins than RDA amounts.

G6PD (glucose-6-phosphate-1-dehydrogenase) is an enzyme involved in the chemical transformation of glucose or sugar. A common polymorphism of G6PD causes a condition called favism.[19] You may never have heard of it, but it is the most common metabolic defect in the world and affects more than 400 million people. To help overcome this defective gene, five times the normal concentration of niacin (vitamin B_3) is needed to activate the abnormal enzymes. For almost half a billion people, RDA amounts of niacin are inadequate.

MTHFR (methylenetetrahydrafolate reductase), which we discussed in chapter 13, is a critically important enzyme that turns folic acid into its active form so it can do its job of lowering homocysteine levels, fighting cancer cells, preventing heart disease, and many other functions. Up to 40 percent of people over 60 in the Caucasian and Asian populations have an MTHFR polymorphism and need up to *50 times* more folic acid than the RDA of 400 micrograms per day. For the 10 to 20 percent of the world's population that carry this polymorphism, taking the RDA amount is rarely adequate to control abnormal methylation, a major cause of many serious diseases and accelerated aging.

Aldehyde dehydrogenase enzyme helps metabolize alcohol, among other functions. A polymorphism of this enzyme is particularly common among Asians, affecting 50 percent or more of this population. People with this genetic defect don't metabolize alcohol properly and are also at increased risk of Alzheimer's disease. The defective enzyme doesn't bind its niacin-based cofactor very well. People with this polymorphism often need 20 times the RDA of niacin to activate the enzyme properly.

CURRENT THINKING ABOUT VITAMIN AND MINERAL SUPPLEMENTATION

Vitamin and antioxidant research has exploded in recent years. Thousands of articles a year are published in scientific journals on the subject of antioxidants alone.[20] While there is not complete agreement, current nutrition research shows that higher levels of vitamins and minerals are associated with better health and more efficient cellular function, while lower levels are linked with suboptimal cellular function.

This research has also increased our understanding of how nutrient supplements work. For example, we now know that folic acid can reduce cancer risk by repairing damaged DNA, and that vitamin E reduces heart attack risk by maintaining blood vessel flexibility and preventing free-radical damage to cholesterol.

Conventional medical opinion is slowly but surely coming around to the opinion that vitamins can do more than prevent deficiency diseases. The two most widely read American medical journals, the *New England Journal of Medicine* and the *Journal of the American Medical Association*, published articles in the past few years recommending that healthy adults take multiple vitamin/mineral supplements.[21] This represents an enormous shift in conventional medical thought, since for many years mainstream physicians believed supplements accomplished little more than helping create expensive urine.

So what are the nutritional supplements you need to take? There are 13 *essential* vitamins, 17 *essential* minerals, and 2 *essential* fatty acids. *Essential* means that you must get them from outside of the body because you can't manufacture them on your own. Many other nutrients that you can make are *conditionally* essential, meaning that under certain *conditions*, such as illness or stress, you need more than your body is able to make.

Vitamins are divided into two groups: the water-soluble vitamins, which include vitamin C and the B vitamins, and the fat-soluble vitamins, A, D, E, and K. Water-soluble vitamins are easily lost in the urine and need to be replenished daily. Fat-soluble vitamins, on the other hand, are stored in your fatty tissues and do not require such frequent dosing. However, because they are stored for longer periods in the body, you need to exercise more caution regarding dosages of fat-soluble vitamins, since toxicity is more easily achieved by overconsumption.

Among the 17 essential minerals, the more common are calcium, copper, iodine, iron, magnesium, manganese, molybdenum, phosphorus, selenium, and zinc. Of the 100 or so naturally occurring mineral elements, only these 17 are required for health. Most of the others found in nature (such as aluminum, lead, cadmium, and mercury) are not needed by the body at all and are simply toxins.

The two EFAs (essential fatty acids) are linoleic acid and alpha-linolenic acid.

RECOMMENDED DIETARY ALLOWANCES (RDA) VERSUS OPTIMAL NUTRITIONAL ALLOWANCES (ONA)

The National Academy of Sciences has established guidelines for intake of nutrients, which they now call the DRIs, or Dietary Reference Intakes. They have further subdivided the DRIs into four subgroups.

Estimated Average Requirement (EAR) is the average daily nutrient intake level estimated to meet the requirement of half the healthy individuals in a particular life stage and gender group.

Recommended Dietary Allowance (RDA) is the average daily nutrient intake level sufficient to meet the nutrient requirement of nearly all (97 to 98 percent) healthy individuals in a particular life stage and gender group.

Adequate Intake (AI) is a recommended average daily nutrient intake level based on observed or experimentally determined approximations or estimates of nutrient intake by a group (or groups) of apparently healthy people that are assumed to be adequate—used when an RDA cannot be determined.

Tolerable Upper Intake Level (UL) is the highest average daily nutrient intake level likely to pose no risk of adverse health effects to almost all individuals in the general population. As intake increases above the UL, the potential risk of adverse effects increases.[22]

Although more precise than the RDAs of the past, the DRI subgroups still fail to provide a solution to the fundamental problem of telling you how much of each nutrient *you* need to overcome *your* individual genetic variations. And since everyone has some of these variations, the RDAs still fall short of our goal of creating conditions for optimal health. Having adequate or even optimal amounts of most nutrients isn't good enough. The RDAs

More Is Better

To optimize your health, maximize healthy longevity, and prevent adverse consequences from your inherited genetic polymorphisms that increase your individual need for one or more specific nutrients, the only surefire way to cover your bases is to take a bit more of all of them. We will be the first to admit that this may mean you end up taking many vitamins and minerals you don't absolutely need, but at this stage of our knowledge, this is the best that Bridge One therapeutics has to offer. Take them all, and let your body use what it needs.

for most nutrients may be applicable to the *majority* of people, but they still fail to provide *you* with a useful mechanism for correcting *all* the metabolic problems you have as a result of your genetic polymorphisms.

There are other reasons why the DRI/RDAs don't provide the real information you need. As defined above, the RDA for a nutrient is "the average daily nutrient intake level sufficient to meet the nutrient requirement of nearly all (97 to 98 percent) *healthy* individuals in a particular life stage and gender group." The first problem with this is that these figures are based on the flawed assumption that the American population is healthy. Where did the National Academy of Sciences researchers who established the RDAs come up with this 97 to 98 percent of "healthy" individuals? Did they exclude the 55 percent of Americans who are overweight, the 47 million with the metabolic syndrome, or the 40 percent who will someday be diagnosed with cancer? Were the tens of millions of Americans who are on medication for high blood pressure, diabetes, asthma, arthritis, depression, and erectile dysfunction considered healthy? Since they began with this flawed assumption that such people are "healthy," their only possible conclusion was the flawed recommendation that relatively small RDA amounts of vitamins are adequate to maintain "health."

Therefore, rather than referring to RDAs, we prefer ONAs (Optimal Nutritional Allowances). In some cases, the amounts of vitamins we recommend are considerably higher than RDA amounts. In others, our recommendations are the same. But our ONAs are designed to optimize health, not maintain the status quo of a population that spends over $1.5 trillion (about 15 percent of the gross national product) on what is euphemistically referred to as "health care."

Our ONA recommendations encompass a wide dosage range. The reason for this is the wide range of individual variability and underlying need for antioxidant protection, detoxification, or energy production. Individuals with genetic polymorphisms that require higher amounts of vitamin/mineral cofactors are at the upper end of this scale or even higher. Unless you know you have a specific disease, condition, hereditary predisposition, or polymorphism that suggests the need for higher supplementation, we suggest you begin at the lower end of the ranges and experiment with dosages. Research into vitamin therapy is ongoing, however, and there may be long-term effects or vitamin interactions that remain unknown.

Doing some of the tests recommended in other chapters of this book may suggest that you need higher amounts of certain supplements. If you know you have elevated homocysteine, for example, increasing your intake of B vitamins and folic acid to the upper limits of our ONA recommendations (or even higher) may be needed. If you have an elevated hs-CRP indicating increased inflammation, more EPA and DHA are needed. If you have elevated cholesterol, megadoses of niacin may be helpful.

Also, if your lifestyle puts you at higher risk of free-radical exposure, you should increase your overall consumption of antioxidants. If you work during the day as a crop duster and go home at night to your apartment next to a heavily trafficked interstate highway, obviously you'll need more antioxidant protection and detoxification than an organic farmer who lives on a mountainside far from sources of pollution. Stress and travel are also frequent contributors to free radicals, so if this describes your lifestyle, you should increase your antioxidant consumption proportionately.

CHOOSING WHICH SUPPLEMENTS ARE RIGHT FOR YOU

In addition to essential nutrients, there are many nonessential but highly desirable nutritional supplements (supernutrients) that you can take to optimize your body functions and slow the aging process. These supernutrients are not necessary for survival, but they can significantly alter the rate and appearance of aging for many people. We recommend several supernutrients in our ONA protocol.

One problem is the sheer number of such nutrients available. If you are like us, your mailbox (and even more so, your e-mail inbox) is regularly filled with literature touting some new and revolutionary breakthrough product to help you look younger, become a sexual dynamo, improve your memory, and so on. It's difficult to sort through all the information and misinformation you receive to decide what supplements to take. In this book, we have tried to focus on the best of these supplemental nutrients. The nutrients we mention are not the only ones available, nor do you need to take all those we discuss. To help clarify the vital question of just what you should take, we have divided supplements into three groups:

- "Universal" supplements—recommended for everyone
- "Supernutrient" supplements—not *necessary* for optimal health, but may be highly useful for most people
- "Specific" supplements—typically taken for treatment or prevention of certain diseases or symptoms

Most supplements that we discuss are not expensive and are almost free of side effects when taken in proper doses, yet each has profound anti-aging benefits. But you will have to experiment and discover for yourself which are best for you. You can also take some of the supernutrient or specific supplements intermittently, switching among them from time to time. In chapters 10 and 17 we authors have each detailed our personal supplement programs to give you some idea of what we take as part of our individualized anti-aging programs.

We will be the first to admit that in these last days of the Bridge One era, if you follow our advice you will find yourself taking a lot of pills every day. But be patient—the newer Bridge Two and Bridge Three therapies will soon allow for easier methods of drug and nutrient delivery, such as through skin creams and inhalers, as well as more potent formulations, drastically reducing the number of pills you'll need to swallow.

In addition, your access to the precise genomic information you need to make accurate decisions about what specific nutrients you need remains somewhat limited. But in the next few years, you'll be able to obtain a personalized genetic map that will alert you to the optimal amount of each nutrient you require, based on your individual biochemical makeup. In the meantime, we suggest your best course is simply to take a little more of all the nutrients that have beneficial effects for most people. Because these supplements have a low level of toxicity, our Bridge One approach of taking extra amounts of all of them is the best way to head off potential problems. You can fine-tune and individualize your program by performing the tests recommended throughout this book, which will enable you to target the specific nutrients you need in greater amounts.

If you perform a homocysteine stress test and discover you have a problem with adequate methylation, you will know that you need to increase your consumption of vitamins B_6, B_{12}, and folic acid well beyond RDA amounts. If your hs-CRP (test for silent inflammation) is elevated, you should take more essential fatty acids than the RDA of 1.2 to 1.6 grams.

You should also add amounts of specific nutrients based on your known medical problems or inherited tendencies. If cancer or heart disease runs in your family, take the supplements recommended in the appropriate protocols. Do *not* rely on RDAs, but supplement aggressively and take amounts of the nutritional supplements to optimize enzyme function and minimize your risks. You now have the ability to diagnose your predisposition to numerous diseases by performing the testing we recommend. Combine the results of these tests along with your family history, the results of any genomics testing you perform, and the information provided by Ray & Terry's Longevity Program, and you will be able to effectively reprogram the expression of your genetic code to avoid entirely the diseases to which you are predisposed.

UNIVERSAL SUPPLEMENTS[23]

Vitamin A

Vitamin A promotes healing and the integrity of your epithelial tissues, which includes the skin, respiratory, and gastrointestinal tracts. It also plays an important role in immune-system function and can help improve resistance to infection. Vitamin A is needed for growth and maintenance of the bones and skin and is essential for proper function of the eye. Vitamin A reduces free-radical tissue damage from UV light and protects the skin from cancer and age-related damage, like age spots and wrinkles. The RDA for vitamin A is 2,660 IU for women and 3,330 IU for men. For most people, the ONA (optimal nutritional allowance) is 5,000 to 10,000 IU of vitamin A and a similar dose of beta-carotene.

Vitamin D

Vitamin D is important in the formation of healthy bones. When not enough vitamin D is consumed, children get rickets, a condition of deformed bones, and adults get osteomalacia, which manifests as soft bones. In the elderly, vitamin D deficiency is common and predisposes one to bone loss and fractures.[24]

You get vitamin D from either your diet (fortified milk is a common source) or exposure to sunlight, which enables the body to form its own D in the skin. While sunlight exposure has been implicated as a cause of skin cancer, it also protects against other, usually more serious internal cancers, breast and prostate cancer in particular.[25] Most experts worry about the toxicity of excessive doses of D, which can lead to elevated levels of calcium in the blood. Newer research, however, suggests that previous recommendations for both the RDA and UL are far too low.

The best way to determine how much vitamin D you should take is to have your blood level of D measured. The test you want is called 25(OH) D (25-hydroxyvitamin D). The normal range for 25(OH) vitamin D is 20 to 56 nanograms per deciliter but you want your level toward the upper limits of this range, around 45 to 56. The DRI for vitamin D is 200 to 600 IU, but we believe a better ONA dose (as long as blood monitoring is done) is 1,600

IU a day. With testing, we sometimes find doses of 2,000 IU daily or more are needed.[26]

In a recent study done on 150 consecutive patients who presented to a Minneapolis inner-city health center complaining of nonspecific musculoskeletal aches and pains, 93 percent were found to have low levels of 25-hydroxyvitamin D (less than 20). The Mayo Clinic authors of this study suggest that physicians check vitamin D levels on all patients who present with vague aches and pains, a problem commonly seen by physicians.[27]

Vitamin E

Vitamin E is a powerful free-radical scavenger, and it also works in harmony with glutathione to recycle vitamin C. Vitamin E protects the body from various toxins and carcinogens that cause free-radical damage, such as mercury, lead, ozone, and nitrous oxide. As a cardiovascular protectant, vitamin E is helpful in the treatment of angina, arteriosclerosis (hardening of the arteries), and thrombophlebitis (blood clots in the legs). It helps prevent blood clots that can cause strokes, improves blood flow to the extremities, and relieves circulatory problems. Vitamin E increases HDL, or good cholesterol, while decreasing overall blood cholesterol levels.[28] It also helps protect against cancers of the lung, esophagus, colon, cervix, and breast.

Up to 80 percent of older adults fail to get the minimal DRI of 30 IU of vitamin E. Food sources of vitamin E include vegetable oils, whole grains, wheat germ, brown rice, eggs, nuts, and leafy green vegetables, but it is almost impossible to get the optimal dose from diet alone. Our ONA for vitamin E is 400 to 1,200 IU of mixed tocopherols, which include several types of vitamin E such as alpha-, beta-, delta- and gamma-tocopherol. Vitamin E may increase bleeding tendency when taken along with aspirin or other blood thinners, so caution is advised in these cases.

B Vitamins

The B complex of vitamins include B_1 (thiamin), B_2 (riboflavin), B_3 (niacin and niacinamide), B_5 (pantothenic acid), B_6 (pyridoxine), B_{12} (cobalamin), folic acid, biotin, choline, inositol, and PABA (para-aminobenzoic acid). These nutrients are grouped together because they have similar functions and are often found together in nature. They are particularly important as cofactors for the enzymes involved in extracting energy from food. The B vitamins are also important in maintaining proper gastrointestinal function and promoting healthy nerves, hair, skin, and eyes.

B-complex vitamins are important in combating stress. Emotional stress as well as physical stress, such as illness, injury, or surgery, can dramatically increase the body's requirement for B vitamins. B vitamins also act as coenzymes for immune-system cells, thus boosting immune function.

Vitamin supplements frequently contain a combination of the various B-complex vitamins. For example, a common supplement is B-complex 50. This usually contains 50 milligrams each of B_1, B_2, B_3, B_6, choline, inositol, and PABA, and 50 *micro*grams each of B_{12}, folic acid, and biotin. Because the B vitamins are so closely intertwined, it is helpful to use a balanced formula like this to avoid competition for absorption of one vitamin over another in the intestines. Since B-complex vitamins are water soluble, toxicity from their usage is relatively rare. Our ONA recommendations for the B vitamins are listed in the "Summary of Recommendations" on pages 334–335.

Vitamin C

Vitamin C is the premier water-soluble antioxidant. Much of C's beneficial effect appears to be related to its role as an antioxidant and its ability to neutralize free radicals. Many researchers believe that vitamin C's ability to protect against heart disease is related to its antioxidant protection of circulating cholesterol. Vitamin C appears to offer some protection against cancers of the gastrointestinal tract, colon, breast, and lung. Vitamin C is quite safe even in relatively high doses; its main side effect in high doses is diarrhea, which can be minimized by gradually increasing the dose over a period of time and spreading it out throughout the day. Vitamin C may be depleted by aspirin and exposure to tobacco smoke.

The DRI is 60 milligrams, the amount necessary to prevent the deficiency syndrome known as scurvy, initially described in sailors. Doses up to 5,000 milligrams a day for more than three years have been found to be safe and without side effects.[29] Based on several epidemiological studies, we have set our ONA dose for vitamin C at 2,000 milligrams. Studies suggest that people who consume *either more or less* than this amount don't live as long!

Minerals

Some minerals, such as zinc, are cofactors in hundreds of different enzymes, while others like selenium are cofactors in relatively few. According to the National Research Council, 25 minerals or elements have some nutritional value in human beings. Of these, only 17 are essential to health, meaning that a deficiency of any of them will result in a disease condition. Dosages of mineral supplementation must be approached more cautiously than with many vitamins, because the margin of safety is narrower. For zinc, the mineral cofactor used in over 300 enzymes, daily requirements are about 10 milligrams, yet toxicity is often seen at doses above 100 milligrams per day.[30]

Under ideal conditions, mineral supplementation would be unnecessary. Because of "modern" farming techniques, use of artificial fertilizers, "convenient" methods of food preparation, and so on, the mineral content of our food supply has deteriorated over the past 50 years. A significant number of American adults now suffer from deficiencies of one or more essential minerals. To correct these imbalances and restore optimal health, mineral supplementation in some form has become a necessity for almost everyone.

There are more than 100 separate mineral elements. Other than the 17 essential minerals, most of the others are not only unessential and unnecessary to good health, they represent a toxic burden to your body's elimination pathways and are just forms of pollution. By competing with essential minerals for binding sites, these toxic elements can gum up your critical enzyme systems and create the same adverse health consequences that would result from a deficiency of the essential minerals. This condition is referred to as toxic metal syndrome and is an unavoidable feature of modern life because of the pollution that surrounds us.

Optimal supplementation requires adequate replacement of the following essential minerals.

Calcium	Selenium	Fluorine
Magnesium	Manganese	Sodium
Phosphorus	Iodine	Iron*
Potassium	Chromium	
Zinc	Boron	
Copper	Molybdenum	

You notice that iron has an asterisk. Iron is an essential mineral needed for formation of hemoglobin, the molecule in your red blood cells that carries oxygen. Yet iron along with copper is a double-edged sword. While some is good, more is not necessarily better. Both iron and copper are so-called transition metals and, under appropriate circumstances, can lead to excessive free-radical production. Therefore, iron supplementation is not generally recommended because the body has no mechanism for ridding itself of an excess. Some researchers have linked excess iron to heart disease, diabetes, cancer, increased risk of infection, and worsening of rheumatoid arthritis.

Except under conditions of increased need, such as pregnancy or chronic blood loss, as with moderate to heavy menstrual bleeding, we do not generally recommend supplemental iron.[31]

Essential Fatty Acids

As discussed earlier, there are only two essential fatty acids—linoleic acid and alpha linolenic acid. Under optimal conditions, the body is able to synthesize the two other omega-3 fatty acids critical to good health: EPA (eicosapentaneoic acid) and DHA (docosahexaneioc acid). Both EPA and DHA are needed to control the production of a family of chemical messengers called *eicosanoids,* which are short-lived intercellular hormones. One type of eicosanoid is the prostaglandins, which have profound effects on your cells.

Some prostaglandins are anti-inflammatory, while others increase inflammation in the body. By taking supplemental omega-3 EFAs such as those found in fish oil or from eating fish, you decrease inflammation. Limiting consumption of red meat and eggs, high in pro-inflammatory fatty acids, also reduces inflammation. An anti-inflammatory diet reduces your chances of developing serious illnesses associated with chronic inflammation, such as heart disease, Alzheimer's, and various cancers, and can even help reduce the pain, swelling, and stiffness associated with arthritis.

In addition to encouraging consumption of fish several times each week, we recommend additional supplementation with EPA and DHA from fish oil. The RDA for EFAs is 1,100 mg for women and 1,600 mg for men. Our ONA recommendations are for 1,700–5,000 mg daily, of which 1,000–3,000 mg is EPA and 700–2,000 mg is DHA. Individuals with inflammatory diseases such as arthritis, asthma, or inflammatory bowel disorders may need to take 10 to 15 grams a day or more.

SUPERNUTRIENT SUPPLEMENTS

Coenzyme Q_{10}

CoQ_{10} (coenzyme Q_{10}) is a naturally occurring antioxidant that plays an important role in the synthesis of ATP (adenosine triphosphate), the universal energy currency of the body. It does so by helping with the metabolism (burning) of carbohydrates and fats. CoQ_{10} is also a powerful antioxidant that aids in the regeneration and recycling of vitamins C and E.[32]

CoQ_{10} also plays an important role in protecting the body from heart disease and various types of cancer. We know that every cell in the body is con-

BRIDGE THREE

EATING FOR FUN IN THE FUTURE

In chapter 7, "You Are What You Digest," we discussed future nanobot-based technologies that will deliver optimal levels of all nutrients, including all of the "supplements" we've discussed in this chapter, directly into your bloodstream. More advanced versions of this technology will include nanobots that can deliver these nutrients directly to your cells, which is where they are ultimately needed. Other nanobots in the bloodstream will also be able to destroy undesirable substances such as toxins, as well as excessive amounts of certain nutrients such as unhealthy fats and glucose. These technologies should emerge in the 2020s.

Initially, this nanobot digestive system will augment the biological digestive process so that the levels of all substances in your blood are at an optimal level for your personal biochemistry, regardless of what you eat. Once this process has been perfected, however, we will rely on it more and more, so you will ultimately be able to separate the pleasurable, sensory, and social aspects of eating from the biochemical requirements of your body.

tinually subjected to free-radical attack. These free radicals react with oxygen to form reactive oxygen species (ROS). ROS can be highly damaging, particularly to DNA molecules. And damaged DNA is a precondition for malignant transformation of a cell. By helping neutralize ROS before they can damage DNA molecules, CoQ_{10} helps prevent cancer.[33]

Coenzyme Q_{10}, also known as ubiquinone, is of value in treating numerous cardiovascular conditions, including angina, high blood pressure, and congestive heart failure.[34] Statin drugs deplete the body of CoQ_{10}, so it is particularly important for people who take statin drugs to take this supplement.

There is almost no one who would not be helped by taking supplemental Coenzyme Q_{10}. We recommend that healthy people take between 30 and 100 milligrams of CoQ_{10} twice a day with meals. Patients with congestive heart failure or cancer should take 400 to 600 milligrams daily.

Proanthocyanidins

Grapeseed proanthocyanidin extract (GSPE) is another powerful antioxidant that helps protect the body from free-radical damage. GSPE is a more potent scavenger of free radicals than vitamin C, vitamin E, or beta-carotene. GSPE enhances the growth of healthy cells while it simultaneously attacks several types of malignancies.[35] GSPE can protect liver and kidney cells from the harmful effects of acetaminophen (Tylenol) overdose. Similarly, it helps prevent the heart damage seen with the chemotherapy drug doxorubicin and the lung damage associated with the cardiac drug amiodarone.[36]

Numerous beneficial effects have been ascribed to GSPE:[37]

- Acts as an exceptional antioxidant
- Prevents heart disease, strokes, and cancer
- Increases elasticity and strength of blood vessels
- Inhibits inflammation
- Enhances the body's ability for collagen repair
- Reverses appearance of aging
- Removes amyloid (cause of age spots and Alzheimer's, for example)

We recommend a dose of 50 to100 milligrams of GSPE twice a day.

Alpha Lipoic Acid

Vitamin C is our premier water-soluble antioxidant, and vitamin E is our primary fat-soluble antioxidant. But alpha lipoic acid (LA) serves as both a water-soluble and fat-soluble antioxidant. It also resembles CoQ_{10} and GSPE in helping recycle other antioxidants, such as vitamin C, vitamin E, glutathione, and even CoQ_{10} itself. Alpha lipoic acid is able to neutralize the most dangerous ROS of all, the OH (hydroxyl) free radical, as well as other harmful free radicals.[38]

LA increases utilization of glucose and the efficiency by which insulin moves sugar into cells.[39] In the presence of free radicals, excess glucose is attached to protein molecules to form AGEs (advanced glycation end products). AGEs are the cause of such diverse conditions as cataracts and age spots on the skin. LA slows or reverses their formation as well as many other types of age-related AGEs.

Because of the safety of LA, there is a wide range of suggested dosages. For healthy adults, 50 to 100 milligrams once or twice a day should be sufficient. For individuals trying to correct glucose intolerance or the metabolic syndrome (see chapter 9), we recommend 100 to 300 milligrams. For diabetics, the recommended dose is 300 to 600 milligrams, while diabetics with neuropathy (nerve damage) should take at least 600 to 900 milligrams a day until their symptoms resolve.

Carnosine

Carnosine is a small protein molecule found naturally in the body that is composed simply of one molecule of alanine and one of histidine. Despite its small size, carnosine is a powerful inhibitor of AGE formation. Like lipoic acid, carnosine can also protect cells from DNA-protein cross-linkages. Carnosine can extend the life span of cells in culture and rejuvenate dying (senescent) cells. Carnosine inhibits the toxic buildup from amyloid peptide of waste products, which leads to numerous disease processes such as Alzheimer's and type 2 diabetes, so it may be of value in the prevention or treatment of these diseases. Cross-linkage of protein with sugar (AGEs) or protein with DNA is a sign of the aging process, so some scientists speculate that carnosine may be a powerful anti-aging nutrient.[40] We recommend that healthy adults take 500 to 1,500 milligrams of carnosine daily.

Resveratrol

Resveratrol is found naturally in red wine. Its antioxidant properties help explain the French Paradox: why the French diet, typically very high in unhealthful saturated fats found in butter and cheese, is *not* associated with a high incidence of coronary heart disease.[41] Just like the antioxidants previously discussed, resveratrol can protect the body from free-radical attack by preventing formation of highly dangerous ROS, even after cells are exposed to tobacco smoke. Resveratrol protects DNA from free-radical damage, so it can play an important role in protecting cells from environmentally induced malignant transformation.[42]

Most cancers and chronic diseases in general are the end result of years of environmental damage in genetically predisposed individuals. Resveratrol, CoQ_{10}, proanthocyanidins, and alpha lipoic acid can serve as powerful chemoprotective nutrients to help the cells of the body withstand the constant assault of free-radical exposure. We recommend 400 milligrams of resveratrol daily as an anti-aging, chemoprotective supplement.

SPECIFIC SUPPLEMENTS

Lutein

Our clearest vision occurs when light is focused on a small, circular yellow region in the center of the retina known as the *macula lutea*. This tiny area, where light energy is changed to electrical signals for transmission to the brain, is a region of intense metabolic activity. A high concentration of antioxidants is required to neutralize the damage from stray high-energy electrons that unavoidably leak from the various electron-transport mechanisms within the macula. Over decades, this damage accumulates and can lead to age-related macular degeneration (AMD), the leading cause of blindness in developed countries.

Lutein is a yellow bioflavonoid found naturally in many fruits and vegetables. Eating a diet high in lutein and other naturally occurring bioflavonoids, which are found in brightly colored produce, can help prevent or delay the occurrence of AMD.[43] Corn and green leafy vegetables are particularly rich sources of these nutrients. In some groups studied there was as much as a tenfold difference in the occurrence of soft drusen (early deposits of debris on the retina that typically result in AMD) between people who had the highest and lowest lutein concentrations.[44] For macular degeneration protection and eye health, we recommend 6 milligrams of supplemental lutein daily.

I3C (Indole-3-carbinol)

Much of the cancer risk associated with estrogen results from an estrogen breakdown product, 16∝OHE1 (16-alpha-hydroxyestrone), while another estrogen breakdown product, 2OHE1 (2-hydroxyestrone) is cancer-preventive. As discussed in chapter 20, the ratio between these two (the 2:16 ratio) can serve as a useful biomarker for estrogen-related cancer risk. Whether estradiol turns into the more beneficial or dangerous type is determined by the level of activity of a liver enzyme known as CYP1A1. Eating cruciferous vegetables such as broccoli, cabbage, and brussels sprouts activates CYP1A1, leading to a safer 2:16 ratio. The active ingredient responsible for this beneficial effect is known as indole-3-carbinol (I3C).[45]

For a healthier 2:16 ratio, you should include generous amounts of cruciferous vegetables in your daily diet. Other dietary nutrients that will improve the 2:16 ratio include omega-3 polyunsaturated fatty acids, found in fish, and lignans found in foods like flaxseed.[46] I3C is also available as a supplement.

By performing genomics (gene) testing, you can also find out if you have a common genetic variation of your CYP1A1 enzyme that makes you more susceptible to estrogen-related risk. Plastics and pesticides are environmental pseudoestrogens (or *xenoestrogens*, as they are called) that must be metabolized by the CYP1A1 pathway. People who find they carry this genetic variation should limit their exposure to environmental xenoestrogens and consume a diet rich in cruciferous vegetables and omega-3 PUFAs.[47]

Supplemental I3C can protect against estrogen-sensitive cancers such as breast cancer in women and prostate cancer in men. Most individuals should take 200 milligrams daily. People with CYP1A1 genetic variations or other evidence of a genetic tendency to estrogen-sensitive cancers (such as a positive family history) should increase this to 300 to 400 milligrams a day.

Lycopene and Saw Palmetto

Lycopene is a bioflavonoid mainly derived from red fruits and vegetables, such as tomatoes and tomato products. In a large prospective study of health care professionals, lycopene intake was associated with a slightly reduced risk of prostate cancer.[48] Tomato sauce seemed more protective than fresh tomatoes, although the risk reduction was modest. Men who ate tomato sauce twice a week were 23 percent less likely to develop prostate cancer than men who consumed tomato sauce less than once a month.[49]

Lycopene is available as a nutritional supplement. Because of its ready availability when tomato sauce is eaten, supplementation is not specifically recommended unless you are at increased risk of prostate cancer, such as having a positive family history, in which case 10 to 30 milligrams a day is suggested.

After age 50, more men notice annoying urinary symptoms such as increased frequency, decreased stream, dribbling, and nocturia (need to urinate during the night). These symptoms are due to harmless enlargement of the prostate known as BPH (benign prostatic hyperplasia) or prostatism. Conventional nonsurgical treatment options rely mainly on the use of prescription drugs such as finasteride (Proscar) or tamsulosin (Flomax). These medications are often effective at improving symptoms of BPH, but they are expensive and may have side effects. An extract of the American saw palmetto or dwarf palm plant, *Serenoa repens*, has been found in several clinical studies to work as well as the prescription drugs for treatment of BPH at a greatly decreased price and lesser incidence of side effects.[50]

It is unclear if saw palmetto can prevent the onset of symptoms of BPH or if it protects against prostate cancer, but it is certainly of value in treatment of prostatism symptoms once they occur. Saw palmetto is taken in a dose of 160 milligrams twice a day.

Garlic

High blood pressure, or hypertension, is a major cause of death and disability in the United States. While 140/90 is the usual upper limit for "normal" blood pressure, the ideal may be more like 110/70. Even modest decreases of 3 to 5 millimeters in blood pressure have been associated with significant health benefits. Animal experiments have clearly shown that garlic is able to lower blood pressure.[51]

Garlic also protects against free-radical-mediated diseases, damage related to the aging process, radiation and chemical exposure, and long-term exposure to environmental toxins. Consumption of garlic extract can lower the risk of heart disease, stroke, cancer, and Alzheimer's disease.[52] We recommend 1,600 milligrams of aged, odor-free (fresh, not packaged) garlic daily.

Summary of Recommendations

UNIVERSAL SUPPLEMENTS (NEEDED BY EVERYONE)[53]

Nutrient	RDA (Recommended Dietary Allowance)	ONA (Optimal Nutritional Allowance)
Vitamin A (IU)	2,660 (women); 3,330 (men)	5,000
Vitamin D (IU)	200–600	600–2,000
Vitamin E (IU)	22–33	400–800
Vitamin K (mcg)	90 (women); 120 (men)	90–120
B_1 (Thiamine) (mg)	1.1 (women); 1.2 (men)	10–200
B_2 (Riboflavin) (mg)	1.1 (women); 1.3 (men)	10–100
B_3 (Niacin) (mg)	14 (women); 16 (men)	20–100
B_6 (Pyridoxine) (mg)	1.3–1.5 (women); 1.3–1.7 (men)	50–100
B_{12} (Cobalamin) (mcg)	2.4	10–25
Folic acid (mcg)	400	400–800
Vitamin C (mg)	75 (women); 90 (men)	500–2,000
Calcium (mg)	1,000–1,200	1,000–1,500
Magnesium (mg)	320 (women); 420 (men)	400–600
Iron (mg)	15–18 (premenopausal women); 8 (postmenopausal women); 8 (men)	15 (premenopausal women); 0 (postmenopausal women); 0 (men)
Zinc (mg)	12 (women); 15 (men)	15–30
Copper (mg)	0.9	0.5–4
Selenium (mcg)	55	100–250
Manganese (mg)	1.8 (women); 2.3 (men)	2–5
Chromium (mcg)	20–25 (women); 30–35 (men)	120–200
Omega-3 EFAs (mg)	1,100 (women); 1600 (men)	EPA 1,000–3,000 DHA 700–2,000

Arginine

Arginine is an essential amino acid that serves as the primary nitrogen donor for many biochemical reactions in the body, such as the formation of nitric oxide (NO), a gas with a very short half-life of only 5 seconds in the body. During this brief interval, NO directs the smooth muscle cells in the walls of the arteries to dilate. By taking supplemental arginine, you can ensure that adequate raw material is on hand for NO synthesis.[54]

By dilating arteries, more bloodflow is available to tissues. Beneficial ef-

SUPERNUTRIENT SUPPLEMENTS (HELPFUL FOR ALMOST EVERYONE)

Supplement	Amount
Coenzyme Q_{10}	30–100 mg 2/day
Grapeseed extract	50–100 mg 2/day
Alpha lipoic acid	50–100 mg 2/day
Carnosine	250–500 mg 2–3/day
Resveratrol	200 mg 2/day

SPECIFIC SUPPLEMENTS (RECOMMENDED FOR SPECIFIC CONDITIONS)

Supplement	Indication	Daily Dose
Lutein	Eye health	6 mg
I3C	Breast, prostate cancer prevention	200 mg
Lycopene	Prevents prostate disease	10–30 mg
Saw palmetto	Prevents prostate disease	320 mg
Garlic extract	Heart, blood pressure	1,600 mg
Arginine	Heart, blood pressure	6,000–9,000 mg
Vinpocetine	Memory	10–20 mg

fects of arginine supplementation include better bloodflow to the heart and brain, lower blood pressure, and better erections in men.

Typical doses of arginine are between 2 and 18 grams. Because of the large volume involved, it is often consumed as a powder. For most people, we recommend 6 to 9 grams a day.[55]

Vinpocetine

Vinpocetine is a powerful memory enhancer. It improves cerebral bloodflow, increasing brain cell ATP production (energy) and utilization of glucose and oxygen. Vinpocetine is derived from an extract of the periwinkle plant.

Vinpocetine is popular in Europe as a treatment for memory problems, acute stroke, and many other neurological conditions. Just 10 to 20 milligrams of vinpocetine can improve short-term memory, as confirmed in numerous scientific studies.

22

KEEP MOVING: THE POWER OF EXERCISE

A vast body of evidence has been accumulating for decades on the profound benefits of both aerobic (active, "with oxygen") and anaerobic (resistance) exercise.[1] Rates of disease and death are dramatically reduced for all of the major progressive diseases such as heart disease, stroke, type 2 diabetes, and cancer. Humans evolved to be physically active, and part of the modern epidemic of degenerative disease results from our society's excessively sedentary lifestyle. The most consistent message that we are discovering from the intensifying research into both body and brain is simply this: Use it or lose it.

The bottom line is that most of our ancestors spent their lives as hunter-gatherers only a few dozen centuries ago. The DNA of 21st-century humans is more than 99.99 percent the same as that of our hunter-gatherer progenitors. These people exercised a lot. They ran, they climbed, they walked. Their bodies—read that, *our* bodies—thrived on this vigorous regimen. The biotechnology and nanotechnology revolutions will eventually provide us the means to change our genetic heritage, but in the meantime we remain locked into bodies that require regular exercise for ideal function.

THE BENEFITS OF EXERCISE

Dr. Ron Klatz, president of the American Academy of Anti-Aging Medicine, and Dr. Robert Goldman, chairman of the board of the American Academy of Anti-Aging Medicine and president of the National Academy of Sports Medicine, have identified the following diverse benefits of exercise:

Lowers heart disease risk
Reduces blood pressure
Increases the strength of ligaments and tendons

Reduces stress and helps relieve depression
Improves sleep
Reduces the risk of several types of cancer (including colon, prostate, and breast)
Enhances physical appearance
Improves self-esteem
Strengthens bones, reducing the risk of osteoporosis
Increases energy

Consider the results of an eight-year study in the *Journal of the American Medical Association.*[2] Researchers divided the 13,344 participants into five fitness categories according to their own exercise habits: category one, sedentary (no regular exercise program); categories two and three, medium fitness (walking 30 to 60 minutes per day, four to five times per week); and categories four and five, high fitness (walking or running 20 to 30 miles per week or more).

The results were unexpectedly dramatic. There were significant gains between low and medium fitness, and again between medium and high fitness. The greatest gains were between low and medium, indicating that even moderate regular exercise is of immense benefit. Overall death rates for the medium exercisers were 60 percent less than those of the sedentary group. Death from cardiovascular disease for men was down by more than two-thirds. There was further benefit for both men and women in the high-fitness category, particularly with regard to cardiovascular disease.

Fitness reduces the health risk associated with many other mortality predictors, including obesity, smoking, and elevated blood pressure. One 1999 study reported in the *American Journal of Clinical Nutrition*, for example, found that fit, obese men were *less* likely to die of heart disease and other causes than unfit, lean men.[3] Another study followed more than 25,000 men and 7,000 women to quantify the relationship between fitness and cardiovascular mortality within "strata of other personal characteristics that predispose to early mortality."[4] The results showed only two statistically independent predictors of mortality for both men and women: smoking and low fitness. In addition, "fit persons with any combination of smoking, elevated blood pressure, or elevated cholesterol level had lower adjusted death rates than low-fit persons with none of these characteristics," according to the researchers.

Even with this body of evidence as a spur, 72 percent of American women and 64 percent of American men do not participate in any regular physical activity. As a result, much debate has focused on how *little* exercise is needed to create a beneficial effect. The lower the goal, the more likely that a largely sedentary population might be inspired to work toward it. In a 2003 editorial discussing some of this research, a Harvard Medical School researcher pointed out that the "modest and achievable level" of 30 minutes per day can have a significant effect for women, and even one to two hours *per week* of moderate activity can decrease the rates of coronary heart disease and premature mortality.[5]

AEROBIC EXERCISE

The mainstay of your exercise program should be regular aerobic exercise. Aerobic exercise is intended to raise both your heart and breathing rates and, as the name suggests, results in increased air (oxygen) consumption. Aerobic exercise such as walking, swimming, cycling, rowing, and cross-country skiing significantly lowers the risk of cardiovascular disease, cancer, and other diseases, as well as providing immediate benefits in terms of weight loss, reduced hypertension, improved sleep, and better mood. Regular aerobic exercise can also reduce elevated triglyceride levels and boost levels of HDL, the good cholesterol.[6]

A key aspect of aerobic exercise is that it involves continuous, rhythmic exertion of the large upper or lower body muscles for at least 20 minutes. It increases your heart rate and your demand for oxygen but is sustainable for

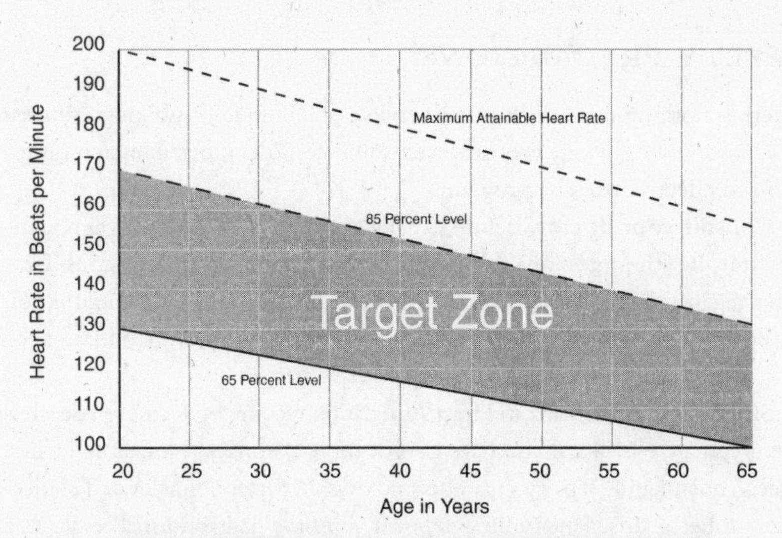

The Maximum Heart Rate shows the peak heart rate for average individuals. Any individual may vary from these averages, which apply to about two-thirds of the population. Shown below the maximum heart rate are the corresponding training heart-rate ranges for each age.

FIGURE 22-1. ADAPTED FROM CHART BY THE AMERICAN HEART ASSOCIATION

an extended period of time. Although there is some cardiac benefit from the significant exertion involved in stop-and-go sports such as tennis and basketball, these are not optimal forms of aerobic exercise. We would consider these types of sports more as supplements to a regular aerobic exercise program, not the primary component of one.

During aerobic exercise, you should be at your training heart rate. That's between 50 and 75 percent of your maximum attainable heart rate, which you can estimate as 220 minus your age, according to the American College of Sports Medicine (ACSM) and Mayo Clinic.[7] So, for example, if you are age 40, your theoretical maximum heart rate is 180, and your training range would be between 90 and 135 beats per minute.[8]

The best way to determine your heart rate is with a sports watch that provides a pulse readout. Otherwise, you need to briefly interrupt your exercise and count your pulse for 15 seconds, then multiply this figure by four. It is important to take your pulse as soon as you stop exercising, because your heart rate will slow down immediately. Once you have estimated your heart rate, continue your exercise routine.

The ideal aerobic exercise is walking. Virtually everyone can do it, almost anywhere. You should have little difficulty elevating your heart rate into your training range on a sustained basis, and it does not put undue strain on any of your joints.

SAFETY PRECAUTIONS

There is nothing unusual about the style or technique of walking required to achieve results. There are, however, some considerations in establishing a safe and effective exercise program.

Consult your doctor. Before starting any regular exercise program, consult your health professional, who can advise you on any special considerations regarding your physical condition and health. This is essential if you have indications of heart disease or other serious illness, you are middle-aged or older, or you haven't been physically active before.

Stress test. The American Heart Association recommends an exercise stress test if you are over 40; if you have two or more coronary risk factors such as male gender, family history, cigarette smoking, hypertension, elevated cholesterol, diabetes, or sedentary lifestyle; or if you have had abnormal results at a physical. The ACSM differentiates between men and women in its guidelines for stress tests at the start of an exercise program as well as between levels of exertion. According to the ACSM, if you are a male under 40 or a woman under

50 without coronary artery disease symptoms, you can begin a moderate regimen without a test. However, the ACSM recommends a stress test for all men 40 or over and women 50 or over who plan to start a vigorous program.[9]

As we discussed in chapter 15, a stress test is an electrocardiogram (ECG) monitored continuously for 10 to 15 minutes while you exercise on a treadmill or stationary bicycle, with progressively increasing difficulty. Your blood pressure is also monitored. The test ends when you are too tired to continue or if any symptoms are noted, such as an abnormally high blood pressure reading, abnormal pulse rate, shortness of breath, chest pain or other discomfort, or an abnormality in the ECG. There is some risk in a stress test, although this is minimal if administered properly by a trained health professional.

The test is far from perfect. About 10 to 20 percent of stress tests give false positives, incorrectly indicating a heart or artery abnormality, and 20 percent to 40 percent yield false negatives, failing to detect an abnormality.[10] So the exercise stress test will not capture all instances of artery blockage, and any positive indications that it does give, need to be confirmed by further medical tests. Nonetheless, it can be a useful screening procedure that your doctor or fitness instructor may request you undergo before starting an exercise program.

Begin slowly. Once you have the go-ahead from your health professional, the next step is to ease into exercise. The objective is to exercise on a regular basis and build this activity into a predictable routine. Endurance and speed will come naturally as your fitness improves.

Don't overdo it. You should feel like you are exerting yourself, but if you feel pain in your legs—shin splints, for example—slow down and rest. *If you ever feel pain in your chest, stop immediately and consult your doctor*, because you may be experiencing angina pain indicative of advanced atherosclerosis. (If the pain persists, you could be having a heart attack.)

Wear the right shoes. Do not use ordinary sneakers or even running shoes. Get a good-quality pair of shoes designed specifically for walking.

Build up gradually. If you are very out of shape, even walking one mile may be strenuous. Just aim to do a bit more each day. Making a real and permanent commitment to a regular and predictable program is the most important step you can take, whereas doing too much too soon could make you so frustrated that you give up.

Walking goal. We recommend walking three miles per day or more, five or more days per week, although even four days per week of regular aerobic exercise offers significant benefit. Once you gain experience and fitness, you'll require about 45 to 50 minutes for each session to continue to make gains.

BRIDGE THREE

THE FUTURE OF EXERCISING?

In chapter 15, "The Real Cause of Heart Disease and How to Prevent It," we discussed Rob Freitas's design for robotic replacements for your red blood cells ("respirocytes"), which will be thousands of times more effective than the biological variety. The implication for exercise will be profound—your endurance will be vastly enhanced. You will be able to run long distances at high speed while taking only occasional breaths. This scenario is a couple of decades in the future, but Intel is planning to construct hemoglobin molecules that transport 10 times more oxygen than natural hemoglobin.[11]

People who are paralyzed face an obvious challenge in exercising, but new developments provide the promise of reconnecting broken neural pathways for people with nerve damage and spinal-cord injuries. It has long been thought that re-creating these pathways would be feasible only for recently injured patients because nerves gradually deteriorate when unused. A recent discovery, however, shows the feasibility of a neuroprosthetic system for patients with long-standing spinal-cord injuries. Researchers at the University of Utah asked a group of long-term quadriplegic patients to move their limbs in a variety of ways and then observed the response of their brains, using magnetic resonance imaging (MRI). Although the neural pathways to their limbs had been inactive for many years, the pattern of their brain activity when attempting to move their limbs was very close to that observed in nondisabled persons.[12]

This provides the opportunity to place sensors in the brain of a paralyzed person. They will be programmed to recognize the brain patterns associated with intended movements and then stimulate the appropriate sequence of muscle movements through wireless communication (from the brain sensors to the muscles). For those patients whose muscles no longer function, there are already designs for nanoelectromechanical systems (NEMS) that can expand and contract to replace damaged muscles and that can be activated by either real or artificial nerves. Rob Freitas also envisions "nanomotors [that] could be implanted in muscles to make them more powerful."[13] We expect to see systems along these lines emerging over the next decade.

THE FIVE PHASES OF EXERCISE

1. Stretching. It is important to start out with a few minutes of stretching, which will help improve coordination and range of motion and help relax the body.[14] The most important targets of stretching are the hamstrings, lower back, quadriceps, shins, calves, and Achilles tendons. See the section below on stretching.

2. Warm-up. The second phase is warm-up, which involves walking at an easy pace, about 3 miles per hour, for a few minutes. This allows you to build momentum, thereby gradually reducing the stress on your muscles and on your heart.

3. Aerobic exercise. The third phase, which should be the bulk of your exercise routine, is the aerobic phase, in which you exercise rigorously enough to bring your heart rate into your training range. This should last at least 20 minutes to get a training effect on your heart. To cover at least 15 miles per week (once you are sufficiently fit to do this), you will probably want to walk in your training range for 30 to 40 minutes each session. In general, you will need to walk at least 4 miles per hour to achieve your training-heart-rate range.

4. Cooldown. Following the aerobic phase, cool down by again walking about 3 miles per hour for a few minutes. This allows your heart rate to return to normal gradually and prevents a pooling of blood in your legs and feet.

5. Recovery (stretching). The fifth and final phase involves another several minutes of stretching to maintain limber joints, muscles, and tendons.

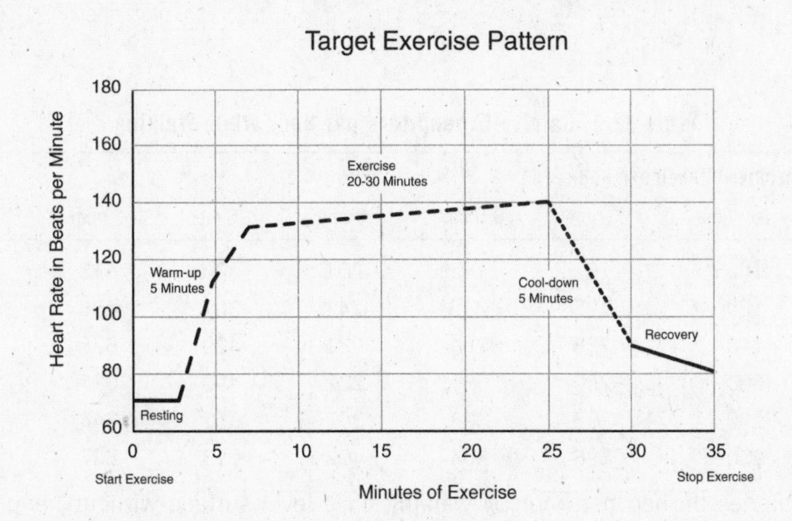

The diagram shows a desirable pattern of heart rate for an aerobic exercise session.

FIGURE 22-2.

WAYS TO INTENSIFY LOW-IMPACT AEROBIC EXERCISE

There are a variety of ways to intensify aerobic exercise while still maintaining a low level of impact. You can vigorously swing your arms, which will increase your heart rate and also boost your calorie consumption by 5 to 10 percent.

You can do interval walking: alternate several minutes of very brisk striding with several minutes at a more moderate pace. This also increases your heart rate and makes the activity more interesting. If you can walk on sand, you'll increase the calories expended by as much as 30 percent.

Or walk at a 10 percent incline, and you'll nearly double your expenditure of calories. Of course, it is only in an Escher drawing that you can walk uphill indefinitely without ever going down.[15] To achieve this effect in real life, you might use a treadmill. The ultimate in walking uphill is stair-climbing. You should be careful, however—walking up metal or concrete stairs carries a high risk of serious injury. A safer option is a stair-climbing machine. By walking two steps per second, a 150-pound person can burn more than 1,000 calories per hour.

Another very effective way to increase both heart rate and calorie burn is by carrying hand weights (unless you have heart disease, hypertension, or back problems of any kind). Use 1- or 2-pound weights and a controlled pattern of arm movements. Do not swing the weights wildly.

Table 22-1. Calorie Expenditure per Hour While Walking

CALORIES BURNED PER HOUR					
Weight (lbs)	3 mph	3.5 mph	4 mph	4.5 mph	5 mph
100	162	181	201	306	413
120	195	218	241	367	496
140	228	254	281	429	578
160	260	291	322	490	661
180	293	327	362	552	744
200	326	364	402	613	827

Calories burned per hour by walking on a level surface without hand weights

Table 22-2. Calorie Expenditure per Hour While Walking, with Incline

CALORIES BURNED PER HOUR			
Weight (lbs)	Flat surface	5 percent incline	10 percent incline
100	162	229	296
110	179	252	326
120	195	275	355
130	212	298	385
140	228	321	141
150	244	345	445
160	260	367	474
170	276	389	503
180	293	413	533
190	309	435	562
200	326	459	593

Calories burned per hour walking 3 miles per hour nearly doubles with a 10 percent incline.

LOW-IMPACT OPTIONS

While we would never discourage someone who runs on a regular basis from continuing this excellent aerobic activity, it is important to point out some disadvantages to jogging or running. These are, in essence, a series of jumps, each conveying a load to your feet and legs equal to three times your weight or more. The force on your feet is even greater, up to 30 times the force of gravity. This sends a shock wave through your body at 200 miles per hour, which is absorbed by your bones and soft tissues.[16]

All of this puts cumulative stress on your body, particularly the feet, ankles, lower legs, and knees. Over time, this can cause shin splints, tendinitis, stress fractures, orthopedic difficulties, and other problems. A large percentage of runners develop injuries, some of which can be serious.[17] To some extent, the potential for injury has been lessened by the development of well-designed running shoes. We strongly recommend that you get appropriate footwear if you wish to consider this form of exercise. It is also best to run on a relatively soft surface, such as dirt, grass, or a running track, as opposed to concrete. There is no question, however, that you can get intense aerobic benefits from running, and it certainly is efficient.

Swimming is low impact too, although it is important to use proper form. One advantage of swimming is that it is easier to move your limbs in the water because of the significant reduction in your apparent weight. That's why swimming and water aerobics (exercise in the water, using a flotation ring) are ideal forms of exercise for the elderly and others who suffer from muscle or joint infirmities and injuries. Swimming is also excellent for active people who, because of injury, cannot continue their usual workout. Because swimming is a non-weight-bearing form of aerobic activity, it is much less stressful on the body than other forms of exercise.

Table 22-3. The Calorie-Burning Value of Different Forms of Exercise

PER HOUR Activity	120-lb person	150-lb person	170-lb person	190-lb person
Skiing, cross-country uphill	897	1,121	1,270	1,420
Squash	694	867	983	1,009
Running (6.7 mph)	632	790	895	1,000
Jumping rope (80 per minute)	537	671	760	850
Swimming (fast crawl)	511	638	723	808
Basketball	452	565	640	715
Aerobic dancing (intense)	442	552	626	700
Walking (climbing hills with 11-lb load)	422	528	598	668
Skiing (moderate cross-country)	389	487	552	617
Tennis	357	446	505	565
Aerobic dancing (medium)	337	421	478	534
Cycling (9.4 mph)	327	409	464	518
Walking (normal pace)	262	327	371	415
Canoeing (leisure)	144	180	204	228
Writing (sitting)	95	119	134	150
Standing	88	110	125	140
Eating (sitting)	75	94	107	119
Resting (lying down)	72	90	102	114

Data from: W. D. McArdle, F. I. Katch, and V. L. Katch. 1991. *Exercise Physiology: Energy Nutrition and Human Performance*. 3rd ed. Philadelphia: Lea and Febiger.

Aerobic classes can be ideal, but we would encourage the low-impact variety. A competent instructor will guide you through the appropriate stages of stretching, warm-up, aerobic exercise, and cooldown. Aerobic dancing is also enjoyable; moving to music can help keep up both momentum and motivation.

BRIDGE THREE

STRENGTHENING THE SKELETON

Your skeleton is the body system that takes the most punishment from exercise. Today's technology includes anti-inflammatory drugs and, in more serious cases, surgery. Both approaches have serious limitations. In an impressive demonstration, a research team at Northwestern University has designed molecules that self-assemble into a structure that has the basic features of human bone at the nanoscale.[18] Jeffrey D. Hartgerink, the lead author of a paper in the journal *Science* describing the research, said, "Re-creating natural bone structure at the nanoscale level . . . is what we set out to do with our experiments, and we succeeded." The three-dimensional structure includes features such as collagen nanofibers that promote infusion with calcium and other minerals, like normal bone.

A promising application for repairing broken bones is to replace surgery with a simple injection of self-assembling gel-like material into the damaged area. Professor Sam Stupp, who leads the project, describes the goal: "You start with something that's liquid, that is injectible, but through self-assembling and mineralization processes becomes a hard, bonelike material."

Ultimately, nanotechnology will allow you to continually maintain and enhance your skeleton. Interlinking nanobots will provide the ability to augment and ultimately replace your bones, bypassing the aging process. Replacing portions of the skeleton today requires painful surgery, but replacing it through nanobots from within can be a gradual and noninvasive process. The human skeleton version 2.0 will be very strong, stable, and self-repairing.

The compelling benefits in overcoming profound diseases and disabilities will keep these technologies on a rapid course, but medical applications represent only the early adoption phase. As the technologies become established, there will be no barriers to using them to realize a vast expansion of human capabilities. In our view, expanding our potential is precisely the primary distinction of our species.

Bicycling is both aerobic and low impact if proper form is used. One disadvantage of bicycling is the rapid change from high intensity to low exertion. Experienced cyclists are able to maintain a fairly even expenditure of energy through appropriate gear changes, but this takes practice. Continuous aerobic exercise is easy to achieve on indoor stationary bikes. When cycling outdoors, be sure to use a safety helmet and avoid roads with heavy traffic.

Finally, cross-country skiing is an ideal form of exercise, providing a workout for both the upper and lower body. It is vigorous and invigorating, particularly in a lovely, snowy field. The only disadvantage is that it is not always available. Another exercise that uses both the upper and lower body is rowing (in a boat or on a machine). Here again, proper form is essential to avoid lower-back injury.

GETTING INTO THE HABIT

After safety, the most important consideration in an exercise program is regularity. We suggest you pick a time of day to exercise that can become a normal part of your routine. It is not a good idea to exercise within 30 to 45 minutes of eating, so many people find it convenient to exercise right before breakfast, lunch, or dinner. It is also not ideal to exercise just prior to bedtime, particularly if you tend to have difficulty sleeping.

Since it is desirable to exercise almost every day, and it isn't always possible to exercise outdoors given the vagaries of weather, a membership at a health club can be both instructive and motivational.

It is also a good idea to have some equipment at home. For many people, this is the most efficient way to make exercise a regular part of their day. Although not inexpensive, the one piece of equipment that we strongly recommend is a treadmill. If your primary aerobic activity is walking, having your own treadmill is an ideal way to assure a continuous program. Otherwise, maintaining a regular walking program may be impossible, unless the weather in your area is particularly accommodating. Most at-home exercisers prefer a treadmill to a stationary bicycle, which can become tedious.

In fact, tedium is a major reason that people fail to maintain their exercise routines. If you find the right activity for you, it can be an enjoyable part of your daily life. Walking can be a refreshing way to tour your neighborhood. Other ways to add interest include exercising with friends, listening to music, or watching television. Or use your exercise period as a time of reflection, an opportunity to let your thoughts wander. Physical exertion combined with mental relaxation is an ideal combination to reduce stress.

You may also wish to vary your routine—walk one day, bicycle another—as a way of staving off boredom. What is most important is that you develop a program that you enjoy and look forward to. Many people report that once they become accustomed to an exercise routine, it feels awful to stop. Aerobic exercise results in the release of endorphins, pleasurable hormones that represent an additional reward and an incentive to maintain a regular program.[19]

Another reason people give up exercise is time pressure from other responsibilities. We all have many obligations: time with family, eating right, work demands. It is important to note, though, that exercise will pay for itself. The dramatic improvements in your energy level, ability to sleep, and sense of well-being mean that you are not really "losing" time. In fact, the busiest and most successful professionals and business executives are frequently the most diligent about their exercise routines.

Traveling need not derail you either. One reason we recommend walking as a mainstay of any exercise program is that, while it is not always possible to find a bicycle or a swimming pool, let alone a location for cross-country skiing, opportunities for walking are usually readily available. However, with the increasing interest in fitness, many hotels do provide exercise rooms, which you can make a requirement when selecting a place to stay.

As with a weight-loss program, it's important to get on the right track and make a commitment to continue exercising. If you keep up the program, you cannot help but advance each week in your endurance and fitness. Seeing a record of your progress will motivate you to keep going. This

No Pain, No Injury

Forget "no pain, no gain"—exercise should definitely *not* hurt. There is no harm in breaking a sweat. In fact, it is difficult to achieve your training-heart-rate range without perspiring. But you should avoid pain. Pain is demoralizing and, depending on the source, can be an indication of injury, angina, or other conditions that should not be tolerated. A simple test to see if you are overexerting yourself is the talk test. If you are too short of breath to carry on a conversation, you are working too hard and should slow down.

doesn't need to be tedious. If you are walking, jot down how far you went and how long it took. You can even do this just once a week to see how your capability improves over time.

Don't exercise if you are ill, particularly if you have a fever. If you miss a week or more of your routine, restart your program cautiously once you are well. You lose fitness twice as quickly as you develop it, so don't be discouraged if it takes time to get back to the level of fitness you reached before the illness.

One last tip regarding aerobic exercise: Take advantage of every opportunity to use your body rather than the array of labor-saving devices that surround us. Use the stairs instead of the elevator. Walk or bicycle to your destination instead of hopping into the car for short trips. These are not substitutes for your regular routine, but this is a great way to work in extra exercise.

ANAEROBIC EXERCISE

The second type of exercise that we recommend is anaerobic, or strength training. This is not only for men; women can benefit too by reducing fat and improving their muscle strength without significant increase of muscle mass. Strength training is the process by which you increase strength and muscular endurance.

You have a choice of free weights or weight machines. Initially, we recommend you start with machines. Proper lifting technique and balance are crucial in weight training to avoid injury and train the correct muscles.[20] Free weights require more technique and balance, and sometimes a partner.

Most health clubs provide introductory training sessions on weight machines and display diagrams and explanations of the exercises. Most modern machines can be used for multiple exercises. Finding a friend or two with whom to train might make the experience more fun and provide a check on your technique.

Do not worry. Women who weight train will not become muscle-bound. They can expect to gain only about 10 percent in muscle size after three to six months, according to the ACSM, but their strength may increase 30 to 50 percent.[21] In fact, the ACSM points out that women often have more to gain functionally from weight training than men because they tend to start from a weaker baseline. A weight-lifting program will tone your whole body and assist with weight loss. Also, for an older woman, weight training can be a vital lifestyle choice by enabling her to maintain the strength to walk, climb stairs, or lift groceries.

Weight-Training Tips

- Work out at least three times a week on alternate days.
- Ensure that you know and use proper lifting technique. Don't compromise form for a higher weight.
- Remember to breathe properly. Rest if you are out of breath.
- Perform at least one set of exercises on each major muscle group. First work the large muscle groups, such as the chest and back, and then the smaller muscles, such as the biceps.[22]
- Use slow, careful, controlled movements.
- Do 8 to 12 repetitions per set.
- Vary your program and increase weights as you progress.

"Generally, sedentary people can lose up to 10 percent of their lean muscle mass each decade after age 30," says Edward Laskowski, M.D., a physical medicine and rehabilitation specialist and co-director of the Sports Medicine Center at the Mayo Clinic in Minnesota. "If you don't do anything to replace that loss, you're losing muscle and increasing fat," says Dr. Laskowski. "But if you do weight training, you can preserve and enhance your muscle mass. It's like having a V-8 engine instead of a 4-cylinder. You have a bigger engine to burn more calories because it takes more calories to keep that [larger] engine running."[23]

STRETCHING

Stretching increases the range of motion in your joints. Stretching does not, as was once thought, prevent injury. As we age, our muscles, tendons, and ligaments shorten. Flexibility training or stretching can slow down that process. In addition, hours behind a desk can lead to a bevy of aches and pains, all due to tight muscles and poor posture. Once again, stretching throughout the day is the answer.

The ACSM identifies many other benefits to flexibility training, including better physical performance, better circulation, improved posture, stress relief, and enhanced coordination and balance.[24]

Here are the ACSM guidelines for setting up a flexibility program:[25]

- Warm up first to make the muscles supple and easier to stretch.
- Focus on the major muscle groups (front and back of the legs, shoulders, chest, and so on).
- Perform the stretches at least three times a week.
- Stretch muscles slowly until you feel a slight pull, not pain.
- Hold each stretch for 10 to 30 seconds. Don't bounce.
- Start slowly and work up.

DON'T FORGET DIET

Exercise alone is not sufficient. For example, exercise will not reduce levels of LDL, the bad cholesterol in your blood.[26] There is a myth that you can eat whatever you want if you just exercise. But the list of marathon runners who have had sudden and often fatal heart attacks due to advanced atherosclerosis is a clear indication that exercise alone is not an antidote for the devastating effects of a diet high in sugar, simple starches, and unhealthy fats and low in healthy nutrients. The dietary and exercise recommendations of Ray & Terry's Longevity Program work together synergistically. The sum of the benefits of applying both principles is greater than the benefits of either one alone.

23

STRESS AND BALANCE

"Don't just do something. . . . Sit there!"

—meditation poster

"We don't stop playing because we grow old; we grow old because we stop playing."

—George Bernard Shaw

Fear of terrorist attacks, proliferation of weapons of mass destruction, international tensions, not to mention the aggravations of the morning commute—it's no wonder everyone is a nervous wreck. But stress is hardly a new phenomenon. Our ancestors lived with the danger of animals that might attack at any moment, hostile neighboring clans, and their biggest concern—the uncertainties of the next hunting or planting season. As recently as one or two centuries ago, human life was extremely precarious. A single misfortune, such as an all-too-common infectious disease (remember that antibiotics are a relatively recent innovation), brought disaster. More recently, one might recall that World War II was hardly a period of stability, and remember the fear of all-out nuclear conflict during the cold war. As Charles Dickens wrote, the human condition—at *any* time—inherently resides in the best of times *and* the worst of times.

We can always find reasons to feel stressed, but the predominant source of stress in our lives is internally generated. Next to the substances we put into our bodies, constructively handling the inevitable pressures of life is the most important objective in maintaining our health and enjoying the journey. To paraphrase Franklin D. Roosevelt, we have nothing to fear but fear itself—and other stressful emotions.

It is an oversimplification to say that stress is bad for you. In many circumstances, certain forms of stress can be very damaging to one's health.

Other situations, situations that we would consider very stressful, may not be damaging and can even be energizing. Consider the experience of the European populations during World War II—some hiding in subway tunnels while their homes were firebombed, others fleeing their homes when caught between opposing armies. Yet the rate of heart disease did not increase during those terrible years. On the contrary, there was a dramatic decrease in heart disease in those countries in which food rationing was imposed during the exact period of time in which rationing was in effect.[1] These people were forced to eat rations of vegetables and forgo their usual diets of meat, cake, butter, and cream. Although war is by definition a period of great conflict, we find suicide rates are low during times of war and rise during periods of peace and stability.[2]

On the other hand, there are many examples of life situations—and perhaps most important, our reactions to them—that appear to have a profound negative effect on our ability to resist or overcome disease. Many studies have shown that chronic stress is a major contributor to disease.[3] Conversely, studies have shown the ability of the mind to assist in both resisting and overcoming disease through relaxation methods such as meditation.[4]

WHAT IS STRESS?

Essentially, stress is the arousal of the body and mind to demands and challenges. That may not sound like such a bad thing, and indeed we need a certain amount of challenge to avoid apathy and boredom. Even positive changes in our lives represent stress. The term *eustress* refers to our reaction to constructive change: a job promotion, an award, getting married, even going on a vacation. The type of stress that appears to be harmful to our health is the excessive and persistent activation of our ancient fight-or-flight response or *distress*.

When our paleolithic ancestors confronted a menacing foe, whether animal or human, they had the choice of confronting the danger or fleeing; hence the term *fight or flight*. The process starts with perceiving danger. Once that judgment is made, the rest is automatic. A perception of danger by the amygdala, a brain region responsible for dealing with danger and fear, triggers a chain reaction of neural and hormonal changes that puts the body into a state of readiness for action. The hypothalamus signals the pituitary gland to produce ACTH (a stress hormone), which in turn stimulates the adrenal cortex to produce cortisol. Cortisol is carried in the bloodstream and causes a dramatic but temporary increase in metabolism (energy) and stimulation of the brain's memory centers to work more quickly. A spinal reflex signals the adrenal glands to produce adrenaline and noradrenaline, also known as epinephrine and norepinephrine.

These hormones have a dramatic effect on the body. They nearly halt the digestive process and increase blood pressure, blood sugar, cholesterol levels, fibrinogen levels (which speed up clotting), and the rates of heartbeat and breathing. Other effects include dilation of the pupils and the mobilization of internal energy stores for the possibility of extreme physical exertion.[5]

The fight-or-flight mechanism is one of those adaptations that were useful to the survival of our species during the times that our bodies were evolving. It remains a useful mechanism from time to time, but the continual and persistent activation of this mechanism is a major contributor to heart disease, type 2 diabetes, stroke, cancer, rheumatoid arthritis, depression, and accelerated aging.[6] If your body is in a constant state of emergency, the temporary effects, such as increased blood pressure and cholesterol and decreased blood flow to the liver and digestive organs, become chronic.[7] As we reviewed in chapter 19, high levels of cortisol accelerate a wide range of aging processes.

A Case Study

At age 71, Woody Strong was diagnosed with inoperable cancer by specialists in Denver. He was told that he had one year to live. After reflecting on the situation, he decided that there was nowhere else he would rather spend his last year than in Nepal, among his many friends. He had been given the name of "father" by many Nepalese because of his kindness and the frequent help he had provided them by bringing in medical and school supplies.

While in Nepal, his Nepalese "family" convinced him to visit a renowned healer in the remote Everest region. With a deep mixture of skepticism but respect for his friends' wishes, Woody consented. For the next five days he underwent an intensive healing ceremony. He laughed and cried and broke into profuse sweats "for no reason." At the end of the ceremony, the lama told Woody that he was "cured."

Shortly thereafter, he returned to Denver for a routine exam. To the amazement of his oncologists, his cancer had undergone a "spontaneous regression." The lama had told the truth—the cancer was gone! Woody rededicated his life to building schools and hospitals in Nepal.[8]

THE TYPE A PERSONALITY REVISITED

The so-called type A personality, which gained publicity in the mid-1970s, describes a person who is hard-driving, overly ambitious, impatient, competitive, aggressive, always working toward a deadline, and generally a workaholic.[9] In contrast, type B personalities are described as relaxed, easygoing, accepting, and complacent. Early studies suggested that having a type A personality was a risk factor for the development of heart disease. More recent studies have cast doubt on that assertion, at least as it was originally defined. Contemporary research indicates that of all the aspects that make up the classic type A pattern, the only ones that appear to be related to an increased risk of heart disease are those involving anger, cynicism, and hostility.[10] People with hot tempers and/or suspicious, angry, hostile natures are more likely to die from heart disease. Other type A characteristics, such as competitiveness, ambition, even workaholism, were not found to be risk factors.

A study in the journal *Circulation* reported that people who become angry easily had a threefold increase in risk of a heart attack or sudden cardiac death, compared with people who scored low on the anger scale.[11] As another example, a long-term study was conducted on a group of 118 lawyers who had taken the Minnesota Multiphasic Personality Inventory, a standard personality test, 25 years earlier while in law school.[12] Those who had higher scores for hostility had a death rate from heart disease that was more than four times higher over the following 25-year period than those with low scores. Another dramatic study was a 25-year follow-up study of 255 physicians.[13] Here the hostile physicians were six times more likely to die than the group who scored low.

Researchers have discovered a similar link between suspicious personalities and increased mortality rate, although suspiciousness is linked to hostility and anger. A study reported in 1987 at Duke University followed 500 men and women at an average starting age of 59 for a period of 15 years.[14] Men who had a suspicious personality were twice as likely to die as their more trusting peers. The suspicious women were 29 percent more likely to die than their more trustful peers. There are many other studies that demonstrate the healthful benefits of a positive and trusting outlook on life.[15]

WOMEN DO STRESS DIFFERENTLY

A study at UCLA asserts that the different levels of hormones between men and women affect the stress cascade.[16] The hormones in women released as part of the stress response include oxytocin (a female reproductive hormone), which has a calming effect and tends to encourage "tend and befriend" behaviors, such as protecting children and gathering with friends. Estrogen enhances this effect of oxytocin, whereas testosterone, which men produce in higher amounts than women, causes hostility. This appears to be another way a premenopausal woman's relatively high estrogen levels may provide protection from heart disease and other negative effects of long-term stress.

THE FOUR C'S: CHALLENGE, COMMITMENT, CURIOSITY, AND CREATIVITY

However, there are several *constructive* reasons why you might be eager to achieve a set of goals. These can be characterized by the four C's: challenge, commitment, curiosity, and creativity. A challenge is a goal that, while difficult to achieve, is worthwhile and meaningful to the individual. Commitment is the ability to place an overriding priority on attaining a challenging goal, to see progress toward a goal as more important than sac-

The Strength of the Spirit

Meyer Friedman, a cardiologist and one of the originators of the type A concept, has spent more than 30 years studying the link between behavior and personality and heart disease. Friedman describes the "negative type A" as a one-dimensional personality, someone with a profound absence of a spiritual life. By spiritual life, Friedman isn't referring specifically to a life with strong religious beliefs, but rather a life that has meaning, that attaches importance to human relationships and to other social and cultural concerns that enrich our lives.[17] If the hard work and apparent impatience of the type A person emerges from concerns and beliefs that are deeply rooted in their own structure of values, this commitment to achievement appears to support cardiac health. If the pattern is the result of the "erosion of personality" that results from chronic suspiciousness, it is destructive.

This perspective sheds light on why the years of World War II, as stressful as they were, did not cause an increase in heart disease. The populations in these wars were not passive bystanders. In previous centuries, war was an activity engaged in primarily by professional armies. But in this century, war has been a struggle of entire societies. The first two C's, challenge and commitment, certainly characterized the attitude of these populations. And since we might regard war as the father of invention and a major impetus to the creation of technology, we can include creativity as well.

rifices that may be required. Curiosity is a desire for knowledge and an openness to what life offers. Creativity is the ability to create knowledge, to harness one's curiosity to discover new wonders.

People who are characterized by the four C's often appear to be type A, since their high level of commitment and willingness to take on challenges make them appear driven and hardworking. But their work ethic may be rooted in a strong sense of self and purpose. The negative type A pattern is driven by something different—by cynicism, anger, and hostility, by a persistent sense of being treated unfairly and a need to be aggressive to get what is due.

STRESSFUL EVENTS AND HOW THEY RATE

Medical researchers Thomas Holmes and Richard Rahe studied how 5,000 individuals reacted to the events in their lives. They developed the Holmes-Rahe Social Readjustment Rating Scale, a stress scale ranking various events ranging from Christmas to the death of a spouse. They found that the higher your total stress score in any particular year, the more likely you were to become ill.

Table 23-1: Stressful Events and How They Rate[18]

Event	Rating
Death of a spouse	100
Divorce	73
Death of an immediate family member	63
Personal injury	53
Personal illness	53
Marriage	50
Fired from job	47
Retirement	45
Change in health of immediate family member	44
Death of a close friend	37
Change in work	29
Children leave home	29
Outstanding achievement	28
Trouble with boss	23
Vacation	13
Christmas	12
Minor law violations	11

These ratings represent average stress levels, which Holmes and Rahe were able to statistically associate with levels of disease. We each have different capacities to cope with stressful change. Stress is inherently an internal phenomenon. How we experience events depends on our outlook and personality. It might seem that the easygoing type B personality would have an easier time accepting change than the type A. However, the passivity of some type Bs may allow difficult situations to fester and become worse, whereas a secure and well-grounded type A person may act sooner to address problems before they erupt into acute danger—such as the largely invisible degenerative health processes we address in this book. A type A person may also see change as a challenge and use his or her creativity to effect a positive result from an otherwise difficult situation.

SYMPTOMS OF STRESS

Physical symptoms often accompany the failure to deal constructively with stress. These include high blood pressure, headaches, rapid heartbeat, aches and pains, muscle tension, and gastrointestinal discomfort. Behavioral indications include difficulty sleeping; compulsive behavior involving food, drugs, alcohol, sex, or gambling; concentration problems; accident proneness; and social withdrawal. Emotional signs include nightmares, crying spells, feelings of worthlessness, excessive or compulsive worrying, mood swings, restlessness, and anxiety. Spiritual signals include a sense of emptiness, loss of life's meaning, excessive confusion, and doubt about one's direction in life.

These may appear to be fairly general symptoms, since there is no simple test to determine how well you are dealing with the stress in your life; it is a matter of judgment. But it is a crucial judgment because the link between our health and our ability to deal constructively with our lives is now strongly supported by a growing body of scientific literature. Most of us can benefit from improving the balance of our lives and our ability to cope with life's challenges.

FALSE STRESS RELIEVERS

Before we discuss how to beneficially manage stress in our lives, let's first review what not to do, starting with food. Compulsive eating isn't an effective way to reduce stress. Even if the foods you overeat are healthful, this habit will contribute to excess weight and will stress your gastrointestinal system.

Moreover, the foods that people eat when combating tension and anxiety tend to be high in sugar, high-glycemic-load starches, unhealthy fats, and calories. As we have all experienced, this approach just doesn't work. You may feel some temporary satisfaction while you are eating, but when you're done you are likely to be physically uncomfortable and have associated feelings of guilt, along with your original anxiety.

Next on the list are the three drugs most abused in American and most other advanced societies: nicotine, alcohol, and caffeine.

Nicotine appears to ease anxiety and promote a sense of alertness by stimulating the production of a variety of hormones, including adrenaline. Smoking also provides oral gratification associated with our earliest feelings of satisfaction from sucking. Yet cigarette smoking is linked to more than 442,000 American deaths a year from heart disease; lung, larynx, and other cancers; emphysema; and other respiratory and circulatory problems.[19] Even putting these devastating diseases aside, smoking substantially reduces the oxygen available to the body's tissues. And the constant assault of carbon monoxide, nicotine, tar, and dozens of other poisonous gases dramatically lowers one's sense of well-being. This deterioration of virtually all of the body's systems clearly adds to the overall level of stress.

Alcohol follows closely after nicotine as our most abused drug, accounting for more than 100,000 deaths per year.[20] Alcohol abuse is a major risk factor in heart disease. It can cause severe liver damage, hypertension, gastrointestinal disorders, and brain damage, and it contributes to a variety of cancers.[21]

Moderate consumption of alcohol is a reasonable stress reliever. Five to 10 drinks per week is associated with reduced rates of heart disease. Alcohol thins the blood, making the formation of an artery-blocking blood clot less likely. It can foster relaxation and the connectedness of social interaction, which are

Alcohol: A Little Goes a Long Way

A study conducted at the Harvard School of Public Health followed 50,000 men for two years and reported a significant protective effect from moderate use of alcohol.[22] Those who drank the equivalent of one or two glasses of wine a day had a 26 percent reduction in the risk of heart disease, compared with those who drank no alcohol. The researchers attributed this benefit to alcohol's blood-thinning effect, as well as the ability of alcohol, when it is metabolized in the liver, to cause an increase in HDL-C levels. At higher levels of consumption, the researchers found the usual harmful effects of alcohol, such as cirrhosis of the liver, high blood pressure, and behavioral problems.

also of value. The dangers of alcohol abuse, however, are well known. Alcohol also contains calories, and relatively high-glycemic-load ones at that.

Although occasional use of alcohol can provide a feeling of relaxation and euphoria, alcohol is basically a depressant. It isn't an answer for chronic feelings of stress and tension. Any attempt to use it for relief of deep-seated feelings of anxiety and isolation are likely only to deepen these feelings, as well as damage one's ability to maintain relationships.

Modest use of **caffeine** can be helpful for improving concentration and alertness. However, the high levels of caffeine found in coffee, colas, and other soft drinks make it our most commonly abused drug. Virtually the entire adult population uses it, and tens of millions of Americans abuse it: more than 22 million Americans drink five or more cups of coffee per day.[23] It is commonly used to combat chronic sleep deprivation, although, ironically, it is a major contributor to sleeplessness. Although small amounts of caffeine improve alertness, excessive amounts create a jumpy yet still tired person. Caffeine can cause headaches, restlessness, digestive problems, heart arrhythmias, and hypertension. Drinking several cups of coffee can significantly elevate blood pressure and adrenaline for more than two hours, compounding the effects of stress.[24]

Caffeine is surprisingly addictive, more so than most people realize. Even very small amounts can contribute significantly to problems of chronic anxiety and panic disorders in some people.[25] One psychiatrist who specializes in panic disorders indicated that at least half his patients were able to eliminate their symptoms by cutting out caffeine. He also found that even a single cup of coffee or one caffeinated soft drink could reactivate panic disorders and sleeplessness. While this is an anecdotal report rather than a controlled scientific study, people suffering from chronic anxiety, panic disorders, and sleeplessness should at least test the impact of eliminating caffeine from their diet.

Our recommendation is to switch from coffee to tea, preferably green tea, which has about one-quarter of the caffeine level of coffee. Tea contains L-theonine, which promotes healthful relaxation. Black and green teas also contain powerful antioxidants. Given the relatively low concentration of caffeine, it's difficult to overdose on caffeine when drinking tea.

Another group of drugs on our list of false stress relievers are the **benzodiazepines**. These drugs are classified as minor tranquilizers and in larger doses as hypnotics, or sleeping pills. Benzodiazepines appear to relieve anxiety in the short term, so they are among the most commonly prescribed drugs in the United States and Canada.[26] However, they also create a chronic pattern of heightened anxiety and drug dependence in the long term. Valium (diazepam), Xanax (alprazolam), and Tranxene (chlorazepate) are benzodi-

BRIDGE TWO

STRESS AND ADDICTION

In the Bridge One material in this chapter, we talk about the false stress relievers you should avoid. We know that for the tens of millions of people in the United States who have an addiction problem, this is easier said than done. As we have seen throughout this book, we are rapidly learning about the detailed molecular pathways that underlie all disease and aging processes, and the same observation holds true for addiction. For example, researchers at the University of Washington at Seattle have described the important role of a specific chemical called neuropeptide Y (NY) in alcohol addiction. Reporting in the journal *Nature*, they showed how blocking NY in genetically engineered mice dramatically exacerbated alcohol abuse, whereas increasing NY ameliorated it.[27] In related research, Emory University's Yerkes Primate Research Center in Atlanta has described the role of the CART (cocaine and amphetamine regulated transcript) peptide, which appears to interact with NY in the addiction cycle.

The common wisdom on addiction is that addicts start out by self-medicating with addictive drugs such as alcohol for feelings of tension, depression, and anxiety. The drugs provide temporary relief but end up deepening the lack of well-being. George Koob, a scientist with Scripps Research Institute and one of the leading addiction researchers, has written hundreds of articles on drug addiction that support this view.[28] Heavy drinking, according to Koob, reduces serotonin, dopamine, GABA, opioids, and other chemicals in the brain that provide pleasure and related feelings of satisfaction. At the same time, this behavior increases levels of CRF (cortisol releasing factor), a stress chemical that deepens feelings of depression and anxiety. The result is the vicious cycle of addiction. Over time, the amount of alcohol or other drugs required to avoid strong feelings of distress increases. Koob has suggested that elevated CRF levels and diminished dopamine levels indicate possible addiction.

azepines, as is the antiseizure medicine Klonopin (clonazepam) and the sleeping pill Halcion (triazolam).

While these drugs may have some value in assisting someone through a brief period of acute stress, sustained use can be both ineffective and dangerous. Benzodiazepines frequently lose their effectiveness as sleep-inducing agents. Sometimes they perpetuate insomnia, thus turning an acute problem into a chronic one. Other common effects include restless or fragmented sleep, nightmares, lethargy, and daytime fatigue.[30]

There are specific genes that appear to cause a person to be susceptible to drug addiction. Research at the Oregon Health Sciences University in Portland demonstrated that mice genetically engineered to lack the dopamine receptor D2 (DRD2) gene were much less likely to become addicted to alcohol. The DRD2 receptor is one of the brain receptors involved in the regulation of feelings of satisfaction and reward. "Taking the DRD2 receptor away cut alcohol consumption in half," reported Tamara Phillips, lead author of the study.[29]

We are not likely to find a single gene or chemical that by itself controls addictive behavior. David Grandy, Phillips's coauthor, says that "research doesn't point toward a single 'alcoholic' gene but rather toward a complex interaction between several receptors and systems in the brain." However, with the increasing power of our tools to track precise molecular interactions, we are closing in on a detailed understanding of the biochemistry of addiction. With a growing arsenal of methods to create and inhibit molecules such as peptides and enzymes, to block gene expression, and ultimately to create entirely new genes in adults, we can expect effective drugs to combat addiction to emerge over the next 5 to 10 years. If we include nicotine, caffeine, and even certain unhealthy foods (such as high-glycemic-load carbohydrates) on our list of addictive substances, the importance of this development is hard to overstate.

We reported that a safe and effective sleeping or antianxiety medication for more than short-term use does not yet exist. The widely used benzodiazepine drugs, for example, are included among drugs that foster addiction. A new generation of sleep aids is expected to be approved soon that appears to be relatively nonaddictive and also more effective. Because of the addiction concern, current sleep drugs are approved for only short-term use. Neurocrine Biosciences Inc. has asked the FDA to approve its drug Indiplon to be used for periods of several months. Sepracor's Estorra drug is also looking for approval for use longer than current sleep drugs. Pfizer is preparing a new sleep drug, 200,390, claiming that it will lengthen the important slow-wave, or delta, stage, the deepest phase of sleep.

Perhaps most serious, these drugs can be addictive and can cause chronic depression. Each antianxiety medication has its pros and cons, but there is widespread agreement in the medical community that a completely safe and effective sleeping or antianxiety medication for more than short-term use simply does not exist. These drugs should be used with caution in acute situations, although there are many cases where certain psychotropic drugs are appropriate, such as for clinical depression, bipolar syndrome, and other psychiatric conditions.

ILLEGAL DRUGS

We discussed the problems associated with legally available drugs first because it is the larger problem in terms of health. Probably because illegal drugs are illegal, their use and abuse is less frequent. However, we don't mean to minimize the problem of illegal drugs, particularly cocaine. Cocaine upsets the regulation of dopamine, norepinephrine, and other neurotransmitters in the brain, which is vital for one's ability to think as well as for one's sense of well-being. The first biochemical effect of cocaine is caused by a surge of dopamine in the brain, experienced as a strong sense of pleasure. As dopamine and other neurotransmitters are depleted from the brain, the effect becomes unpleasant.

People often attempt to restore the euphoric state by taking more cocaine. This cycle can quickly lead to dependency, particularly by individuals with a genetic susceptibility to addiction. Ultimately, extensive repeated use can diminish the brain's supply of dopamine on a sustained basis. At this stage, a person requires a large dose of the drug just to achieve neurotransmitter levels that are merely not agonizing. The situation of the cocaine addict is quite desperate; his or her regulation of dopamine and other neurotransmitters is out of control. Unfortunately, regaining that control is very difficult.[31]

The social urgency of the problem stems from the psychological and behavioral impact of cocaine addiction. Unlike heroin, which produces a relatively withdrawn state (at least for those four-hour periods during which an addict's need for heroin is satisfied), cocaine produces a state of paranoia, irritability, and aggressiveness. This can be a dangerous state of mind, particularly when combined with one other attribute of cocaine-induced psychosis: loss of certain inhibitions, such as those concerning interpersonal violence. Crime related to heroin addiction is typically economic: the heroin addict is desperate for his or her next fix. But the potential violence of a cocaine addict is worse: he or she is dangerous with or without the drug.

Which problem is worse—legal or illegal drugs? In terms of lives lost and sheer impact on health and well-being, one would have to say legal drugs. Fifty million Americans smoke, and any regular use of tobacco has to be considered abuse.[32] About 30 million Americans abuse alcohol.[33] More than 20 million Americans abuse caffeine.[34] About 20 million abuse prescription drugs, with some estimates running much higher. There is, of course, a great deal of overlap in these numbers, since, on average, one drug abuser abuses about two drugs. But the number of Americans that abuse one or more legal drugs is approximately 75 million—nearly half the adult population!

There are about 6 million individuals addicted to "hard" illegal drugs, such as cocaine and heroin. About 16 million Americans, or 7.1 percent of the population over 12 years of age, use illicit drugs on occasion.[35] So the health implications are comparably smaller. But when assessing the problems

Ideas for Living More Fully

- Be aware of seasons.
- Get some sun each day (about 5 to 10 minutes).
- Practice lucid dreaming (see Ray's use of lucid dreaming described in chapter 10).
- Seek quiet environments.
- Seek out beauty.
- Don't be unduly attached to things (consider how many wars are fought over rocks and sand).
- Don't be unduly concerned with what other people think of you (except that you may be able to learn something useful about yourself by listening to valid criticism and taking it to heart).
- Give criticism very sparingly (and only if you really need to help someone).
- Keep learning.
- Keep challenging yourself.
- Be optimistic.
- Be grateful (and express it).
- Give yourself to someone.
- Have integrity.
- Take responsibility for your well-being.
- Keep a journal (it helps put concerns into perspective).
- Never retire, but do change the nature of your work from time to time.
- Keep an open mind.
- Most important, practice the four C's: challenge, commitment, curiosity, and creativity.

of illegal drugs, you have to add the compounding social dislocation, crime, and violence.

Regardless of how one ranks these two aspects of our drug problem, the more important perspective is to see the abuse of drugs, both legal and illegal, as part of the same problem, rooted in the same quest, however misguided, for a quick fix to chronic anxiety. Any discussion of the "drug problem" that ignores the abuse of legal drugs is bound to be ineffectual. One observer compared our national war on drugs, which largely ignores the overarching problems of tobacco and alcohol addiction, to a naval strategy that ignores the Atlantic and Pacific oceans.

RAY & TERRY'S 12-POINT PROGRAM FOR MANAGING STRESS CONSTRUCTIVELY

Having reviewed what *not* to do, let's consider how to transform stress into a constructive challenge rather than a destructive disturbance.

1. Food. Our recommendation is to follow the Ray & Terry Longevity Program nutritional guidelines. Ending the assault of a poisonous diet on one's body and mind will avoid an enormous source of stress. The resulting sense of well-being and health will go a long way in reducing stress. You will sleep better. Your brain and other tissues will be better oxygenated. These and the many other benefits we have reviewed in this book all have a major impact on your level of stress. Following our nutritional recommendations will also enable you to achieve and maintain your ideal weight. Avoiding excess weight will eliminate another chronic source of stress. Of course, worrying a lot less about heart disease, cancer, stroke, and other serious diseases is undoubtedly helpful as well.

2. No addictive drugs. Closely related to the first is to ingest a "diet" that is also low in addictive drugs. For nicotine and for cocaine and other illegal drugs "low" means none. Benzodiazepines should be used with caution in acute situations of anxiety under a doctor's care. Five to 10 drinks of alcohol per week may be beneficial, but moderation is the key. The same guidance holds for caffeine, particularly by substituting green tea for coffee.

3. Exercise. Both aerobic exercise and weight training have health benefits, but aerobic exercise directly and immediately provides tangible benefits in reducing stress and promoting relaxation. The natural release of endorphins (brain chemicals that reduce stress) from continuous exertion is a healthy alternative to artificial stimulants and depressants. We explore this exercise in chapter 22.

4. Adequate sleep. Inadequate or poor-quality sleep not only contributes to increasing the level of stress but is an independent risk factor for heart disease, as we discussed in chapter 15. (Of course, excessive stress may cause an inability to sleep in the first place.) Our program for attaining a natural and healthy sleep cycle includes the following.

- Give a high priority to getting a healthy quantity and quality of sleep. If you have not slept adequately, even minor problems feel quite stressful and your entire outlook is colored negatively. Researchers have estimated that more than a third of the population is chronically sleep-deprived.[36] Consuming a lot of caffeine in the morning will only aggravate the inherent stressfulness of inadequate sleep.
- The optimal amount of sleep varies from individual to individual, although 7 to 8 hours a night is typical. If you stop consuming a toxic diet and get in touch with your body and its feelings, you will know when you have obtained adequate rest.
- The other guidelines in this book will help you to get a good night's sleep, especially those concerning nutrition, exercise, and (in this chapter) stress guidelines. Exercise promotes a natural cycle of sleep. Anyone with difficulty sleeping should get more aerobic exercise—but not just before retiring for the day.
- Maintain a regular routine, especially at night. Make it a practice to slow down and engage in relaxing activities, such as reading for pleasure, before you go to sleep. Try to stick to the same sleep schedule as much as possible.
- If you have difficulty sleeping, cut down on caffeine or eliminate it altogether. Don't consume caffeine after midday.
- Supplements that are helpful to healthy sleeping include the following (the first two—L-theonine and GABA—are useful for managing stress in general):
  - L-theonine, the ingredient in tea that promotes relaxation, is available as a supplement.
  - GABA (not kava, which we don't recommend), a neurotransmitter available as a supplement, induces a natural sense of calm and can promote sleep. Recommended dosage is 500 milligrams.

- Melatonin, a hormone that controls the body's sleep clock, normally surges at bedtime, triggering other body systems to shut down and prepare for sleep. As we get older, melatonin levels decline, which contributes to increasing difficulty in falling and staying asleep. Supplementing with small amounts of melatonin can be very helpful in achieving a restful night's sleep. The usual dosage sold over the counter—3 milligrams—is generally too high, however, and is likely to cause drowsiness the next day. We recommend a dosage of 200 to 1,000 micrograms (0.2 to 1 milligram) in a sublingual form, which goes to the brain quicker and avoids going through the GI tract, lowering effects during the day.

5. Balance. Stress isn't an isolated issue that you can deal with once a week. Handling stress effectively is a matter of gaining balance in your life. By balance, we are referring to keeping the three poles of one's life—work, family/friends, and self—in balance. If your work, for example, is raising your children, the poles of work and family overlap, but there is still a distinction between the work of child rearing and the opportunity to experience moments of love and sharing. Freud said the two great issues in life are work and love, which are the first two of the three poles we have mentioned. The third pole—one's self—refers to the importance of living a life that is satisfying and meaningful, of taking the time to assess your values and goals to understand your own needs and give them priority.

One obvious form of imbalance is represented by the workaholic, but excessive dependence on social forms of gratification, even your relationships with family and friends, or excessive concern with your own needs and desires (self) to the exclusion of others, can represent a lack of balance. A healthy balance in your life provides a time for challenge—to work hard—and a time for relaxation—to play hard. This guideline is different from many of the others in this book in that it is impossible to quantify, but it is important nonetheless to assess this issue for yourself periodically.

6. Time management. How you spend your time reflects your priorities. As an exercise, write down your priorities in terms of work, family, friends, exercise, sports, relaxation, and so on. Then, for the following week, write down how you actually spend your time. How does your allocation of time match your stated priorities? In terms of this sixth guideline, how well does your time management reflect an optimal and comfortable balance between work, family/friends, and self? With all of the pressures of modern life, it is easy for your allocation of time to stray substantially from what you desire, and also from what is healthy. If you find a discrepancy, consider how you would prefer your time be spent, and then develop a strategy for making the change. It's often not possible to make significant changes overnight. After all, you have responsibilities and obligations that cannot just be dropped. But with a well-thought-out plan, most people have a greater ability to control their destiny, not to mention their daily schedule, than they realize.

A worthwhile exercise is to develop a schedule that will accommodate your various objectives. This is particularly important if you are attempting to make a change—adding a regular aerobic exercise program, for example. To avoid unnecessary stress, try not to overschedule and overcommit—projects and plans often take more time than anticipated. Also, leave time free for problems and opportunities that arise. An overscheduled life does not leave time for spontaneity.

The value of a schedule is that you avoid procrastination that will only worsen stressful situations. You have the opportunity to establish your priorities. You decide what responsibilities you need or want to accommodate. You also learn when and how to say no. After all, you can't please everyone.

7. Take vacations. Take time periodically to change your routine. A vacation does not have to mean lying in the sun (although that may be what you enjoy). It could involve taking a week to work in your yard, paint the garage, take a course, even sort out old files. Just do something different from your usual routine.

8. Talk with someone. By talking we mean expressing your true feelings and regularly sharing your fears, worries, hopes, and delights. It is important to have someone you can really talk to without worrying about being embarrassed, making a good impression, or appearing silly.

Of course, it isn't always obvious who can fulfill that role. Sometimes it is a relative, although family members are often the source of the feelings you need to talk about, and they may not possess the kind of nonjudgmental attitude required. A good friend, perhaps, can provide the necessary trust and confidence. A spouse, partner, or lover can offer the requisite intimacy, and it is certainly desirable that you talk about your feelings in such a relationship. As with a relative, however, you may still need someone you can talk to about your partner. Other possibilities include a pastor or teacher, maybe even your boss.

Very often, this role can be filled by a therapist, psychiatrist, psychologist, social worker, or counselor trained to provide exactly this kind of supportive and understanding relationship. Unfortunately, the idea of talking to a therapist has negative associations for many people. Some people mistakenly attach a feeling of shame to the idea of seeing a therapist. It is a common perspective that there has to be something wrong with you to have regular sessions with a therapist—that you must be mentally ill, or at least neurotic. Yet, for most of us, our lives are sufficiently complex and demanding that having a professionally trained person, capable of being objective, with whom to share our feelings and important life decisions can be helpful. We don't hesitate to hire assistance for taxes, legal problems, house selection, money management, and many similar issues, so why not professional consultation on the most important issues facing us: dealing with our emotions? Our internal life is at least as complex as our tax returns and even more confusing. Most people can benefit from a relationship of this type.

Regardless of who can fill this role for you, everyone has the need to share their most intimate feelings. Just the act of articulating one's feelings to another human being has an enormously beneficial impact—provided, of course, that the person you are talking to is truly listening. It helps put difficult issues into perspective. Even painful subjects can begin to be seen in a constructive context once you articulate them. One's perspective can become increasingly distorted if this type of intimate sharing is routinely ignored in one's life.

9. Listen. It is very therapeutic to listen, to truly listen to what others have to say. When people feel that you are really listening, they will start to open up and share their inner feelings. It's just a natural human response and need. Don't be so enamored with what you have to say that you fail to listen to others. They say a wise man can learn more from a fool than the other way around.

Being a good listener is more difficult than it may appear. The first challenge is to simply allow the other person to talk, by not interrupting and by paying close attention. The more important challenge, however, concerns what you do with the information you receive. You need to keep an open mind, to try to perceive the world from the other person's perspective. Even if you don't agree with everything that is being said, provide feedback that lets the other person know you understand his or her words, feelings, and thoughts. The most critical aspect of creative listening is empathy. You'll benefit too: it's often said that you can make more friends by listening to others than by talking about yourself.

10. Regular massages. There are many schools of massage therapy, and all forms are beneficial in relieving physical and emotional stress. Shiatsu and acupressure massage are based on principles of Chinese medicine and are intended to correct imbalances according to principles of energy flow between different organ systems in the body.

11. Have a life partner. People with successful, long-term marriages or committed relationships live longer and healthier lives. Although easier said than done, we offer a few of the ingredients of a successful marriage.[37]

- Keep in mind that in disputes and arguments, you can't win—it's either win-win, or lose-lose.
- Be like the wise bamboo and bend.
- Keep your partner "special."
- Devote time to your relationship. You won't have the quality times without devoting the quantity.
- Have a "life" outside your marriage.
- Healthy flirting is okay.
- Strive to learn new things together.
- Don't stop having sex.

12. Evoke the relaxation response. The relaxation response was discovered when Dr. Herbert Benson, then the director of the hypertension section of Boston's Beth Israel Deaconess Hospital, and other researchers at the Harvard Medical School and Beth Israel Hospital studied the physical and mental effects of a variety of methods of evoking a calm state, including yoga and several forms of meditation. They discovered a hypothalamic response that was the converse of the fight-or-flight response: reduced levels of epinephrine and norepinephrine and, in turn, lowered levels of blood pressure and blood sugar and breathing and heart rates.[38] Moreover, they discovered that regular use of these techniques and regular elicitation of this response were able to produce permanently lowered blood pressure, improved sleep patterns, improved gastrointestinal functioning, improved blood flow, and other benefits.[39] A study of the elderly found that regular use of a meditation technique that elicited the relaxation response resulted in a dramatic reduction in deaths during the three-year period of the study, as well as substantially improved mental acuity and mental outlook.[40]

Research directed by Benson has cataloged a number of techniques that demonstrably and reliably produce the relaxation response. One such class of techniques is yoga, which combines meditation with stretching exercises and controlled breathing. Benson has also documented the health benefits of people experiencing the relaxation response on a regular basis.[41]

Another researcher who has developed a series of techniques in what he calls "mindfulness-based stress reduction" is Jon Kabat-Zinn. More than 10,000 patients at the University of Massachusetts Medical Center have completed Kabat-Zinn's program, which he has popularized in a series of best-selling books.[42]

Yoga involves an extensive body of knowledge and skill, and one can certainly devote many years of study and practice to mastering this school of thought. Indeed, there are several different schools of yoga that comprise this ancient tradition. However, to simply evoke the relaxation response does not require becoming a master of these techniques. A beginner can learn enough in a relatively short period of time to begin to achieve some of the benefits.

Another technique is **biofeedback**, which involves the use of equipment to provide visual or auditory feedback reflecting internal states of tension, such as blood pressure and heartbeat. Usually performed at a clinic, biofeedback techniques have been shown to be effective in treating some cases of hypertension, headaches, and other stress disorders.[43]

Another method is called **visualization**, which involves using all of the senses to imagine a desired result. This is often used by athletes to improve performance. When using this as a method for treating anxiety, one visualizes a situation that is peaceful and serene.

Learning yoga, biofeedback, or visualization from a book is certainly not as effective as obtaining the guidance of a skilled practitioner. However, we describe here one technique that Benson has extensively studied. It is a simple method he derived from transcendental meditation and other sources, with a concentration on its applicability as a treatment for stress.[44] Benson and his colleagues at Harvard have spent more than 20 years studying the physiological changes produced by this and other relaxation techniques and the health benefits of experiencing the *relaxation response*, a term that Benson popularized, on a regular basis. By studying the practices of both Eastern and Western religions and other lay practices that produce the response, Benson sought to find and describe a simple method that would capture the essential components necessary to produce the physiological changes involved.[45]

HOW TO MEDITATE

First, find a quiet and comfortable environment, preferably one in which you feel safe and where you enjoy spending time. It is also desirable that you not be disturbed by other people, the phone, or other distractions.

Second, sit comfortably and close your eyes.

Relax your muscles, starting with your feet and working up to your face.

BRIDGE THREE

EXPANDING HUMAN POTENTIAL

Ultimately, we will have the ability through drugs and using nanobots to overcome the dysfunction in the regulation of the brain's centers of pleasure and satisfaction that cause addictive behavior. The ability to directly control your mood raises obvious philosophical issues. If you have the means to control your sense of satisfaction independent of your actions and situation, where does that leave the positive aspects of motivation and purpose? Our perspective is that these emerging technologies, like all technology through the ages, can and probably will be abused. However, the opportunity to relieve and overcome the agony of addiction and severe mood disorders will inherently be an immensely positive development. Although intricate, these pleasure-center controls are relatively crude compared with the deeper, more complex, and subtle satisfaction to be gained from the creative act: creating works of music, art, literature, technology, and science—including health books! By alleviating the distractions of addiction and other self-defeating behaviors that result from malfunctions of the pleasure and reward centers in our brains, greater human effort will be freed for this meaningful pursuit.

We stated earlier that the human species is unique in seeking to extend its horizons and to reach beyond limitations. We are also the only species that creates knowledge. Our civilization-wide knowledge base is expanding exponentially in size, and the Bridge Three developments we've described in this book will enable this noble pursuit to continue throughout the remarkable century ahead.

Now become aware of your breathing. As you breathe out, say a particular word or sound to yourself. This is the heart of the technique, so this aspect bears some discussion. Most any sound that you like will do, although we suggest a one- or two-syllable sound that contains no hard consonants (such as b, d, g, k, p, q, t, x). A suitable sound would be "oh one" or "ah one." (Compare the "om" used by Zen Buddhists, "amen" of Christians and Moslems, "omain" in Judaism.) Another possibility, which Benson himself recommends, is simply the word "one." When you say the word to yourself, don't actually say it aloud, just think it. Let the sound say itself. Just start the sound off in your mind and let it repeat itself with your breath. It is important to let your mind feel free to wander where it may, so you don't want repeating this sound to be difficult. Once you start it up, it should just repeat itself naturally. If, after a while, you notice that it has stopped, gently start it up again. Don't force the sound to repeat itself. Just let it happen.

A vital aspect of the method is a passive attitude. The technique used here is essentially the opposite of a mental discipline. This technique is considered nonconcentrative as opposed to concentrative. For people who are used to disciplining themselves both physically and mentally, this can be confusing. Don't worry about how well you are doing. Let thoughts come and go. Some will be pleasant; some may be distressing. Both will pass and lead to other thoughts. The repetition of the sound with your breathing should also not be a discipline; just let it happen, and if you notice that it's stopped, gently start imagining the sound again repeating itself with your breath.

As you gain experience meditating, the sound (sometimes called a mantra, although, technically, a mantra is a Sanskrit word from a specific tradition) will become more subtle and less clearly articulated. This is desirable—eventually the sound will become just a feeling of the sound. To assist with this process, it is desirable not to say the word(s) out loud, even when not meditating.

If you find certain thoughts to be disturbing, let them pass. If that fails, try returning to the sound that is repeating itself with your breath. If necessary, stop the sound, wait a few seconds, and end the meditation by gradually opening your eyes.

Continue this process for 15 to 30 minutes. You can open your eyes briefly to check the time if you wish, but don't use an alarm. When you are done, stop the sound and sit quietly with your eyes closed for a couple of minutes. Open your eyes gradually and sit for a few minutes more. Then stand up. If you are tired, you may find yourself falling asleep. While the purpose of meditating isn't to get additional sleep, if you do find yourself nodding off, that's okay.

To obtain a health benefit, the technique should be practiced once or twice a day. In general, avoid the two-hour period after each meal, since the digestive process may interfere with your ability to elicit the desired response. It makes sense, therefore, to practice this meditation technique prior to a meal.

The key to the technique is the passive attitude. This includes not worrying about how well you are doing and letting thoughts, positive or negative, wash over you.

Benson's research has uncovered a wide range of subjective experiences that occur during meditation, although a feeling of peacefulness and tranquility isn't uncommon. However, because the method specifically does not include trying to elicit peaceful thoughts or feelings, your experience may vary from session to session and from minute to minute. The research has demonstrated that the physiological changes associated with the relaxation response are elicited regardless of the subjective experience, whether tranquil or otherwise.

This method contrasts with other mental techniques that involve disciplining the mind to concentrate, whether on a particular sound or even on the idea of relaxation. On the book's Web site (Fantastic-Voyage.net), we provide information about where you can receive training in this type of meditation.

Would it do just as well to use these 20 minutes to get some extra sleep? Isn't sleep a relaxation technique? As we've discussed, getting adequate quantity and quality of sleep is very important as one element of a lifestyle that deals effectively with stress. Not sleeping adequately is indeed very stressful. Both sleep and the relaxation response involve significant and measurable physiological changes, which have been extensively studied. They are not the same, however. And while sleep is necessary, getting more will not achieve the beneficial endocrine changes that are achieved through regular elicitation of the relaxation response.

Between following the nutritional guidelines described in this book—exercising, spending time with your family, getting adequate sleep, and now practicing relaxation techniques—how are you supposed to find the time to do anything else, like earn a living?

This is a reasonable question. Let's take these issues one at a time.

The nutritional guidelines may take some time at first, in terms of learning nutritional breakdown of foods and exploring the world of foods that comply with the guidelines and that you enjoy. But we can share with you from our own experience and that of many others that, after a period of learning and adjustment, following the guidelines does not involve an ongoing time investment. Spending time with your family or friends is, presumably, something you want to do. If you really believe you have no desire or need for interaction with others, you might wish to examine this priority. As for sleep, this is also something you need, and it is counterproductive to try to cut back on it. Similarly, we have found that exercise more than pays for itself in terms of greater energy, better sleep patterns (meaning you will get more sleep without spending more time in bed), and a more positive attitude about each day's challenges.

Relaxation techniques offer the same promise. The time spent will repay itself in terms of demonstrable physiological, mental, and emotional benefits, seen in greater effectiveness in other spheres and an enhanced sense of well-being. That is why we emphasized time management earlier. You will be surprised at how much you can accomplish and how many things you have time for if you carefully consider the management of your time. The authors put a priority on practicing these guidelines. Most people waste enormous amounts of time and personal energy in ways that are both unproductive and don't contribute to the sense of well-being that everyone seeks. The different elements of the program we have outlined—diet, exercise, balance, relaxation—work synergistically. Rather than interfering with your life, a healthy and well-balanced lifestyle will make you more effective in achieving your personal goals and enhancing your life. There is really no alternative.

EPILOGUE

While this is the end of our *Fantastic Voyage*, we hope it is just the beginning of your personal journey of discovery. If there is one message we have tried to impart, it is the unprecedented opportunity you have to greatly improve your health. Health is not simply the absence of disease; rather, it refers to the effectiveness of every level of your existence, something you can always improve. The effort you put into this endeavor will be repaid many times and will assist you in whatever life goals you may have.

Society has a number of powerful and widely held but misleading ideas that we have tried to counter:

- *The health care system will take care of me if I have a problem.* The reality: Our medical system is largely geared toward dealing with health issues once they erupt as full-blown disease. Very little attention is paid in our medical schools to nutrition and disease prevention. Waiting until symptoms of disease appear is often too late. The first symptom of heart disease may be a heart attack; cancer may not be evident until it has already metastasized. The knowledge to avoid the degenerative diseases that cause more than 90 percent of all deaths and vast suffering is available, but the responsibility to apply this knowledge is yours.

- *Taking a supplement or drug is a last resort.* The reality: When our bodies evolved tens of thousands of years ago, it was in the interest of the species for humans (and other animals) not to live much beyond their child-rearing days. Now that we live in an era of abundance rather than scarcity, this evolutionary program is no longer relevant. We have the means to dramatically slow down and in many cases halt and reverse degenerative disease and aging processes, but these require reprogramming your biochemistry through nutrition, exercise, and lifestyle as well as taking advantage of supplements and drugs. This process will become easier with the more powerful Bridge Two and Bridge Three therapies and interventions being developed, but today it requires understanding and effort.

- *The only things we can count on are death and taxes.* The reality: We'll leave the issue of taxes for another book, but the means to extend longevity indefinitely are in our grasp. Although we do not yet have all the tools we need to stop and reverse all aging processes, we do have the means right now to stay in good health and spirits until the full blossoming of the biotechnology and nanotechnology/artificial intelligence revolutions, which will indeed provide radical life extension.

Health knowledge is expanding at an accelerating pace. While this book was being written, dramatic new developments continued to become available on a weekly basis. We kept reworking the text to include many of these, but realized that if we kept doing this, the book would never be finished. So this book is a snapshot in time. Our primary mission has been not to provide an unchanging set of rules, but rather to describe an overall attitude and approach to improving your health, one based on a continued search for insight in an era of expanding knowledge. This search will indeed be a fantastic voyage. We hope that you will join us.

RESOURCES AND CONTACT INFORMATION

FANTASTIC-VOYAGE.NET

New medical information is accumulating so quickly that we find ourselves making changes to our own personal health programs at least once a month. To help you keep abreast of the rapidly changing health and longevity field, we invite you to visit Fantastic-Voyage.net where you will find:

- The latest updates to Ray & Terry's Longevity Program
- News of the most recent medical breakthroughs in longevity medicine
- A Short Guide to a Long Life: a summary of the principles of Ray & Terry's program
- Information on the tests recommended throughout this book, and how you can perform many of these tests at home
- Sources for finding organic foods and low-carbohydrate substitutes for many popular foods
- Resources for finding high-quality nutritional supplements
- Listings of physicians and medical and health organizations who are knowledgeable about the principles of our program

OTHER WEB RESOURCES

We have also developed a line of nutritional supplements and healthy low-carbohydrate foods that complement our longevity recommendations, which can be viewed at www.RayandTerry.com.

Additional resources on future technologies, including a free e-newsletter, can be found at KurzweilAI.net.

CONTACTING THE AUTHORS

The authors are committed to helping spread the word about Ray & Terry's Longevity Program. Ray can be reached at ray@RayandTerry.com, Terry at terry@RayandTerry.com. Terry's longevity clinic (wwww.fmiclinic.com) in the Denver area can be reached at info@fmiclinic.com or toll-free at 877-548-4387.

NOTES

CHAPTER 1

1 A. M. Cunningham. 2003. "BioBots." *ScienCentralNews*; www.sciencentral.com/articles/view.php3?article_id=218391960&language=English; A. Moore. 2001. "Of silicon and submarines." *EMBO Reports*. 2(5): 367–370; www.nature.com/cgi-taf/DynaPage.taf?file=/embor/journal/v2/n5/full/embor411.html.

2 "Purdue researchers connect life's blueprints with its energy source." *Purdue News*, February 4, 2003; http://news.uns.purdue.edu/html4ever/030204.Guo.ATP.html.

3 "Today at UCI." May 8, 2003; http://today.uci.edu/news/release_detail.asp?key=995.

4 R. Kurzweil. "The Law of Accelerating Returns." *KurzweilAI.net*. www.kurzweilai.net/meme/frame.html?main=/articles/art0134.html; R. Kurzweil. 2005 (upcoming). *The Singularity Is Near: When Humans Transcend Biology*. New York: Viking Press.

5 Ray Kurzweil's theory of the "law of accelerating returns," and its social and economic impact, was introduced in *The Age of Spiritual Machines* (Viking, 1999) and will be further explored in his upcoming book *The Singularity Is Near: When Humans Transcend Biology* (Viking, 2005).

6 R. N. Anderson. The Ten Leading Causes of Death in the U.S., Final 2000 Data. Heart Disease: 710,760, Cancer: 553,091, Stroke: 167661, Chronic Lower Respiratory Disease: 122,009, Accidents: 97,900, Diabetes: 69,301, Pneumonia/Influenza: 65,313, Alzheimer's Disease: 49,558, Nephritis, nephrotic syndrome, and nephrosis: 37,251, Septicemia: 31,224.

7 J. C. Riley. 2001. *Rising Life Expectancy: A Global History*. Cambridge: Cambridge University Press.

8 F. Fukuyama. 2002. *Our Posthuman Future: Consequences of the Biotechnology Revolution*. New York: Farrar Straus.

9 The U.S. Department of Agriculture's "Food Pyramid" can be viewed at www.nal.usda.gov:8001/py/pmap.htm. The emphasis on starches and grains at the base of the pyramid has been linked with the current "epidemic" of obesity plaguing our country.

10 R. N. Anderson. National Vital Statistics Report, 2002 (Sept 16);50:16: 1–86. Also, according to the Minneapolis Heart Institute Foundation, "Approximately two-thirds of heart attacks are first heart attacks and one-third of all heart attacks are fatal. The first symptom of heart attack is often sudden death." See www.mplsheartfoundation.org.

11 L. A. Ries et al., eds. SEER Cancer Statistics Review, 1973–1999, National Cancer Institute, Bethesda, Maryland. For example, the Alliance for Lung Cancer Fact Sheet states that, due to a lack of screening, lung cancer is diagnosed in the late stages up to 85 percent of the time; www.alcase.org/factsabout_lungcancer.html. Late-stage diagnosis occurs in close to half of all cervical cancer occurrences. J. M. Ferrante et al. 2000. "Clinical and Demographic Predictors of Late-Stage Cervical Cancer, *Arch Fam Med*. 9: 439–445. And, over 50 percent of all cases of ovarian cancer are diagnosed in late stages. A. Srikameswaran. "Experts discuss promising new test for ovarian cancer." *Pittsburgh Post-Gazette*. May 7, 2002.

12 The Recommended Dietary Allowances were first issued in 1968 by the National Academy of Sciences and were last revised in 1989. These standards vary depending on age, gender, and whether a woman is pregnant or lactating. They are not designed to be "optimal" but rather to avoid specific nutritional deficiency diseases. They are expressed as average daily intakes over time. They rely on dietary sources rather than vitamin or mineral supplementation, and do not account for unusual requirements due to disease or environmental stress. See www.blionline.com/HDB/NutritionalStandardsRDAUSRDAAndRDIAntioxidantsBooklet.htm.

13 O. W. Rasmussen et al. 1993. "Effects on blood pressure, glucose and lipid levels of a high-monounsaturated fat diet compared with a high-carbohydrate diet in NIDDM subjects." *Diabetes Care.* 16: 1565–1571.

14 On the Web site for the American Diabetes Association is found the "Diabetic Food Pyramid." Interestingly, it is essentially identical to the Department of Agriculture's Food Pyramid recommended for the general public. The same reliance on a starch- and grain-based diet with 30 percent of calories coming from fat is recommended. See www.diabetes.org/main/health/nutrition/article031799.jsp.

15 D. Ornish. "Can lifestyle changes reverse coronary heart disease?" 1990. *Lancet.* 336: 129–133.

16 E. G. Vermeulen et al. 2000. "Effect of homocysteine-lowering treatment with folic acid plus vitamin B_6 on progression of sub clinical atherosclerosis: a randomized, placebo-controlled trial." *Lancet.* Feb 12;355(9203): 517–522. Also, in an editorial accompanying A. D. Korczyn. 2002. "Homocysteine, Stroke, and Dementia," *Stroke.* 33: 2343–2344, Dr. Korczyn of Tel-Aviv University Medical School in Ramat-Aviv, Israel, says, "Since dietary habits are so different among people, it may be appropriate to recommend 2 to 5 mg folic acid and a similar dose of vitamin B_{12} daily. This recommendation is based on the known safety of both vitamins, which do not have side effects even if used in excessive amounts, and their low cost."

CHAPTER 2

1 Nanotechnology is "thorough, inexpensive control of the structure of matter based on molecule-by-molecule control of products and byproducts; the products and processes of molecular manufacturing, including molecular machinery." (E. Drexler and C. Peterson. 1991. *Unbounding the Future: The Nanotechnology Revolution.* New York: William Morrow and Company.) According to the authors (chapter 1): "Technology has been moving toward greater control of the structure of matter for millennia . . . [P]ast advanced technologies—microwave tubes, lasers, superconductors, satellites, robots, and the like—have come trickling out of factories, at first with high price tags and narrow applications. Molecular manufacturing, though, will be more like computers: a flexible technology with a huge range of applications. And molecular manufacturing won't come trickling out of conventional factories as computers did; it will replace factories and replace or upgrade their products. This is something new and basic, not just another twentieth-century gadget. It will arise out of twentieth-century trends in science, but it will break the trend-lines in technology, economics, and environmental affairs."

Drexler and Peterson outline the following possible scenarios to explain the scope of the revolution: efficient solar cells "as cheap as newspaper and as tough as asphalt," molecular mechanisms that can kill cold viruses in six hours before biodegrading, immune machines that destroy malignant cells in the body at the push of a button, pocket supercomputers, the end of the use of fossil fuels, space travel, and restoration of lost species. Also see another book by K. E. Drexler, *Engines of Creation* (Anchor Books, 1986). Foresight Institute has a useful list of nanotechnology FAQs (www.foresight.org/NanoRev/FIFAQ1.html) and other information. Other Web resources include the National Nanotechnology Initiative (www.nano.gov), www.nanotechweb.org, Dr. Ralph Merkle's Nanotechnology page (www.zyvex.com/nano/), and *Nanotechnology* (an online journal: www.iop.org/EJ/journal/0957-4484).

Extensive material on nanotechnology can be found on Ray Kurzweil's Web site, www.kurzweilai.net.

2 Nanotechnology is technology in which objects are built from individual atoms or molecules, or where one or more dimensions are on a scale of nanometers (billionths of meter). For further information, see K. E. Drexler's 1986 classic *Engines of Creation*; www.kurzweilai.net/meme/frame.html?m=8.

3 Besides the functions of different types of cells, two other causes for cells to control the expression of genes are environmental cues and developmental processes. Even simple organisms such as bacteria can turn on and off the synthesis of proteins, depending on environmental cues. E. coli, for example, can turn off the synthesis of proteins that allow it to fix nitrogen gas from the air when there are other, less-energy-intensive sources of nitrogen in its environment. A recent study of 1,800 strawberry genes found that the expression of 200 of those genes varied during different stages of development (E. Marshall. 1999. "An array of uses: expression patterns in strawberries, Ebola, TB, and mouse cells." *Science.* 286(5439): 445).

4 Along with a protein-encoding region, genes include regulatory sequences called promoters and enhancers that control where and when that gene is expressed. Promoters are located "upstream" (on base pairs nearby the transcription site) on the DNA molecule. An enhancer activates a promoter, thereby controlling the rate of gene expression. To be expressed, most genes require enhancers; enhancers determine when genes are expressed and for which target protein cell type. Each gene can have several different enhancer sites linked to it (S. F. Gilbert. 2000. *Developmental Biology*, 6th ed. Sunderland, Massachusetts: Sinauer Associates; searchable online at www.ncbi.nlm.nih.gov/books/bv.fcgi?call=bv.View..ShowTOC&rid=dbio.TOC&depth=2.

By binding to enhancer or promoter regions, transcription factors start or repress the expression of a gene. New knowledge of transcription factors has transformed our understanding of gene expression. Per S. F. Gilbert in the chapter "The Genetic Core of Development: Differential Gene Expression," "The gene itself is no longer seen as an independent entity controlling the synthesis of proteins. Rather, the gene both directs and is directed by protein synthesis. Natalie Anger (1992) has written, 'A series of discoveries suggests that DNA is more like a certain type of politician, surrounded by a flock of protein handlers and advisors that must vigorously massage it, twist it and, on occasion, reinvent it before the grand blueprint of the body can make any sense at all.'"

5 Many antisense RNAs "have shown convincing in vitro reduction in target gene expression and promising activity against a wide variety of tumors." A. Biroccio, C. Leonetti, and G. Zupi. 2003. "The future of antisense therapy: combination with anticancer treatments." *Oncogene*. Sep 29;22(42): 6579–6588. See also "Subtle gene therapy tackles blood disorder." October 11, 2002, *NewScientist.com*; www.newscientist.com/news/news.jsp?id-ns99992915; X. Jiang et al. 2003. "Inhibition of MMP-1 expression by antisense RNA decreases invasiveness of human chrondrosarcoma." *J Orthop Res*. Nov;21(6): 1063–1070.

6 B. Holmes. "Gene therapy may switch off Huntington's." *NewScientist.com*, March 13, 2003; www.newscientist.com/news/news.jsp?id=ns99993493.

"Emerging as a powerful tool for reverse genetic analysis, RNAi is rapidly being applied to study the function of many genes associated with human disease, in particular those associated with oncogenesis and infectious disease." J. C. Cheng, T. B. Moore, and K. M. Sakamoto. 2003. "RNA interference and human disease." *Mol Genet Metab*. Oct;80(1–2): 121–128. RNAi is a "potent and highly sequence-specific mechanism" (L. Zhang, D. K. Fogg, and D. M. Waisman. 2003. "RNA interference-mediated silencing of the S100A10 gene attenuates plasmin generation and invasiveness of Colo 222 colorectal cancer cells." *J Biol Chem*. Oct 21 [e-pub ahead of print]).

7 Gene transfer to somatic cells affects a subset of cells in the body for a period of time. It is theoretically possible to also alter genetic information in egg and sperm (germ line) cells, for the purpose of passing on those changes to the next generations. Such therapy poses many ethical concerns and has not yet been attempted.

8 Genes encode proteins, which perform vital functions in the human body. Abnormal or mutated genes encode proteins that are unable to perform those functions, resulting in genetic disorders and diseases. The goal of gene therapy is to replace the defective genes so that normal proteins are produced. This can be done in a number of ways, but the most typical way is to insert a therapeutic replacement gene into the patient's target cells using a carrier molecule called a vector. "Currently, the most common vector is a virus that has been genetically altered to carry normal human DNA. Viruses have evolved a way of encapsulating and delivering their genes to human cells in a pathogenic manner. Scientists have tried to take advantage of this capability and manipulate the virus genome to remove the disease-causing genes and insert therapeutic genes." (Human Genome Project, "Gene Therapy," www.ornl.gov/TechResources/Human_Genome/medicine/genetherapy.html). See the Human Genome Project site for more information about gene therapy and links. Gene therapy is an important enough area of research that there are currently six scientific peer-reviewed gene therapy journals and four professional associations dedicated to this topic.

9 K. Smith. 2002. "Gene transfer in higher animals: theoretical considerations and key concepts." *J Biotechnol*. Oct 9;99(1): 1–22.

10 "'Miracle' gene therapy trial halted." *NewScientist.com*, October 3, 2003; www.newscientist.com/news/news.jsp?id=ns99992878; Human Genome Project. "Gene therapy," www.ornl.gov/TechResources/Human_Genome/medicine/genetherapy.html.

11 L. Wu, M. Johnson, and M. Sato. 2003. "Transcriptionally targeted gene therapy to detect and treat cancer." *Trends Mol Med*. Oct;9(10): 421–429.

12 S. Westphal. "Virus synthesized in a fortnight." *NewScientist.com*, November 14, 2003; www.newscientist.com/news/news.jsp?id=ns99994383.

13 A. Ananthaswamy. "Undercover genes slip into the brain." *NewScientist.com*, March 20, 2003; www.newscientist.com/news/news.jsp?id=ns99993520.

14 A. E. Trezise et al. 2003. "In vivo gene expression: DNA electrotransfer." *Curr Opin Mol Ther*. Aug;5(4): 397–404.

15 S. Westphal. "DNA nanoballs boost gene therapy." *NewScientist.com*, May 12, 2002; www.newscientist.com/news/news.jsp?id=ns99992257.

16 B. Dekel et al. 2003. "Human and porcine early kidney precursors as a new source for transplantation." *Nature Med*. Jan 1;(9): 53–60.

17 Here is one possible explanation: "In mammals, female embryos have two X-chromosomes and males have one. During early development in females, one of the X's and most of its genes are normally silenced or inactivated. That way, the amount of gene expression in males and females is the same. But in cloned animals, one X-chromosome is already inactivated in the donated nucleus. It must be reprogrammed and then later inactivated again, which introduces the possibility of errors." "Genetic defects may explain cloning failures." *CBCNews*, May 27, 2002; www.cbc.ca/storview/CBC/2002/05/27/cloning_errors020527. That story reports on F. Xue et al. 2002. "Aberrant patterns of X chromosome inactivation in bovine clones." *Nat Genet.* Jun;31(2): 216–220.

18 J. B. Gurdon and A. Colman. 1999. "The future of cloning." *Nature.* 402: 743–746; G. Stock and J. Campbell, eds. 2000. *Engineering the Human Germline: An Exploration of the Science and Ethics of Altering the Genes We Pass to Our Children.* New York: Oxford University Press.

19 W. S. Hwang. 2004. "Evidence of a Pluripotent Human Embryonic Stem Cell Line Derived from a Cloned Blastocyst." *Science.* Mar 12;303(5664): 1669–1674.

20 G. Vince. "Nanotechnology may create new organs." *NewScientist.com*, July 8, 2003; www.newscientist.com/news/news.jsp?id=ns99993916.

21 S. Westphal. "'Virgin birth' method promises ethical stem cells." *NewScientist.com*, April 3, 2003; www.newscientist.com/news/news.jsp?id=ns99993654.

22 Liver stem cells have been transformed into pancreatic cells (L. Yang et al. 2002. "In vitro trans-differentiation of adult hepatic stem cells into pancreatic endocrine hormone-producing cells." *Proc Natl Acad Sci USA.* Jun 11;99(12): 8078–8083). Adult muscle stem cells can be transformed into muscle, neural tissue, and blood vessels. Z. Qu-Petersen et al. 2002. "Identification of a novel population of muscle stem cells in mice: potential for muscle regeneration." *J Cell Biol.* May;157: 851–864.

23 A. M. Hakelien et al. 2002. "Reprogramming fibroblasts to express T-cell functions using cell extracts." *Nature Biotechnology.* May;20: 460–466.

24 See the description of transcription factors in note 3, page 380.

25 R. P. Lanza et al. 2000. "Extension of cell life-span and telomere length in animals cloned from senescent somatic cells." *Science.* Apr 28;288(5466): 665–669. See also J. C. Ameisen. 2002. "On the origin, evolution, and nature of programmed cell death: a timeline of four billion years." *Cell Death & Differentiation.* Apr;9(4): 367–393; M. E. Shay. "Transplantation without a donor." *Dream: The magazine of possibilities*, Children's Hospital Boston, Fall 2001; www.childrenshospital.org/about/dreamfall01.pdf.

26 S. Bhattacharya. "Stem cell 'immortality' gene found." *NewScientist.com*, May 30, 2003; www.newscientist.com/news/news.jsp?id=ns99993786.

27 A. D. de Grey. 2003. "The foreseeability of real anti-aging medicine: focusing the debate." *Exp Gerontol.* Sep;38(9): 927–934; A. D. de Grey. 2003. "An engineer's approach to the development of real anti-aging medicine." *Sci SAGE KE.* Jan 8;2003(1): VP1; A. D. de Grey et al. 2002. "Is human aging still mysterious enough to be left only to scientists?" *Bioessays.* Jul;24(7): 667–676.

28 A. D. de Grey. "Engineering negligible senescence: rational design of feasible, comprehensive rejuvenation biotechnology," at www.gen.cam.ac.uk/sens/sensov.ppt.

29 A. D. de Grey et al. 2004. "Total deletion of in vivo telomere elongation capacity: an ambitious but possibly ultimate cure for all age-related human cancers." *Annals NY Acad Sci.* 1019: 147–170.

30 O. J. Finn. 2003. "Cancer vaccines: between the idea and the reality." *Nat Rev Immunol.* Aug;3(8): 630–641; R. C. Kennedy and M. H. Shearer. 2003. "A role for antibodies in tumor immunity." *Int Rev Immunol.* Mar–Apr;22(2): 141–172.

31 A. D. de Grey. 2002. "The reductive hotspot hypothesis of mammalian aging: membrane metabolism magnifies mutant mitochondrial mischief." *Eur J Biochem.* Apr;269(8): 2003–2009; P. F. Chinnery et al. 2002. "Accumulation of mitochondrial DNA mutations in ageing, cancer, and mitochondrial disease: is there a common mechanism?" *Lancet.* Oct 26;360(9342): 1323–1235; A. D. de Grey. 2000. "Mitochondrial gene therapy: an arena for the biomedical use of inteins." *Trends Biotechnol.* Sep;18(9): 394–399.

32 S. Graham. "Methuselah worm remains energetic for life." *ScientificAmerican.com*, October 27, 2003; www.sciam.com/article.cfm?chanID=sa003&articleID=000C601F-8711-1F99-86FB83414B7F0156.

33 P. Ball and H. Pearson. "Drug may give cells a fresh start." *Nature Science Update*, January 30, 2004; www.nature.com/nsu/040126/040126-14.html.

34 H. Pearson. "Instant stem cells—just add water." *Nature Science Update*, December 19, 2003; www.nature.com/nsu/031215/031215-11.html.

35 R. A. Freitas Jr. *Nanomedicine, Volume I: Basic Capabilities*, first in an anticipated four-volume *Nanomedicine* technical book series. Freitas offers a pioneering and fascinating glimpse into a molecular-nanotechnology future with far-reaching implications for the medical profession—and ultimately for the radical improvement and extension of natural human biological structure and function.

36 Sensors and diagnostic tools are important applications of nanotechnology because the devices can be placed in direct contact with cells and the molecules in it. Another option is nanoimaging, in which nanocrystals would seek out different types of molecules, such as cancer cells. When stimulated by a laser, the crystals would emit light. These applications are just the "tip of the nano-iceberg" (P. Balasubramanian and S. Japa. 2003. "Nanosensing," *Stanford Biomedicine Quarterly*. Spring, p. 13).

37 For activities of the International Society for BioMEMS and Biomedical Nanotechnology, see its site (www.bme.ohio-state.edu/isb/). You can also find BioMEMS conferences listed on the SPIE site (www.spie.org/Conferences).

38 As reported in the *Stanford Biomedicine Quarterly* article in note 36 above, researchers used a gold nanoparticle to monitor blood sugar in diabetics. Y. Xiao et al. 2003. "'Plugging into enzymes': Nanowiring of redox enzymes by a gold nanoparticle." *Science*. Mar 21;299(5614): 1877–1881. "One might even speculate on the design of an 'artificial pancreas,' an implant that would release appropriate levels of insulin into the blood from moment to moment according to the blood sugar readings provided by this nanosensor system" (p. 12).

39 Dr. Michael Cima at MIT is one researcher examining in vivo drug release from implantable MEMS arrays. He is one of the authors on G. Voskerician et al. 2002. "Biocompatibility and biofouling of MEMS drug delivery devices." *Biomaterials*. 24: 1959–1967.

40 According to Wise, one reason for relatively slow advances over the past thirty years is because of the "aggressive saltwater environment" of living tissue. (Quoted in D. Lammers. "Micro medical devices could transform health care." *EE Times*, June 21, 2002; www.eetimes.com/at/news/OEG20020620S0060.) See also the discussion of Wise's work in J. DeGaspari. "Tiny, tuned, and unattached." *Mechanical Engineering*, July 1, 2001; www.memagazine.org/backissues/july01/features/tinytune/tinytune.html.

41 "A team of scientists from Japan have developed tiny spinning screws that can swim along veins. The screws could then be used to ferry drugs to infected tissues or even burrow into tumours to kill them off with a hot lance." "'Microbots' hunt down disease." *BBC News*, June 13, 2001; http://news.bbc.co.uk/1/hi/health/1386440.stm. The micromachines are based on cylindrical magnets. (K. Ishiyama, M. Sendoh, and K. I. Arai. 2002. "Magnetic micromachines for medical applications." *J Magnetism Magnetic Materials*. 242–245(P1): 41–46.)

42 See the Sandia National Laboratories August 15, 2001, press release "Pac-Man-like microstructure interacts with red blood cells," www.sandia.gov/media/NewsRel/NR2001/gobbler.htm. For an industry trade article in response, see D. Wilson. "Microteeth have a big bite." August 17, 2001; www.e4engineering.com/item.asp?ch=e4_home&type=Features&id=42543.

43 P. Ball. "Chemists build body fluid battery." *Nature Science Update*, November 12, 2002; www.nature.com/nsu/021111/021111-1.html.

44 M. Bernstein. "Tiny nanowire could be next big diagnostic tool for doctors." *EurekAlert*, December 16, 2003; www.eurekalert.org/pub_releases/2003-12/acs-nc121603.php.

45 J. Sliwa. Researchers envision intelligent implants." *EurekAlert*, July 8, 2003; www.eurekalert.org/pub_releases/2003-07/asfm-rei070303.php.

46 J. Whitfield. "Lasers operate inside single cells," *Nature Science Update*, October 6, 2003; www.nature.com/nsu/030929/030929-12.html.

47 Ron Weiss's home page at Princeton University (www.ee.princeton.edu/~rweiss/) lists his publications, such as 2003. "Genetic circuit building blocks for cellular computation, communications, and signal processing." *Natural Computing, an International Journal*. (2): 47–84.

48 S. L. Garfinkel. "Biological computing." *Technology Review*, May/June 2000; www.simson.net/clips/2000.TR.BiologicalComputing.htm.

49 Ibid. See also the list of current research on the MIT Media Lab Web site; www.media.mit.edu/research/index.html.

CHAPTER 3

1 Great Smokies Diagnostic Laboratory. "Integrative Medicine," at www.gsdl.com/gsdl/functional_med.html.

2 Terry Grossman's personal experience is by no means unique. In a recent article in *JAMA*, the official journal of the American Medical Association, 46.3 percent of American people consulted a practitioner of alternative medicine in 1997. (D. M. Eisenberg et al. "Trends in alternative medicine use in the United States, 1990–1997: results of a follow-up national survey." 1998. *JAMA*. Nov 11;280(18): 1569–75.)

3 N. M. Bressler et al. 2003. "Potential public health impact of Age-Related Eye Disease Study results: AREDS report no. 11." *Arch Ophthalmol*. Nov;121(11): 1621–1624.

4 Approximately 70 percent of Terry Grossman's macular-degeneration patients have experienced some degree of visual improvement with his treatment protocol, which also includes electrical stimulation of the eyes, while 25 percent stabilized their existing vision. See also E. L. Paul. 2002. "The Treatment of Retinal Diseases with Micro Current Stimulation and Nutritional Supplementation." Presentation to the International Society for Low-Vision Research and Rehabilitation (ISLRR), Göteborg University, Faculty of Medicine, Göteborg, Sweden; and also L. D. Michael and M. J. Allen. 1993. "Nutritional Supplementation, Electrical Stimulation and Age-Related Macular Degeneration." *J Orthomol Med.* 8: 168–171.

5 For further information on the nutritional treatment of autistic disorders, see the DAN! (Defeat Autism Now!) protocols, available through the Autism Research Institute; www.autism.com/ari/contents.html.

CHAPTER 4

1 Charles Darwin recognized that one of the primary factors in the size of an animal population is food, which all species require to survive. Species are linked in food chains, starting with producers, which create their own food by converting inorganic compounds into organic compounds. When plants (producers) create organic compounds through photosynthesis, they are storing energy that will then be passed up the food chain as those plants are eaten and then the eaters of those plants are eaten, and so on. Decomposers break down the complex organic compounds created by energy conversion and return the nutrients to the soil, where producers use them once again.

2 Plant cultivation began much sooner than originally thought: there are signs that squash was cultivated in Ecuador and rice in China 10,000 to 11,000 years ago (H. Pringle. 1998. "Neolithic agriculture: the slow birth of agriculture." *Science*. 282(5393): 1446–1449). Since then, humans have spread into almost every ecosystem on earth, eating widely ranging diets. Until recently, the maintenance energy required to acquire food was a high proportion of all humans' energy budgets. One reason may be our brains, which require 20 to 25 percent of our energy needs, compared with 8 to 10 percent for nonhuman primates and 3 to 5 percent for other mammals. Animals with larger brains typically seek richer diets (W. Leonard. 2002. "Food for thought: dietary change was a driving force in human evolution." *Sci Am*. Nov 13: 106–115).

Many researchers see a link between obesity in developed countries and the amount of energy that humans have, until recently, spent on food acquisition. "It's only been maybe the last 30 years, certainly after the Industrial Revolution, since food stopped being scarce," says Ann Kelley, a neuroscientist at the University of Wisconsin at Madison. "No way have the brain and the physiological systems that regulate body weight had a chance to catch up." (C. T. Hall. "Caveman history blamed for U.S. obesity." *San Francisco Chronicle*, January 12, 2003.)

3 According to the U.S. Department of Agriculture ("A history of American agriculture 1776–1990," www.usda.gov/history2/text3.htm), in 1790, farmers made up 90 percent of the labor force in a population of almost 4 million. By 1840, that percentage dropped to 69 percent. By 1900, out of a population of 76 million, 38 percent of the labor force worked on farms, which averaged 147 acres in size. By 1990, only 2.6 percent of the labor force worked on farms, which averaged 460 acres in size, out of a total U.S. population of 246 million.

4 In 2000, the Centers for Disease Control (CDC) defined poor nutrition and lack of exercise as the second leading "actual" cause of death in the United States, behind tobacco. "Actual causes of death are defined as lifestyle and behavioral factors such as smoking and physical inactivity that contribute to this nation's leading killers including heart disease, cancer and stroke" ("Physical Inactivity and Poor Nutrition Catching up to Tobacco as Actual Cause of Death." March 9, 2004; www.cdc.gov/od/oc/media/pressrel/fs040309.htm). In that same year, "fewer than one-fourth of U.S. adults reported eating recommended amounts of fruits and vegetables daily." ("Chronic disease prevention: the burden of chronic diseases and their risk factors." National Center for Chronic Disease Prevention and Health Promotion; www.cdc.gov/nccdphp/burdenbook2002/03_nutriadult.htm.)

5 The producers at the bottom of a food chain build complex energy-rich compounds from four atoms (carbon, nitrogen, oxygen, and hydrogen). For example, proteins are chains of amino acids, each of which contains an amino group (NH_2) and a carboxyl group (COOH). Proteins are broken down into their amino acids during digestion, and these amino acids pass into your bloodstream, from which they are absorbed by cells. Your body uses 20 out of the approximately 100 amino acids in nature as building blocks.

Plants also produce carbohydrates, which your cells absorb and convert into energy to drive all your bodily functions. Glucose, composed of six carbon atoms and six water molecules, is one of the simplest carbohydrates so it can pass directly into the bloodstream. More complex carbohydrates, made up of chains of glucose molecules, need to be broken down into glucose molecules before they can be absorbed. Hydrolysis is the enzymatic reaction that breaks chemical bonds in food through the addition of water.

6 "The apparent simplicity of the water molecule belies the enormous complexity of its interactions with other molecules, including other water molecules" (A. Soper. 2002. "Water and ice." *Science*. 297: 1288–1289). There is much that is still up for debate, as shown by the numerous articles still being published about this most basic of molecules, H_2O. For example, D. Klug. 2001. "Glassy water." *Science*. 294: 2305–2306; P. Geissler et al. 2001. "Autoionization in liquid water." *Science* 291(5511): 2121–2124; J. K. Gregory et al. 1997. "The water dipole moment in water clusters." *Science*. 275: 814–817; K. Liu et al. 1996. "Water clusters." *Science*. 271: 929–933.

A water molecule has slightly negative and slightly positive ends, which means water molecules interact with other water molecules to form networks. The partially positive hydrogen atom on one molecule is attracted to the partially negative oxygen on a neighboring molecule (hydrogen bonding). Three-dimensional hexamers involving six molecules are thought to be particularly stable, though none of these clusters lasts longer than a few picoseconds.

The polarity of water results in a number of anomalous properties. One of the best known is that the solid phase (ice) is less dense than the liquid phase. This is because the volume of water varies with the temperature, and the volume increases by about 9 percent on freezing. Due to hydrogen bonding, water also has a higher-than-expected boiling point.

7 M. S. Jhon. 1989. "Water and health." Korea Applied Science Research Center for Water, Seoul, Korea. Other articles include M. S. Jhon and J. D. Andrade. 1973. "Water and hydrogels." *J Biomed Mater Res*. Nov;7(6): 509–522; and J. D. Andrade et al. 1973. "Water as a biomaterial." *Trans. Am Soc. Artif. Intern Organs*. 19: 1–7.

8 The following study cites many benefits from an alkalinizing diet. L. A. Frassetto et al. 1998. "Estimation of net endogenous noncarbonic acid production in humans from diet potassium and protein contents." *Am J Clinical Nutrition*. 68: 576–83. "Normal adult humans eating Western diets have chronic, low-grade metabolic acidosis, the severity of which is determined in part by the net rate of endogenous noncarbonic acid production (NEAP), which varies with diet. . . . Normal adult humans eating typical American diets characteristically have chronic, low-grade metabolic acidosis. . . . With advancing age, the severity of diet-dependent acidosis increases independently of diet. That occurs because kidney function ordinarily declines substantially with age, resulting in a condition similar to that of chronic renal insufficiency. Renal insufficiency induces metabolic acidosis by reducing conservation of filtered bicarbonate and excretion of acid. Failure to recognize the respective and independent roles of age-related impaired renal acid-base regulatory capacity and diet net acid load has until recently prevented the recognition that low-grade metabolic acidosis is characteristically present and worsens with age in otherwise healthy adults. . . . Potassium bicarbonate is a natural base that the body generates from the metabolism of organic acid salts of potassium (e.g. potassium citrate) (8), whose density (i.e., mmol K/kJ food item) is greatest in fruit and vegetables. Long-term supplementation of the diet with potassium bicarbonate has numerous anabolic effects. In postmenopausal women, for example, calcium and phosphorus balances improve (1), bone resorption markers decrease (1), bone formation markers increase (1), nitrogen balance improves (9), and serum growth hormone concentrations increase (10). These findings suggest that the adverse effects of chronic, low-grade, diet-dependent acidosis are not inconsequential and may contribute to such age-related disturbances as bone mass decline, osteoporosis, and muscle wasting. One way to reduce or eliminate diet-dependent metabolic acidosis is by eating diets that impose little or no net acid load."

9 The body maintains the pH of blood at around 7.4. pH. The pH measure, first used by the Danish biochemist S. P. L. Sorensen (1868–1939), expresses the concentration of the hydrogen ion as a number between 1 and 14. A solution with a pH less than 7 is considered acidic, while a solution with a pH of 7 is considered basic, or alkaline. Thus, human blood is slightly alkaline.

"The concentration of H+ in blood plasma and various other body solutions is among the most tightly regulated variables in human physiology. . . . Acute changes in blood pH induce powerful regulatory effects at the level of the cell, organ, and organism" (J. Kellum. 2000. "Determinants of blood pH in health and disease." *Crit. Care*. 4: 6–14). In other words, if the pH level changes by even a few tenths of a pH unit, serious problems can result.

Disturbances in the acid-base balance in the blood can cause either acidosis (too much acid, resulting in a decrease in blood pH) or alkalosis (too much base, resulting in an increase in blood pH). There is still much debate regarding how to treat the metabolic disorders that cause these imbalances (see, for example, M. A. Shafiee et al. 2002. "A conceptual approach to the patient with metabolic acidosis." *Nephron*. 92 Suppl 1: 46–55).

10 CH_3COOH. Acetic acid, one of the carboxylic acids, is a metabolic intermediate in the body. Vinegar is a dilute solution of acetic acid produced by fermenting and oxidizing carbohydrates. See, for example, S. Weinhouse. 1995. "The acetyl group in fatty acid metabolism." *FASEB J.* Jun;9(9): 820–821; L. R. Empey et al. 1991. "Fish oil-enriched diet is mucosal protective against acetic acid-induced colitis in rats." *Can J Physiol Pharmacol.* Apr;69(4): 480–487.

11 $CH_3CHOHCOOH$. Lactic acid, one of the carboxylic acids, is found in the blood as a salt (lactate). The body creates lactic acid by exercising muscles. This acid is also found in fermented milk products such as sour milk, cheese, and buttermilk. Certain bacteria create lactates during fermentation. For more information, see J. S. Pringle and A. M. Jones. 2002. "Maximal lactate steady state, critical power and EMG during cycling." *Eur J Appl Physiol.* Dec;88(3): 214–216; B. S. Dien et al. 2002. "Fermentation of sugar mixtures using Escherichia coli catabolite repression mutants engineered for production of L-lactic acid." *J Ind Microbiol Biotechnol.* Nov;29(5): 221–227; H. Pitkanen et al. 2002. "Serum amino acid responses to three different exercise sessions in male power athletes." *J Sports Med Phys Fitness.* Dec;42(4): 472–480.

12 H_2CO_3. The carbonic acid–bicarbonate buffering system helps maintain blood pH (see note 11 above). Two types of salt created from carbonic acid are hydrogen carbonate, which contains HCO_3-, and carbonates, which contain CO_{32}-. For more information, see, for example, S. Kimura et al. 2003. "Enzymatic assay for determination of bicarbonate ion in plasma using urea amidolyase." *Clin Chim Acta.* Feb;328(1–2): 179–184; A. Vesela and J. Wilhelm. 2002. "The role of carbon dioxide in free radical reactions of the organism." *Physiol Res.* 2002;51(4): 335–339; D. A. Bushinsky et al. 2002. "Acute acidosis-induced alteration in bone bicarbonate and phosphate." *Am J Physiol Renal Physiol.* Nov;283(5): F1091–1097.

13 $C_5H_4N_4O_3$. In the purine group, uric acid is created as the body digests proteins. As with other acidic by-products of digestion, uric acid must be excreted at sufficient levels to avoid health problems such as gout. For more information, see T. Nakámura et al. 2003. "Serum fatty acid levels, dietary style and coronary heart disease in three neighboring areas in Japan: the Kumihama study." *Br J. Nutr.* Feb;89(2): 267–272; F. Perez-Ruiz et al. 2002. "Renal underexcretion of uric acid is present in patients with apparent high urinary uric acid output." *Arthritis Rheum.* Dec 15;47(6): 610–613.

14 CnH_2nO_2. Fatty acids are components of lipids and composed of chains of carbon and hydrogen atoms. The carboxyl group (-COOH) at one end of a fatty acid makes it a carboxylic acid. Single carbon-to-carbon bonds make the acid saturated, while double and triple bonds make it unsaturated. Oleic acid is the most common fatty acid; you can find it in vegetable oils such as olive, palm, and peanut oil. Oleic acid also makes up 46 percent of human fat. See, for example, M. Nydahl et al. 2003. "Achievement of dietary fatty acid intakes in long-term controlled intervention studies: approach and methodology." *Public Health Nutr.* Feb;6(1):31-40; G. R. Hynes et al. 2003. "Effects of dietary fat type and energy restriction on adipose tissue fatty acid composition and leptin production in rats." *J Lipid Res.* May;44(5): 893–901; S. F. Knutsen et al. 2003. "Comparison of adipose tissue fatty acids with dietary fatty acids as measured by 24-hour recall and food frequency questionnaire in black and white adventists." *Ann Epidemiol.* Feb;13(2): 119–127.

15 H_3PO_4. Phosphoric acid is used in fertilizers, in dental cements, and in the sugar and textile industries; it is also used in food products to provide a fruitlike flavoring. Most of the peer-reviewed literature focuses on the effects of phosphoric acid on tooth enamel (see, for example, B. Dincer et al. 2002. "Scanning electron microscope study of the effects of soft drinks on etched and sealed enamel." *Am J Orthod Dentofacial Orthop.* Aug:122(2): 135–141) and on bone density, particularly in girls. (See, for example, J. Fisher et al. 2001. "Maternal milk consumption predicts the tradeoff between milk and soft drinks in young girls' diets." *J Nutr.* Feb:131(2): 246–250; F. Carcia-Contreras et al. 2000. "Cola beverage consumption induces bone mineralization reduction in ovariectomized rats." *Arch Med Res.* Jul–Aug 31(4): 360–365.)

16 $NaHCO_3$. Sodium bicarbonate is often called the most important pH blood buffer. Typically, the concentration of bicarbonate in the blood plasma is 25 millimoles per liter. This level is called the bicarbonate threshold. The body produces sodium bicarbonate from the carbon dioxide (CO_2) formed in the cells as a by-product of chemical reactions.

After the carbon dioxide filters into the capillaries, it combines with an enzyme of red blood cells called carbonic anhydrase to form carbonic acid (H_2CO_3). This acid quickly separates into hydrogen ions (H+) and bicarbonate ions (HC_3-). The reaction can also reverse, yielding carbon dioxide and water from bicarbonate and hydrogen ions, with the carbon dioxide eliminated through the lungs.

Sodium bicarbonate is used as a medicine to relieve heartburn, sour stomach, or acid indigestion by neutralizing excess stomach acid.

17 Na_2HPO_4. Sodium phosphate is an important nonbicarbonate base in the renal system. Monobasic phosphate (NaH_2PO_4) forms when this base accepts hydrogen ions.

18 The balance of bases to hydrogen ions is key to how the renal system eliminates wastes from the metabolism of our food. The kidneys regulate the blood by filtering 20 percent of the plasma and noncell elements from the blood, reabsorbing key components (fluid, ions, small molecules) as needed, and secreting unwanted components in the urine. The entire blood volume of an adult is typically filtered 20 to 25 times a day.

Bicarbonate is one of the components filtered from the blood and then reabsorbed. When the concentration of bicarbonate falls below the threshold of 25 millimoles per liter, no bicarbonate is excreted, which means all of it is reabsorbed into the blood. When the concentration is higher than the threshold, bicarbonate is passed into the urine.

19 H_2SO_4. Sulfuric acid is a strong acid that ionizes to form hydronium ions (H_3O+) and hydrogen sulfate ions (HSO_4-). See, for example, T. Ubuka. 2002. "Assay methods and biological roles of labile sulfur in animal tissues." *J Chromatogr B Analyt Technol Biomed Life Sci.* Dec 5;781(1-2): 227–249.

20 H_3PO_4. See note 15 on page 387 on H_3PO_4 for more detail.

21 The kidneys are an effective mechanism for maintaining the blood pH. To control the concentration of hydrogen ions, for example, the kidneys can excrete 2,500 times more ions in the urine than are found in the blood. Likewise, the kidneys can excrete more or less bicarbonate.

The human body, however, creates many organic and inorganic acids as it breaks down food; and the more acidic the diet, the more time it takes for the kidneys to restore the pH balance in the blood. When you eat an acidic diet, the bicarbonate concentration in the blood is reduced (as is the pH). The kidneys compensate by secreting more hydrogen ions in the urine and secreting more bicarbonate back to the blood than it filtered out. This process continues until the concentrations of hydrogen and bicarbonate ions are returned to normal. For more detail, see C. Freudenrich. "How your kidneys work" (http://science.howstuffworks.com/kidney.htm).

Western diets rich in meats and other acid sources, such as colas, produce a heavy acid load. An increasing level of attention is being paid to the resulting health effects. See, for example, M. Maurer et al. 2003. "Neutralization of Western diet inhibits bone resorption independently of K intake and reduces cortisol secretion in humans." *Am J Physiol Renal Physiol.* Jan;284(1): F32–40; U. S. Barzel. 1995. "The skeleton as an ion exchange system: implications for the role of acid-base imbalance in the genesis of osteoporosis." *J Bone Miner Res.* Oct;10(10): 1431–1436; L. A. Frassetto et al. 2001. "Diet, evolution and aging—the pathophysiologic effects of the post-agricultural inversion of the potassium-to-sodium and base-to-chloride ratios in the human diet." *Eur J Nutr.* Oct;40(5): 200–213.

22 J. Shuster et al. 1992. "Soft drink consumption and urinary stone recurrence: a randomized prevention trial." *Journal Clinical Epidemiology*, Aug;45(8): 911–6. This study demonstrated a significant increase in the risk of stone formation for those who consumed phosophoric acid (found in colas): "those who reported at the time of the index stone that their most consumed drink was acidified by phosphoric acid but not citric acid, the experimental group had a 15 percent higher 3 yr recurrence-free rate than the controls, $p = 0.002$." For those who consumed primarily citric acid, no increase was found in risk.

Similar results were found in J. Shuster et al. 1985. "Primary liquid intake and urinary stone disease." *Journal Chronic Disease.* 38(11): 907–14. "This investigation indicates that there are important associations between urinary stone disease and a person's primary liquid intake. . . . an important (p less than 0.01) positive association was found between urinary stone disease and soda (carbonated beverage) consumption. . . . no important associations exist between urinary stone disease and any of milk, water, or tea, when these beverages represent a person's primary liquid intake. Moreover, soda can be viewed almost synonymously as sugared cola, since few subjects had diet sodas or sugared non-cola soda as primary fluid."

The following study concludes with a warning to avoid cola consumption: A. Rodgers. 1999. "Effect of cola consumption on urinary biochemical and physiocochemical risk factors associated with calcium oxalate urolithiasis." *Urology Research.* 27(1): 77–81. "Since stone formers are advised to increase their intake of fluid, the present study was undertaken to determine the effect of cola beverage consumption on calcium oxalate kidney stone risk factors. . . . Several risk factors changed unfavourably following consumption of cola. In males, oxalate excretion, the Tiselius risk index and modified activity product increased significantly ($P < 0.05$). In females, oxalate excretion increased significantly while magnesium excretion and pH decreased significantly ($P < 0.05$). Scanning electron microscopy showed that urines obtained from both sexes after cola consumption supported calcium oxalate crystallization to a greater extent than the control urines. It is concluded that consumption of cola causes unfavourable changes in the risk factors associated with calcium oxalate stone formation and that therefore patients should possibly avoid this soft drink in their efforts to increase their fluid intake."

The following study demonstrated benefit in avoiding urinary stones from a high fluid intake. R. Siener and A. Hesse. 2003. "Fluid intake and epidemiology or utolithiasis." *Eur J Clin Nutr*, Dec;57 Suppl 2:

S47–51. "A review of the literature shows that an increased urine volume achieved by a high fluid intake exerts an efficacious preventive effective on the onset and recurrence of urinary stones." The following study demonstrated the value of consumption of alkalinizing mineral water: T. Kessler and A. Hesse. 2000. "Cross-over study of the influence of bicarbonate-rich mineral water on urinary composition in comparison with sodium potassium citrate in healthy male subjects." *Br J Nutr.* 84(6): 865–87. "The aim of the present study on healthy male subjects aged 23–38 years was to evaluate the influence of bicarbonate-rich mineral water (1715 mg bicarbonate/l) on urinary-stone risk factors in comparison with sodium potassium citrate, a well-established treatment in that case. The results showed that the effect of the bicarbonate-rich mineral water was similar to that of the sodium potassium citrate, which suggests that it could be useful in the prevention of the recurrence of calcium oxalate and uric acid stones."

The following study concludes that people with a history of calcium-containing kidney stones should not avoid calcium and should drink adequate liquids: G. C. Curhan and S. G. Curhan. 1994. "Dietary factors and kidney stone formation." *Compr. Ther.* 20(9): 485–9. "Specifically, for individuals who have a history of a calcium-containing kidney stone, important dietary recommendations should include the following: Achieve adequate fluid intake to produce at least 2 liters of urine per day. Avoid calcium restriction (except in the rare instances of excessive intake of greater than several grams per day). A dietary intake of elemental calcium of at least 800 mg/day (the current RDA for adults) is recommended to prevent a negative calcium balance, bone mineral loss, and increased intestinal absorption of oxalate. At present, there is no evidence to support the belief that calcium restriction is beneficial and current data suggest that it may in fact be harmful."

See also P. M. Hall. 2002. "Preventing kidney stones: calcium restriction not warranted." *Cleve Clin J Med.* Nov;69(11): 885–888; B. Shekarraiz and M. L. Stoller. 2002. "Uric acid nephrolithiasis: current concepts and controversies." *J Urol.* Oct;168(4 Pt 1): 1307–1314; S. T. Reddy et al. 2002. "Effect of low-carbohydrate high-protein diets on acid-base balance, stone-forming propensity, and calcium metabolism." *Am J Kidney Dis.* Aug;40(2): 265–274; N. A. Breslau et al. 1988. "Relationship of animal protein-rich diet to kidney stone formation and calcium metabolism." *J Clin Endocrinol Metab.* Jan;66(1): 140–146; F. Grases et al. 1998. "Biopathological crystallization: a general view about the mechanisms of renal stone formation." *Adv Colloid Interface Sci.* Feb;74: 169–194; J. M. Aguado and J. M. Morales. 1993. "The pathogenesis and treatment of kidney stones." *N Engl J Med.* Feb 11;328(6): 444.

23 NIH. "Kidney Stones in Adults." http://kidney.niddk.nih.gov/kudiseases/pubs/stonesadults/index.htm.

24 V. Radosavljevic, S. Jankovic, J. Marinkovic, and M. Djokic. 2003. "Fluid intake and bladder cancer. A case control study." *Neoplasma.* 50(3): 234–8. The study states, "Multivariate logistic regression model showed consumption of: soda (OR=8.32; 95%CI=3.18-21.76), coffee (OR=1.46; 95%CI=1.05-2.01) and spirits (OR=1.15; 95%CI=1.04-1.28) as statistically significant risk factors, while mineral water (OR=0.52; 95%CI=0.34-0.79), skim milk (OR=0.38; 95%CI=0.16-0.91), yogurt (OR=0.34; 95%CI=0.12-0.97) and frequency of daily urination (OR=0.27; 95%CI=0.18-0.41) were statistically significant protective variables. In our study no statistically significant association was observed for total fluid intake. The findings suggest consumption of soda, coffee and spirits were indicated as risk factors for bladder cancer, while mineral water, skim milk, yogurt and frequency of urination as protective factors for bladder cancer."

25 That is the conclusion of G. R. Fernando, R. M. Martha, and R. Evangelina. 1999. "Consumption of soft drinks with phosphoric acid as a risk factor for the development of hypocalcemia in postmenopausal women." *Journal Clin Epidemiol.* Oct;52(10): 1007–10. "The objective of this study was to determine the relationship between the consumption of phosphoric acid-containing soft drinks and hypocalcemia in postmenopausal women . . . In the multivariate regression analysis consumption of one or more bottles per day of cola soft drinks showed association with hypocalcemia (1.28, CI 95% 1.06-1.53). The consumption of soft drinks with phosphoric acid should be considered as an independent risk factor for hypocalcemia in postmenopausal women."

The following study compared diets with primarily "acid precursors" to diets with primarily "base precursors" and concluded that diets that promote an alkaline body environment reduce "the rate of bone loss and the risk of fracture in postmenopausal women." D. E. Sellmeyer et al. 2001. "A high ratio of dietary animal to vegetable protein increases the rate of bone loss and the risk of fracture in postmenopausal women." *Am J Clin Nutr.* Jan;73(1): 118–122. "Different sources of dietary protein may have different effects on bone metabolism. Animal foods provide predominantly acid precursors, whereas protein in vegetable foods is accompanied by base precursors not found in animal foods. Imbalance between dietary acid and base precursors leads to a chronic net dietary acid load that may have adverse consequences on bone. . . . Elderly women with a high dietary ratio of animal to vegetable protein intake have more rapid femoral neck bone loss and a greater risk of hip fracture than do those with a low ratio. This suggests that an increase in vegetable protein intake and a decrease in animal protein intake may decrease bone loss and the risk of hip fracture."

26 M. Bertoni et al. 2002 "Effects of a bicarbonate-alkaline mineral water on gastric functions and functional dyspepsia: a preclinical and clinical study." *Pharmacol Res.* Dec;46(6): 525–31. "The present study was performed in order to evaluate: (1) the influence of a bicarbonate-alkaline mineral water (Uliveto) on digestive symptoms in patients with functional dyspepsia; (2) the effects of Uliveto on preclinical models of gastric functions. . . . These findings indicate that a regular intake of Uliveto favors an improvement of dyspeptic symptoms." The preclinical study suggests that the clinical actions of Uliveto water depend mainly on its ability to enhance gastric motor and secretory functions.

27 L. A. Frassetto et al., ibid. See note 8 on page 386.

28 Water can dissociate into hydroxide (OH-) ions, which makes it alkaline (basic), and hydrogen ions (H+). As a result, water can act as a base or an acid. Drinking alkaline water has been claimed to help with constipation, diarrhea, high or low blood pressure, and diabetes.

29 C. L. Wabner and C. Y. Pak. 1993. "Effect of orange juice consumption on urinary stone risk factors." *Journal Urology*. Jun;149(6): 1405–8. The study demonstrates the alkalinizing effect of orange juice: "Compared to potassium citrate, orange juice delivered an equivalent alkali load and caused a similar increase in urinary pH (6.48 versus 6.75 from 5.71) and urinary citrate (952 versus 944 from 571 mg. per day)." The study concludes that orange juice reduces two underlying processes in urinary stone formation: "Overall, orange juice should be beneficial in the control of calcareous and uric acid nephrolithiasis."

A similar protective effect was found for grapefruit juice and apple juice in R. Honow et al. 2003. "Influence of grapefruit, orange, and apple juice consumption on urinary variables and risk of crystallization." *Br J Nutrition*. Aug;90(2): 295–300. "Alkalizing beverages are highly effective in preventing the recurrence of calcium oxalate (Ox), uric acid and cystine lithiasis. The aim of the present study was to evaluate the influence of grapefruit-juice and apple-juice consumption on the excretion of urinary variables and the risk of crystallization in comparison with orange juice. . . . We showed that both grapefruit juice and apple juice reduce the risk of CaOx stone formation at a magnitude comparable with the effects obtained from orange juice."

30 A free radical is a molecule that, in contrast to most molecules, contains at least one unpaired electron and as a result is usually highly reactive. A considerable body of literature explores the role of oxygen free radicals in aging as well as in disease processes such as heart disease and cancer. According to some theories, mitochondrial DNA is a major target of free radical attack. See, for example, A. Ishchenko et al. 2003. "Age-dependent increase of 8-oxoguanine-, hypoxanthine-, and uracil-DNA glycosylase activities in liver extracts from OXYS rats with inherited overgeneration of free radicals and Wistar rats." *Med Sci Monit.* Jan;9(1): BR16–24; Y. Okatani et al. 2003. "Acutely administred melatonin restores hepatic mitochondrial physiology in old mice." *Int J Biochem Cell Biol.* Mar;35(3): 367–375; J. Sastre. 2002. "Ginkgo biloba extract EGb 761 protects against mitochondrial aging in the brain and in the liver." *Cell Mol Biol.* Sep;48(6): 685–692; A. Anantharaju. 2002. "Aging Liver: A review." *Gerontology*. Nov–Dec;48(6): 343–353.

31 H. Valtin. 2004. Upcoming article in *Journal of Physiology: Regulatory, Integrative and Comparative Physiology*. See American Physiological Society press release at www.the-aps.org/press/journal/release8-13-02.htm.

CHAPTER 5

1 Glucose, fructose, and galactose are isomers (molecules with the same number and types of atoms as another molecule, but with different properties). The different arrangement of atoms gives these sugars different properties.

2 M. Bloomfield and L. Stephens. 1996. *Chemistry and the Living Organism*, 6th ed. New York: John Wiley and Sons; W. Tamborlane et al., eds. 1997. *The Yale Guide to Children's Nutrition*. New Haven and London: Yale University Press.

3 E. Westman. 2002. "Is dietary carbohydrate essential for human nutrition?" *Am J Clin Nutr.* 75(5): 951–953. The established human nutrients are water, energy, amino acids, essential fatty acids, vitamins, minerals, trace minerals, electrolytes, and ultratrace minerals. (A. E. Harper. "Defining the essentiality of nutrients." In M. D. Shils et al., eds. 1993. *Modern Nutrition in Health and Disease*, 9th ed. Boston: William and Wilkins, pp. 3–10.)

4 Lactose intolerance varies by age and by race. The activity of the enzyme lactase declines after babies are weaned, so most of the human adult population is lactose-intolerant (J. L. Vilotte. 2002. "Lowering the milk lactose content in vivo." *Reprod Nutr Dev.* Mar–Apr 42: 127–132). Depending on race, the deficiency occurs in 50–90 percent of most populations. "White, western Europeans are the exception." A. Ferguson. 1995. "Mechanisms in adverse reactions to food." *Allergy.* 50: 32–38.

5 W. Willett and M. Stampfer. 2003. "Rebuilding the food pyramid." *Sci Amer.* Jan: 64–71; "The basics of good nutrition: Essential nutrients and their functions." In D. Tapley et al., eds. 1995. *Columbia University College of Physicians and Surgeons Complete Home Medical Guide.* New York: Crown Publishers (or available online at http://cpmcnet.columbia.edu/texts/guide).

6 Our primate ancestors also could not digest fiber (K. Milton. 1993. "Diet and primate evolution." *Sci Amer.* Aug: 86–93). See also "The basics of good nutrition: Essential nutrients and their functions." In D. Tapley et al., eds. 1995. *Columbia University College of Physicians and Surgeons Complete Home Medical Guide.* New York: Crown Publishers (or available online at http://cpmcnet.columbia.edu/texts/guide).

7 Fructose has a glycemic index in the 30s. Glucose and sucrose, along with white bread and potatoes, have index values over 85. In fact, glucose is sometimes used as the reference food for the scale, with an index of 100. For more information, see K. Foster-Powell et al. 2002. "International table of glycemic index and glycemic load values: 2002." *Am J Clin Nutr.* 76(1): 5–56.

8 Insulin surges lead to overeating and also foster the deposition of fat. D. S. Ludwig et al. 1999. "High glycemic index foods, overeating, and obesity." *Pediatrics.* Mar;103: E26; D. S. Ludwig. 2001. "Relation between consumption of sugar-sweetened drinks and childhood obesity: a prospective, observational analysis." *Lancet.* Feb 17;357: 505–508. This last study found that "each additional sugar-sweetened drink consumed" per day significantly increased a child's chance of developing obesity later.

9 F. S. Facchini et al. 2001. "Insulin resistance as a predictor of age-related diseases." *J Clin Endocrinol Metab.* Aug;86(8): 3574–3578; J. Salmeron et al. 1997. "Dietary fiber, glycemic load, and risk of non-insulin-dependent diabetes mellitus in women." *JAMA.* Feb 12; 277(6): 472–477.

10 The bulk provided by insoluble fiber increases stool size and shortens stool transit time through the intestine. Shorter transit times are better for bowel function. There is less time, for example, for "bad" bacteria to proliferate and produce toxins. In addition, insoluble fiber may inhibit the metabolism of carcinogens in the gut. In ascertaining the contribution of insoluble fiber to preventing colon cancer, it has been difficult to distinguish between the potential benefits of the fiber and the benefits of other cancer-inhibiting nutrients found in fiber-rich foods such as vegetables. S. A. Bingham et al. 2003. "Dietary fibre in food and protection against colorectal cancer in the European Prospective Investigation into Cancer and Nutrition (EPIC): an observational study." *Lancet.* May 3;361(9368) :1496–1501; U. Peters et al. 2003. "Dietary fibre and colorectal adenoma in a colorectal cancer early detection programme." *Lancet.* May 3;361(9368): 1491–1495; S. Gråsten et al. 2000. "Rye bread improves bowel function and decreases the concentrations of some compounds that are putative colon cancer risk markers in middle-aged women and men." *J Nutr.* 130: 2215–2221.

11 N. M. Avena and B. G. Hoebel. 2003. "Amphetamine-sensitized rats show sugar-induced hyperactivity (cross-sensitization) and sugar hyperphagia." *Pharmacol Biochem Behav.* Feb;74(3): 635–639; C. Colantuoni et al. 2002. "Evidence that intermittent, excessive sugar intake causes endogenous opioid dependence." *Obes Res.* Jun;10(6): 478–488; C. Colantuoni et al. 2001. "Excessive sugar intake alters binding to dopamine and muopioid receptors in the brain." *Neuroreport.* Nov 16;12(16): 3549–3552.

12 C. B. Ebbeling and D. S. Ludwig. 2001. "Treating obesity in youth: should dietary glycemic load be a consideration?" *Adv Pediatr.* 48: 179–212; S. B. Roberts. 2000. "High-glycemic index foods, hunger, and obesity: is there a connection?" *Nutr Rev.* 58: 163–169.

13 S. Higgenbotham et al. 2004. Dietary glycemic load and risk of colorectal cancer in the Women's Health Study. *J Natl Cancer Inst.* Feb 4; 96(3): 229–233.

14 K. Foster-Powell et al. 2002. "International table of glycemic index and glycemic load values: 2002." *Am J Clin Nutr.* Jul;76(1): 5–56.

15 For more complete lists, see www.lifelonghealth.us/mhc_home/pdf_docs/GLYCEMIC_INDEX.pdf; C. T. Netzer. 2000. *The Complete Book of Food Counts.* New York: Dell.

16 These findings were especially pronounced in overweight women. S. Liu et al. 2001. "Dietary glycemic load assessed by food-frequency questionnaire in relation to plasma high-density-lipoprotein cholesterol and fasting plasma triacylglycerols in postmenopausal women." *Am J Clin Nutr.* Mar;73(3): 560–566.

17 G. M. Reaven. 2003. "Age and glucose intolerance." *Diabetes Care.* 26: 539–540; G. M. Reaven. 1998. "Insulin resistance and human disease: a short history." *Basic Clin Physiol Pharmacol.* 9(2–4): 387–406. "The number of adults in the United States with diabetes increased by 49 percent between 1991 and 2000 . . . and Type II diabetes accounts for practically all of that increase" (J. Marx. 2002. "Unraveling the causes of diabetes." *Science.* 296(5568): 686–689, summarizing Centers for Disease Control and Prevention data).

18 M. Blüher et al. 2003. "Extended longevity in mice lacking the insulin receptor in adipose tissue." *Science*. Jan 24;299(5606): 572–574; S. H. Golden et al. 2002. "Risk factor groupings related to insulin resistance and their synergistic effects on subclinical atherosclerosis: the atherosclerosis risk in communities study." *Diabetes*. Oct;51: 3069–3076; F. Wollesen et al. 2002. "Insulin resistance and atherosclerosis in diabetes mellitus." *Metabolism*. Aug;51: 941–948.

19 J. Marx. 2002. "Unraveling the causes of diabetes." *Science*. 296(5568): 686–689; R. K. Campbell and J. R. White. 2002. "Insulin therapy in type 2 diabetes." *J Am Pharm Assoc*. Jul–Aug;42: 602–611; G. M. Reaven. 1999. "Insulin resistance: a chicken that has come to roost." *Ann NY Acad Sci*. Nov 18;892: 45–57.

20 F. Abbasi et al. 2002. "Relationship between obesity, insulin resistance, and coronary heart disease risk." *J Am Coll Cardiol*. Sep 4;40(5): 944–945; S. Liu and W. C. Willett. 2002. "Dietary glycemic load and atherothrombotic risk." *Curr Atheroscler Rep*. Nov 4: 454–461; G. M. Reaven. 2000. "Diet and Syndrome X." *Curr Atheroscler Rep*. Nov;2(6): 503–507; S. Liu et al. 2000. "A prospective study of dietary glycemic load, carbohydrate intake, and risk of coronary heart disease in U.S. women." *Am J Clin Nutr*. 71(6):1455–1461; J. Yip et al. 1998. "Resistance to insulin-mediated glucose disposal as a predictor of cardiovascular disease." *J Clin Endocrinol Metab*. Aug;83(8): 2773–2776.

21 High glycemic load has been noted as a risk factor for pancreatic, breast, and colon cancers. See, for example, D. S. Michaud. 2002. "Dietary sugar, glycemic load, and pancreatic cancer risk in a prospective study." *J Natl Cancer Inst*. 94(17): 1293–1300; L. S. Augustin et al. 2001. "Dietary glycemic index and glycemic load, and breast cancer risk: a case-control study." *Ann Oncol*. 12(11): 1533–1538; E. Giovannucci. 2001. "Insulin, insulin-like growth factors and colon cancer: a review of the evidence." *J Nutr*. Nov;131(11): 3109S–20S.

22 Discovered in 1879, saccharin is 300 times sweeter than sugar. In 1977, when the Food and Drug Administration (FDA) proposed to ban saccharin, it was the only alternative sweetener. The public outcry prompted Congress to pass the Saccharin Study and Labeling Act, which required foods containing saccharin to display a warning label. That requirement remained in place for more than two decades. Even though the government now claims to have exonerated saccharine through the Saccharin Warning Elimination via Environmental Testing Employing Science and Technology Act of 2000, many scientists remain concerned about the tens of millions of people consuming the sweetener and evidence of carcinogenesis. See, for example, W. Bell et al. 2002. "Carcinogenicity of saccharin in laboratory animals and humans." *Int J Occup Environ Health*. Oct–Dec;8: 387–393; Y. Sasaki et al. 2002. "The comet assay with 8 mouse organs: results with 39 currently used food additives." *Mutation Research*. 519(1–2): 103–119.

23 H. J. Roberts. 1992. *Aspartame (Nutrasweet): Is It Safe?* New York: The Charles Press. As with saccharin, controversy continues to rage over the safety of aspartame. The FDA defends the sweetener, although research shows a variety of possible heath effects (see, for example, S. K. van den Eeden et al. 1994. "Aspartame ingestion and headaches: a randomized crossover trial." *Neurology*. Oct;44(10): 1787–1793).

24 R. J. Wurtman. 1983. "Neurochemical changes following high-dose aspartame with dietary carbohydrates." *N Engl J Med*. Aug 18;309(7): 429–30; "Migraine provoked by aspartame." 1986. *N Engl J Med*. Aug 14;315(7): 456; S. E. Moller. 1991. "Effect of aspartame and protein, administered in phenylalanine-equivalent doses, on plasma neutral amino acids, aspartate, insulin and glucose in man." *Pharmacol Toxicol*. 68(5): 408–412.

25 Soluble fiber can be digested by the body as opposed to insoluble fiber, which cannot. D. L Sprecher and G. L. Pearce. 2002. "Fiber-multivitamin combination therapy: a beneficial influence on low-density lipoprotein and homocysteine." *Metabolism*. Sep;51(9): 1166–1170; B. M. Davy et al. 2002. "High-fiber oat cereal compared with wheat cereal consumption favorably alters LDL-cholesterol subclass and particle numbers in middle-aged and older men." *Am J Clin Nutr*. Aug;76(2): 351–358.

Eating soluble fiber may also be beneficial for individuals with syndrome X. (B. M. Davy and C. L. Melby. 2003. "The effect of fiber-rich carbohydrates on features of Syndrome X." *J Am Diet Assoc*. Jan;103(1): 86–96.)

26 Insoluble fiber cannot be digested by humans and is found in foods such as wheat bran, vegetables, and whole grains. M. Hill. 2003. "Dietary fibre and colon cancer: where do we go from here?" *Proc Nutr Soc*. Feb;62(1): 63–65; American Dietetic Association and Dietitians of Canada. 2003. "Position of the American Dietetic Association and Dieticians of Canada: Vegetarian Diets." *Can J Diet Pract Res*. Summer;64(2): 62–81.

27 A. Mukherjee and J. Chakrabarti. 1997. "In vivo cytogenetic studies on mice exposed to acesulfame-K-a non-nutritive sweetener." *Food Chem. Toxicol*. Dec;35(12): 1177–1179.

28 See www.ffcr.or.jp/zaidan/FFCRHOME.nsf/pages/e-kousei-sucra for Japanese studies on the safety of sucralose.

29 Stevia is a Paraguayan plant. Each leaf of stevia "contains 9 to 13 percent stevioside, which is 300 times sweeter than sugar." Stevia has been used as a sweetener in Japan for over three decades; this may be a reason for the number of Japanese studies on the plant. (See, for example, E. Koyama et al. 2003. "In vitro metabolism of the glycosidic sweeteners, stevia mixture and enzymatically modified stevia in human intestinal microflora." *Food and Chem Toxicol.* 41(3): 359–374; M. Matsui et al. 1996. "Evaluation of the genotoxicity of stevioside and steviol using six in vitro and one in vivo mutagenicity assays." *Mutagenesis.* Nov;11(6): 573–579.)

CHAPTER 6

1 Diet is key because humans do not synthesize either essential fatty acid. See, for example, A. P. Simopoulos. 2002. "The importance of the ratio of omega-6/omega-3 essential fatty acids." *Biomed Pharmacother.* Oct;56(8): 365–379; M. Crawford et al. 2000. "Role of plant-derived omega-3 fatty acids in human nutrition." *Ann Nutr Metab.* 44(5–6): 263–265.

2 F. B. Hu and M. J. Stampfer. 1999. "Nut consumption and risk of coronary heart disease: a review of epidemiologic evidence." *Curr Atherscler Rep.* Nov;1(3): 204–209. Other epidemiological studies also support this finding. See J. L. Ellsworth et al. 2001. "Frequent nut intake and risk of death from coronary heart disease and all causes in postmenopausal women in the Iowa Women's Health Study." *Nutr Metab Cardiovasc Dis.* Dec;11(6): 372–377. A plethora of important insights about nutrition have come from this study (see www.channing.harvard.edu/nhs/pub.html).

3 "There is increasing evidence that inflammation is also involved in the atherogenic process." J. T. Kuvin and R. H. Karas. 2003. "The effects of LDL reduction and HDL augmentation on physiologic and inflammatory markers." *Curr Opin Cardiol.* Jul;18(4): 295–300. See also T. Pischon et al. 2003. "Habitual dietary intake of n-3 and n-6 fatty acids in relation to inflammatory markers among US men and women." *Circulation.* Jul 15;108(2): 155–160.

4 Omega-3 fatty acids, for example, affect myocardial contractility, blood pressure, and coagulation factors. They have also been shown to prevent "sudden death after myocardial infarction." D. Bhatnagar and P. N. Durrington. 2003. "Omega-3 fatty acids: their role in the prevention and treatment of atherosclerosis related risk factors and complications." *Int J. Clin Pract.* May;57(4): 305–314; F. Thies. 2003. "Association of n-3 polyunsaturated fatty acids with stability of atherosclerotic plaques: a randomized controlled trial." *Lancet.* Feb 8;361(9356): 477–485.

5 An appraisal of "33 published case-control and cohort studies that examined the relationship between prostate cancer and dietary fat or specific fatty food types" found eight studies that "suggested a statistically significant association, and many studies noted significant associations for specific types of fatty foods (e.g., milk or meat) and prostate cancer." N. Fleshner et al. 2004. "Dietary fat and prostate cancer." *J Urol.* Feb;171(2 Pt 2): S19–24; L. N. Kolonel, A. M. Y. Nomura, and R. V. Cooney. 1999. *J Natl Cancer Inst.* 91(5): 414–428; L. M. Newcomer et al. 2001. "The association of fatty acids with prostate cancer risk." *Prostate.* Jun 1;47(4): 262–268.

6 Y. Park and W. S. Harris. 2003. "Omega-3 fatty acid supplementation accelerates chylomicron triglyceride clearance." *J Lipid Res.* Mar;44(3): 455–463; W. S. Harris. 1997. "n-3 fatty acids and serum lipoproteins: human studies." *Am J Clin Nutr.* May;65(5 Suppl): 1645S–1654S.

7 A. H. Lichtenstein. 2003. "Dietary fat and cardiovascular disease risk: quantity or quality?" *J Womens Health (Larchmt).* Mar;12(2): 109–114.

8 H. Chen et al. 2003. "EPA and DHA attenuate ox-LDL-induced expression of adhesion molecules in human coronary artery endothelial cells." *J Mol Cell Cardiol.* Jul;35(7): 769–775; S. Renaud and D. Lanzmann-Petithory. 2002. "Dietary fats and coronary heart disease pathogenesis." *Curr Atheroscler Rep.* Nov;4(6): 419–424.

9 A. Nordoy et al. 2001. "n-3 polyunsaturated fatty acids and cardiovascular diseases." *Lipids.* 36 Suppl: S127–129; K. Imaizumi et al. 2000. "Role of dietary lipids in arteriosclerosis in experimental animals." *Biofactors.* 13(1-4): 25–28; J. A. Conquer et al. 1999. "Effect of supplementation with dietary seal oil on selected cardiovascular risk factors and hemostatic variables in healthy male subjects." *Thromb Res.* 1;96(3): 239–250.

10 P. C. Calder. 2002. "Dietary modification of inflammation with lipids." *Proc Nutr Soc.* Aug;61(3): 345–358; P. Yang et al. 2002. "Quantitative high-performance liquid chromatography/electrospray ionization tandem mass spectrometric analysis of 2- and 3-series prostaglandins in cultured tumor cells." *Anal Biochem.* Sep 1;308(1): 168–177.

11 L. A. Sauer et al. 2000. "Mechanism for the antitumor and anticachectic effects of n-3 fatty acids." *Cancer Res.* Sep 15;60(18): 5289–5295.

12 P. M. Kris-Etherton et al. 2002. "Fish consumption, fish oil, omega-3 fatty acids, and cardiovascular disease." *Circulation.* Nov 19;106(21): 2747–2757. An article by two of the authors (W. S. Harris and L. J. Appel) is also available on the AMA site (www.americanheart.org).

13 M. E. Surette. 2003. "Inhibition of leukotriene synthesis, pharmacokinetics, and tolerability of a novel dietary fatty acid formulation in healthy adult subjects." *Clin Ther.* Mar;25(3): 948–971; L. S. Harbige. 2003. "Fatty acids, the immune response, and autoimmunity: question of n-6 essentiality and the balance between n-6 and n-3." *Lipids.* Apr;38(4): 323–341.

14 D. Bagga et al. 2003. "Differential effects of prostaglandin derived from omega-6 and omega-3 polyunsaturated fatty acids on COX-2 expression and IL-6 secretion." *Proc Natl Acad Sci USA.* Feb 18;100(4): 1751–1756.

15 T. van Vliet and M. B. Katan. 1990. "Lower ratio of n-3 to n-6 fatty acids in cultured than in wild fish." *Am J Clin Nutr.* Jan;51(1): 1–2. There are a variety of other issues associated with farming salmon. See, for example, M. D. Eason et al. 2002. "Preliminary examination of contaminant loadings in farmed salmon, wild salmon and commercial salmon feed." *Chemosphere.* Feb;46(7): 1053–1074.

16 M. Massaro et al. 2002. "Quenching of intracellular ROS generation as a mechanism for oleate-induced reduction of endothelia activation and early atherogenesis." *Thromb Haemost.* Aug;88(2): 335–344; C. M. Williams. 2001. "Beneficial nutritional properties of olive oil: implications for postprandial lipoproteins and factor VII." *Nutr Metab Cardiovasc Dis.* Aug;11(4 Suppl): 51–56.

17 "Progression of CAD over 39 mo, measured by a decrease in minimum absolute width of coronary segments (MinAWS) on angiography, was highly correlated with intakes of palmitic, stearic (18:0), palmitoleic, and elaidic (t-18:1) acids ($P < 0.001$) . . . "

"Our results indicate that three nonessential fatty acids—stearic acid, palmitoleic acid, and omega 9 eicosatrienoic acid, and one essential fatty acid—dihomogammalinolenic acid, are independent correlates of blood pressure among middle-aged American men at high risk of coronary heart disease."

Center for Science in the Public Interest, July/August *Nutrition Action Newsletter,* using the USDA Nutrient Database for Standard Reference (Release 14) as a source.

18 T. Thostrup et al. 1994. "Fat high in stearic acid favorably affects blood lipids and factor VII coagulant activity in comparison with fats high in palmitic acid or high in myristic and lauric acids." *Am J Clin Nutr.* Feb;59(2): 371–377.

19 F. Joffre et al. 2001. "Kinetic parameters of hepatic oxidation of cyclic fatty acid monomers formed from linoleic and linolenic acid." *J Nutr Biochem.* Oct;12(10): 554–558; B. Potteau. 1976. "Influence of heated linseed oil on reproduction in the female rat and on the composition of hepatic lipids in young rats." *Ann Nutr Aliment.* 30(1): 67–88.

20 J. K. Donnelly and D. S. Robinson. 1995. "Free radicals in food." *Free Radic Res.* Feb;22(2): 147–176.

21 Methods of commercial oil production include expeller pressing (which can generate temperatures as high as 185°F); cold pressing; and solvent extraction, in which oils are extracted using petroleum solvents.

22 A. H. Lichtenstein et al. 1999. "Effects of different forms of dietary hydrogenated fats on serum lipoprotein cholesterol levels." *N Engl J Med.* Jun 24;340(25): 1933–1940.

23 N. de Roos et al. 2001. "Consumption of a solid fat rich in lauric acid results in a more favorable serum lipid profile in healthy men and women than consumption of a solid fat rich in trans-fatty acids." *J Nutr.* Feb;131(2): 242–245.

24 A. Ammouche et al. 2002. "Effect of ingestion of thermally oxidized sunflower oil on the fatty acid composition and antioxidant enzyme of rat liver and brain in development." *Ann Nutr Metab.* 46(6); 268–275.

25 S. M. Marcovina and M. L. Koschinsky. 2003. "Evaluation of lipoprotein(a) as a prothrombotic factor: progress from bench to bedside." *Curr Opin Lipidol.* Aug;14(4): 361–366; M. Koruk et al. 2003. "Serum lipids, lipoproteins, and apolipoproteins levels in patients with nonalcoholic steatohepatitis." *J Clin Gastroenterol.* Aug;37(2): 177–182.

26 R. Rosmond and P. Björntorp. 1998. "The interactions between hypothalamic-pituitary-adrenal axis activity, testosterone, insulin-like growth factor I and abdominal obesity with metabolism and blood pressure in men." *Int. J Obes Relat Metab Disord.* Dec;22(12): 1184–1196.

27 S. M. Grundy et al. 2002. "Diet composition and the metabolic syndrome: what is the optimal fat intake?" *Am J Med.* Dec 30;113 Suppl 9B: 25S–29S.

28 "The more hydrogenated an oil is, the harder it will be at room temperature. For example, a spreadable tub margarine is less hydrogenated and so has fewer trans fats than a stick margarine." "Fats and cholesterol," Harvard School of Public Health (www.hsph.harvard.edu/nutritionsource/fats.html). See also F. D. Kelly et al. 2001. "A stearic acid-rich diet improves thrombogenic and atherogenic risk factor profiles in healthy males." *Eur J Clin Nutr.* Feb;55(2): 88–96; C. M. Nieuwenhuys and G. Hornstra. 1998. "The effects of purified eicosapentaneoic and docosahexaneoic acids on arterial thrombosis tendency and platelet function in rats." *Biochem Biophys Acta.* Feb 23;1390(3): 313–322.

29 J. W. Ju and M. Y. Jung. 2003. "Formation of conjugated linoleic acids in soybean oil during hydrogenation with a nickel catalyst as affected by sulfur addition." *J Agric Food Chem.* May 7;51(10): 3144–3149; M. A. De Oliveira et al. 2003. "Method development for the analysis of trans-fatty acids in hydrogenated oils by capillary electrophoresis." *Electrophoresis.* May;24(10): 1641–1647.

30 S. Vincent et al. 2003. "Targeting of proteins to membranes through hedgehog auto-processing." *Nature Biotech* (advance online publication July 13); C. Thiele et al. 2000. "Cholesterol binds to synaptophysin and is required for biogenesis of synaptic vesicles." *Nature Cell Biol.* Jan;2: 42–49.

31 L. Ellegård et al. 2000. "Will recommended changes in fat and fibre intake affect cholesterol absorption and sterol excretion?" *Eur J Clin Nutr.* Apr;54(4): 306–313.

32 I. S. Cowin and P. M. Emmett. 2001. "Associations between dietary intakes and blood cholesterol concentrations at 31 months." *Eur J Clin Nutr.* Jan;55(1): 39–49.

33 Cortisol is one of several steroids (others include progesterone, estradiol, and testosterone) synthesized from cholesterol. See H. Lodish et al. 2000. "Cell-to-cell signaling: hormones and receptors." *Molecular Cell Biology.* New York: W. H. Freeman and Company.

34 See Step by Step: Eating to Lower Your High Blood Cholesterol. National Institutes of Health: National Heart, Lung, and Blood Institute; www.limcpc.com/Medical%20Info/cholest/eattolowerchol.htm.

35 J. Scott. 1999. "Heart disease: good cholesterol news." *Nature.* Aug 26;400: 816–819.

36 For further discussion on the importance of limiting dietary cholesterol consumption, see, for example, *High Blood Cholesterol: What You Need to Know*, National Cholesterol Education Program; www.nhlbi.nih.gov/health/public/heart/chol/hbc_what.htm; *Taking Charge of Your Health: The Harvard Medical School Family Health Guide* (available online at www.health.harvard.edu/fhg/fhgupdate/A/A2.shtml); R. M. Weggemans et al. 2001. "Dietary cholesterol from eggs increases the ratio of total cholesterol to high-density lipoprotein cholesterol in humans: a meta-analysis." *Am J Clin Nutr.* May;73(5): 885–891.

37 Like antioxidants, ketone bodies are attracting a great deal of research interest, often around treatment for diabetes, obesity, and epilepsy (ketogenic diets for epilepsy are primarily based on fats). The following citation states that the brain can use both glucose and ketones for energy: A. E. Greene et al. 2003. "Perspectives on the metabolic management of epilepsy through dietary reduction of glucose and elevation of ketone bodies." *J Neurochem.* Aug;86(3): 529–537.

38 National Institutes of Health: National Heart Lung and Blood Institute. 2001. "High blood cholesterol: what you need to know," at www.nhlbi.nih.gov/health/public/heart/chol/wyntk.pdf (NIH Pub. No. 01-3290).

39 R. M. Anderson et al. 1996. "Transmission dynamics and epidemiology of BSE in British cattle." *Nature.* 382: 779–788.

40 J. X. Kang. 2004."Transgenic mice: fat-1 mice convert n-6 to n-3 fatty acids." *Nature.* Feb 5;427(6974): 504.

41 A. Baguisi et al. 1999. "Production of goats by somatic cell nuclear transfer." *Nature Biotechnology.* May 17(5): 456–461. For more information on the partnership between Genzyme Transgenics Corporation, Louisiana State University, and Tufts University School of Medicine that produced this work, see the press release on the GTC Biotherapeutics Web site, www.transgenics.com/pressreleases/pr042799.html.

42 This five-year project was announced in December 1999. Why protein folding? "The life sciences have benefited from computational capabilities and will be driving the requirements for data, network, and computational capabilities in the future. . . . The understanding of the protein folding phenomenon is a recognized 'grand challenge problem' of great interest to the life sciences." F. Allen et al. 2001. "Blue Gene: A vision for protein science using a petaflop supercomputer." *IBM Sys J.* 40(2): 310–327.

43 C. L. Scott. 2003. "Diagnosis, prevention, and intervention for the metabolic syndrome." *Am J Cardiol.* Jul 3;92(1A): 35i–42i; F. B. Hu and W. C. Willett. 2002. "Optimal diets for prevention of coronary heart disease." *JAMA.* Nov 27;288(20): 2569–2578.

44 In November 1989, FDA recalled all dietary supplements containing more than 100 mg of L-tryptophan due to over 1,500 cases of a rare, sometimes fatal condition known as EMS (eosinophilia-myalgia syndrome). See CDC. 1990. "Update: Eosinophilia-Myalgia Syndrome Associated with Ingestion of L-Tryptophan—United States, through August 24, 1990." *MMWR.* Aug 31;39(34); 587–589. Subsequent investigation suggested that these cases of disease were due to contaminants introduced in the manufacture of this amino acid and not due to the amino acid itself. It is now available again in the U.S., but either by prescription or at a much increased price.

45 S. Moncada and A. Higgs. 1993. "The L-arginine-nitric-oxide pathway." *N Engl J Med.* Dec 30;329(27): 2002–2012.

46 The following book cites extensive research documenting the value of arginine supplementation in maintaining healthy cardiac arteries and in avoiding heart disease and stroke: J. Zimmer and J. P. Cooke. 2002. *The Cardiovascular Cure: How to Strengthen Your Self-Defense Against Heart Attack and Stroke.* New York: Broadway. Also A. Lerman et al. 1998. "Long-term L-arginine supplementation improves small-vessel coronary endothelial function in humans." *Circulation.* 97: 2123–2128; B. Y. Wang et al. 1999. "Regression of atherosclerosis: role of nitric oxide and apoptosis." *Circulation.* 99: 1236–1241.

47 A. L. Jenkins. 2002. "Depression of the glycemic index by high levels of â-glucan fiber in two functional foods tested in type 2 diabetes." *Eur J Clin Nutr.* Jul;56(7): 622–628.

48 J. W. Helge. 2002. "Prolonged adaptation to fat-rich diet and training; effects on body fat stores and insulin resistance in man." *Intl J Obesity.* Aug;26(8): 1118–1124.

49 D. J. Jenkins et al. 2000. "Dietary fibre, lente carbohydrates and the insulin-resistant diseases." *Br J Nutr.* Mar;83(Suppl 1): S157–163.

50 Of the many recent articles on this subject, see E. Södergren et al. 2001. "A diet containing rapeseed oil-based fats does not increase lipid peroxidation in humans when compared to a diet rich in saturated fatty acids." *Eur J Clin Nutr.* Nov;55(11): 922–931.

CHAPTER 7

1 D. A. Drossman et al. 1993. "U.S. householder survey of functional gastrointestinal disorders. Prevalence, sociodemography, and health impact." *Dig Dis Sci.* Sep;38(9): 1569–1580. According to the American Gastroenterological Association (www.gastro.org/public/brochures/yourdigest.html), "each month 44% of adults take antacids or other medicines" to treat a single gastrointestinal problem—heartburn.

2 Hypochlorhydria has often been misdiagnosed, either by doctors or by patients, because its symptoms, such as bloating, flatulence, and burning, resemble those of hyperchlorhydria (too much acid). As a result, some patients take antacids when they have the opposite problem. Many studies show that atrophic gastritis (little or no acid secretion in the stomach) is an increasing problem with age. As many as 20 to 30 percent of those over 60 in the U.S. have this condition. See, for example, S. D. Krasinski et al. 1986. "Fundic atrophic gastritis in an elderly population." *J Am Geriatr Soc.* Nov;34(11): 800–806. Atrophic gastritis is a "predisposing factor for gastric cancer." (M. Inoue et al. 2000. "Severity of chronic atrophic gastritis and subsequent gastric cancer occurrence." *Cancer Lett.* Dec 8;161(1): 105–112.) Other problems associated with too little stomach acid include rheumatoid arthritis, anemia, coronary disease, asthma, anemia, and gallstones. See J. Wright and L. Lenard. 2001. *Why Stomach Acid Is Good for You.* New York: M. Evans & Co.

3 "In addition, 75% of adults worldwide are said to be lactose maldigesters or have low lactase levels." "Lactose Maldigestion/Lactose Intolerance," National Dairy Council (www.nationaldairycouncil.org/lv104/nutrilib/calccounsel/06_ccr_rev.htm). "In Africa, Asia, and Latin America, prevalence rates range from 15–100%, depending on the population studied." N. S. Scrimshaw and E. B. Murray. 1988. "The acceptability of milk and milk products in populations with a high prevalence of lactose intolerance." *Am J Clin Nutr.* Oct;48(Suppl 4): 1079–1159. See also D. L. Swagerty Jr. et al. 2002. "Lactose intolerance." *Am Fam Physician.* May 1;65(9): 1845–1850.

4 M. Morotomi and S. Kado. 2003. "Intestinal microflora and cancer prevention." *Gan To Kagaku Ryoho.* Jun;30(6): 741–747; F. Guarner and J. R. Malagelada. 2003. "Gut flora in health and disease." *Lancet.* May 24;361(9371): 1831.

5 Recent studies suggest that the cut and type of meat consumed may influence the risk for colon cancer. L. Ferguson. 2002. "Meat consumption, cancer risk and population groups within New Zealand." *Mut Res.* Sep 30;506–507: 215–224; E. L. Matos and A. Brandani. 2002. "Review on meat consumption and cancer in South America." *Mut Res.* Sep 30;506–507: 243–249.

6 One recent study found a reduction in cancer risk of 40 percent by doubling fiber intake from food. S. Bingham et al. 2003. "Dietary fibre in food and protection against colorectal cancer." *Lancet.* May 3;361(9368): 1496–1501. These results contradict earlier research. "People in the top 20% who had the biggest reduction were eating far more fibre than in other studies which have not shown a relationship," said S. Bingham, interviewed by P. Reaney for *News in Science* (www.abc.net.au/science/news). See also L. Ferguson and P. Harris. 2003. "The dietary fibre debate: more food for thought." *Lancet.* May 3;361(9368): 1487–1488.

7 P. D'Adamo. 1997. *Eat Right for Your Type*. New York: G. P. Putnam's & Sons.

8 For information about this test, see data from diagnostic laboratories, such as Great Smokies Diagnostic Laboratory (www.gsdl.com/assessments/cdsa) or Doctors' Data Laboratory (www.doctorsdata.com). See also a naturopathic text such as S. Barrie. 1999. "Comprehensive digestive stool analysis." In M. T. Murray and J. E. Pizzorno. *Textbook of Natural Medicine*, 2nd ed. New York: Churchill Livingstone, pp. 107–116.

9 J. Bland. 1999. *The 20-Day Rejuvenation Diet Program*. New York: McGraw-Hill.

10 "Demonstration of the potential health benefits of short-chain fructooligosaccharides on colon cancer risk is an active field of research in animal and human nutrition," according to F. R. Bornet and F. Brouns. "Immune-stimulating and gut health-promoting properties of short-chain fructo-oligosaccharides." *J Nutr Oct*. 60(10 Pt 1): 326–334. See also C. Cherbut et al. 2003. "The prebiotic characteristics of fructooligosaccharides are necessary for reduction of TNBS-induced colitis in rats." *J Nutr*. Jan;133(1): 21–27.

11 A. Ferrar. 2003. "Metal poisoning." *An Sist Sanit Navar*. 26(Suppl 1): 141–153; L. Patrick. 2003. "Toxic metals and antioxidants." *Altern Med Rev*. Apr;8(2): 106–128.

12 Antibodies are immunoglobulins (Ig). These are produced in response to an antigen, which is a substance the body perceives as a threat. IgG is one class of antibody, and it is normally present in the body at relatively high levels (10 mg/ml). IgG responses to food are typically delayed by as much as 48 hours and thus the symptoms, such as wheezing, bloating, loss of energy, and headaches, are often not associated with the triggering food. Most negative food reactions involve IgG.

13 IgE, like IgG, is a class of immunoglobulin. IgE is normally present in the body at low levels (0.5 μg/ml). With an IgE allergic reaction to a food, symptoms appear within seconds to a few hours. The IgE triggers cells that orchestrate immune responses, called mast cells, to start an inflammatory response. An extreme IgE allergic response can result in anaphylactic shock and death. For more information, see C. Janeway et al. 2001. "Allergy and hypersensitivity" in *Immunobiology*, 5th. ed. New York: Garland Publishing.

14 *Toxoplasma gondii* is a parasite prevalent in wild and domestic animals worldwide. According to an expert at the National Institute of Allergy and Infectious Diseases (NIAID), "Many parasitic diseases such as giardiasis and cryptosporidiosis are not always reported to health authorities, so that we suspect that the extent and impact of parasitic diseases in the United States is underestimated." In addition, "up to three million women have acquired sexually transmitted T. vaginalis." News from NIAID, November 1, 1993 (www.aegis.com/news/niaid/1993/CDC93081.html).

15 D. Karsenti et al. 2001. "Small intestine bacterial overgrowth: six case reports and literature review." *Rev Med Interne*. Jan;22(1): 20–29; S. M. Riordan et al. 2001. "Small intestinal bacterial overgrowth and the irritable bowel syndrome." *Am J Gastroenterol*. Aug;96(8): 2506–2508.

16 "Helicobacter pylori infection is one of the most common in man," according to S. A. Dowsett and M. J. Kowolik. 2003. "Oral Helicobacter pylori: can we stomach it?" *Crit Rev Oral Biol Med*. 14(3): 226–233.

17 A. Gewirtz et al. 2002. "Intestinal epithelial pathobiology: past, present, and future." *Best Practice & Res Clin Gastroenterol*. Dec;16(6): 851–867; D. Hollander. 1999. "Intestinal permeability, leaky gut, and intestinal disorders." *Curr Gastroenterol Rep*. Oct;1(5): 410–416.

18 "The word 'auto' is the Greek word for self. The immune system is a complicated network of cells and cell components (called molecules) that normally work to defend the body and eliminate infections caused by bacteria, viruses, and other invading microbes. If a person has an autoimmune disease, the immune system mistakenly attacks self, targeting the cells, tissues, and organs of a person's own body. A collection of immune system cells and molecules at a target site is broadly referred to as inflammation." Understanding Autoimmune Diseases, National Institute of Allergy and Infectious Diseases; www.niaid.nih.gov/publications/autoimmune.htm.

19 See note 17, above.

20 S. Holt, ed. 2000. *Natural Ways to Digestive Health: Interfaces Between Conventional and Alternative Medicine*. New York: M. Evans and Company.

21 Y. Ringel et al. 2001. "Irritable bowel syndrome." *Annu Rev Med*. 52: 319–338; C. M. Porth. 1998. "Irritable bowel syndrome." In *Pathophysiology: Concepts of Altered Health States*, 5th ed. Philadelphia: Lippincott, pp. 729–730.

22 A. R. Gaby. 2003. "Treatment with enteric-coated peppermint oil reduced small-intestinal bacterial overgrowth in a patient with irritable bowel syndrome." *Altern Med Rev*. Feb;8(1):3; R. M. Kline et al. 2001. "Enteric-coated, pH-dependent peppermint oil capsules for the treatment of irritable bowel syndrome in children." *J Pediatr*. Jan;138(1): 125–128.

23 Produced by Proper Nutrition; www.propernutrition.com.

24 See A. Picard. "Today's fruits, vegetables lack yesterday's nutrition." *Globe and Mail*, July 6, 2002; www.globeandmail.com/special/food/wxfood.html.

25 Pesticides are classified when they are registered on the basis of animal and epidemiological tests. See www.epa.gov/pesticides/health/tox_categories.htm.

26 The extoxnet is a good source of information on these agricultural chemicals. It is a pesticide information project of the Cooperative Extension Offices of Cornell University, Michigan State University, Oregon State University, and University of California at Davis. For terbutryn, see http://pmep.cce.cornell.edu/profiles/extoxnet/pyrethrins-ziram/terbutryn-ext.html. Also see the index of cleared science reviews under the Freedom of Information Act; www.epa.gov/pesticides/foia/reviews/080813.htm.

27 I. Kimber and R. J. Dearman. 2002. "Factors affecting the development of food allergy." *Proc Nutr Soc.* Nov;61(4): 435–439; E. Fernandez et al. 2000. "Diet diversity and colorectal cancer." *Prev Med.* Jul;31(1): 11–14.

28 M. Zimmerman. 2001. *Eat Your Colors: Maximize Your Health by Eating the Right Foods for Your Body Type.* New York: Henry Holt & Co.

29 K. Mukamal et al. 2002. "Tea consumption and mortality after acute myocardial infarction." *Circulation.* May 6;105: 2476.

30 Green tea contains a high level of catechins, which are a type of polyphenol. Polyphenols are antioxidants. The catechins in black tea are lost during processing. J. D. Lambert and C. S. Yang. 2003. "Cancer chemopreventative activity and bioavailability of tea and tea polyphenols." *Mutat Res.* Feb–Mar;523–524: 201–208; K. Maeda et al. 2003. "Green tea catechins inhibit the cultured smooth muscle cell invasion through the basement barrier." *Atherosclerosis.* Jan;166(1): 23–30.

31 A. Sierksma et al. 2002. "Moderate alcohol consumption reduces plasma C-reactive protein and fibrinogen levels; a randomized, diet-controlled intervention study." *Eur J Clin Nutr.* Nov;56(11): 1130–1136.

32 *Atkins for Life: The Complete Controlled Carb Program for Permanent Weight Loss and Good Health* and *Dr. Atkins' Age-Defying Diet.*

33 WHO. 2002. FAO/WHO Consultation on the Health Implications of Acrylamide in Food. Summary report of a meeting held in Geneva, 25–27 June 2002 (available at www.who.int/fsf/). Also see "Acrylamide in food." European Commission, Scientific Committee on Food; http://europa.eu.int/comm/food/fs/sfp/fcr/acrylamide/acryl_index_en.html. FDA Action Plan for Acrylamide in Food, March 2004, FDA/Center for Food Safety & Applied Nutrition, www.cfsan.fda.gov/~dms/acrypla3.html; Exploratory Data on Acrylamide in Food, March 2003, FDA/Center for Food Safety & Applied Nutrition, www.cfsan.fda.gov/~dms/acrydata.html; Exploratory Data on Acrylamide in Food, March 2004, FY 2003 Total Diet Study Results, FDA/Center for Food Safety & Applied Nutrition, www.cfsan.fda.gov/~dms/acrydat2.html.

34 *Eat More, Weigh Less: Dr. Dean Ornish's Life Choice Program for Losing Weight Safely While Eating Abundantly* and *Everyday Cooking with Dr. Dean Ornish: 150 Easy, Low-Fat, High Flavor Recipes.*

35 Information about the USDA food pyramid is available at www.nal.usda.gov/fnic/Fpyr/pyramid.html and www.nal.usda.gov:8001/py/pmap.htm.

36 See W. Willett. 2001. *Eat, Drink, and Be Healthy*. New York: Simon & Schuster. Also see "Food pyramids." Harvard School of Public Health; www.hsph.harvard.edu/nutritionsource/pyramids.html.

CHAPTER 8

1 According to this study's results, a woman who is obese at age 20 can expect a reduction in life expectancy of 8 years, while an obese 20-year-old man can anticipate a loss of 13 years compared to the life span of his normal-weight peers. When weight gain does not occur until later in life, the results are still significant, though not as dramatic. Merely being overweight (not obese) at age 40 shortens average life span by 3.1 years. People who are overweight and smoke can anticipate living 7 years less. The loss of life consequent to being overweight is about equal to that from cigarette smoking. K. R. Fontaine et al. 2003. "Years of life lost due to obesity." *JAMA.* Jan 8;289(2): 187–193.

2 T. E. Strandberg et al. 2003. "Impact of midlife weight change on mortality and quality of life in old age." *Int J Obes.* Aug;27(8): 950–954; S. A. French et al. 1997. "Weight variability and incident disease in older women: the Iowa Women's Health Study." *Int J Obes.* Mar;21(3): 217–223.

3 A. H. Mokdad et al. "Prevalence of obesity, diabetes, and obesity-related health risk factors, 2001." *JAMA.* Jan 1;289(1): 76–79; National Institutes of Health. 1998. Clinical guidelines on the identification, evaluation, and treatment of overweight and obesity in adults. Bethesda, Maryland: Department of Health and Human Services, National Institutes of Health, National Heart, Lung, and Blood Institute, pp. 12–20; "Overweight and obesity: health consequences." The Surgeon General's Call to Action; www.surgeongeneral.gov/topics/obesity/calltoaction/fact_consequences.htm.

4 "After adjustment for established risk factors, there was an increase in the risk of heart failure of 5 percent for men and 7 percent for women for each increment of 1 in BMI," per S. Kenchaiah et al. 2002. "Obesity and the risk of heart failure." *N Engl J Med.* Aug 1;347(5): 305–313; F. W. Ashley Jr. and W. B. Kannell. 1974. "Relation of weight change to changes in atherogenic traits: the Framingham Study." *J Chronic Dis.* Mar.;27(3): 103–114.

5 According to a recent Centers for Disease Control and Prevention (CDC) survey, "more than two-thirds of Americans—64 percent of men and 78 percent of women—are either dieting to lose weight or watching what they eat," as reported in "Many Americans fed up with diet advice," *New York Times*, January 2, 2001. Yet currently "more than 60 million Americans (a third of the population) are overweight," per the CDC. "The link between physical activity and morbidity and mortality," National Center for Chronic Disease Prevention and Health Promotion; www.cdc.gov/nccdphp/sgr/mm.htm.

6 S. Orenstein. "The pill that will make you thin." *Business 2.0*, March 2004, pp. 108–115.

7 G. K. Goodrick and J. P. Foreyt. 1991. "Why treatments for obesity don't last." *J Am Diet Assoc.* Oct; 91(10): 1243–1247; F. M. Kramer et al. 1989. "Long-term follow-up of behavioral treatment for obesity: patterns of weight regain among men and women." *Int J Obes.* 13(2): 123–136.

8 M. Hendricks. 2003. "Off the scale." *Johns Hopkins Public Health: The Magazine of the Johns Hopkins Bloomberg School of Public Health*, Spring; K. D. Brownell. 1989. "Weight cycling," *Am J Clin Nutr.* May;49(Suppl 5): 937; G. L. Blackburn et al. 1989. "Weight cycling: the experience of human dieters." *Am J Clin Nutr.* May;49(Suppl 5): 1105–1109.

9 Each pound of fat stores about 3,500 calories.

10 Weights at ages 25–59 based on lowest mortality. Weight in pounds according to frame (wearing in-door clothing weighing 3 lbs. and shoes with 1-in. heels). Courtesy of Metropolitan Life Insurance Company.

11 A number of recent studies are exploring the relationship between body mass index (BMI) and the percentage of body fat (%BF) as a means of developing health guidelines. See, for example, U. G. Kyle et al. 2003. "Body composition interpretation: Contributions of the fat-free mass index and the body fat mass index." *Nutrition.* Jul–Aug;19(7–8): 597–604; D. Gallagher et al. 2000. "Healthy percentage body fat ranges: an approach for developing guidelines based on body mass index." *Am J Clin Nutr.* Sep;72(3): 694–701. Age and race have both been cited as factors to consider in setting the normal ranges for %BF. The onset of puberty is linked to the development of energy storage in the form of fat. See, for example, B. Vizmanos and C. Martí-Henneberg. 2000. "Puberty begins with a characteristic subcutaneous body fat mass in each sex." *Eur J Clin Nutr.* Mar;54(3): 203–208.

12 S. P. Weisberg and A. W. Ferrante Jr. 2003. "Obesity is associated with macrophage accumulation in adipose tissue." *Journal of Clinical Investigation.* Dec 15;112: 1796–1808. Available at www.jci.org/cgi/content/full/112/12/1796; K. E. Wellen and G. S. Hotamisligil. 2003. "Obesity-induced inflammatory changes in adipose tissue." *Journal of Clinical Investigation.* Dec 15;112: 1785–1788. Available at www.jci.org/cgi/content/full/112/12/1785; H. Xu and H. Chen. 2003. "Chronic inflammation in fat plays a crucial role in the development of obesity-related insulin resistance." *Journal of Clinical Investigation* Dec 15;112: 1821–1830. Available at www.jci.org/cgi/content/full/112/12/1821; G. S. Hotamisligil et al. 1994. "Tumor necrosis factor alpha inhibits signaling from the insulin receptor." *Proceedings of the National Academy of Sciences* May 24;91: 4854–4858. Available at www.pnas.org/cgi/reprint/91/11/4854.

13 D. V. Schapira et al. 1991. "Upper-body fat distribution and endometrial cancer risk," *JAMA.* Oct 2;266(13): 1808–1811.

14 The link between abdominal obesity and health issues such as diabetes and metabolic syndrome has been made in many different populations around the globe. See, for example, N. K. Vikram et al. 2003. "Anthropometry and body composition in northern Asian Indian patients with type 2 diabetes." *Diabetes Nutr Metab.* Feb;16(1): 32–40; J. A. Lawati et al. "Prevalence of the metabolic syndrome among omani adults." *Diabetes Care.* Jun;26(6): 1781–1785. There is a "modest relationship" between abdominal obesity and coronary heart disease, according to K. M. Rexrode et al. 2001. "Abdominal and total adiposity and risk of coronary heart disease in men." *Int J Obes.* Jul;25(7): 1047–1056.

15 In the Nurses' Health Study II, "most women who lost a clinically significant amount of weight regained it, [however,] they gained less weight over the entire 6 year period than their peers," per A. E. Field et al. 2001. "Relationship of a large weight loss to long-term weight change among young and middle-aged U.S. women." *Int J Obes.* Aug;25(8): 1113–1121. Some recent results are encouraging: "A large proportion of the American population has lost = 10% of their maximum weight and has maintained this weight for at least 1 year." M. T. McGuire et al. 1999. "The prevalence of weight loss maintenance among American adults." *Int J. Obes.* Dec;23(22): 1314–1319.

16 D. E. Cummings et al. 2002. "Plasma ghrelin levels after diet-induced weight loss or gastric bypass surgery." *N Engl J Med.* May 23;346(21): 1623–1630. See also D. E. Cummings and M. W. Schwartz. 2003. "Genetics and pathophysiology of human obesity." *Annu Rev Med.* 54: 453–471.

17 The more muscle you have, the more calories you burn, even while resting. Many fitness trainers recommend weight training to build muscle mass. See, for example, R. Roubenoff et al. 2000. "The effect of gender and body composition method on the apparent decline in lean mass-adjusted resting metabolic rate with age." *J Gerontol A Biol Sci Med Sci.* Dec;55(12): M757–760. There may also be "metabolic demands of resynthesizing glycogen and repairing tissue damage," per R. Andersen. 1999. "Exercise, an active lifestyle, and obesity." *Phys Sportmed.* Oct 1;27(10).

18 "Energy can be expended by performing work or producing heat (thermogenesis). Adaptive thermogenesis, or the regulated production of heat, is influenced by environmental temperature and diet. Mitochondria, the organelles that convert food to carbon dioxide, water and DP, are fundamental in mediating effects on energy dissipation." B. B. Lowell and B. M. Spiegelman. 2000. "Towards a molecular understanding of adaptive thermogenesis." *Nature.* 404: 652–660.

C. R. Kahn points out that exercise represents only 10-20 percent of energy expenditure in most people, with the rest "represented by the basal metabolic rate and thermogenesis." He claims that in mammals, at least 20 percent of the thermogenesis "is due to an 'energy leak' that occurs through movement of protons across the mitochondrial inner membrane of cells." "Triclycerides and toggling the tummy." 2000. *Nature Genetics.* 25(1): 6–7.

19 S. D. Hursting et al. 2003. "Calorie restriction, aging, and cancer prevention: mechanisms of action and applicability to humans." *Annu Rev Med.* 54: 131–152; V.E. Archer. 2003. "Does dietary sugar and fat influence longevity?" *Med Hypotheses.* Jun;60(6): 924–929.

20 "The effect of caloric restriction (CR) on lifespan has been reported in nearly all [short-lived] species tested and has been reproduced hundreds of times under a variety of different laboratory conditions. In addition to prolonging lifespan, CR also prevents or delays the onset of age-related disease and maintains many physiological functions at more youthful levels . . . The studies on nonhuman primates are . . . suggesting that the effect of CR on aging is universal across species." M. A. Lane et al. 2002. "Caloric restriction and aging in primates: relevance to humans and possible CR mimetics." *Microsc Res Tech.* Nov 15;59(4): 335–338.

21 B. P. Yu et al. 1982. "Life span study of SPF Fischer 344 male rats fed ad libitum or restricted diets: longevity, growth, lean body mass, and disease," *J Gerontol.* Mar;37(2): 130–141.

22 Y. Minokoshi et al. "AMP-Kinase regulates food intake by responding to hormonal and nutrient signals in the hypothalamus," *Nature* online, March 17, 2004.

23 S. G. Bouret, S. J. Draper, and R. B. Simerly. 2004. "Trophic Action of Leptin on Hypothalamic Neurons that Regulate Feeding," *Science.* April 2;304.

24 N. Angier. "Diet offers tantalizing clues to a long life." *New York Times*, April 17, 1990, sec. C.

25 Caloric restriction has been shown to inhibit the growth of spontaneous, transplanted, or chemically induced tumors in rats and mice. At 40 percent caloric restriction, growth of chemically induced breast and colon tumors was significantly inhibited. Exercise has also been shown to inhibit tumor growth. Sedentary rats who were allowed to eat freely had 108 percent higher incidence of induced colon tumors than free-eating rats subjected to vigorous treadmill exercise. D. Kritchevsky. 1990. "Influence of caloric restriction and exercise on tumorigenesis in rats," *Proc Soc Exp Biol Med.* Jan;193(1): 35–38.

26 N. Angier. See note 24, above.

27 M. Blüher, B. B. Kahn, and C. R. Kahn. 2003. "Extended longevity in mice lacking the insulin receptor in adipose tissue." *Science.* Jan 24;299(5606): 572–574; M. Blüher et al. 2002. "Adipose tissue selective insulin receptor knockout protects against obesity and obesity-related glucose intolerance." *Dev Cell.* Jul;3(1): 25–38.

28 A. Cerami. 1985. "Hypothesis: glucose as a mediator of aging." *J Am Geriatr Soc.* Sep;33 (9): 626–634; E. J. Masoro et al. "Evidence for the glycation hypothesis of aging from the food-restricted rodent model." *J Gerontol.* Jan;44(1): B20–22.

29 C. K. Ferrari and E. A. Torres. 2003. "Biochemical pharmacology of functional foods and prevention of chronic diseases of aging." *Biomed Pharmacother.* Jul;57(5–6): 251–260; D. T. Chiu and T. Z. Liu. 1997. "Free radical and oxidative damage in human blood cells." *J Biomed Sci.* 4(5): 256–259.

30 A. Koizumi et al. 1987. "Influences of dietary restriction and age on liver enzyme activities and lipid peroxidation in mice." *J Nutr.* Feb;117(2): 361–367.

31 R. Licastro, R. Weindruch, and R. L. Walford. 1986. "Dietary restriction retards the age-related decline of DNA repair capacity in mouse splenocytes," in *Topics in Aging Research in Europe 9*. A. Facchini, J. J. Haaijman, and G. Labo, eds. Rijswijk: EURAGE, pp. 53–61; R. J. Tice and R. B. Setlow. 1985. "DNA repair and replication in aging organisms and cells," in *Handbook of the Biology of Aging*. C. E. Finch and E. L. Schneider, eds. New York: Van Nostrand Reinhold, pp. 173–224.

32 C. Kahn. 1990. "His theory is simple: eat less, live longer. A lot longer." *Longevity*. Oct: 61–66, esp. 64.

33 N. Angier. See note 24 on page 400.

34 "Caloric Restriction without the Restriction" is a registered trademark of Ray & Terry's Longevity Products.

35 Joslin Diabetes Center press release, January 2004, "Study shows it may someday be possible to stay slim," www.joslin.harvard.edu/news/FirkoMouseStudy01.shtml. The study was published in *Science*: M. Blüher, B. Kahn, and C. R. Kahn. 2003. "Extended longevity in mice lacking the insulin receptor in adipose tissue." Jan 24;299(5606): 572–574.

36 Precose is recommended by the Joslin Diabetes Center as "one of six types of diabetes pills currently available to treat type 2 diabetes." www.joslin.harvard.edu/education/library/precose.shtml.

37 Xenical is produced by Roche Pharmaceuticals (www.rocheusa.com/products/xenical/). A number of recent studies have supported the effectiveness of Xenical (orlistat). See, for example, S. A. Harrison et al. 2003. "Orlistat in the treatment of NASH: a case series." *Am J Gastroenterol*. Apr;98(4): 926–930; M. Hanefeld and G. Sachse. 2002. "The effects of orlistat on body weight and glycemic control in overweight patients with type 2 diabetes: a randomized, placebo controlled trial." *Diabetes Obes Metab*. Nov;4(6): 415–423.

38 M. D. Gades and J. S. Stern. 2003. "Chitosan supplementation and fecal fat excretion in men." *Obes Res*. May;11(5): 683–688.

CHAPTER 9

1 U.S. Department of Agriculture, *Agriculture Factbook 2001–2002*, chapter 2, "Profiling food consumption in America," www.usda.gov/factbook/chapter2.htm. This 152-pound figure represents the amount of sugar available wholesale. USDA recommends that "an average person on a 2,000-calorie daily diet" consume no more than 20 teaspoons of sugar per day. The average annual consumption of 152 pounds of sweeteners is equivalent to 52 teaspoonfuls of added sugar per day. Though approximately 20 teaspoonfuls are lost or wasted, Americans are still consuming at least double the recommended amount of sugar. The percentage increase is also alarming, increasing by almost 40 percent between 1960 and 2000 (for a chart of the increase since 1983, see www.cspinet.org/reports/sugar/sugarconsumption.html). "Consumption has risen every year but once since 1983" (www.cspinet.org/new/sugar_limit.html).

In addition, see USDA, *Agricultural Outlook March 1997*, "U.S. sugar consumption continues to grow," www.ers.usda.gov/publications/agoutlook/mar1997/ao238g.pdf.

2 www.nsda.org/softdrinks/History/funfacts.html. Estimates vary on the percentage of sugar consumed in soft drinks, but they account for between 22 percent (*Agriculture Factbook*, see En. 1) and 33 percent (www.cspinet.org/reports/sugar/sugarorigin.html) of Americans' sugar intake. According to a USDA researcher, soft drinks are the prime culprit in the diet of high consumers of sugar (www.cspinet.org/new/sugar_limit.html).

3 See the WHO Technical Report # 916. 2003. "Diet, Nutrition and the Prevention of Chronic Diseases" available at www.who.int/hpr/NPH/docs/who_fao_expert_report.pdf.

4 As quoted in O. Dwyer. 2004. "U.S. government rejects WHO's attempts to improve diet." *BMJ*. Jan 24;328(7433): 185.

5 The U.S. Dept. of Health and Human Services issued a report largely condemning the findings of the WHO Technical Report #916 "Diet, Nutrition and the Prevention of Chronic Diseases" on Jan. 4, 2004, stating their beliefs that there was little proven connection between obesity and consumption of fast food or high glycemic foods. See www.commercialalert.org/bushadmincomment.pdf.

6 American Academy of Pediatrics Committee on School Health. 2004. "Soft drinks in schools." *Pediatrics*. Jan;113(1 Pt 1): 152–154.

7 Dr. Banting was joined in his research by a medical student working with him at the University of Toronto, Mr. C. H. Best. The discovery of insulin by two young unheralded researchers is "one of medicine's great success stories" (www.aventis.com/future/downloads/PDF/fut0203/En_03_2002_the_discovery_of_insulin.pdf). See also F. G. Banting et al. 1991. "Pancreatic extracts in the treatment of diabetes mellitus: preliminary report. 1922." *CMAJ*. Nov 15;145(10): 1281–1286.

8 According to the National Center for Health Statistics, in 2000, 64.5 percent of Americans were overweight and 30.5 percent were obese. K. M. Flegel et al. 2002. "Prevalence and trends in obesity among U.S. adults, 1999–2000." *JAMA*. Oct 9;288(14): 1723–1727.

9 Ten years ago, "only" one out of two adults was overweight and one in five obese. Talk about a growth industry! J. E. Manson and S. S. Bassuk. 2003. "Obesity in the United States: a fresh look at its high toll." *JAMA*. Jan 8;289(2): 229–230.

10 K. R. Fontaine et al. 2003. "Years of life lost due to obesity." *JAMA*. Jan 8;289(2): 187–193.

11 E. S. Ford, W. H. Giles, and W. H. Dietz. 2002. "Prevalence of the metabolic syndrome among US adults: findings from the third National Health and Nutrition Examination Survey." *JAMA*. Jan 16;287(3): 356–359.

12 A. Agatston. 2003. *The South Beach Diet*. Emmaus, Pennsylvania: Rodale, p. 76.

13 Reaven initially called it "metabolic syndrome," see G. M. Reaven. 1988. "Banting Lecture: Role of insulin resistance in human disease." *Diabetes*. 37: 1595–1607. As of October 2003, according to the ICD-9-CM (International Classification of Diseases), it should now be called "dysmetabolic syndrome X" or "dysmetabolic syndrome," but because "metabolic syndrome" is more commonly known and used, we will continue to use the older nomenclature.

14 Executive Summary of the 3rd Report of the U.S. National Cholesterol Education Program (NCEP)—Adult Treatment Panel III (ATP III); www.nhlbi.nih.gov/guidelines/cholesterol/profmats.htm.

15 If people are taking antihypertensives or antidiabetic drugs, they are counted as if they had a high blood pressure or an elevated fasting blood glucose.

16 G. M. Reaven et al. 1993. "Insulin resistance and hyperinsulinemia in individuals with small dense LDL particles." *J Clin Invest*. 92: 141-146.

17 Ford, Giles, and Dietz, op cit.

18 The other 80 percent are able to compensate at least temporarily by increased pancreatic insulin production, creating a long-standing state of elevated insulin levels. L. C. Jones and A. Clark. 2001. "Beta cell neogenesis in type 2 diabetes mellitus." *Diabetes*. 50(Suppl 1): S186–187.

19 J. L. Wautier and P. J. Guillausseau. 2001. "Advanced glycation end products, their receptors and diabetic angiopathy." *Diabetes Metab*. Nov;27(5 Pt 1): 535–542.

20 C. Netzer. 2000. *The Complete Book of Food Counts*. New York: Random House.

21 News release, 2003. "Type 2 diabetes linked to a family of metabolic genes," Joslin Diabetes Center, July, at www.joslin.org/news/GenesType2.shtml.

22 C. Chen et al. 2003. *Nature Biotechnology*. 21: 294–301.

23 J. M. Lehman et al. 1995. "An antidiabetic thiazolidinedione is a high affinity ligand for peroxisome proliferator-activated receptor gamma (PPAR gamma)." *J Biol Chem*. Jun 2;270(22): 12953–12956.

24 A. M. J. 2003. "Clinical islet transplant: current and future directions towards tolerance." *Immun Reviews*. 196: 219–236.

25 "Growing human organs on the farm." *NewScientist*. 180(2426): 4.

26 News release. "Joslin Comments on Diabetes Study Published in *Science* on Nov. 14, 2003"; www.joslin.org/news/ScienceReport1103.shtml.

27 V. K. Ramiya. 2000. "Reversal of insulin-dependent diabetes using islets generated in vitro from pancreatic stem cells." *Ann NY Acad Science*. May; 958: 59–68.

28 L. Knapp. "Diagnosis and medicine in a pill." *Wired News*, July 28, 2003.

29 S. Vasan, P. Foiles, and H. Founds. 2003. "Therapeutic potential of breakers of advanced glycation end product-protein crosslinks." *Arch Biochem Biophys*. Nov 1;419(1): 89–96; D. A. Kass. 2003. "Getting better without AGE: new insights into the diabetic heart." *Circ Res*. Apr 18;92(7): 704–706.

30 D. A. Kass et al. 2001. "Improved arterial compliance by a novel advanced glycation end-product crosslink breaker." *Circulation*. Sep 25;104(13): 1464–70.

31 J. P. Despres et al. 1996. "Hyperinsulinemia as an independent risk factor for ischemic heart disease." *N Engl J Med*. Apr 11;334(15): 952–957.

32 In many centers, even the venerable glucose tolerance test has been abandoned in favor of the more streamlined "hemoglobin A1c," an excellent test to monitor diabetes, but completely inadequate to detect prediabetics or people with a tendency to TMS.

33 Early in the course of type 2 diabetes, insulin levels often remain elevated. Later on, after years of excess insulin production (and gradual replacement of insulin-producing cells with amyloid), the pancreas can "burn out" and insulin levels can fall to "normal" or low levels.

34 See, for example, G. S. Watson and S. Craft. 2003. "The role of insulin resistance in the pathogenesis of Alzheimer's disease: implications for treatment." *CNS Drugs*. 17(1): 27–45. Coronary artery disease is also linked to insulin resistance: "The development of insulin resistance is considered to be a pivotal event in vascular risk" (P. J. Grant. 2003. "The genetics of atherothrombotic disorders: a clinician's view." *J Thromb Haemost.* Jul;1(7): 1381–1390). Yet a third disease linked to insulin resistance is fatty liver disease; both the prevalence and severity of the disease are linked to body mass index and waist circumference. (A. J. Scheen and F. H. Luyckx. 2003). "Nonalcoholic steatohepatitis and insulin resistance." *Acta Clin Belg.* Mar–Apr;58(2): 81–91.)

Insulin resistance is also linked to non-age-related problems such as pregnancy-induced hypertension (E. W. Seely and C. G. Solomon. 2003. "Insulin resistance and its potential role in pregnancy-induced hypertension." *J Clin Endocrinol Metab.* Jun;88(6): 2393–2398).

35 F. S. Facchini et al. 2001. "Insulin resistance as a predictor of age-related diseases." *J Clin Endocrinol Metab.* Aug;86(8): 3574–3578.

36 R. A. Freitas Jr. 1999. *Nanomedicine, Volume 1: Basic Capabilities.* Austin, Texas: Landes Bioscience, pp. 93–122.

37 www.bizjournals.com/columbus/stories/2000/09/25/story2.html?page=2.

38 Medtronic news release, "Research Presented at ADA Annual Meeting Demonstrates Accuracy and Feasibility of Artificial Pancreas Components," at www.medtronic.com/newsroom/news_20020617b.html.

39 S. Jacob et al. 1999. "Oral administration of RAC-alpha-lipoic acid modulates insulin sensitivity in patients with type-2 diabetes mellitus: a placebo-controlled pilot trial." *Free Radic Bio Med.* 27(3–4): 309–314.

40 G. Boden et al. 1996. "Effects of vanadyl sulfate on carbohydrate and lipid metabolism in patients with non-insulin-dependent diabetes mellitus." *Metabolism.* Sep;45(9): 1130–1135.

41 L. H. Storlien et al. 1987. "Fish oil prevents insulin resistance induced by high-fat feeding in rats." *Science.* 237(4817): 885–888.

42 R. B. Singh et al. 1999. "Effect of hydrosoluble coenzyme Q_{10} on blood pressures and insulin resistance in hypertensive patients with coronary artery disease." *J Hum Hypertens.* 13: 203–208.

43 M. F. McCarty. 1999. "High-dose biotin, an inducer of glucokinase expression may synergize with chromium picolinate to enable a definitive nutritional therapy for type II diabetes." *Med Hypotheses.* 52(5): 401–406.

44 P. M. Piatti et al. 2001. "Long-term oral L-arginine administration improves peripheral and hepatic insulin sensitivity in type 2 diabetic patients." *Diabetes Care.* 24(5): 875–880.

45 C. W. Bates. 1995. "DHEA attenuates study-induced declines in insulin sensitivity in postmenopausal women." *Ann NY Acad Sci.* 774: 291–293.

46 W. Dean. "Metformin: The Most Effective Life Extension Drug Is Also a Safe, Effective Weight Loss Drug" on www.antiaging-systems.com/extract/metforminweight.htm.

47 E. L. Barrett-Connor. 1995. "Testosterone and risk factors for cardiovascular disease in men." *Diab Metab.* 21: 156–161.

CHAPTER 10

1 The test for "biological age," called the H scan test, includes tests for auditory reaction time, highest audible pitch, vibrotactile sensitivity, visual reaction time, muscle movement time, lung (forced expiratory volume), visual reaction time with decision, muscle movement time with decision, memory (length of sequence), alternative button tapping time, and visual accommodation.

2 G. S. Rothfeld, S. Levert. 2001. *The Acupuncture Response.* New York: Contemporary Books.

CHAPTER 11

1 See "Race for the $1000 Genome Is On" in www.newscientist.com/news/news.jsp?id=ns99992900.

2 www.research.ibm.com/resources/news/20031114_bluegene.shtml.

3 J. Cohen. "Big-picture biotech." *MIT Technology Review V,* December 2003–January 2004.

4 Ibid.

5 R. J. Williams. 1998. *Biochemical Individuality: The Basis for the Genetotrophic Concept.* New York: Keats.

6 Interestingly, over the past dozen years or so since the Human Genome Project began, about $3 billion has been spent to complete the sequencing or about $1 per base pair.

7 The completion of the Human Genome Project occurred in the "fiftieth anniversary year of the discovery of the double-helical structure of DNA. . . . The genomic era is now a reality." F. S. Collins et al.

2003. "A vision for the future of genomics research." *Nature*. Apr 24;422: 835–847. To celebrate, entire issues of the major scientific journals *Nature* and *Science* were dedicated to discussions of the implications. For more information, see the Human Genome Project Information site (www.ornl.gov/TechResources/Human_Genome/project/50yr.html).

Among the challenges is applying the lessons learned from the Human Genome Project to understanding thousands of other organisms. See, for example, M. E. Frazier et al. 2003. "Realizing the potential of the genome revolution: the Genomes to Life Program." *Science*. 300: 290; F. S. Collins, M. Morgan, and A. Patrinos. "The Human Genome Project: lessons from large-scale biology." *Science*. 300: 286. See also M. Ridley. 1999. *Genome: The Autobiography of a Species in 23 Chapters*. New York: Perennial.

8 Andi Braun, chief medical officer of Sequenon, as quoted in *Wired*, November 2002, p.183.

9 R. Carlson. 2003. "The pace and proliferation of biological technologies." *Biosecurity and Bioterrorism*. 1(3); published online August 20, 2003, at www.molsci.org/~rcarlson/Carlson_Pace_and_Prolif.pdf.

10 D. Weatherall. 2003. "Evolving with the enemy." *NewScientist*. 802(2422): 44.

11 Each chip contains synthetic oligonucleotides that replicate sequences that identify specific genes. "To determine which genes have been expressed in a sample, researchers isolate messenger RNA from test samples, convert it to complementary DNA (cDNA), tag it with fluorescent dye, and run the sample over the wafer. Each tagged cDNA will stick to an oligo with a matching sequence, lighting up a spot on the wafer where the sequence is known. An automated scanner then determines which oligos have bound, and hence which genes were expressed." E. Marshall. 1999. "Do-it-yourself gene watching." *Science*. Oct 15;286(5439): 444–447.

12 Ibid.

13 J. Rosamond and A. Allsop. 2000. "Harnessing the power of the genome in the search for new antibiotics." *Science*. Mar 17;287(5460): 1973–1976.

14 A. Dove. 2002. "Antisense and sensibility." *Nature Biotechnology*. Feb;20: 121–124.

15 K. Philipkoski. "Next big thing in biotech: RNAi." *Wired News*. November 20, 2003; www.wired.com/news/medtech/0,1286,61305,00.html.

16 A. Goho. "Life made to order." *MIT Technology Review*, April 2003; www.technologyreview.com/articles/print_version/goho20403.asp.

17 Pima Indians in Arizona maintained their traditional way of life until the late 19th century. Then farmers diverted their water supply, resulting in many relying on the lard, sugar, and white flour provided by the government. During World War II, many Pimas entered military service or migrated to cities to work in factories. Though many Pimas returned to the reservations in the 1950s, their way of life was "profoundly affected." See "Obesity associated with high rates of diabetes in the Pima Indians," http://diabetes.niddk.nih.gov/dm/pubs/pima/obesity/obesity.htm.

According to a recently published theory, the reason only 2 percent of Europeans suffer from diabetes is that a diabetes epidemic centuries ago killed many people and prevented them from passing on the gene. Other populations, particularly indigenous peoples, carry the genes that make them highly prone to the risk factors found in urbanized settings. This is one reason 50 percent of Native Americans have diabetes. J. Diamond. 2003. "The double puzzle of diabetes." *Nature*. Jun 05;423: 599–602. See also D. L. Coleman. 1978. "Diabetes and obesity: thrifty mutants?" *Nutr Rev*. May;36(5): 129–132.

18 J. Hahm and C. M. Lieber. 2004. "Direct Ultrasensitive Electrical Detection of DNA and DNA Sequence Variations Using Nanowire Nanosensors." *Nano Letters*. 4(1): 51–54. See also http://pubs.acs.org/cgi-bin/sample.cgi/nalefd/2004/4/i01/html/nl034853b.html.

19 Emory Health Sciences news release. March 27, 2003. www.emory.edu/WHSC/HSNEWS/releases/mar03/nanotech.html.

20 R. A. Freitas Jr. 1999. *Nanomedicine, Volume I: Basic Capabilities*. Austin, Texas: Landes Bioscience. Or see www.nanomedicine.com.

21 Currently available risk panels include those for cardiac risk, high blood pressure, osteoporosis, immune function, detoxification capability, alcoholism, obesity, and more. For additional information, see Fantastic-Voyage.net.

22 R. Kurzweil. 1999. *The Age of Spiritual Machines*. New York: Viking, p. 30.

23 See the NIH National Human Genome Institute Web site: www.genome.gov/11511175.

24 Lifetime risk of developing breast cancer in women who test positive for the BRCA 1 mutation has been estimated at 80 percent, while lifetime risk for noncarriers is about 10 percent. See J. M. Lancaster. 1997. "BRCA 1 and 2—A Genetic Link to Familial Breast and Ovarian Cancer." *Medscape Women's Health*. Feb;2(2): 7. Other studies cite a 92 percent total lifetime risk.

In one Dutch study, 50 percent of healthy women whose mothers had breast cancer refused testing for BRCA1, preferring not to know whether they harbored such a potent cancer risk. The National Center for Technology Information (www.ncbi.nlm.nih.gov/entrez/dispomim.cgi?id=113705), with Johns Hopkins University, has summarized studies conducted on the BRCA gene.

25 Using genomics information to adversely prejudice against an individual is now called "genism."

26 J. Zhang et al. 2003. "Strikingly higher frequency in centenarians and twins of mtDNA mutation causing remodeling of replication origin in leukocytes." *Proc Natl Acad Sci USA*. Feb 4;100(3): 1116–1121.

27 Some studies are focusing on the patterns of variations in apolipoproteins across populations. Their focus is to determine "the usefulness of apolipoproteins as genetic markers for clinical, population, and anthropological studies." P. P. Singh, M. Singh, and S. S. Mastana. 2002. "Genetic variation of apolipoproteins in North Indians." *Hum Biol*. Oct;74(5): 673–682.

Other studies are exploring the significance of a particular genetic pattern for a specific disease. X. Li, Y. Du, and X. Huang. 2003. "Association of apoliproprotein E gene polymorphism with essential hypertension and its complications." *Clin Exp Med*. Feb;2(4): 175–179. See also M. Eto et al. 1988. "Familial hypercholesterolemia and apolipoprotein E4." *Atherosclerosis*. Aug;72(2-3): 123–128.

28 R. H. Myers et al. 1996. "Apolipoprotein E epsilon4 association with dementia in a population-based study: The Framingham study." *Neurology*. Mar;46(3): 673–677.

29 M. I. Kamboh. 1995. "Apolipoprotein E polymorphism and susceptibility to Alzheimer's disease." *Hum Biol*. Apr;67(2): 195–215.

30 L. A. Farrer et al. 1997. "Effects of age, sex, and ethnicity on the association between apolipoprotein E genotype and Alzheimer's disease. A meta-analysis. APOE and Alzheimer's Disease Meta Analysis Consortium." *JAMA*. Oct 22–29;278(16): 1349–1356.

31 R. H. Myers et al. 1996. "Apolipoprotein E epsilon4 association with dementia in a population-based study: The Framingham study." *Neurology*. Mar;46(3): 673–677.

32 See, for example, A. J. Slooter et al. 1998. "Risk estimates of dementia by apolipoprotein E genotypes from a population-based incidence study: the Rotterdam Study." *Arch Neurol*. Jul;55(7): 964–968.

33 H. K. Hamdi and C. Keney. 2003. "Age-Related Macular Degeneration: A New Viewpoint." *Frontiers in Bioscience*. May1;8: e305–314.

34 W. Retz et al. 1998. "Free radicals in Alzheimer's disease." *J Neural Transm Suppl*. 54: 221–36.

CHAPTER 12

1 P. M. Ridker et al. 1998. "Prospective study of C-reactive protein and the risk of future cardiovascular events among apparently healthy women." *Circulation*. 98: 731–733.

2 R. N. Kalaria. 2002. "Small vessel disease and Alzheimer's dementia: pathological considerations." *Cerebrovasc Dis*. 13(Suppl 2): 48–52.

3 E. M. Castano et al. 1995. "Fibrillogenesis in Alzheimer's disease of amyloid beta peptides and apolipoprotein E." *Biochem J*. Mar 1;306(Pt 2): 599–604.

4 R. A. Floyd. 1999. "Neuroinflammatory processes are important in neurodegenerative diseases: an hypothesis to explain the increased formation of reactive oxygen and nitrogen species as major factors involved in neurodegenerative disease development." *Free Radic Biol Med*. May;26(9–10): 1346–1355.

5 In fact, the Apo E4 polymorphism is often called the Alzheimer's gene. The connection between the gene and Alzheimer's was discovered in 1993 at Duke University. The risks of harboring the Apo E4 genotype are discussed more fully in chapter 11, "The Promise of Genomics."

6 W. Marz et al. 1996. "Apolipoprotein E polymorphism is associated with both senile plaque load and Alzheimer-type neurofibrillary tangle formation." *Ann NY Acad Sci*. Jan 17;777: 276–280.

7 T. G. Ohm et al. 1999. "Apolipoprotein E isoforms and the development of low and high Braak stages of Alzheimer's disease-related lesions." *Acta Neuropathol (Berl)*. Sep;98(3): 273–280. D. S. Yang et al. 1997. "Characterization of the binding of amyloid-beta peptide to cell culture-derived native apolipoprotein E2, E3, and E4 isoforms and to isoforms from human plasma." *J Neurochem*. Feb;68(2): 721–725.

8 R. B. Pyles. 2001. "The association of herpes simplex virus and Alzheimer's disease: a potential synthesis of genetic and environmental factors." *Herpes*. Nov;8(3): 64–68. R. F. Itzhaki et al. 1997. "Herpes simplex virus type 1 in brain and risk of Alzheimer's disease." *Lancet*. Jan 25;349(9047): 241–244.

9 M. R. Hayden. 2002. "Islet amyloid, metabolic syndrome, and the natural progressive history of type 2 diabetes mellitus." JOP. *J Pancreas* (Online). 3(5): 126–138. See www.joplink.net/prev/200209/02.html.

10 C. Gorman and A. Park. 2004. "The fires within." *Time*. Feb 23; 163(8): 41.

11 B. S. Reddy et al. 1992. "Inhibition of colon carcinogenesis by prostaglandin synthesis inhibitors and related compounds." *Carcinogenesis*. Jun;13(6): 1019–1023.

12 A. Akhmedkhanov et al. 2002. "Aspirin and lung cancer in women." *Br J Cancer.* Jul 1;87(1): 49–53.

13 Y. Y. Fan, K. S. Ramos, and R. S. Chapkin. 1997. "Dietary gamma-linolenic acid enhances mouse macrophage-derived prostaglandin E1 which inhibits vascular smooth muscle cell proliferation." 1997. *J Nutr.* Sep;127(9): 1765–1771. U. N. Das et al. 1989. "Prostaglandins can modify gamma-radiation and chemical induced cytotoxicity and genetic damage in vitro and in vivo." *Prostaglandins.* Dec;38(6): 689–716.

14 J. I. Kreisberg and P. Y. Patel. 1983. "The effects of insulin, glucose and diabetes on prostaglandin production by rat kidney glomeruli and cultured glomerular mesangial cells." *Prostaglandins Leukot Med.* Aug;11(4): 431–442.

15 T. Hishinuma, T. Yamasaki, and M. Mizugaki. 1999. "Effects of long-term supplementation of eicosapentaneoic and docosahexaneoic acid on the 2-, 3-series of prostacyclin production by endothelial cells." *Prostaglandins Other Lipid Mediat.* Jul;57: 333–340; V. E. Kelley et al. 1985. "A fish oil diet rich in eicosapentaneoic acid reduces cyclooxygenase metabolites, and suppresses lupus in MRL-lpr mice." *J Immunol.* Mar;134(3): 1914–1919.

16 The best vegetarian source of preformed EPA is wakame, a type of seaweed that contains 186 mg of EPA per 100 grams of seaweed. Yet vegans and vegetarians are still advised to supplement with flaxseed oil, because to obtain 650 mg of EPA, the minimum daily requirement, over 12 ounces of wakame a day would be needed.

17 M. Laimer et al. 2002. "Markers of chronic inflammation and obesity: a prospective study on the reversibility of this association in middle-aged women undergoing weight loss by surgical intervention." *Int J Obes Relat Metab Disord.* May;26: 659–662; M. Visser. 2001. "Higher levels of inflammation in obese children." *Nutrition.* Jun;17: 480–481.

18 J. K. Kiecolt-Glaser et al. 2003. "Chronic stress and age-related increases in the proinflammatory cytokine IL-6." *Proc Natl Acad Sci USA.* Jul 22;100(15): 9090–9095.

19 H. Bucher et al. 2002. "n-3 Polyunsaturated fatty acids in coronary heart disease: a meta-analysis of randomized controlled trials." *Am J Med.* 112: 298–304.

20 A multicenter double-blind study of 500 patients is an example of the ongoing work to further investigate this link. I. A. Brouwer et al. 2003. "Rationale and design of a randomised controlled clinical trial on supplemental intake of n-3 fatty acids and incidence of cardiac arrhythmia: SOFA." *Eur J Clin Nutr.* Oct;57(10): 1323–1330. See also I. Rosenberg. 2002. "Fish-food to calm the heart." *New Engl J Med.* 346(15): 1102–1103.

21 For more information on gum disease and health, see the American Academy of Periodontology site (www.perio.org/consumer/2a.html). The link between heart disease and gum disease has still not been conclusively established. See S. Abou-Raya, A. Naeem, and K. H. Abou-El. 2002. "Coronary artery disease and periodontal disease: is there a link?" *Angiology.* Mar–Apr;53(2): 141–148; P. Hujoel et al. 2000. "Periodontal disease and coronary heart disease risk." *JAMA.* Sept 20;284(11): 1406–1410.

22 P. P. Zandi, J. C. Breitner, and J. C. Anthony. 2002. "Is pharmacological prevention of Alzheimer's a realistic goal?" *Expert Opin Pharmacother.* Apr;3(4): 365–380; B. M. McLendon, G. G. Chen, and P. M. Doraiswamy. 2000. "Current and future treatments for cognitive deficits in dementia." *Curr Psychiatry Rep.* Feb;2(1): 20–23.

23 W. F. Stewart et al. 1997. "Risk of Alzheimer's disease and duration of NSAID use." *Neurology.* Mar;48(3): 626–632.

24 P. S. Sanmuganathan et al. 2001. "Aspirin for primary prevention of coronary heart disease: safety and absolute benefit related to coronary risk derived from meta-analysis of randomised trials." *Heart.* Mar;85(3): 265–271.

25 Chronic NSAID use is associated with a very high incidence of adverse drug reactions, such as gastrointestinal hemorrhage. Over 16,500 deaths and 100,000 hospitalizations annually have been associated with prescription NSAID usage (and the number would be even higher if over-the-counter usage were included). See M. Wolfe, D. Lichtenstein, and S. Gurkirpal. 1999. "Gastrointestinal toxicity of nonsteroidal anti-inflammatory drugs." *N Engl J Med.* Jun 17;340(24): 1888–1899.

26 F. E. Silverstein et al. 2000. "Gastrointestinal toxicity with celecoxib versus nonsteroidal anti-inflammatory drugs for osteoarthritis and rheumatoid arthritis. The CLASS study: a randomized controlled trial." *JAMA.* 284: 1247–1255; C. Bombardier et al. for the VIGOR Study Group. 2000. "Comparison of upper gastrointestinal toxicity of rofecoxib and naproxen in patients with rheumatoid arthritis." *N Engl J Med.* 343: 1520–1528.

27 P. Libby, op cit, p.55.

28 B. Lindahl et al. 2000. "Markers of myocardial damage and inflammation in relation to long-term mortality in unstable coronary artery disease. FRISC Study Group. Fragmin during instability in coronary artery disease." *N Engl J Med.* Oct 19;343(16): 1139–1147; D. J. Rader. 2000. "Inflammatory markers of coronary risk." *N Engl J Med.* Oct 19;343(16): 1179–1182; C. J. Packard et al. 2000. "Lipoprotein-associated phospholipase A2 as an independent predictor of coronary heart disease. West of Scotland Coronary Prevention Study Group." *N Engl J Med.* Oct 19;343(16): 1148–1155.

29 J. Danesh et al. 2000. "Low grade inflammation and coronary heart disease: prospective study and updated meta-analyses." *BMJ.* Jul 22;321(7255): 199–204.

30 P. M. Ridker et al. 1998. "C-reactive protein adds to the predictive value of total and HDL cholesterol in determining risk of first myocardial infarction." *Circulation.* May 26;97(20): 2007–2011.

31 I. Kushner. 2001. "C-reactive protein elevation can be caused by conditions other than inflammation and may reflect biologic aging." *Cleve Clin J Med.* Jun;68(6): 535–537.

32 N. Rifai and P. M. Ridker. 2001. "High-sensitivity C-reactive protein: a novel and promising marker of coronary heart disease." *Clin Chem.* Mar; 47(3): 403–411; P. M. Ridker et al. 2000. "C-reactive protein and other markers of inflammation in the prediction of cardiovascular disease in women." *N Engl J Med.* Mar 23;342(12): 836–843.

33 In this slight variation from normal, the 31st nucleotide in the DNA chain that codes for IL-1β, one nucleotide, cytosine, is replaced by thymidine (31C→T polymorphism).

34 N. Sueoka et al. 2001. "A new function of green tea: prevention of lifestyle-related diseases." *Ann NY Acad Sci.* Apr; 928: 274–280.

CHAPTER 13

1 The risk of defective homocysteine metabolism rises with age and varies with ethnicity; hence the wide spread between 10 and 44 percent. See G. L. Booth and E. E. Wang. 2000. "Preventive health care, 2000 update: screening and management of hyperhomocysteinemia for the prevention of coronary artery disease events. The Canadian Task Force on Preventive Health Care." *CMAJ.* Jul 11;163(1): 21–29.

2 Medicare does not pay for homocysteine testing, regarding it as neither medically reasonable nor necessary. Although controversial, we feel that by paying a few tens of dollars for routine homocysteine screening, it would help identify many individuals at significant risk of heart attack, stroke, and Alzheimer's diseases that Medicare then pays tens of thousands of dollars to treat.

3 In the same paragraph on their Web site that the American Heart Association doesn't acknowledge homocysteine as a "major risk factor for cardiovascular disease," they also "don't recommend widespread use of folic acid and B vitamin supplements to reduce the risk of heart disease and stroke." See www.americanheart.org/presenter.jhtml?identifier=4677. See also, M. R. Malinow et al. 1999. "Homocyst(e)ine, diet, and cardiovascular diseases: a statement for healthcare professionals from the Nutrition Committee, American Heart Association." *Circulation.* 99: 178–182.

4 Any product containing more than 800 mcg of folic acid requires a prescription.

5 More precisely, cytosine first undergoes another chemical reaction known as "deamination" to form uracil, which is then methylated to form thymine.

6 J. Yokota et al. 2003. "Genetic alterations responsible for metastatic phenotypes of lung cancer cells." *Clin Exp Metastasis.* 20(3): 189–193. According to this study, one gene associated with lung cancer is "inactivated in 50% of lung cancers by deletions, mutations, and methylation." See also K. S. McCully. 1994. "Chemical pathology of homocysteine. II. Carcinogenesis and homocysteine thiolactone metabolism." *Ann Clin Lab Sci.* Jan–Feb;24(1): 27–59.

7 See M. Iscan et al. 2002. "The organochlorine pesticide residues and antioxidant enzyme activities in human breast tumors: is there any association?" *Breast Cancer Res Treat.* Mar;72(2): 173–182; C. Charlier and G. Plomteux. 2002. "Environmental chemical pollution and risk of human exposure: the role of organochlorine pesticides." *Ann Biol Clin* (Paris). Jan–Feb;60(1): 37–46; M. S. Wolff and P. G. Toniolo. 1995. "Environmental organochlorine exposure as a potential etiologic factor in breast cancer." *Environ Health Perspect.* Oct;103(Suppl 7): 141–145.

8 K. Nilsson et al. 1996. "Hyperhomocysteinaemia—a common finding in a psychogeriatric population." *Eur J Clin Invest.* Oct;26(10): 853–859.

9 S. R. Maxwell. 2000. "Coronary artery disease—free radical damage, antioxidant protection and the role of homocysteine." *Basic Res Cardiol.* 95(Suppl 1): 165–171.

10 W. P. Castelli. 1996. "Lipids, risk factors and ischaemic heart disease." *Atherosclerosis.* Jul;124(Suppl): S1–9.

Many other studies have also shown such a connection. For example, an Irish study showed a fivefold increase in the risk of stroke with elevated homocysteine levels. The lead author suggested that the unavailability of fortified foods, particularly cereals, in the United Kingdom made supplementation even more important. S. P. McIlroy et al. 2002. "Moderately elevated plasma homocysteine, methylenetetrahydrofolate reductase geneotype, and risk for stroke, vascular dementia, and Alzheimer disease in Northern Ireland." *Stroke*. Oct;33(10): 2351–2356.

11 M. J. Stampfer et al. 1992. "A prospective study of plasma homocyst(e)ine and risk of myocardial infarction in U.S. physicians." *JAMA*. Aug 19;268(7): 877–881.

12 R. Clarke et al. 1998. "Folate, vitamin B_{12}, and serum total homocysteine levels in confirmed Alzheimer disease." *Arch Neurol*. Nov;55(11): 1449–1455.

13 While elevated homocysteine confers a cardiovascular risk equal to smoking, the combination is even worse. One study "suggests that smokers with high plasma homocysteine are at greatly increased risk of cardiovascular disease and should therefore be offered intensive advice to help them cease smoking." P. O'Callaghan et al. 2002. "Smoking and plasma homocysteine." *Eur Heart J*. Oct;23(20): 1580–1586; S. Tonstad and P. Urdal. 2002. "Does short-term smoking cessation reduce plasma total homocysteine concentrations?" *Scand J Clin Lab Invest*. 62(4): 279–284. See also I. M. Graham et al. 1997. "Plasma homocysteine as a risk factor for vascular disease. The European Concerted Action Project." *JAMA*. Jun 11;277(22): 1775–1781.

14 J. M. Ellis and K. S. McCully. 1995. "Prevention of myocardial infarction by vitamin B_6." *Res Commun Mol Pathol Pharmacol*. Aug;89(2): 208–220.

15 Functional levels refer to amounts in the blood that prevent biochemical abnormality and are distinct from the absolute bloodstream level. D. G. Savage et al. 1994. "Sensitivity of serum methylmalonic acid and total homocysteine determinations for diagnosing cobalamin and folate deficiencies." *Am J Med*. Mar;96(3): 239–246.

16 D. J. DeRose et al. 2000. "Vegan diet-based lifestyle program rapidly lowers homocysteine levels." *Prev Med*. Mar 30: 225–33.

17 L. L. Husemoen et al. 2004. "Effect of lifestyle factors on plasma total homocysteine concentrations in relation to MTHFR(C677T) genotype." *Eur J Clin Nutr*. Advance online publication March 31, 2004.

18 M. S. van der Gaag et al. 2000. "Effect of consumption of red wine, spirits, and beer on serum homocysteine." *Lancet*. Apr 29;355(9214): 1522.

19 J. F. Toole et al. 2004. "Lowering homocysteine in patients with ischemic stroke to prevent recurrent stroke, myocardial infarction, and death: the Vitamin Intervention for Stroke Prevention (VISP) randomized controlled trial." *JAMA*. Feb 4;291(5): 565–575.

20 Homocysteine Lowering Trialists' Collaboration. 2000. "Lowering blood homocysteine with folic acid-based supplements: meta-analysis of randomised trials." *Indian Heart J*. Nov–Dec;52(7 Suppl): S59–64.

21 I. M. Graham et al. 1997. "Plasma homocysteine as a risk factor for vascular disease. The European Concerted Action Project." *JAMA*. Jun 11;277(22): 1775–1781.

22 See, for example, M. R. Malinow et al. 1998. "Reduction of Plasma Homocyst(e)ine Levels by Breakfast Cereal Fortified with Folic Acid in Patients with Coronary Heart Disease." *N Engl J Med*. 338: 1009–1015.

23 E. Arnesen et al. 1995. "Serum total homocysteine and coronary heart disease." *Int J Epidemiol*. Aug;24(4): 704–709.

24 www.labcorp.com/datasets/labcorp/html/chapter/mono/sr021700.htm.

25 I. M. Graham et al. 1997, op cit.

26 E. K. Amouzou et al. 2004. "High prevalence of hyperhomocysteinemia related to folate deficiency and the 677C—>T mutation of the gene encoding methylenetetrahydrofolate reductase in coastal West Africa." *Am J Clin Nutr*. Apr;79(4): 619–624.

27 L. D. Botto and Q. Yang. 2000. "5,10-Methylenetetrahydrofolate reductase (MTHFR) Gene Variants and Congenital Anomalies." *Am J Epidemiol*. 151(9): 862–877. W. Herrman et al. "Homocysteine, methylenetetrahydrofolate reductase C677T polymorphism and the B-vitamins: a facet of nature-nurture interplay." *Clin Chem Lab Med*. Apr;41(4): 547–553; S. S. Kang et al. 1991. "Intermediate hyperhomocysteinemia resulting from compound heterozygosity of methylenetetrahydrofolate reductase mutations." *Am J Hum Genet*. Mar;48(3): 546–551.

28 L. A. Kluijtmans et al. 1996. "Molecular genetic analysis in mild hyperhomocysteinemia: a common mutation in the methylenetetrahydrofolate reductase gene is a genetic risk factor for cardiovascular disease." *Am J Hum Genet*. Jan;58(1): 35–41.

29 M. Goodman et al. 2001. "Association of methylenetetrahydrofolate reductase polymorphism C677T and dietary folate with the risk of cervical dysplasia." *Cancer Epidemiol Biomarkers Prev*. Dec;10(12): 1275–1280.

30 S. Matsushita et al. 1997. "The frequency of the methylenetetrahydrofolate reductase-gene mutation varies with age in the normal population [letter]." *Am J Hum Genet*. 61: 1459–1460.

CHAPTER 14

1 "Louisiana led the nation in toxic waste generated, with more than nine billion pounds generated, or approximately one quarter of the nation's toxic waste. Nevada led the nation in direct releases, with 14 percent of the nation's pollution, mostly from the mining industry." U.S. PIRG news release, "Toxic waste production increased by eight billion pounds in 2000: New dioxin data show high amounts of hazardous pollution," May 23, 2002, www.uspirg.org/uspirgnewsroom.asp?id2=7030&id3=USPIRGnewsroom&.

2 B. C. Wolverton et al. "Interior Landscape Plants for Indoor Air Pollution Abatement," NASA/ALCA Final Report, *Plants for Clean Air Council*, Davidsonville, Maryland, 1989.

3 The U.S. EPA maintains information on air and radiation at www.epa.gov/air/concerns. See information on environmental sites such as the Rainforest Action Network (www.ran.org/info_center/factsheets/04a.html) and Environmental Defense (www.environmentaldefense.org/system/templates/page/focus.cfm?focus=3).

NOVA Online (www.pbs.org/wgbh/nova/ice/greenhouse.html) provides useful background: "Greenhouse gas concentrations in the atmosphere have been naturally rising and falling for billions of years, creating cold and warm periods in the Earth's history. For example, as the Ice Age progressed, scientists believe the amount of natural carbon dioxide in the atmosphere dropped over thousands of years, reducing the greenhouse effect, and making the Earth cooler. But many disagree on how that change in carbon dioxide occurred. Today, scientists are looking at effects of global warming as they debate the long-term impact of man-made carbon dioxide and CFCs entering the atmosphere. Many climatologists argue that we are artificially increasing the greenhouse effect, warming the Earth faster than would occur naturally, which could cause problems for the Earth in the future."

4 T. J. Woodruff et al. 1998. "Public health implications of 1990 air toxics concentrations across the United States." *Environ Health Perspect.* May;106(5): 245–251.

5 EPA Office of Air and Radiation. 1993. "Targeting Indoor Air Pollution: EPA's Approach and Progress." EPA 400R 92012; EPA Office of Air and Radiation. 2001. "Healthy Buildings, Healthy People: A Vision for the 21st Century." EPA 402K01003, p. 8.

6 B. O. Brooks et al. 1991. "Indoor air pollution: an edifice complex." *J Toxicol Clin Toxicol.* 29(3): 315–374. Both the EPA (www.epa.gov/iaq/pubs/sbs.html) and the National Safety Council (www.nsc.org/ehc/indoor/sbs.htm) maintain information about sick buildings.

7 www.epa.gov/ogwdw/dwh/health.html.

8 J. A. Varner et al. 1998. "Chronic administration of aluminum-fluoride or sodium-fluoride to rats in drinking water: alterations in neuronal and cerebrovascular integrity." *Brain Res.* Feb 16; 784(1–2): 284–298.

9 A. Hoshi et al. 2001. "Concentrations of trace elements in sweat during sauna bathing." *Tohoku J Exp Med.* Nov;195(3): 163–169.

10 B. C. Kross et al. 1996. "Proportionate mortality study of golf course superintendents." *Am J Ind Med.* May;29(5): 501–506.

11 Interestingly, this same group of refinery workers was noted to have significantly decreased mortality from respiratory tuberculosis (29 percent), esophageal cancer (45 percent), rectal cancer (49 percent), and cancers of the bladder and other urinary organs (40 percent), suggesting a multifactorial cause for the expression of exposure to petrochemicals at refineries and genetic expression of cancer-causing potential. J. M. Dement et al. 1998. "Proportionate mortality among union members employed at three Texas refineries." *Am J Ind Med.* Apr;33(4): 327–340.

12 B. A. Evanoff, P. Gustavsson, and C. Hogstedt. 1993. "Mortality and incidence of cancer in a cohort of Swedish chimney sweeps: an extended follow up study." *Br J Ind Med.* May;50(5): 450–459.

13 "Estimated daily doses of dieldrin alone exceed US Environmental Protection Agency and US Agency for Toxic Substances Disease Control reference dose for children. Given the widespread occurrence of POPs in the food supply and the serious health risks associated with even extremely small levels of exposure, prevention of further food contamination must be a national health policy priority in every country." K. S. Schafer and S. E. Kegley. 2002. "Persistent toxic chemicals in the U.S. food supply." *J Epidemiol Community Health.* Nov;56(11): 813–817.

14 See www.foodnews.org/reportcard.php.

15 G. Hyland. 2001. "The Physiological and Environmental Effects of Non-Ionising Electromagnetic Radiation." European Parliament Directorate General for Research.

16 H. Lai and N. P. Singh. 1996. "Single- and double-strand DNA breaks in rat brain cells after acute exposure to radiofrequency electromagnetic radiation." *Int J Radiat Biol.* Apr;69(4): 513–521.

17 D. Leszczynski et al. 2002. "Non-thermal activation of the hsp27/p38MAPK stress pathway by mobile phone radiation in human endothelial cells: Molecular mechanism for cancer- and blood-brain barrier-related effects." *Differentiation.* May;70(2–3): 120–129.

18 R. O. Becker. 1990. *Cross Currents: The Promise of Electromedicine, the Perils of Electropollution*. New York: J. P. Tarcher.

19 Ibid.

20 S. Boseley. "Hands-Free Mobiles Increase Radiation Risk." *The Guardian*. April 4, 2000.

21 S. Overell. "Scientists Believe a Ferrite Choke Clipped to the Wire of a Hands-Free Set Could Dramatically Lower Radiation." *Financial Times*, February 12, 2001.

22 The Institute for Genomic Research (TIGR) deciphered the genome. The "extraordinary" capabilities of Geobacter come from over 100 genes that code c-type cytochromes, which are proteins involved in electron transfer and metal reduction. This is the largest number of this type of gene yet found in a bacterial species. "Scientists decipher genome of bacterium that remediates uranium contamination and generates electricity through its metabolism." TIGR news release at www.tigr.org/new/press_release_12-11-03.shtml, referring to B. A. Methe et al. 2003. "Genome of Geobacter sulfurreducens: metal reduction in subsurface environments." *Science*. Dec 12(302): 1967–1969.

23 A. Goho. "Life made to order." *MIT Technology Review*, April 2003; www.technologyreview.com/articles/print_version/goho20403.asp.

24 S. Duke. 2003. "Weeding with transgenes." *Trends in Biotechnology*. 21(5): 192–195.

25 E. Baard. "Plants have a way with metals." *Wired News*, September 5, 2003; www.wired.com/news/print/0,1294,60302,00.html.

26 S. Duke, op cit.

27 H. Y. Ha et al. 2003. "Chronic restraint stress massively alters the expression of genes important for lipid metabolism and detoxification in liver." *Toxicol Lett*. Dec(146): 49–63.

28 L. Carroll. 2004. "Genes, toxins, and Parkinson's." *International Herald Tribune*, February 12, at www.iht.com/articles/129155.html.

29 A. D. de Grey. 2003. "An Engineer's Approach to the Development of Real Anti-Aging Medicine." *Sci SAGE KE*. Jan 8: VP1. See http://sageke.sciencemag.org/cgi/content/full/2003/1/vp1 and also A. D. de Grey. 2002. "Bioremediation meets biomedicine: therapeutic translation of microbial catabolism to the lysosome." *Trends Bioltechnol*. 20(11): 452–455.

30 "The notion of 'vaccinating' individuals against a neurodegenerative disorder such as Alzheimer's disease is a marked departure from classical thinking about mechanism and treatment, and yet therapeutic vaccines for both Alzheimer's disease and multiple sclerosis have been validated in animal models and are in the clinic. Such approaches, however, have the potential to induce unwanted inflammatory responses as well as to provide benefit." H. L. Weiner and D. J. Selkoe. 2002. "Inflammation and therapeutic vaccination in CNS diseases." *Nature*. Dec 19–26;420(6917): 879–884. These researchers showed that a vaccine in the form of nose drops could slow the brain deterioration of Alzheimer's. H. L. Weiner et al. 2000. "Nasal administration of amyloid-beta peptide decreases cerebral amyloid burden in a mouse model of Alzheimer's disease." *Ann Neurol*. Oct;48(4): 567–579.

31 D. Beyersmann. 2002. "Effects of carcinogenic metals on gene expression." *Toxicol Lett*. Feb 28;127(1–3): 63–68.

32 P. Weihe et al. 2002. "Neurobehavioral performance of Inuit children with increased prenatal exposure to methylmercury." *Int J Circumpolar Health*. Feb;61(1): 41–49.

33 S. A. Thompson et al. 1998. "Alterations in immune parameters associated with low level methyl mercury exposure in mice." *Immunopharmacol Immunotoxicol*. May;20(2): 299–314.

34 "Coal-fired power plants are the largest industrial emitters of mercury, producing over one third of all mercury pollution in the U.S.," per the Clear the Air public education campaign (http://cta.policy.net/mercury/).

"In very small quantities, [mercury] conducts electricity, measures temperature and pressure, and forms alloys with almost all other metals. With these and other unique properties, mercury plays an important role as a process or product ingredient in several industrial sectors." Background Information on Mercury Sources and Regulations, available along with other mercury information at www.epa.gov/mercury/information.htm#fact_sheets.

35 A 2001 report by U.S. PIRG (the national lobbying office for the state Public Interest Research Groups, which are nonprofit, nonpartisan public interest advocacy groups) and the Environmental Working Group found that mercury contamination of fish is so great that 25 percent of pregnant women who eat fish regularly expose their unborn babies to levels of mercury that could threaten a developing fetus. The situation will only get worse with full enactment of the Bush administration's "Clear Skies Initiative," which would allow three times more mercury pollution than full enforcement of the current Clean Air Act.

36 Farm-raised salmon are fed fish food containing high levels of pollutants. Unlike wild ocean salmon that consume phytoplankton, which they then turn into EPA, farm-raised salmon have little EPA. A recent report by the Environmental Working Group suggests farmed salmon may also be high in PCBs due to contamination of their food (www.ewg.org/news/story.php?id=1871).

37 The Environmental Working Group's fish list is at www.ewg.org/reports/BrainFood/sidebar.html.

38 The health benefits of amalgam removal are more theoretical than proven, although it would seem that having mercury inside one's mouth is less than ideal. To learn more about mercury toxicity, read *It's All in Your Head*, Hal Huggins (New York: Avery Penguin Putnam, 1993).

39 A. Szutowicz. 2001. "Aluminum, NO, and nerve growth factor neurotoxicity in cholinergic neurons." *J Neurosci Res*. Dec 1;66(5): 1009–1018.

40 P. Fairley. "Saving Lives with Living Machines." *Technology Review*. July/August 2003. www.technologyreview.com/articles/print_version/fairley0703.asp.

41 Ibid.

42 R. Zacks. "The Liver Chip." *Technology Review*. March 2003. www.technologyreview.com/articles/demo0303.asp?x=38&y=11.

43 R. A. Freitas Jr. "Death is an Outrage." KurzweilAI.net Jan. 9, 2003. www.kurzweilai.net/articles/art0536.html. (Based on a lecture by the author at the Fifth Alcor Conference on Extreme Life Extension.)

44 This test is called the Comprehensive Detoxification Profile and is available through your health care practitioner from Great Smokies Diagnostic Laboratory (www.gsdl.com).

45 N. Song et al. 2001. "CYP 1A1 polymorphism and risk of lung cancer in relation to tobacco smoking: a case-control study in China." *Carcinogenesis*. Jan;22(1): 11–16.

46 H. Payami et al. 2001. "Parkinson's disease, *CYP2D6* polymorphism, and age." *Neurology 2001*; May 22;56(10): 1363–1370.

47 T. Konishi et al. 2003. "The ADH3*2 and CYP2E1 c2 alleles increase the risk of alcoholism in Mexican American men." *Exp Mol Pathol*. Apr;74(2): 183–189.

48 H. Zheng et al. 2004. "Tacrolimus dosing in adult lung transplant patients is related to cytochrome P4503A5 gene polymorphism." *J Clin Pharmacol*. Feb;44(2): 135–140.

49 V. Fonte et al. 2002. "Interaction of intracellular beta amyloid peptide with chaperone proteins." *Proc Natl Acad Sci USA*. Jul 9;99(14): 9439–9444.

50 P. Hammarstrom, F. Schneider, and J. W. Kelly. 2001. "Trans-suppression of misfolding in an amyloid disease." *Science* Sep 28;293(5539): 2459–2462.

51 P. M. Harrison et al. 1999. "Thermodynamics of model prions and its implications for the problem of prion protein folding." *J Mol Biol*. Feb 19;286(2): 593–606.

CHAPTER 15

1 "Chronic Disease Overview," National Center for Chronic Disease Prevention and Health Promotion (www.cdc.gov/nccdphp/overview.htm). See also the *2003 Heart Disease and Stroke Statistical Update*, American Heart Association (www.americanheart.org/presenter.jhtml?identifier=3000090).

2 Even though women's comparable risk trails that of men by 10 years, "more than half of persons who die each year of heart disease are women." "Chronic Disease Overview," op. cit. Also, "38 percent of women compared to 25 percent of men will die within one year after a heart attack." "Statistics You Need to Know," American Heart Association (www.americanheart.org/presenter.jhtml?identifier=107).

3 In the United States, for example, approximately 75,000 bypass surgeries were performed in 1979 and over 520,000 in 2000. Angioplasties were first tracked by the AHA in 1986; by 2002, close to 570,000 were performed per year in the United States. "Trends in Cardiovascular Operations and Procedures," in "2003 Heart Disease and Stroke Statistical Update," American Heart Association (www.americanheart.org/presenter.jhtml?identifier=3009972). See also A. Michaels and K. Chatterjee. 2002. "Angioplasty versus bypass surgery for coronary artery disease." *Circulation*. Dec 3;106(23):187–190.

4 Recent data suggest that the volume of cases handled by the hospital and the surgeon in particular has an impact on the mortality rate from bypass surgery. See E. L. Hannan et al. 2003. "Do hospitals and surgeons with higher coronary artery bypass graft surgery volumes still have lower risk-adjusted mortality rates?" *Circulation*. Aug 19;108(7):795–801.

5 Public information materials often claim that the mental decline is temporary; see, for example, "Coronary bypass surgery," MayoClinic.com (www.mayoclinic.com/invoke.cfm?id=HB00022). However, recent research has found "early improvement followed by later decline"; see, for example, M. F. Newman et al. 2001. "Longitudinal assessment of neurocognitive function after coronary-artery bypass surgery." *N Engl J Med*. Feb 8;344(6):395–402.

6 Bypass surgery and angioplasty cannot fix the injured heart tissue or the microvessel obstructions that are most likely to cause post-heart-attack complications. *Johns Hopkins Magazine*, April 1998, "Treating heart attacks through MRI," www.jhu.edu/~jhumag/0698web/health.html. This study showed a higher risk of death for patients with invasive treatment: L. F. Wexler et al. 2001. "Non-Q-wave myocardial infarction following thrombolytic therapy: a comparison of outcomes in patients randomized to invasive or conservative post-infarct assessment strategies in the Veterans Affairs Non-Q-Wave Infarction Strategies In-Hospital (VANQWISH) Trial." *J Am Coll Cardiol.* Jan;37(1):19–25.

In this study, patients treated invasively experienced less angina but significantly more critical events, including death. H. C. Bucher et al. 2000. "Percutaneous transluminal angioplasty versus medical treatment for non-acute coronary heart disease: meta-analysis of randomized controlled trials." *BMJ.* Jul 8;321(7253): 73–77. Another study showed many repeat operations in patients treated invasively and a 22-year cumulative survival rate of 20 percent in the surgically treated group compared with 25 percent in the medically treated group. "This trial provides strong evidence that initial bypass surgery did not improve survival for low-risk patients and that it did not reduce the overall risk of myocardial infarction." P. Peduzzi, A. Kamina, and K. Detrie. 1999. "Twenty-two-year follow-up in the VA Cooperative Study of coronary artery bypass surgery for stable angina." *Am J Cardiol.* Jan 15;83(2): 301–304.

In yet another study, "patients undergoing coronary angioplasty had twice the rate of adverse outcomes as normal subjects, seven times the rate of angina, almost four times the number of heart attacks and twice the rate of congestive heart failure." G. A. Van Norman and K. Posner. 2000. "Coronary stenting or percutaneous transluminal coronary angioplasty before noncardiac surgery increases adverse events: the evidence is mounting." *J Am Coll Cardiol.* Dec;36(7): 2351–2352 (as described on the Noninvasive Heart Center site, www.heartprotect.com/comparison-studies.shtml). See also note 7 below and note 82 on page 418.

7 Note 6 above cites several studies that show improved outcomes for treatment with medication as compared to treatment with surgery. Two studies that show a slightly better outcome with surgery are: C. Espinoza-Klein et al. 2000. "Ten-year outcome after coronary angioplasty in patients with single-vessel coronary artery disease and comparison with the results of the Coronary Artery Surgery Study (CASS)." *Am J Cardiol.* Feb 1;85(3): 321–326. This study showed relative rates of survival after 10 years was 86 percent after balloon angioplasty, 85 percent after bypass surgery, and 82 percent with medical treatment alone.

M. M. Graham et al. 2002. "Survival after coronary revascularization in the elderly." *Circulation.* May 21: 105(20): 2378–2384. This study showed 17 percent better survival in older patients (greater than 80 years old) who had surgery versus those who did not. This study also showed a small survival increase in patients under 70 years old who had surgery versus those who did not (95 percent versus 91 percent).

Studies have shown that angioplasty improves survival when applied during a heart attack to clear away the thrombus (blood clot).

Additional studies on treatment with medication include the following:

R. Conti. 1999. "Single-Vessel Disease: What is the Evidence Favoring Medical Versus Interventional Therapy?" *Clin. Cardiol.* 22, 3–5 (1999) at www.clinicalcardiology.org/briefs/9901briefs/22-003.html; "Treatment with a combination of statin and niacin can slash the risk of a fatal or non-fatal heart attack or hospitalization for chest pain by 70 percent among patients who are likely to suffer heart attacks and/or death from coronary heart disease, according to a study by University of Washington researchers in the Nov. 29 *New England Journal of Medicine*."; B. G. Brown et al. 2001. "Simvastatin and niacin, antioxidant vitamins, or the combination for the prevention of coronary disease." *N Engl J Med.* Nov 29;345(22): 1583–92. Summary for patients in: *Curr Cardiol Rep.* Nov 2002;4(6): 486. See also notes 6 and 82 on pages 411 and 418.

8 L. L. Demer et al. 1994. "Mechanism of calcification in atherosclerosis." *Trends Cardiovasc Med.* 4: 45–49; L. L. Demer. "Effect of calcification on in vivo mechanical response of rabbit arteries to balloon dilation." *Circulation.* June 1;83(6): 2083–2093.

9 In addition to the new model of heart disease presented in the next section, another primary reason for the failure of invasive procedures is that these procedures address symptoms of the problem, not the problem itself. "There is a common misconception that most of the excess risk accumulated over many years can be erased by aggressive short-term prevention introduced later in life." S. Grundy et al. 1999. "Assessment of cardiovascular risk by use of multiple-risk-factor assessment equations." *Circulation.* 100: 1481–1492. The importance of behavioral changes in preventing future cardiovascular events and mortality has been shown by many studies, including N. C. Campbell et al. 1998. "Secondary prevention in coronary heart disease: a randomized trial of nurse-led clinics in primary care." *Heart.* Nov;80(5): 447–452. With regard to the elderly, see last reference in note 7 above.

"Insulin resistance with or without frank type 2 diabetes has emerged as a major determinant of accelerated coronary artery disease and its sequelae." Thus, a number of randomized clinical trials have been performed to compare the efficacy of these procedures for diabetic patients who fall into the subset of patients with severe multivessel disease who benefit from surgery. B. E. Sobel et al. 2003. "Burgeoning dilemmas in the management of diabetes and cardiovascular disease: rationale for the Bypass Angioplasty Revascularization Investigation 2 Diabetes (BARI 2D) Trial." *Circulation.* Feb 4;107(4): 636–642.

10 K. S. Prediman. 2003. "Mechanisms of plaque vulnerability and rupture." *J Am Coll Cardiol.* Feb 19;41(4 Suppl 1): S15–S22; M. Takano et al. 2001. "Mechanical and structural characteristics of vulnerable plaques analysis by coronary angioscopy and intravascular ultrasound." *J Am Coll Cardiol.* Jul;38(1): 99–104.

11 New Studies Question Value of Opening Arteries, www.cse.buffalo.edu/~rapaport/510/nyt-heart-printerfriendly.htm; B. G. Brown et al. 1986. "Incomplete Lysis of Thrombus in the Moderate Underlying Atherosclerotic Lesion during Intracoronary Infusion of Streptokinase for Acute Myocardial Infarction: Quantitative Angiographic Observations," *Circulation.* 73: 653–661.

12 S. E. Nissen and J. C. Curley. 1991. "Application of intravascular ultrasound for detection and quantitation of coronary atherosclerosis." *Int J Card Imaging.* Jan; 6(3–4): 165–177; P. Schoenhagen, E. S. McErlean, and S. E. Nissen. 2000. "The vulnerable coronary plaque." *J Cardiovasc Nurs.* Oct;15(1): 1–12.

13 D. D. Waters. 2000. "Medical therapy versus revascularization: the atorvastatin versus revascularization treatment AVERT trial." *Can J Cardiol.* Jan;16(Suppl A): 11A–3A.

14 G. Kolata. "New Heart Studies Question the Value of Opening Arteries," *New York Times,* March 21, 2004.

15 Ibid.

16 R. J. Aiello et al. 2002. "Leukotriene B4 receptor antagonism reduces monocytic foam cells in mice," *Arteriosclerosis Thromb Vasc Biology.* Mar;22(3): 361–363. The study states, "Compared with age-matched controls, lipid accumulation and monocyte infiltration were significantly reduced in treated apoE(-/-) mice at all time points tested. Lesion area reduction was also demonstrated in LDLr(-/-) mice maintained on a high-fat diet."

17 "Accumulating evidence indicates that the arteries of a cardiac patient become inflamed with white blood cells and other immune system agents in much the same way as arthritic joints and asthmatic airways." R. Langreth. 2004. "Prevention Puzzle." *Forbes.com,* Feb. 9, at www.forbes.com/global/2004/0209/060_print.html; J. H. Dwyer et al. 2004. "Arachidonate 5-lipoxygenase promoter genotype, dietary arachidonic acid, and atherosclerosis." *N Engl J Med.* Jan 1;350(1): 29–37; I. Wickelgren. 2004. "Heart disease. Gene suggests asthma drugs may ease cardiovascular inflammation." *Science.* Feb 13;303(5660): 941.

18 J-C Tardif et al. 2003. "Effects of AGI-1067 and probucol after percutaneous coronary interventions." *Circulation.* 107: 552.

For more information on how AGI-1067 was developed, see M. Herper. 2003. "Inflamed Hearts." *Forbes.com,* July 23, www.forbes.com/forbes/2003/0623/168_print.html; G. Coté et al. 1999. "Effects of probucol on vascular remodeling after coronary angioplasty." *Circulation.* 99: 30–35.

19 M. Herper and R. Langreth. 2004. "Cardiovascular drugs to watch." *Forbes.com,* April 27, www.forbes.com/2004/01/22/cx_mh_rl_cardiotear_9.html; Y. Hirakawa and H. Shimokawa. 2001. "Lipid-lowering drugs." *Nippon Yakurigaku Zasshi.* Dec 1;118(6): 389–395.

20 J. A. Blackie et al. 2003. "The identification of clinical candidate SB-480848: a potent inhibitor of lipoprotein-associated phospholipase A2." *Bioorg Med Chem Lett.* 13(6) (Mar 24): 1067–1070; D. P. Rotella. 2004. "SB-480848. GlaxoSmithKline." *Curr Opin Invest Drugs.* Mar;5(3): 348–351; M. Herper and R. Langreth. 2004. "Cardiovascular drugs to watch." *Forbes.com,* April 27, www.forbes.com/2004/01/22/cx_mh_rl_cardiotear_2.html.

21 G. Kolata, ibid.

22 G. Kolata, ibid.

23 National Cholesterol Education Program, Adult Treatment Panel III Report, 2001 (www.nhlbi.nih.gov/guidelines/cholesterol/atp3_rpt.pdf); J. Berliner et al. 1995. "Atherosclerosis: basic mechanisms." *Circulation.* May 1;91(9): 2488–2496.

24 M. R. Naghavi et al. 2001. "New developments in the detection of vulnerable plaque." *Curr Ather Rep.* 3(2): 125–135; M. R. Naghavi et al. 2001. "MRI detection of atherosclerotic vulnerable plaque using superparamagnetic iron oxide contrast media." *Am J Cardiol.* July 19;88(2 Suppl 1): 82.

25 ". . . [T]he conditions provided by a chronic inflammatory environment are so essential for the progression of the neoplastic process that therapeutic intervention aimed at inhibiting inflammation . . . and stimulating cell-mediated immune responses may have a major role in reducing the incidence of common cancers." K. J. O'Byrne and A. G. Dalgleish. 2001. *Br J Cancer.* Aug;85(4): 473–483.

Another of the many possible examples: "Mild chronic inflammation may play a significant role in the incidence of HBP [high blood pressure]." L. E. Bautista. 2003. "Inflammation, endothelial dysfunction and the risk of high blood pressure: epidemiologic and biological evidence." *J Hum Hypertens.* April;17(4): 223–230.

26 "Heart and Stroke Facts." American Heart Association (www.americanheart.org/presenter.jhtml?identifier=3000333); National Cholesterol Education Program, Adult Treatment Panel III Report, 2001 (www.nhlbi.nih.gov/guidelines/cholesterol/atp3_rpt.pdf).

27 The role of iron in atherosclerosis is still controversial. Some studies support such a role: "These results provide direct evidence for a key role of iron in initiating atherogenesis" (D. Ponraj et al. 1999. "The onset of atherosclerotic lesion formation in hypercholesterolemic rabbits is delayed by iron depletion." *FEBS Lett.* Oct 8;459(2): 218: 222). Others do not: "Overall the results do not support the hypothesis that positive body iron stores, as measured by serum ferritin, are associated with an increased risk of cardiovascular diseases (CVD), coronary heart disease (CHD), or myocardial infarction (MI) . . . " (C. T. Sempos et al. 2000. "Serum ferritin and death from all causes and cardiovascular disease: The NHANES II mortality study." *Ann Epidemiol.* Oct 1;10(7): 441–448).

28 E. Falk et al. 1995. "Coronary plaque disruption." *Circulation.* Aug 1;92(3): 657–671; A. P. Schroeder and E. Falk. 1995. "Vulnerable and dangerous coronary plaques." *Atherosclerosis.* Dec;118 (Suppl): S141–149.

29 A. C. van der Wal and A. E. Becker. 1999. "Atherosclerotic plaque rupture: pathologic basis of plaque stability and instability." *Cardiovasc Res.* 41: 334–344; E. Falk et al. 1995. "Coronary plaque disruption." *Circulation.* Aug 1;92(3): 657–671; E. Falk. 1992; "Why do plaques rupture?" *Circulation.* 86(Suppl III): III-30-III-42.

30 G. Chiesa. 2002. "Recombinant apolipoprotein A-I(Milano) infusion into rabbit carotid artery rapidly removes lipid from fatty streaks." *Circ Res.* May 17;90(9): 974–980; P. K. Shah et al. 2001. "High-dose recombinant apolipoprotein A-I milano mobilizes tissue cholesterol and rapidly reduces plaque lipid and macrophage content in apolipoprotein e-deficient mice." *Circulation.* Jun 26;103(25): 3047–3050.

31 S. E. Nissen et al. 2003. "Effect of recombinant ApoA-I Milano on coronary atherosclerosis in patients with acute coronary syndromes: a randomized controlled trial." *JAMA.* Nov 5;290(17): 2292–2300.

32 A recent Phase 2 study reported in the *New England Journal of Medicine* "markedly increased HDL cholesterol levels and also decreased LDL cholesterol levels . . . " M. E. Brousseau et al. 2004. "Effects of an inhibitor of cholesteryl ester transfer protein on HDL cholesterol." *N Engl J Med.* Apr 8; 350(15): 1505–1515. Global Phase 3 trials began in late 2003.

Information on Torcetrapib is available on the Pfizer site: www.pfizer.com/are/investors_reports/annual_2003/review/p2003ar14_15.htm.

33 G. Etgen et al. 2002. "A tailored therapy for the metabolic syndrome." *Diabetes.* 51: 1083–1087.

Information on the PPAR alpha agonist is available in the 2003 Eli Lilly annual report: www.lilly.com/investor/annual_report/lillyar2003complete.pdf; M. Herper and R. Langreth. 2004. "Cardiovascular drugs to watch." *Forbes.com*, April 27, www.forbes.com/2004/01/22/cx_mh_rl_cardiotear_10.html.

34 Coronary calcium scores courtesy of Dr. Melvin E. Clouse of Boston's Beth Israel Deaconess Medical Center and Imatron. Data based on 13,073 asymptomatic men and 5,227 asymptomatic women.

35 B. G. Brown. 2002. "Measurement of coronary calcification: a new clinical tool." *University of Washington Regional Heart Center Consult.* Issue #3, Winter; W. Stanford. "Coronary artery calcification: significance and methods of detection." Society of Thoracic Radiology (www.thoracicrad.org/STR_Archive/PostGraduatePapers/StandfordW.html), based on W. Stanford et al. 1993. "Coronary artery calcification." *AJR Am J Roentgenol.* Dec;161(6): 1139–1146.

Dropping calcium from the diet will not reduce your calcium score. "It's not the milk you're drinking that's causing the problem," says Dr. Larry Dean, professor of medicine at the University of Washington Medical Center. "It's the butterfat in the milk that's causing the cholesterol problem, which is then causing the inflammatory process in the blood vessels. The calcium is a marker of the underlying disease process" (quoted in "Calcium scoring: A new technique useful for some with heart risk factors," *University Week*, University of Washington. Vol. 19(25), May 2, 2002; (www.depts.washington.edu/uweek/archives/2002.05.MAY_02/hs_e.html).

36 Calcium scoring is a new technique used to help identify patients at risk of heart disease. A series of quick images of the heart taken by a computed tomography (CT) scanner allows doctors to "score" the level of calcium deposits in the coronary arteries. Numerous studies are now examining the relationship between high calcium scores and other risk factors in predicting illness. N. D. Wong et al. 1994. "Coronary calcium and atherosclerosis by ultrafast computed tomography in asymptomatic men and women." *Am Heart J.* Feb;127(2): 422–430; A. S. Agatston et al. 1990. "Quantification of coronary artery calcium using ultrafast computed tomography." *J Am Coll Cardiol.* Mar 15;15(4): 827–832; A. S. Fiorino. 1998. "Electron-beam computed tomography, coronary artery calcium and evaluation of patients with coronary heart disease." *Ann Int Med.* May 15;128: 839–847.

37 C. Francis et al. 2003. "Comparison of ximelagatran with warfarin for the prevention of venous thromboembolism after total knee replacement." *N Engl J Med.* Oct 30;349: 1703–1712; S. B. Olsson et al. 2003. "Stroke prevention with the oral direct thrombin inhibitor ximelagatran compared with warfarin in patients with non-valvular atrial fibrillation (SPORTIF III): randomised controlled trial." *Lancet.* Nov 22;362(9397): 1691–1698.

38 A. Eisenberg. "An Ultrasound that Navigates Every Nook and Cranny." *New York Times*, January 15, 2004.

39 D. Ornish. 1996. *Dr. Dean Ornish's Program for Reversing Heart Disease: The Only System Scientifically Proven to Reverse Heart Disease Without Drugs or Surgery.* New York: Ballantine Books.

40 See discussion in the text of this chapter "Elevated Cholesterol, LDL, and Triglyceride Levels and Diminished HDL Levels." See also C. P. Cannon et al. 2004. "Comparison of intensive and moderate lipid lowering with statins after acute coronary syndromes." *N. Engl. J. Med.* 350(15): 1495–1504. The research compared the experimental group, which took 80 mg per day of Lipitor, to the control group, which took 40 mg a day of Pravachol. The LDL-C comparison was 62 for the experimental group versus 95 for the control group. The experimental group had substantially fewer heart attacks and recommendations for surgery.

41 "The study provides further evidence that genes play a large role in early-onset coronary heart disease (CHD) and that it clusters in families, regardless of environmental factors," according to M. Laakso, senior author of A. Kareinen et al. 2001. "Cardiovascular risk factors associated with insulin resistance cluster in families with early-onset coronary heart disease." *Arterioscler Thromb Vasc Biol.* Aug;21(8): 1346–1352. Quotation from AMA August 9, 2001, news release (www.americanheart.org/presenter.jhtml?identifier=10964).

42 G. H. Gibbons and V. J. Dzau. 1994. "The emerging concept of vascular remodeling." *N Engl J Med.* May 19;330(20): 1431–8.

43 See www.womens-health.org/press/Releases/prheartstudy.htm. The survey, conducted by International Communications Research, included 1,019 women. Studies such as this have served as catalysts for the *Heart Truth* campaign sponsored by the National Heart, Lung and Blood Institute, the National Institutes of Health, and the U.S. Department of Health and Human Services.

44 "Chronic Disease Overview," National Center for Chronic Disease Prevention and Health Promotion (www.cdc.gov/nccdphp/overview.htm). See also the *2003 Heart Disease and Stroke Statistical Update*, American Heart Association (www.americanheart.org/presenter.jhtml?identifier=3000090).

45 "Smoking costs Americans over $157 billion annually in medical care." *2003 Heart Disease and Stroke Statistical Update*, American Heart Association, op cit. Stroke is as much a concern as heart attacks; see, for example, G. A. Colditz et al. 1988. "Cigarette smoking and risk of stroke in middle-aged women." *N Engl J Med.* Apr 14;318(15): 937–941. Even regular exposure to secondhand smoke has been shown to nearly double a woman's risk of a heart attack (I. Kawachi et al. 1997. "A prospective study of passive smoking and coronary heart disease." *Circulation.* 95: 2374–2379).

46 S. Kenchaiah et al. 2002. "Obesity and the risk of heart failure." *N Engl J Med.* Aug 1;347(5): 305–313; P. W. Wilson et al. 2002. "Overweight and obesity as determinants of cardiovascular risk: the Framingham experience." *Arch Intern Med.* Sep 9;162(16): 1867–1872.

W. B. Kannel et al. 1988. "Cardiac failure and sudden death in the Framingham Study." *Am Heart J.* Apr;115(4): 869–875.

47 A number of studies citing low cholesterol as a factor in hemorrhagic stroke have been reported widely (see, for example, www.cnn.com/HEALTH/9902/06/strokes/). According to the American Heart Association, these results should be considered cautiously: study sizes have often been small and no cause-and-effect mechanism has been identified. Furthermore, "there is no trend for an increase in total mortality unless the total cholesterol level is less than 160 mg/dL. It is estimated that in the United States less than 10% of middle-aged men and women have serum cholesterol levels below this range." M. Criqui. 1994. "A statement for healthcare professionals from the American Heart Association Task Force on Cholesterol Issues" (www.americanheart.org/presenter.jhtml?identifier=1208).

48 T. Partonen et al. 1999. "Association of low serum total cholesterol with major depression and suicide." *Br J Psych.* 175: 259–262.

As with the link to stroke, however, some researchers are cautious about results that might deter clinicians from "prescribing cholesterol-lowering drugs, to reduce the risk of death from coronary heart disease." "Many confounding factors, e.g., poor health, depression and loss of appetite may play a role in the apparent relationship between serum cholesterol levels and suicide." R. Manfredini et al. 2000. "The association of low serum cholesterol with depression and suicidal behaviors: new hypotheses for the missing link." *J Int Med Res.* Nov 1;28(6): 247–257.

49 C. P. Cannon et al. "Comparison of Intensive and Moderate Lipid Lowering with Statins after Acute Coronary Syndromes." *New England Journal of Medicine,* March 8, 2004 (http://content.nejm.org/cgi/content/abstract/NEJMoa040583). The research compared the experimental group, which took 80 mg per day of Lipitor, to the control group, which took 40 mg a day of Pravachol. The LDL-C comparison was 62 for the experimental group versus 95 for the control group. The experimental group had substantially fewer heart attacks and recommendations for surgery.

50 HDL2 and HDL3 are the two major HDL subclasses. See, for example, M. C. Bakogianni et al. 2001. "Clinical evaluation of plasma high-density lipoprotein subfractions (HDL2, HDL3) in insulin-dependent diabetics with coronary artery disease." *J Diabetes Complications.* Sep–Oct;15(5): 265–269.

Regarding the debate over the more protective subfraction, see, for example, "Antioxidative activity of HDL subfractions increased with increment in density, as follows: HDL2b <HDL2a <HDL3a <HDL3b <HDL3c . . . " A. Kontush et al. 2003. "Small, dense HDL particles exert potent protection of atherogenic LDL against oxidative stress." *Arterioscler Thromb Vasc Biol,* published online before print, August 14, 2003.

In addition, "from a statistical standpoint, the present data suggest that the HDL2 subfraction may be more closely related to the development of IHD than the HDL3 subfraction. However, the qualitative difference in the relative predictive value of each subfraction was trivial, since it only corresponded to a modest quantitative difference. Thus, the possibility that a significant proportion of the cardioprotective effect of elevated HDL cholesterol levels may be mediated by the HDL3 subfraction still cannot be excluded." B. Lamarche et al. 1997. "Associations of HDL2 and HDL3 subfractions with ischemic heart disease in men: Prospective results from the Quebec Cardiovascular Study." *Arterioscler Thromb Vasc Biol.* Jun 1;17(6): 1098–1105.

51 A number of recent news stories have highlighted the importance of cholesterol tests "that look beyond the usual definitions of good and bad cholesterol, that separate the bad from the really bad and the mildly good from the angelic." D. Franklin. 2001. "What this CEO didn't know about his cholesterol almost killed him." *Fortune,* March 19 (and reported on www.berkeleyheartlab.com/GENERAL/news_fortune.html).

52 LDL particle size and density gained attention in the early 1990s, when their role in heart disease was uncovered (see R. M. Krauss et al. 1994. "A prospective study of LDL particle diameter and risk of myocardial infarction." *Circulation.* 90:I–460; described in D. Gilbert. 1994. "Small dense cholesterol particles worse for your heart." November 29, www.lbl.gov/Science-Articles/Archive/cholesterol-particles.html). See also P. T. Williams et al. 2003. "Smallest LDL particles are most strongly related to coronary disease progression in men." *Arterioscler Thromb Vasc Biol.* Feb 1:23(2): 314–321; and note 39 on page 415.

53 At a 1999 colloquium, R. M. Krauss, a researcher at the E. O. Lawrence Berkeley National Laboratory, explained: "Small LDL bind more tightly to the artery wall, they are oxidized more rapidly and they may cause greater endothelial dysfunction. There is an increasing body of evidence that suggest that some of the damage caused by higher levels of triglyceride is mediated by this effect on LDL, as well as some of the other metabolic conditions that are associated with high triglyceride, including low HDL and insulin resistance. We call this an atherogenic phenotype. It is a collection of abnormalities that together comprise a real significant coronary disease risk profile that is above and beyond what we can detect just by measuring the levels of LDL." Sante Fe Colloquium on Preventive Cardiovascular Therapy, October 7–9, 1999; Sante Fe, New Mexico (www.acc.org/education/online/sante_fe/krauss.htm).

R. Krauss is also an author on a study that supports triglycerides as an independent risk factor for myocardial infarction. See M. J. Stampfer et al. 1996. "A prospective study of triglyceride level, low-density lipoprotein particle diameter and risk of myocardial infarction." *JAMA.* Sep 18;276(11): 882–888.

54 "Policosanol is a mixture of higher primary aliphatic alcohols isolated from sugar cane wax, whose main component is octasanol. This mixture has been shown to lower cholesterol in animal models, healthy volunteers and patients with type II hypercholesterolemia." I. Gouni-Berthold and H. K. Berthold. 2002. "Policosanol: clinical pharmacology and therapeutic significance of a new lipid-lowering agent." *Am Heart J.* Feb;143(2): 356–365.

55 Ibid. "Because higher doses have not been tested up to now, it cannot be excluded that effectiveness may be even greater. Daily doses of 10 mg of policosanol have been shown to be equally effective in lowering total or LDL cholesterol as the same dose of simvastatin or pravastatin. Triglyceride levels are not influenced by policosanol."

56 S. Nityanand et al. 1989. "Clinical trials with gugulipid: a new hypolipidaemic agent." *J Assoc Physicians India*. 37: 323–328.

57 N. G. Stephens et al. 1996. "Randomised controlled trial of vitamin E in patients with coronary disease: Cambridge Heart Antioxidant Study (CHAOS)." *Lancet*. Mar 23;347(9004): 781–786.

58 M. N. Nanjee et al. 2001. "Intravenous apo A-I/lecithin discs increase pre-Beta-HDL concentration in tissue fluid and stimulate reverse cholesterol transport in humans," *Journal of Lipid Research*. Oct(42): 1586–1593. The study examined "intravenous infusion of apolipoprotein A-I/phosphatidylcholine discs in humans" and concluded "Intravenous apoA-I/lecithin discs increase pre-Beta-HDL concentration in tissue fluid and stimulate reverse cholesterol transport in humans."

59 A. Gotto Jr. 2003. "Safety and statin therapy: reconsidering the risks and benefits." *Arch Intern Med*. 163: 657–659; J. Tobert. 2003. "Lovastatin and beyond: the history of the HMG-CoA reductase inhibitors." *Nature Rev Drug Discov*. 2: 517–526.

60 Lipitor is produced by Pfizer, Inc. (www.lipitor.com). See also, for example, B. R. Krause and R. S. Newton. 1995. "Lipid-lowering activity of atorvastatin and lovastatin in rodent species: triglyceride-lowering in rats correlates with efficacy in LDL animal models." *Atherosclerosis*. Oct 1;117(2): 237–244.

61 P. M. Ridker et al. 1998. "Prospective study of C-reactive protein and the risk of future cardiovascular events among apparently healthy women." *Circulation*. 98: 731–733; P. M. Ridker et al. 1997. "Inflammation, aspirin and the risk of cardiovascular disease in apparently healthy men." *N Engl J Med*. Apr 3;336(14): 973–979. For a later report, see P. M. Ridker. 2001. "High-sensitivity C-reactive protein." *Circulation*. 103: 1813–1818.

62 K. Miura et al. 2001. "Relationship of blood pressure to 25-year mortality due to coronary heart disease, cardiovascular diseases and all causes in young adult men: The Chicago Heart Association Detection Project in Industry." *Arch Int Med*. Jun 25;161(12): 1501–1508.

63 For the different types of drugs for hypertension and how they work, see the list on the American Heart Association site: "Blood pressure-lowering drugs" (www.americanheart.org/presenter.jhtml?identifier=159). One study observed that losartan (Cozaar) had fewer side effects than calcium channel blockers in a community-based setting, which meant that patients would be more likely to adhere to their treatment regimens (J. P. Grégoire et al. 2001. "Tolerability of antihypertensive drugs in a community-based setting." *Clin Ther*. May;23(5): 715–726). Another research compared angiotensin II antagonists and calcium channel blockers and suggested a role for each. (M. Weir. 2001. "Appropriate use of calcium antagonists in hypertension." *Hosp Pract* (Off Ed). Sep 15;36(9): 47–48, 53–55.)

64 S. Jacob et al. 1998. "Antihypertensive therapy and insulin sensitivity: do we have to redefine the role of beta-blocking agents?" *Am J Hypertens*. Oct;11(10): 1258–1265. Impotence may also be a problem with beta-blockers: R. Fogari. 1998. "Sexual function in hypertensive males treated with lisinopril or atenolol: a cross-over study." *Am J Hypertens*. Oct;11(10): 1244–1247.

65 "Hostility is an independent risk factor for coronary heart disease (CHD)." T. Q. Miller et al. "A meta-analytic review of research on hostility and physical health." *Psychol Bull*. Mar;119(2): 322–348. For the link between adrenaline and inflammation, see, for example, P. H. Black. 2002. "Stress and the inflammatory response: a review of neurogenic inflammation." *Brain Behav Immun*. Dec;16(6): 622–653.

66 J. Denollet and D. Brutsaert. 1998. "Personality, disease severity and the risk of long-term cardiac events in patients with a decreased ejection fraction after myocardial infarction." *Circulation*. Jan;97: 167–173; J. Denollet. 2000. "Type D personality. A potential risk factor defined." *J Psychosom Res*. Oct;49(4): 255–266.

67 I. Wilcox et al. 1998. "'Syndrome Z': The interaction of sleep apnoea, vascular risk factors and heart disease." *Thorax*. Oct;53(Suppl 3): S5–S28; J. E. Muller et al. 1997. "Mechanisms precipitating acute cardiac events." *Circulation*. 96: 3233–3239.

68 Companies developing heart simulations include Artesian Therapeutics and Immersion Medical, both in Gaithersburg, Maryland; Insillicomed in La Jolla, California; and Predix Pharmaceuticals in Woburn, Massachusetts.

69 D. H. Freedman. "The Virtual Heart." *Technology Review*, March 2004.

70 "Plaque rupture was significantly associated with high fibrinogen levels." A. Mauriello et al. 2000. "Hyperfibrinogenemia is associated with specific histocytological composition and complications of atherosclerotic carotid plaques in patients affected by transient ischemic attacks." *Circulation*. 101: 744. See also A. Maseri and V. Fuster. 2003. "Is there a vulnerable plaque?" *Circulation*. Apr;107: 2068–2071.

71 P. Lotufo et al. 2000. "Male pattern baldness and coronary heart disease." *Arch Int Med*. Jan 24;160(2): 165–71.

72 For general information about hemochromatosis, see "Hemochromatosis," National Digestive Diseases Information Clearinghouse (http://digestive.niddk.nih.gov/ddiseases/pubs/hemochromatosis/index.htm).

For recent research, see, for example, M. Rasmussen et al. 2001. "A prospective study of coronary heart disease and the hemochromatosis gene (HFE) C282Y mutation: the Atherosclerosis Risk in Communities (ARIC) study." *Atherosclerosis.* Feb 15;154(3): 739–746.

73 Contradictory results have been reported on the link between periodontal disease and coronary heart disease. P. P. Pussinen et al. (2003) reported that "men with antibodies to the dental bacteria were 50 percent more likely to have heart disease than men without these antibodies," per a Reuters Health news report ("Bugs in mouth bad for heart," http://12.31.13.29/HealthNews/Reuters/NewsStory0717200318.htm). "Antibodies to periodontal pathogens are associated with coronary heart disease." *Arterioscler Thromb Vasc Biol.* Apr 24;23: 1250. However, P. P. Hujoel et al. (2000) had different results: "This study did not find convincing evidence of a causal association between periodontal disease and CHD risk." "Periodontal disease and coronary heart disease risk." *JAMA.* 284(11): 1406–1410.

74 M. Christ-Crain et al. 2003. "Elevated C-reactive protein and homocysteine values: cardiovascular risk factors in hypothyroidism? A cross-sectional and a double-blind placebo-controlled trial." *Atherosclerosis.* Feb;166(2):379–386; I. Klein. 2003. "Thyroid hormone and cardiac contractility." *Am J Cardiol.* Jun;91(11): 1331–1332.

75 See his book *Nanomedicine* (vol. 1, 1999, and vol. 2, 2003; Georgetown, Texas: Landes Bioscience). Also see the Foresight Institute's "Nanomedicine" page by Robert Freitas Jr., which lists his current technical works (www.foresight.org/Nanomedicine/index.html#MedNanoBots).

76 One of the authors of this book, Ray Kurzweil, and his company, Kurzweil Technologies, Inc., is working with Medicomp (a subsidiary of United Therapeutics, Inc.), a leader in Holter and event monitoring, to create a new generation of computer-based pattern recognition to automatically evaluate ECG recordings from holster and event monitors.

77 For general information about external counterpulsation, see "Enhanced external counterpulsation (EECP)" (www.americanheart.org/presenter.jhtml?identifier=4577). Clinic sites also have descriptions; see, for example, http://cardiology/ucsf.edu/clinical/eecp/. For studies supporting the technique's efficacy, see, for example, A. D. Michaels et al. 2002. "Left ventricular systolic unloading and augmentation of intracoronary pressure and Doppler flow during enhanced external counterpulsation." *Circulation.* Aug 19;106: 1237.

78 Many labs and universities have received funding to research tiny fuel cells. See the University of Notre Dame news release, "Team receives $1.6 million grant for fuel cell research," http://newsinfo.nd.edu/content.cfm?topicId=3311. To follow government support for fuel cells, see the U.S. Department of Energy Hydrogen, Fuel Cells & Infrastructure Technologies Program Web page, www.eere.energy.gov/hydrogenandfuelcells/). Also see A. V. Chadwick. 2000. "Nanotechnology: solid progress in ion conduction." *Nature.* Dec 21;408: 925–926.

79 "Angiogram: what risks are there from the test?" *Harvard Medical School Family Health Guide* (www.health.harvard.edu/fhg/diagnostics/angiogram/angiogramRisks.shtml).

80 No significant differences in outcome were noted between:

Veterans treated medically and surgically (R. J. Scott et al. 1987. "Comparison of medical and surgical treatment for unstable angina pectoris." *N Engl J Med.* Apr 16;316(16): 977–984).

Hospitalized patients in Sweden receiving dramatically less surgical intervention than in the U.S. (P. G. McGovern et al. 1997. "Comparison of medical care and one and 12-month mortality of hospitalized patients with acute myocardial infarction." *Am J Cardiol.* Sept 1;80(5): 557–562).

Patients in different parts of the U.S. receiving radically different types of care (L. Pilote et al. 1995. "Regional variation across the United States in the management of acute myocardial infarction." *N Engl J Med.* Aug 31;333(9): 589–590).

At the same time, significant differences in outcome have been associated with better oversight of patients and lifestyle changes. Note, however, that the selection of patients can influence the results from comparative studies. Studies that screen out sicker patients will inevitably show fewer differences between surgical and medical treatment.

81 C. M. Winslow et al. 1988. "The appropriateness of performing coronary artery bypass surgery." *JAMA.* Jul 22–29;260(4): 505–509; R. Lange and D. L. Hillis. "Use and overuse of angiography and revascularization for acute coronary syndromes." *N Engl J Med.* 338(25): 1838–1839.

While the health care system in the United States is set up to encourage the overuse of expensive treatments, patients and patients' families play their part as well by assuming that more expensive options are necessarily better.

82 S. G. Ellis et al. 1992. "Randomized trial of late angioplasty versus conservative management for patients with residual stenosis after thrombolytic treatment of myocardial infarction." *Circulation.* Nov;86(5): 1400–1406. This study "strongly suggests" patients who had an "uncomplicated myocardial infarction" should be treated medically (with drugs) rather than with surgery.

Another study concluded, "because conservative strategy achieves equally good short and long term outcomes with less morbidity and a lower use of [angioplasty], it seems to be the preferred initial management strategy." W. J. Rogers et al. "Comparison of immediate invasive, delayed invasive and conservative strategies after tissue-type plasminogen activator." *Circulation*. May;81(5): 1457–1476. For the guidelines the medical profession uses to grade the seriousness of occluded arteries, see the report from the American College of Cardiology Foundation and American Heart Association, "ACC/AHA 2002 Guideline Update for the Management of Patients with Chronic Stable Angina."

E. Schneider et al. 2001. "Overuse of coronary artery bypass graft surgery and percutaneous transluminal coronary arngioplasty." *Annals of Internal Medicine*. Sept 4;135(5): S35; E. Schneider et al. 2001. "Racial differences in cardiac revascularization rates: does 'overuse' explain higher rates among white patients?" *Annals of Internal Medicine*. Sept 4;135(5): 328–337; W. E. Boden et al. 1998. "Outcomes in patients with acute non-Q-wave myocardial infarction randomly assigned to an invasive as compared with a conservative management strategy." *N Engl J Med*. 338: 1785; E. Braunwald. 1988. "Evolution of the management of acute myocardial infarction: a 20th-century saga." *Lancet*. 352: 1771–1774.

The following articles speak to the "sickest subset" issue:

www.clevelandclinic.org/heartcenter/pub/news/archive/2004/survival4_29.asp and

www.dukemednews.org/news/article.php?id=6479.

See also notes 6 and 7 on page 412.

83 M. F. Newman et al. 2001. "Longitudinal assessment of neurocognitive function after coronary-artery bypass surgery." *N Engl J Med*. Feb 8;344(6): 395–402.

84 "No development in interventional cardiology has created a stir like the drug-eluting stent for preventing restenosis. . . . Finally, in our excitement about the potential for interventional cardiology, we must remember that atherosclerosis will not be cured by drug-eluting stents. Prevention of progression of this disease requires changing the metabolic milieu of the patient who has it. Interventional procedures are superb for alleviating the current ischemia and related symptoms, but a concerted effort by the healthcare team and the patient are necessary to change the ultimate outcome. Although the restenosis mouse 'has roared,' it may not be necessary in all cases to use an elephant gun to eliminate him." S. King. "Restenosis: the mouse that roared." *Circulation*. 108: 248.

In addition, "sirolimus-eluting stent edge restenosis is frequently associated with local trauma outside the stent." P. A. Lemos et al. "Coronary restenosis after sirolimus-eluting stent implantation: morphological description and mechanistic analysis from a consecutive series of cases." *Circulation*. Jul 22;108(3): 256–260.

CHAPTER 16

1 Only heart disease kills more people than cancer. E. Arias and B. L. Smith. 2003. "Deaths: preliminary data for 2001." *National Vital Statistics Reports* 51(5), www.cdc.gov/nchs/data/nvsr/nvsr51/nvsr51_05.pdf. and R. Davis. 2004. "Cancer stats cite new danger." *USA Today* online edition, Jan 14. See www.usatoday.com/news/health/2004-01-15-cancer-obesity_x.htm.

2 Infection with *Helicobacter pylori* bacteria is one of the primary factors leading to stomach cancer. J. Parsonnet et al. 1994. "Helicobacter pylori infection and gastric lymphoma." *N Engl J Med*. May 5;330(18): 1267–1271. Methods of storing food may also be a factor. See S. H. Landis et al. 1999. "Cancer statistics, 1999." *Ca-A Cancer J Clin*. 49(1): 8–31.

"Several reasons may have led to this drop in stomach cancer rates, such as improved detection and treatment as well as improved dietary habits, such as eating more fruits, vegetables, and fiber. Many studies show that a diet rich in fruits and vegetables lowers the risk for many cancers. But if stomach cancer is not caught early before it has spread, the prognosis is poor and the disease may be fatal." (Cleveland Clinic, "*Helicobacter pylori* and stomach cancer," www.clevelandclinic.org/health/health-info/docs/1800/1816.asp?index=8107&src=news).

"Cigarette smoking is the most important risk factor for lung cancer, accounting for 68 to 78 percent of lung cancer deaths among females and 88 to 91 percent of lung cancer deaths among males." CDC. 1990. Cigarette smoking–attributable mortality and years of potential life lost—United States, 1990. *Morbidity and Mortality Weekly Report* 42(33): 645–649, reported in CDC and NIH, Healthy People 2010, www.healthypeople.gov/document/html/volume1/03cancer.htm.

"Unexplained cancer-related health disparities remain among population subgroups. For example, Blacks and people with low socioeconomic status have the highest overall rates for both new cancers and deaths." National Cancer Institute, 2001 Progress Report, http://progressreport.cancer.gov/highlights.asp?coid=17.

3 American Cancer Society, Cancer Facts & Figures 2002, www.cancer.org/downloads/STT/CancerFacts&Figures2002TM.pdf.

4 Genomic Health information page, at www.genomichealth.com/oncotype/faq/pat.aspx; Biospace, CCIS, "The National Surgical Adjuvant Breast and Bowel Project (NSABP) and Genomic Health, Inc. Announce Positive Results from Large-Scale, Prospective Validation Study to Quantify Breast Cancer Recurrence in Newly Diagnosed Patients," www.biospace.com/ccis/news_story.cfm?StoryID=14550020&full=1.

5 "10 emerging technologies that will change your world." *MIT Technology Review*, February 2004; www.technologyreview.com/articles/print_version/emerging0204.asp; H. Brody. "Taming the terahertz." *MIT Technology Review*, June 2003; www.technologyreview.com/articles/innovation40603.asp.

6 "Spotting cancer before it sickens." *Wired News*, April 9, 2003; www.wired.com/news/medtech/0,1286,58407,00.html.

7 Ibid. See also, T. Parker-Pope. "Ten major advances you're likely to see in the coming year." *Wall Street Journal*, January 26, 2004; Y. Yu et al. 2004. "Visualization of tumors and metastases in live animals with bacteria and vaccinia virus encoding light-emitting proteins." *Nature Biotech.* 22(Mar 01): 313–320.

8 See T. J. Key et al. 2004. "Diet, nutrition and the prevention of cancer." *Public Health Nutr.* Feb;7(1A): 187–200.

Yet this view remains controversial. In 2003, the U.S. Preventive Services Task Force (USPSTF) concluded "that the evidence is insufficient to recommend for or against the use of supplements of vitamins A, C, or E; multivitamins with folic acid; or antioxidant combinations for the prevention of cancer or cardiovascular disease." See www.ahrq.gov/clinic/3rduspstf/vitamins/vitaminsrr.htm.

9 Instead of preventing cancer, radiation from mammograms has been identified as the cause of 3,000–5,000 additional cases of breast cancer each year. See J. W. Gofman and E. O'Connor. 1999. *Radiation from Medical Procedures in the Pathogenesis of Cancer and Ischemic Heart Disease: Dose-Response Studies with Physicians per 100,000 Population.* San Francisco: CNR Books.

10 We say there is "often" value in early detection because many cases of malignancy are already metastasized by the time they can be detected and their early detection has little predictive value.

11 Take the case of prostate cancer, for example. "Available screening tests (for example prostate specific antigen) can detect early stage disease but there is no evidence that clinical outcomes are improved by early detection. The potential harms of screening 28 million men older than 50 years include unnecessary interventions for thousands of men without disease or with clinically insignificant cancer. The billions of dollars required for this effort could displace resources away from health care services of proved benefit." S. H. Woolf. 1994. "Public health perspective: the health policy implications of screening for prostate cancer." *J Urol.* Nov;152(5 Pt 2): 1685–1688.

12 L. A. Koutsky.2002. "A controlled trial of a human papilloma virus type 16 vaccine." *N Engl J Med.* Nov 21;347(21): 1645–1651.

13 O. J. Finn. 2003. "Cancer vaccines: between the idea and the reality." *Nat Rev Immunol.* Aug;3(8): 630–641; R. C. Kennedy and M. H. Shearer. 2003. "A role for antibodies in tumor immunity." *Int Rev Immunol.* Mar–Apr;22(2): 141–172.

14 E. Jonietz. "DNA drugs." *MIT Technology Review*, November 2002; www.technologyreview.com/articles/innovation51102.asp?p=1.

15 "In normal development, [the surface molecule] 5T4 is involved in helping cells move around in a regulated way" according to Peter Stern at Cancer Research UK. "In cancer, it's not regulated—it's out of control." S. Bhattacharya. "Stem cell mobility linked to cancer's spread." *NewScientist.com*. October 24, 2003; www.newscientist.com/news/news.jsp?id=ns99994309.

16 Interleukin-12 is not normally produced by cancer cells. "Mount Sinai School of Medicine conducting clinical trials with gene therapy for colorectal cancer." *Bio.com*, January 22, 2004; www.bio.com/newsfeatures/newsfeatures_research.jhtml?cid=129742418&page=1.

17 J. K. Gohagan et al. 2000. "The prostate, lung, colorectal and ovarian (PLCO) cancer screening trial of the National Cancer Institute: history, organization, and status." *Control Clin Trials.* Dec;21(6 Suppl): 251S–272S.

18 National Cancer Institute, "Screening and Testing for Cancer;" www.nci.nih.gov/cancerinfo/screening.

19 For more information about the DR-70 test, see the AMDL, Inc. site; www.amdl.com/Products/DR-70/index.html.

20 Stanford University Medical Center, Office of Communication & Public Affairs, "Stanford researchers weigh risks vs. benefits of self-referred body scanning." July 29, 2003; http://mednews.stanford.edu/news_releases_html/2003/julyrelease/scanning.htm. The news release refers to J. Illes et al. 2003. "Self-referred whole-body CT imaging: current implications for health care consumers." *Radiology.* Aug;228(2): 346–351.

21 Quoted in G. Kolata. "Questions grow over usefulness of some routine cancer tests." *New York Times*, December 30, 2001.

22 The only food that cancer cells can eat is sugar, which is another reason we emphasize avoiding simple sugars in the diet as well as high-glycemic-index foods that raise blood sugar levels quickly.

23 T. Boehm et al. 1997. "Antiangiogenic therapy of experimental cancer does not induce acquired drug resistance." *Nature*. 390: 404–407.

24 Angiogenesis Foundation, "Understanding Angiogenesis". www.angio.org/understanding/content_understanding.html.

25 National Cancer Institute, Clinical Trial Results: "Bevacizumab (Avastin(tm)) Improves Survival in Metastatic Colorectal Cancer"; www.cancer.gov/clinicaltrials/results/bevacizumab-and-colorectal-cancer0601. With these trials showing benefit in colon cancer, trials are now looking at Avastatin in the treatment of renal cell cancer, prostate cancer, non-Hodgkin's lymphoma, and many others.

26 J. Perkel. 2002. "Telomeres as the key to cancer." *The Scientist*. 16(11):38; www.the-scientist.com/yr2002/may/profile_020527.html.

27 J. Alam. 2003. "Apoptosis: target for novel drugs." *Trends in Biotechnology*. 21(11): 479–483.

28 There are over 700 gene therapy treatments being tested worldwide. There have been a number of highly publicized setbacks for some of these trials in the U.S. and Europe. Shenzhen SiBiono Gene Technologies Co. has been criticized for commercially releasing Gendicine too early without sufficiently large human trials. J. Hepeng. "First gene-therapy medicine commercialized." *Business Weekly*, December 9, 2003; www1.chinadaily.com.cn/en/doc/2003-12/09/content_289867.htm; "Cancer gene therapy is first to be approved." *NewScientist.com*, November 28, 2003; www.newscientist.com/news/news.jsp?id=ns99994420.

29 S. Bhattacharya. "Deadly spread of cancer halted." *NewScientist.com*, June 5, 2003; www.newscientist.com/news/news.jsp?id=ns99993801.

30 Cancer Research UK press release. "Scientists overpower cancer's drug defenses," February 19, 2004; www.cancerresearchuk.org/news/pressreleases/cancer_drugdefences_19feb04.

31 S. Bhattacharya. "GM blood kills human cancer cells." *NewScientist.com*, April 1, 2003; www.newscientist.com/news/news.jsp?id=ns99993574.

32 See P. J. Pickhardt et al. 2003. "Computed tomographic virtual colonoscopy to screen for colorectal neoplasia in asymptomatic adults." *N Engl J Med*. Dec 4;349: 2191–2200; M. M. Morrin and J. T. LaMont. 2003. "Screening virtual colonoscopy—Ready for prime time?" *N Engl J Med*. Dec 4;349: 2261–2264.

33 P. B. Cotton et al. 2004. "Computed tomographic colonography (virtual colonoscopy): a multicenter comparison with standard colonoscopy for detection of colorectal neoplasia." *JAMA*. Apr 14;291(14): 1713–1719.

34 P. Lichtenstein et al. 2002. "The Swedish Twin Registry: a unique resource for clinical, epidemiological and genetic studies." *J Intern Med*. Sep;252(3): 184–205; N. L. Pedersen et al. 2002. "The Swedish Twin Registry in the third millennium." *Twin Res*. 5(5): 427–432.

35 See, for example, J. D. Hayes and R. C. Strange. 2000. "Glutathione S-transferase polymorphisms and their biological consequences." *Pharmacology*. Sep;61(3): 154–166.

36 See our Web site Fantastic-Voyage.net for further information on available genomics tests.

37 Up to 48 percent of the cases of stomach cancer may be due to the GSTM 1 null polymorphism combined with mutations of the IL-1B and NAT1 genes, as demonstrated by C. A. Gonzalez et al. 2002. "Genetic susceptibility and gastric cancer risk." *Int J Cancer*. Jul 20;100(3): 249–260.

38 Women often refuse to have the test for a myriad of reasons, such as concern over insurance coverage for preexisting conditions. "Misuse of genetic information can have devastating consequences—job loss, social stigmatization, loss of health and life insurance or inability to obtain them—and all these must be guarded against whenever possible." "Genetic testing for breast and ovarian cancer susceptibility." *DukeMed Magazine*, Summer 2001; http://dukemednews.duke.edu/news/controversy.php?id=1733. See also, A. Berchuck and M. G. Muto. 1996. "Status of testing for genetic predisposition to ovarian cancer." *SCO Issues* Fall;16(3); www.sgo.org/publications/SGOIssues/fall96/Science.html.

39 T. Soucci. 2000. "The p53 tumor suppressor gene: from molecular biology to clinical investigation." *Ann NY Acad Sci*. Jun;910: 121–137; discussion 137–139.

40 T. J. Key et al. 2004, op cit.

41 The National Cancer Institute defines a serving as any of the following: one medium-size fruit (such as apple, orange, banana, pear), ½ cup raw, cooked, canned or frozen fruits or vegetables, ¾ cup (6 oz.) 100 percent fruit or vegetable juice, ½ cup cut-up fruit, ½ cup cooked or canned legumes (beans and peas), 1 cup raw, leafy vegetables (lettuce, spinach), or ¼ cup dried fruit (raisins, apricots, mango). For more information, visit the National Cancer Institute's 5 A Day program page: www.5aday.gov.

42 Behavior risk factors surveillance system CD-ROM (1984–1995, 1996, 1998) and public use data tape (2000), National Center for Chronic Disease Prevention and Health Promotion, Centers for Disease Control and prevention, 1997, 1999, 2000, 2001.

43 C. N. Holick et al. 2002. "Dietary carotenoids, serum beta-carotene, and retinol and risk of lung cancer in the alpha-tocopherol, beta-carotene cohort study." *Am J Epidemiol.* Sep 15;156(6): 536–547.

44 C. La Vecchia et al. 2001. "Nutrition and health: epidemiology of diet, cancer and cardiovascular disease in Italy." *Nutr Metab Cardiovasc Dis.* Aug;11(4 Suppl): 10–15.

45 A. Trichopoulou et al. 2003. "Adherence to a Mediterranean diet and survival in a Greek population." *N Engl J Med.* Jun 26;348(26): 2599–2608.

46 S. Watanabe, S. Uesugi, and Y. Kikuchi. 2002. "Isoflavones for prevention of cancer, cardiovascular diseases, gynecological problems and possible immune potentiation." *Biomed Pharmacother.* Aug;56(6): 302–312.

47 Y. C. Wang and U. Bachrach. 2002. "The specific anti-cancer activity of green tea (-)-epigallocatechin-3-gallate (EGCG)." *Amino Acids.* 22(2): 131–143.

48 V. Rice. "University Health Network researchers discover new class of human stem cells." University of Toronto news release, June 8, 2003; www.eurekalert.org/pub_releases?2003-06/uot-uhn060503.php.

49 R. W. Owen et al. 2000. "Olive-oil consumption and health: the possible role of antioxidants." *Lancet Oncol.* Oct;1: 107–112.

50 E. Giovannucci et al. 2002. "Importance of lycopene and tomato products to prevent prostate cancer. A prospective study of tomato products, lycopene, and prostate cancer risk." *J Natl Cancer Inst.* Mar 6;94(5): 391–398.

51 O. Warburg. 1956. "On the origin of cancer cells." *Science.* Feb 24;123(3191): 309–314.

52 S. Higgenbotham et al. 2004. "Dietary glycemic load and risk of colorectal cancer in the Women's Health Study." *J Natl Cancer Inst.* Feb 4;96(3): 229–233. Interestingly, the same researchers did not find a high-glycemic-load diet to be associated with an increased risk of breast cancer. S. Higgenbotham et al. 2004. "Dietary glycemic load and breast cancer risk in the Women's Health Study." *Cancer Epidemiol Biomarkers Prev.* Jan;13(1): 65–70.

53 D. S. Michaud et al. 2002. "Dietary sugar, glycemic load, and pancreatic cancer risk in a prospective study." *J Natl Cancer Inst.* Sep 4;94(17): 1293–1300.

54 American Cancer Society. "Fitting in Fitness." www.cancer.org/docroot/PED/content/PED_6_1X_Be_Physically_Active_Achieve_and_Maintain_a_Healthy_Weight.asp?sitearea=PED.

55 Much of the concern over sunlight exposure seems misdirected. While excessive exposure to bright sunlight will damage skin and increase the incidence of skin cancer, moderate exposure to sunlight may be cancer protective, as shown in the study by W. B. Grant. 2002. "An estimate of premature cancer mortality in the U.S. due to inadequate doses of solar ultraviolet-B radiation." *Cancer.* Mar 15;94(6): 1867–1875.

56 A. P. Albino et al. 2000. "Cell cycle arrest and apoptosis of melanoma cells by docosahexaneoic acid: association with decreased pRb phosphorylation." *Cancer Res.* Aug 1;60(15): 4139–4145.

57 L. Settimi et al. 2001. "Cancer risk among male farmers: a multi-site case-control study." *Int J Occup Med Environ Health.* 14(4): 339–347.

58 These results come from the 1999–2000 National Health and Nutrition Examination Survey (NHANES). The report, "Prevalence of Overweight and Obesity Among Adults: United States, 1999–2000," is available at the CDC Web site www.cdc.gov/nchs/products/pubs/pubd/hestats/obese/obse99.htm. See also K. M. Flegal et al. 2002. "Prevalence and trends in obesity among U.S. adults, 1999–2000." *JAMA.* 288: 1723–1727.

59 E. E. Calle et al. 2003. "Overweight, obesity, and mortality from cancer in a prospectively studied cohort of U.S. adults." *N Engl J Med.* Apr 24;348(17): 1625–1638.

60 R. A. Freitas Jr. 2002. "The future of nanofabrication and molecular scale devices in nanomedicine." *Studies in Health Technology and Informatics.* 80: 45–59.

61 X. H. Gao et al. 2002. "Quantum-dot nanocrystals for ultrasensitive biological labeling and multicolor optical encoding." *Journal of Biomedical Optics.* 7(4): 532–537.

62 Fred Hutchinson Cancer Research Center press release. "Intel and Fred Hutchinson to explore the use of nanotechnology tools for early disease detection." Fred Hutchinson Cancer Research Center. October 23, 2003; www.eurekalert.org/pub_releases/2003-10/fhcr-iaf102303.php.

63 "Optical biopsies on horizon using noninvasive biomedical imaging technique developed by Cornell-Harvard group." *Cornell News Service*, June 11, 2003; www.news.cornell.edu/releases/June03/Intrinsic.Fluor.hrs.html.

64 M. Kelly. "Startups seek perfect particles to search and destroy cancer." *Small Times*, April 18, 2003; www.smalltimes.com/document_display.cfm?document_id=5867.

65 Ibid.

66 J. Couzin. "Nanoparticles Cut Tumors' Supply Lines," *Science*, June 27, 2002; http://sciencenow.sciencemag.org/cgi/content/full/2002/627/3.

67 E. Cameron and L. Pauling. 1992. *Cancer and Vitamin C: A Discussion of the Nature, Causes, Prevention, and Treatment of Cancer with Special Reference to the Value of Vitamin C.* Philadelphia: Camino Books.

68 Q. S. Zheng and R. L. Zheng. 2002. "Effects of ascorbic acid and sodium selenite on growth and redifferentiation in human hepatoma cells and its mechanisms." *Pharmazie*. Apr;57(4): 265–269.

69 A. J. Duffield-Lillico et al. 2002. "Baseline characteristics and the effect of selenium supplementation on cancer incidence in a randomized clinical trial: a summary report of the Nutritional Prevention of Cancer Trial." *Cancer Epidemiol Biomarkers Prev*. Jul;11(7): 630–639.

70 O. Portakal et al. 2000. "Coenzyme Q_{10} concentrations and antioxidant status in tissues of breast cancer patients." *Clin Biochem*. Jun;33(4): 279–284.

71 S. P. Verma et al. 1997. "Curcumin and genistein, plant natural products, show synergistic inhibitory effects on the growth of human breast cancer MCF-7 cells induced by estrogenic pesticides." *Biochem Biphy Res Comm*. 233: 692–696.

72 R. Rashmi, T. R. Santhosh Kumar, and D. Karunagaran. 2003. "Human colon cancer cells differ in their sensitivity to curcumin-induced apoptosis and heat shock protects them by inhibiting the release of apoptosis-inducting factor and caspases." *FEBS Lett*. Mar 13;538(1-3): 19–24; T. Kawamori et al. 1999. "Chemopreventive effect of curcumin, a naturally occurring anti-inflammatory agent, during the promotion/progression stages of colon cancer." *Cancer Res*. 59: 597–601.

73 R. A. Freitas Jr. "Robots in the bloodstream: the promise of nanomedicine." *KurzweilAI.net*, February 26, 2002; www.kurzweilai.net/meme/frame.html?main=/articles/art0410.html.

74 R. Smith. "Lung cancer cluster bombs created by researchers." *Medical News Today*, January 31, 2004; www.medicalnewstoday.com/index.php?newsid=5604.

75 J. Mason. "Coatings and arrays help put medication where it's needed." *Small Times*, June 27, 2003; www.smalltimes.com/document_display.cfm?document_id=6288.

76 J. Gorman. "Buckymedicine: Coming soon to a pharmacy near you?" *Science News*, July 13, 2002; www.sciencenews.org/20020713/bob10.asp.

77 D. Penman. "Carbon nanotubes show drug delivery promise." *NewScientist.com*, December 16, 2003; www.newscientist.com/news/news.jsp?id=ns99994485.

78 M. R. McDevitt et al. 2001. "Tumor Therapy with Targeted Atomic Nanogenerators." *Science*. Nov 16;294(5546): 1537–1540.

79 "Nanoprobe to be developed for a 'Fantastic Voyage' in the human body, finding and treating deadly tumors." *Today@UCI*, May 8, 2003; http://today.uci.edu/news/release_detail.asp?key=995.

80 "Lasers operate inside single cells." *Nature Science Update*, October 6, 2003; www.nature.com/nsu/030929/030929-12.html.

81 "Microbeams have big impact on cancer cells." *Reuters*, December 2, 2003; www.cnn.com/2003/HEALTH/conditions/12/02/cancer.microbeams.reut/index.html.

82 A. Salkever. "How High Tech Is Operating on Medicine." *Business Week*, October 15, 2002; www.businessweek.com/technology/content/oct2002/tc20021015_8842.htm.

83 R. A. Freitas Jr., op cit.

84 A. Panzer and M. Viljoen. 1997. "The validity of melatonin as an oncostatic agent." *J Pineal Res*. May;22(4):184–202.

85 S. Cos and E. J. Sanchez-Barcelo. 2000. "Melatonin, experimental basis for a possible application in breast cancer prevention and treatment." *Histol Histopathol*. Apr;15(2): 637–647.

86 M. Eichholzer et al. 2001. "Folate and the risk of colorectal, breast and cervix cancer: the epidemiological evidence." *Swiss Med Wkly*. Sep 22;131(37–38): 539–549.

87 H. Tapiero et al. 2002. "Polyunsaturated fatty acids (PUFA) and eicosanoids in human health and pathologies." *Biomed Pharmacother*. Jul;56(5): 215–222.

88 J. Virtamo et al. 2003. "Incidence of cancer and mortality following alpha-tocopherol and beta-carotene supplementation: a postintervention follow-up." *JAMA*. Jul 23;290(4): 476–485.

89 M. Caraballoso et al. 2003. "Drugs for preventing lung cancer in healthy people." *Cochrane Database Syst Rev*. (2):CD002141; A. Arora, C. A. Willhite, and D. C. Liebler. 2001. "Interactions of beta-carotene and cigarette smoke in human bronchial epithelial cells." *Carcinogenesis*. Aug;22(8): 1173–1178.

CHAPTER 17

1 You can also use the "Life Expectancy Calculator" I used to calculate how long you might expect to live. It may be found at http://gosset.wharton.upenn.edu/~foster/mortality/perl/CalcForm.html.

2 Anne Collins's Web site, "Ideal weight for men," www.annecollins.com/weight-loss/ideal-weight-men.htm.

3 See Fantastic-Voyage.net.

4 The Ray & Terry Meal Replacement Shake is available at www.RayandTerry.com.

CHAPTER 18

1 See www.lloydwatts.com and the site of his company www.audience-inc.com. The tagline on the audience site is "We let machines hear."

2 The University of Texas cerebellum simulation included 10,000 granule cells, 900 Golgi cells, 500 mossy fiber cells, 20 Purkinje cells, and 6 nucleus cells.

J. L. Raymond et al. 1996. "The cerebellum: a neuronal learning machine?" *Science*. 272: 1126–1131; J. J. Kim and R. F. Thompson. 1997. "Cerebellar circuits and synaptic mechanisms involved in classical eyeblink conditioning." *Trends Neuroscience*. 20: 188–191; J. F. Medina et al. 2000. "Timing mechanisms in the cerebellum: testing predictions of a large-scale computer simulation." *Journal Neuroscience*. 20: 5516–5525; D. V. Buonomano and M. D. Mauk. 1994. "Neural network model of the cerebellum: temporal discrimination and the timing of motor responses." *Neural Computation*. 6: 38–55.

3 B. Fischl. 2000. "Measuring the thickness of the human cerebral cortex from magnetic resonance images." *Proc Natl Acad Sci USA*. Sep 26;97(20): 11050–11055.

4 G. Huang. "Mind-machine merger." *MIT Technology Review*, May 2003; www.technologyreview.com/articles/print_version/huang0503.asp.

5 Ibid. In 2004, a research team plans to have a monkey in St. Louis control a robot in Ann Arbor as it moves through an obstacle course. The monkey will watch the robot's movements on a screen, and the monkey's commands and feedback from the robot will be sent via the Internet. This is an early step toward remote human control of robots by thought alone. See K. Philipkoski. "Transforming thoughts into deeds." *Wired News*, January 14, 2004; www.wired.com/news/medtech/0,1286,61889,00.html; Reuters. "Monkey thinks, makes his moves." *Wired News*, October 13, 2003; www.wired.com/news/medtech/0,1286,60803,00.html.

6 The BrainBrowser Internet software is under development at Georgia State University. "When a user focuses his attention on a button, it becomes highlighted, and when the user successfully focuses on clicking the button, it emits a low tone." "Browser boosts brain interface." *MIT Technology Review*, May 22, 2003; www.technologyreview.com/articles/rnb_052203.asp?p=1; see also R. Brooks. "Toward a brain-Internet link." *MIT Technology Review*, November 2003; www.technology.review.com/articles/print_version/brooks1103.asp.

7 "Microchip promises smart artificial arms." *BBC News*, June 15, 2003; http://news.bbc.co.uk/2/hi/health/2975828.stm.

8 J. Hogan. "Synapse chip taps into brain chemistry." *NewScientist.com*, March 24, 2003; www.newscientist.com/news/news.jsp?id=ns99993523.

9 D. H. Hubel and T. N. Wiesel. 1965. "Binocular interaction in striate cortex of kittens reared with artificial squint." *Journal of Neurophysiology*. 28(6): 1041–1059.

10 M. A. Packer et al. 2003. "Nitric oxide negatively regulates mammalian adult neurogenesis." *Proc Natl Acad Sci USA*. Aug 5;100(16): 9566–9571.

11 C. Lie Dieter et al. "Neurogenesis in the Adult Brain: New Strategies for CNS Diseases." In *Annual Reviews of Pharmacology and Toxicology* (in press).

12 "Bat spit drug aids stroke victims." *BBC News*, February 6, 2004; http://news.bbc.co.uk/go/pr/fr/-/1/hi/health/3465419.stm.

13 L. Spinney. 2004. "Tea strainer in the neck 'stops strokes.'" *New Scientist*. 181(2432): 12.

14 "Neurologists create a font of human nerve cells." *Science Daily*, adapted from a University of Rochester news release, February 16, 2004; www.sciencedaily.com/print.php?url=/releases/2004/02/040216083710.htm.

15 S. Westphal. "Re-implanted stem cells tackle Parkinson's." *NewScientist.com*, April 8, 2002; www.newscientist.com/news/news.jsp?id=ns99992139.

16 G. Stix. 2003. "Ultimate self-improvement." *Sci Amer*. Sept: 44.

17 E. Jonietz. "7 hot projects." *MIT Technology Review*, December–January 2004; www.techologyreview.com/articles/print_version/jonietz1203.asp.

18 "Key advance reported in regenerating nerve fibers." *Science Daily*, based on news release from Children's Hospital Boston, February 18, 2004; www.sciencedaily.com/print.php?url-/releases/2004/02/040218075713.htm; D. Fischer, Z. He, and L. I. Benowtiz. 2004. "Counteracting the Nogo receptor enhances optic nerve regeneration if retinal ganglion cells are in an active growth state." *J Neurosci*. Feb 18;24(7): 1646–1651. See also, S. Seethaler. 2004. "Scientists discover new gene essential for the development of normal brain connections resulting from sensory input." UCSD news release; http://ucsdnews.ucsd.edu/newsreel/science/screst.asp.

19 R. Dotinga. "Cool new ways to save brains." *Wired News*, Feb. 10, 2004; www.wired.com/news/medtech/0,1286,62224,00.html.

20 One fRMI research group claims to be able to use fMRI to catch "a word or concept as it forms itself in the brain." Marcel Just at Carnegie Mellon University has run tests on volunteers using a small number of concepts. "We have 12 categories and can determine which of the 12 the subjects are thinking of with 80 to 90 percent accuracy." P. Ross. 2003. "Mind readers." *Sci Amer*. Sept: 77.

21 fMRI image of Ray Kurzweil's brain is courtesy of *Inc. Magazine*, "Your Brain on Innovation," by T. Singer, September 2002.

22 Buddhist monks have also shown changes in brain activity when they were asked to "induce a state of compassion in themselves." C. Newton. "Meditation and the brain." *MIT Technology Review*, February 2004; www.technologyreview.com/articles/print_version/newton0204.asp; T. Singer. "The Innovation Factor: Your Brain on Innovation." *Inc. Magazine*, September 2002.

23 R. E. Callaway and R. Yuste. (The Salk Institute for Biological Studies, Systems Neurobiology Laboratory). 2002. "Stimulating Neurons with Light," *Curr. Opin. Neurobiology*. October 1; 12(5): 587.

24 B. L. Sabatini and K. Svoboda. 2000. "Analysis of calcium channels in single spines using optical fluctuation analysis." *Nature*. 408: 589–593.

25 J. L. Etnier and D. M. Landers. 1995. "Brain Function and Exercise: Current Perspectives." *Sports Medicine*. 19(2): 81–85.

26 "Direct brain-to-brain communication and the transfer of minds between bodies are among the advances forecast in a recent report by the U.S. National Science Foundation and Department of Commerce." G. Brumfiel. 2002. "Futurists predict body swaps for planet hops." *Nature*. Jul 25;418: 359.

Deep brain stimulation, by which electric current from implanted electrodes influences brain function, is one possible neural implant. See A. Abbott. 2002. "Brain implants show promise against obsessive disorder." *Nature*. Oct 17;419: 658; B. Nuttin et al. 1999. "Electrical stimulation in anterior limbs of internal capsules in patients with obsessive-compulsive disorder." *Lancet*. Oct 30;354(9189): 1526.

27 R. S. Hong et al. 2003. "Dynamic range enhancement for cochlear implants." *Otol Neurotol*. Jul;24(4): 590–595; R. S. Tyler et al. 2002. "Three-month results with bilateral cochlear implants." *Ear Hear*. Feb;23(1 Suppl): 80S–89S.

28 See the Retinal Implant Project Web site (www.rle.mit.edu/retinaweb/), which contains a range of resources including recent papers. Here is one such recent paper from the team: R. J. Jensen et al. 2003. "Thresholds for activation of rabbit retinal ganglion cells with an ultrafine, extracellular microelectrode." *Invest Ophthalmal Vis Sci*. Aug;44(8): 3533–3543.

29 The FDA approved the Medtronic implant for this purpose in 1997 for only one side of the brain; it was approved for both sides of the brain on January 14, 2002. S. Snider. "FDA approves expanded use of brain implant for Parkinson's disease." U.S. Food and Drug Administration FDA Talk Paper, January 14, 2002; www.fda.gov/bbs/topics/ANSWERS/2002/ANS01130.html.

30 Medtronic also makes an implant for cerebral palsy. See S. Hart. "Brain implant quells tremors." *ABCNews.com*, December 23, 1997; http://more.abcnews.go.com/sections/living/brainstim1223. Also see the Medtronic site, www.medtronic.com.

31 This prosthesis, already 10 years in development, would perform the short-term memory tasks of the hippocampus rather than simply stimulating brain activity. To develop it, the developers had to "devise a mathematical model of how the hippocampus performs under all possible conditions, build that model into a silicon chip, and then interface the chip with the brain." D. Graham-Rowe. 2003. "The world's first brain prosthesis revealed." *NewScientist.com* 177(2386): 4; www.newscientist.com/news/news.jsp?id=ns99993488.

32 G. Zeck and P. Fromherz. 2001. "Noninvasive neuroelectronic interfacing with synaptically connected snail neurons immobilized on a semiconductor chip." *Proc Natl Acad Sci USA*. Aug 28;98(18): 10457–10462.

33 Quantum dots are nanosize crystals based on photosensitive semiconductor materials that detect photons or fluoresce (light up) in specific colors based on their size. See R. C. Johnson. "Scientists activate neurons with quantum dots." *EE Times*, December 6, 2001; www.eetimes.com/story/OEG20011204S0068.

34 M. George. 2003. "Stimulating the brain." *Sci Amer*. Sept: 67–73; F. M. Mottagy et al. 1999. "Facilitation of picture naming after repetitive transcranial magnetic stimulation." *Neurology*. 53: 1806–1812.

35 S. Pridmore et al. 2000. "Comparison of unlimited numbers of rapid transcranial magnetic stimulation (rTMS) and ECT treatment sessions in major depressive episode." *Int J Neuropsychopharmacol*. Jun; 3(2): 129–134; www.wireheading.com/rtms/.

36 E. Stockstad. 2001. "New hints into the biological basis of autism." *Science*. 294: 34–37.

37 D. J. Gerber et al. 2003. "Evidence for association of schizophrenia with genetic variation in the 8p21.3 gene, PPP3CC, encoding the calcineurin gamma subunit." *PNAS*. 100: 8993–8998.

38 J. S. Rhodes et al. 2003. "Exercise increases hippocampal neurogenesis to high levels but does not improve spatial learning in mice bred for increased voluntary wheel running." *Behav Neurosci*. Oct;117(5): 1006–1016.

39 B. Draganski et al. 2004. "Neuroplasticity: Changes in grey matter induced by training." *Nature*. Jan;427: 311–312.

40 D. F. Hultsch et al. 1999. "Use it or lose it: engaged lifestyle as a buffer of cognitive decline in aging?" *Psychol Agin*. Jun;14(2): 245–263.

41 E. P. Noble. 2000. "Addiction and its reward process through polymorphisms of the D2 dopamine receptor gene: a review." *Eur Psychiatry*. Mar; 15(2): 79–89.

42 F. H. Gage. 2003. "Brain: Repair yourself." *Sci Amer*. Sept: 46–53.

43 M. A. McDaneil. 2003. "'Brain-specific nutrients: a memory cure?" *Nutrition*. Nov–Dec; (11–12): 957–75.

44 J. Polich. 2001. "Cognitive effects of a ginkgo biloba/vinpocetine compound in normal adults: systematic assessment of perception, attention and memory." *Hum Psychopharmacol*. Jul;16(5): 409–416.

45 P. Bönöczk et al. 2000. "Role of sodium channel inhibition in neuroprotection: effect of vinpocetine." *Brain Res. Bull*. 53(3): 245–254; S. A. Erdo et al. 1996. "Vincamine and vincanol are potent blockers of voltage-gated Na channels." *Eur J Pharmacol*. Oct 24;314(1–2): 69–73; R. Balestreri, L. Fontana, and F. Astengo. 1987. "A double-blind placebo controlled evaluation of the safety and efficacy of vinpocetine in the treatment of patients with chronic vascular senile cerebral dysfunction." *J Am Geriatr Soc*. May;35(5): 425–430.

46 M. Furushiro et al. 1997. "Effects of administration of soybean lecithin transphosphatidylated phosphatidylserine on impaired learning of passive avoidance in mice." *Jpn J Pharmacol*. Dec;75(4): 447–450.

47 E. H. Sharman et al. 2002. "Reversal of biochemical and behavioral parameters of brain aging by melatonin and acetyl L-carnitine." *Brain Res*. Dec 13;957(2): 223–230.

48 "Researchers at the UCLA Neuropsychiatric Institute found significant improvement in verbal recall among a group of people with age-associated memory impairment who took the herbal supplement ginkgo biloba for six months when compared with a group that received a placebo." R. Champeau. "UCLA researchers find ginkgo biloba may help improve memory." UCLA news release, November 10, 2003; www.eurekalert.org/pub_releases/2003-11/uoc--urf111003.php. See also, R. W. Stackman et al. 2003. "Prevention of age-related spatial memory deficits in a transgenic mouse model of Alzheimer's disease by chronic ginkgo biloba treatment." *Exp Neurol*. Nov;184(1): 510–520.

49 J. M. Bourre et al. 1991. "Essentiality of n-3 fatty acids for brain structure and function." *World Rev Nutr Diet*. 66: 103–117.

50 D. S. Heron et al. 1980. "Lipid fluidity markedly modulates the binding of serotonin to mouse brain membranes." *Proc Natl Acad Sci*. 77: 7463–7467.

51 A. P. Simopoulos. 2001. "Evolutionary aspects of diet and essential fatty acids." *World Rev Nutr Diet*. 88: 18–27; A. P. Simopoulos, A. Leaf, and N. Salem. 1999. "Workshop on the essentiality of and recommended dietary intakes for omega-6 and omega-3 fatty acids." *J Am Coll Nutr*. 18: 487–489.

52 There is not universal agreement about the optimal ratio of omega-6 to omega-3 fatty acid consumption. Dr. Yehuda et al. from Israel feels that a 4:1 ratio is optimal. See S. Yehuda et al. 2002. "The role of polyunsaturated fatty acids in restoring the aging neuronal membrane." *Neurobiol Aging*. Sep–Oct;23(5): 843–853.

53 E. Nemets et al. 2002. "Addition of omega-3 fatty acid to maintenance medication treatment for recurrent unipolar depressive disorder." *Am J Psychiatry*. 159: 477–479; L. B. Marangell et al. 2003. "A double-blind, placebo-controlled study of the omega-3 fatty acid docosahexaneoic acid in the treatment of major depression." *Am J Psychiatry*. 160: 996–998.

54 S. Y. Chung et al. 1995. "Administration of phosphatidylcholine increases brain acetylcholine concentration and improves memory in dementia mice." *J Nutr*. Jun;125(6): 1484–1489.

55 S. L. Ladd et al. 1993. "Effect of phosphatidylcholine on explicit memory." *Clin Neuropharmacol*. Dec;16(6): 540–549.

56 "Memory is a biological process that can be manipulated by modern biology like anything else. Not only can you disrupt it, you can improve it," according to Timothy Tully of Helicon Therapeutics. Another founder of memory research, Nobel laureate Eric Kandel, has explored memory using a marine snail model. Some of the nerve cells in the Aplysia slug are "big enough to be seen with the naked eye." R. Langreth. "Viagra for the brain." *Forbes.com*, February 4, 2002; www.forbes.com/forbes/2002/0202/046_print.html.

57 F. Fagnani et al. 2004. "Donepezil for the treatment of mild to moderate Alzheimer's disease in France: the economic implications." *Dement Geriatr Cogn Disord*. 17(1–2): 5–13; K. R. Krishnan et al. 2003. "Randomized, placebo-controlled trial of the effects of donepezil on neuronal markers and hippocampal volumes in Alzheimer's disease." *Am J Psychiatry*. Nov;160(11): 2003–2011.

CHAPTER 19

1 N. Carvalhaes-Neto et al. 2002. "Urinary free cortisol is similar in older and younger women." *Exp Aging Res*. Apr–Jun;28(2): 163–168; E. Beale et al. 2002. "Changes in serum cortisol with age in critically ill patients." *Gerontology*. Mar–Apr;48(2): 84–92.

2 R. Sapolsky. 1998. *Why Zebras Don't Get Ulcers*. New York: W. H. Freeman.

3 D. S. Khalsa and C. Stauth. 1997. *Brain Longevity*. New York: Warner Books.

4 E. Ferrari et al. 2001. "Age-related changes of the adrenal secretory pattern: possible role in pathological brain aging." *Brain Res Rev*. Nov;37(1–3): 294–300; E. Ferrari et al. 2001. "Age-related changes of the hypothalamic-pituitary-adrenal axis: pathophysiological correlates." *Eur J Endocrinol*. Apr;144(4): 319–329.

5 For information on this test, see Fantastic-Voyage.net.

6 According to one study, "The physiological role of dehydroepiandrosterone (DHEA) and its sulphated ester DHEA(S) has been studied for nearly 2 decades and still eludes final clarification." The authors also suggest that the availability of the supplement is "hampering the rigorous scientific evaluation of its potential." M. Racchi, C. Balduzzi, and E. Corsini. 2003. "Dehydroepiandrosterone (DHEA) and the aging brain: flipping a coin in the 'fountain of youth.'" *CNS Drug Rev* Spring;9(1): 21–40. See also M. Boudarene and J. J. Legros. 2002. "Study of the stress response: role of anxiety, cortisol and DHEAs." *Encephale*. Mar–Apr;28(2): 139–146.

7 S. Shibata. 2000. "A drug over the millennia: pharmacognosy, chemistry, and pharmacology of licorice." *Yakugaku Zasshi*. Oct;120(10): 849–862.

8 B. Singh et al. 2001. "Adaptogenic activity of a novel, withanolide-free aqueous fraction from the roots of Withania somnifera Dun." *Phytother Res*. Jun;15(4): 311–318.

9 Barry Sears has long been preaching the gospel of the hazards of cortisol and insulin. See *The Anti-Aging Zone*. New York: HarperCollins, 1999, p. 138.

10 www.sciam.com/article.cfm?chanID=sa003&articleID=000C601F-8711-1F99-86FB83414B7F0156.

11 P. Zimmet and S. Baba. 1990. "Central obesity, glucose intolerance and other cardiovascular disease risk factors: an old syndrome rediscovered." *Diabetes Res Clin Pract*. 10(Suppl 1): S167–171.

12 J. Nandi et al. 2002. "Central mechanisms involved with catabolism." *Curr Opin Clin Nutr Metab Care*. Jul;5(4): 407–418.

13 According to one study, "Insulin resistance is an important risk factor for type 2 diabetes and coronary heart disease. Our results suggest that genetic factors, intrauterine environment, early childhood, and adult environmental factors are all relevant in determining adult insulin resistance." D. A. Lawlor, G. Smith, and S. Ebrahim. 2003. "Life course influences on insulin resistance: findings from the British Women's Heart and Health Study." *Diabetes Care*. Jan;26(1): 97–103.

See also, J. P. Despres et al. 1996. "Hyperinsulinemia as an independent risk factor for ischemic heart disease." *N Engl J Med*. Apr 11;334(15): 952–957.

14 W. Regelson and C. Colman. 1996. *The Super Hormone Promise*. New York: Simon and Schuster.

15 W. Leowattana. 2001. "DHEA(S): the fountain of youth." *J Med Assoc Thai*. Oct;84(Suppl 2): S605–612.

16 J. A. Lemon, D. R. Boreham, and C. D. Rollo. 2003. "A dietary supplement abolishes age-related cognitive decline in transgenic mice expressing elevated free radical processes." *Exp Biol Med*. 228: 800–810; R. N. Butler et al. 2002. "Is there an anti-aging medicine?" Journals of Gerontology Series A: *Biol Sci Med Sci* 57: B333–B338.

17 E. Barrett-Connor, K. T. Khaw, and S. S. Yen. 1986. "A prospective study of dehydroepiandrosterone sulfate, mortality, and cardiovascular disease." *N Engl J Med*. Dec 11;315(24): 1519–1524.

18 R. H. Straub et al. 2002. "Dehydroepiandrosterone in relation to other adrenal hormones during an acute inflammatory stressful disease state compared with chronic inflammatory disease: role of interleukin-6 and tumour necrosis factor." *Eur J Endocrinol*. Mar;146(3): 365–374.

19 W. Regelson. 1985. "Vitamin A, dehydroepiandrosterone (DHEA) and 5' nucleotidase: regulatory factors in tumor growth." *Cancer Invest*. 3(4): 407–409.

20 A. H. Young, P. Gallagher, and R. J. Porter. 2002. "Elevation of the cortisol-dehydroepiandrosterone ratio in drug-free depressed patients." *Am J Psychiatry*. Jul;159(7): 1237–1239.

21 According to one study, "Increases in mental and physical sexual arousal ratings significantly increased in response to an acute dose of DHEA in postmenopausal women." L. Hackbert and J. R. Heiman. 2002. "Acute dehydroepiandrosterone (DHEA) effects on sexual arousal in postmenopausal women." *J Women's Health Gend Based Med*. Mar;11(2): 155–162.

See also R. F. Spark. 2002. "Dehydroepiandrosterone: a springboard hormone for female sexuality." *Fertil Steril*. Apr;77(Suppl 4): 19–25.

22 W. Leowattana, op cit.

23 D. Rudman et al. 1990. "Effects of human growth hormone in men over 60 years old." *N Engl J Med.* Jul 5;323(1): 1–6.

24 At www.ncbi.nlm.nih.gov, search the PubMed database for "growth hormone" to see these results.

25 A. Vermeulin. 2002. "Aging, hormones, body composition, metabolic effects." *World J Urol.* May;20(1): 23–27.

26 R. D. Murray et al. 2002. "Low-dose GH replacement improves the adverse lipid profile associated with the adult GH deficiency syndrome." *Clin Endocrinol (Oxf).* Apr;56(4): 525–532.

27 A. M. Ahmad et al. 2002. "Effects of GH replacement on 24-h ambulatory blood pressure and its circadian rhythm in adult GH deficiency." *Clin Endocrinol (Oxf).* Apr;56(4): 431–437.

28 J. Svensson et al. 2002. "Effects of seven years of GH-replacement therapy on insulin sensitivity in GH-deficient adults." *J Clin Endocrinol Metab.* May;87(5): 2121–2127.

29 D. E. Cummings and G. R. Merriam. 1999. "Age-related changes in growth hormone secretion: should the somatopause be treated?" *Semin Reprod Endocrinol.* 17(4): 311–325.

Research is also being conducted on GH treatment in adults worldwide. See, for example, Y. B. Sverrisdottir et al. 2003. "The effect of growth hormone (GH) replacement therapy on sympathetic nerve hyperactivity in hypopituitary adults: a double-blind, placebo-controlled, crossover, short-term trial followed by long-term open GH replacement in hypopituitary adults." *J Hypertens.* Oct;21(10): 1905–1914.

30 Other points to consider: this study utilized the "high-dose low-frequency" GH injection protocol. Most anti-aging physicians recommend "low-dose high-frequency" GH therapy to reduce side effects. Also, artificial, not bio-identical forms of estrogen, progestin, and testosterone were used. See M. R. Blackman et al. 2002. "Growth hormone and sex steroid administration in healthy aged women and men." *JAMA.* 288: 2282–2292.

31 M. Shim and P. Cohen. 1999. "IGFs and human cancer: implications regarding the risk of growth hormone therapy." *Horm Res.* 51(Suppl 3): 42–51.

32 A. Beentjes et al. 2000. "One year growth hormone replacement therapy does not alter colonic epithelial cell proliferation in growth hormone deficient adults." *Clin Endocrinol (Oxf).* Apr;52(4): 457–462 and M. Letsch et al. 2003. "Growth hormone-releasing hormone (GHRH) antagonists inhibit the proliferation of androgen-dependent and -independent prostate cancers." *Proc Natl Acad Sci USA.* Feb 4;100(3): 1250–1255; M. H. Torosian. 1993. "Growth hormone and prostate cancer growth and metastasis in tumor-bearing animals." *J Pediatr Endocrinol.* Jan–Mar;6(1): 93–97.

33 J. R. Stout. 2002. "Amino acids and growth hormone manipulation." *Nutrition.* July–Aug;18(7–8): 683–684; R. Savine and P. H. Sonksen. 1999. "Is the somatopause an indication for growth hormone replacement?" *J Endocrinol Invest.* 22(5 Suppl): 142–149.

34 A. J. Morales et al. 1998. "The effect of six months treatment with a 100 mg daily dose of dehydroepiandrosterone (DHEA) on circulating sex steroids, body composition and muscle strength in age-advanced men and women." *Clin Endocrinol (Oxf).* Oct;49(4): 421–432.

35 D. Leger et al. 2004. "Nocturnal 6-sulfatoxymelatonin excretion in insomnia and its relation to the response to melatonin replacement therapy." *Am J Med.* Jan 15;116: 91–95.

36 Studies of women who work night shifts have been used to explore the role of melatonin. The implications of the results, which showed an increase in risk, "'extend beyond women who work at night' . . . 'Women in developing countries have one-fifth the risk of breast cancer compared to women in industrialized nations' . . . It is possible that exposure to more light at night, a common phenomenon in industrialized nations, may account for increased cancer risk in women. 'This has implications that are independent of shift work.'" M. T. Willis. 2001. "Light at night," *ABCNews.com* (abcnews.go.com/sections/living/DailyNews/breastcancer011016.html). See E. Schernhammer et al. 2003. "Night-shift work and risk of colorectal cancer in the Nurses' Health Study." *JNCI.* Jun 4:95(11): 825–828.

See also Y. Touitou. 2001. "Human aging and melatonin. Clinical relevance." *Exp Gerontol.* Jul;36(7): 1083–1100; F. Fraschini et al. 1998. "Melatonin involvement in immunity and cancer." *Biol Signals.* Jan;7(1): 61–72.

CHAPTER 20

1 It is interesting that in the rush to "level the playing field" in the name of sexual equality, many physicians seem to have lost sight of the fact that there are, nevertheless, major physical differences between men and women. Recent research suggests that gender must be taken into account when many types of drugs and other therapies are used. There are significant differences between men and women in the function of their brains, hearts, lungs, and immune and digestive systems, in addition to their obvious differences in reproductive systems. For more information, see M. J. Legato. 2002. *Eve's Rib: The New Science of Gender-Specific Medicine and How It Can Save Your Life.* New York: Harmony Books.

2 BERT isn't really new, but has been around for decades. It is now becoming more widely known, thanks to popular books such as Suzanne Somers's *The Sexy Years.* 2004. New York: Crown.

3 Interestingly, the makers of Premarin seem proud of this fact and promote Premarin as containing "estrogens obtained exclusively from natural sources," even though these natural sources are pregnant mare urine.

4 V. W. Pinn et al. 2002. NIH Research and Other Efforts Related to the Menopausal Transition. April 22. Bethesda, Maryland: Office of Research on Women's Health/National Institutes of Health (www4.od.nih.gov/orwh/MenopauseRpt4-02.pdf).

5 A. Vashisht et al. 2001. "Prevalence of and satisfaction with complementary therapies and hormone replacement therapy in a specialist menopause clinic." *Climacteric.* Sep;4(3): 250–256. See also S. L. Nand et al. 1998. "Menopausal symptom control and side-effects on continuous estrone sulfate and three doses of medroxyprogesterone acetate. Ogen/Provera Study Group." *Climacteric.* Sep;1(3): 211–218.

6 The lack of consensus is not due to lack of effort. Several trials are currently under way or have recently concluded, which attempt to provide better guidance for menopausal women. The main studies include the Women's Health Initiative (WHI), the Postmenopausal Estrogen/Progestin Intervention (PEPI) Trial, the Heart and Estrogen-Progestin Replacement Study (HERS), the Women's International Study of long Duration Oestrogen after Menopause (WISDOM), and the Million Women Study.

7 For an interesting discussion of the types of problems faced by today's clinicians in deciding what types of women and what risk factors are amenable to HRT, see J. E. Manson and K. A. Martin. 2001. "Clinical practice. Postmenopausal hormone-replacement therapy." *N Engl J Med.* Jul 5;345(1): 34–40.

8 J. A. Cauley et al. 2001. "Effects of hormone replacement therapy on clinical fractures and height loss: The Heart and Estrogen/Progestin Replacement Study (HERS)." *Am J Med.* Apr 15;110(6): 442–450.

9 M. P. Warren et al. 2003. "Persistent osteopenia in ballet dancers with amenorrhea and delayed menarche despite hormone therapy: A longitudinal study." *Fertil Steril.* Aug;80: 398–404.

10 J. A. Cauley et al. 2003. "Effects of estrogen plus progestin on risk of fracture and bone mineral density: the Women's Health Initiative randomized trial." *JAMA.* Oct 1;290(13): 1729–1738.

11 C. T. Owens. 2002. "Estrogen replacement therapy for Alzheimer disease in postmenopausal women." *Ann Pharmacother.* Jul;36(7): 1273–1276 and K. Yaffe. 1998. "Estrogen therapy in postmenopausal women: effects on cognitive function and dementia." *JAMA.* Mar 4;279(9): 688–695.

12 Conclusions from the large Heart and Estrogen/Progestin Replacement Study suggest that women at risk of heart disease suffer an even greater cardiac risk by taking HRT. For low-risk women, "There is a risk that women without coronary heart disease might experience even greater net harm from HRT." See J. A. Blakely. 2000. "The heart and estrogen/progestin replacement study revisited: hormone replacement therapy produced net harm, consistent with the observational data." *Arch Intern Med.* Oct 23;160(19): 2897–2900. Also see R. SoRelle. 2002. "Second year of HERS same as the first—no clear benefit or harm for cardiovascular disease." *Circulation.* Feb 26;105(8): e9077–9078; T. W. Meade and M. R. Vickers. 1999. "HRT and cardiovascular disease." *J Epidemiol Biostat.* 4(3): 165–190.

13 G. A. Colditz et al. 1995." The use of estrogens and progestins and the risk of breast cancer in postmenopausal women." *N Engl J Med.* Jun 15;332(24): 1589–1593.

14 C. Rodriguez et al. 1995. "Estrogen replacement therapy and fatal ovarian cancer." *Am J Epidemiol.* May 1;141(9): 828–835.

15 Writing Group for the Women's Health Initiative Investigators. 2002. "Risks and benefits of estrogen plus progestin in healthy postmenopausal women: principal results from the Women's Health Initiative randomized controlled trial." *JAMA.* 288: 321–333.

16 Data taken from the American College Of OB/GYN Web site ("Questions and Answers on Hormone Therapy," www.acog.org/from_home/publications/press_releases/nr08-30-02.cfm). All the data were not bad for women who took Prempro, however. A 37 percent decrease in colon cancer and a 24–34 percent decrease in bone fractures were seen as well.

17 A. L. Hersh, M. L. Stefanick, and R. S. Stafford. 2004. "National use of postmenopausal hormone therapy: annual trends and response to recent evidence." *JAMA.* Jan 7; 291(1): 47–53; J. S. Haas et al. 2004." Changes in the use of postmenopausal hormone therapy after the publication of clinical trial results." *Ann Intern Med.* Feb 3;140(3): 184–188.

18 J. T. Hargrove et al. 1989. "Menopausal hormone replacement therapy with continuous daily oral micronized estradiol and progesterone." *Obstet Gynecol.* Apr;73(4): 606–612.

19 A. DuPont et al. 1991. "Comparative endocrinological and clinical effects of percutaneous estradiol and oral conjugated estrogens as replacement therapy in menopausal women." *Maturitas.* Oct;13(4): 297–311.

20 K. M. Prestwood et al. 2000. "The effect of low dose micronized 17ss-estradiol on bone turnover, sex hormone levels, and side effects in older women: a randomized, double-blind, placebo-controlled study." *J Clin Endocrinol Metab.* Dec;85(12): 4462–4469. See also B. Ettinger et al. 1992. "Low-dosage micronized 17 beta-estradiol prevents bone loss in postmenopausal women." *Am J Obstet Gynecol.* Feb;166(2): 479–88.

21 M. C. Snabes et al. 1997. "Physiologic estradiol replacement therapy and cardiac structure and function in normal postmenopausal women: A randomized, double-blind, placebo-controlled crossover trial." *Obstet Gynecol.* 89: 332–339; G. M. C. Rosano et al. 1993. "Beneficial effect of oestrogen on exercise induced myocardial ischemia in women with coronary artery disease." *Lancet.* 342: 133–136; C. Haines et al. 1996. "Effect of oral estradiol on Lp(a) and other lipoproteins in postmenopausal women. A randomized, double-blind, placebo-controlled crossover study." *Arch Intern Med.* 156: 886–872.

22 H. N. Hodis et al. 2003. "Hormone therapy and the progression of coronary-artery atherosclerosis in postmenopausal women." *N Engl J Med.* Aug 7;349(6): 535–545.

23 E. N. Meilahn et al. 1998. "Do urinary oestrogen metabolites predict breast cancer? Guernsey III cohort follow-up." *Br J Cancer.* Nov;78(9): 1250–1255.

24 The role of estrogen in protecting brain function is still not understood, but it is clearly important for men as well as for women. V. Bisagno, R. Bowman, and V. Luine. 2003. "Functional aspects of estrogen neuroprotection." *Endocrine.* Jun 1;21(1): 33–41; D. F. Swaab et al. 2003. "Sex differences in the hypothalamus in the different stages of human life." *Neurobiol Aging.* May 1;24(Suppl 1): S1–S16, discussion, S17–S19.

25 A. Kamada et al. 2004. "A new series of estrogen receptor modulators: effect of alkyl substituents on receptor-binding affinity." *Chem Pharm Bull (Tokyo).* Jan;52: 79–88.

26 D. Yin et al. 2003. "Pharmacodynamics of selective androgen receptor modulators." *J Pharmacol Exp Ther.* Mar;304(3): 1334–1340.

27 A. Vincent and L. A. Fitzpatrick. 2000. "Soy isoflavones: are they useful in menopause?" *Mayo Clin Proc.* Nov;75(11): 1174–1184.

28 T. Horiuchi et al. 2000. "Effect of soy protein on bone metabolism in postmenopausal Japanese women." *Osteoporos Int.* 11(8): 721–724; Y. Somekawa et al. 2001. "Soy intake related to menopausal symptoms, serum lipids, and bone mineral density in postmenopausal Japanese women." *Obstet Gynecol.* Jan;97(1): 109–115.

29 One randomized crossover study examined the effects of three different soy diets on 18 healthy postmenopausal women. "When compared with baseline values, consumption of all three soy diets . . . decreased the ratio of genotoxic:total estrogens. These data suggest that both isoflavones and other soy constituents may exert cancer-preventative effects in postmenopausal women by altering estrogen metabolism away from genotoxic metabolites toward inactive metabolites." X. Xu et al. 2000. "Soy consumption alters endogenous estrogen metabolism in postmenopausal women." *Cancer Epidemiol Biomarkers Prev.* Aug;9: 781–786. See also L. A. Fitzpatrick. 2003. "Soy isoflavones: hope or hype?" *Maturitas.* March&14;44(Suppl 1): S21–29; T. Kishida et al. 2000. "Effect of dietary soy isoflavone aglycones on the urinary 16alpha-to-2-hydroxyestrone ratio in C3H/HeJ mice." *Nutr Cancer.* 38(2): 209–214.

30 U.S. Soyfoods Directory; www.soyfoods.com/nutrition/isoflavoneconcentration.html.

31 F. Kronenberg and A. Fugh-Berman. 2002. "Complementary and alternative medicine for menopausal symptoms: a review of randomized, controlled trials." *Ann Intern Med.* Nov 19;137(10): 805–813.

32 In the mid-1990s, researchers at Johns Hopkins established a link in animal studies between chemicals found in broccoli and protection against cancer. Since then, a number of studies have looked at the beneficial effects for people of cruciferous vegetables.

One study focused on why Polish women in the Midwest are more likely to develop breast cancer than their relatives in Europe. The answer may be in the cabbage, which is a cruciferous vegetable that the European Poles eat. Researchers from the University of Illinois at Urbana–Champaign "stimulated test-tube colonies of human breast-cancer cells with estrogen, then added extracts of plain cabbage, sauerkraut, or acidified brussels sprouts." At higher concentrations, each extract "not only slowed the growth of estrogen-fed cells but also blocked estrogen's ability to turn on a particular gene." Though the study's findings do not point conclusively to the agent in the vegetables that is at work, the findings do suggest "these foods might offer even more 'potentially important' agents and point toward a new class of drugs to reduce cancer risk." J. Raloff. 2001. "Fighting cancer from the cabbage patch." *Science News.* Mar 3;159(9); www.sciencenews.org/20010303/food.asp. For the study itself, see Y. H. Ju et al. 2000. "Estrogenic effects of extracts from cabbage, fermented cabbage, and acidified brussels sprouts on growth and gene expression of estrogen-dependent human breast cancer (MCF-7) cells." *J Agricultural Food Chem.* Oct;48: 4628.

See also, G. Murillo and R. G. Mehta. 2001. "Cruciferous vegetables and cancer prevention." *Nutr Cancer.* 41(1–2): 17–28; J. H. Fowke, C. Longcope, and J. R. Hebert. 2000. "Brassica vegetable consumption shifts estrogen metabolism in healthy postmenopausal women." *Cancer Epidemiol Biomarkers Prev.* Aug;9(8): 773–779.

33 M. S. Brignall. 2001. "Prevention and treatment of cancer with indole-3-carbinol." *Altern Med Rev.* Dec;6(6): 580–589.

34 J. R. Lee. 1999. *What Your Doctor May Not Tell You About Premenopause.* New York: Warner Books.

35 H. B. Leonetti et al. 1999. "Transdermal progesterone cream for vasomotor symptoms and postmenopausal bone loss." *Obstet Gynecol.* Aug;94(2): 225–228.

36 Ibid. See also B. G. Wren et al. 2003. "Transdermal progesterone and its effect on vasomotor symptoms, blood lipid levels, bone metabolic markers, moods, and quality of life for postmenopausal women." *Menopause.* Jan–Feb;10(1): 13–18.

37 "p53, a tumor suppressor gene, is a target of genetic alternations in many human and animal cancers. Compared to normal tissues, cancer tissues overexpress mutant p53 protein thus allowing their detection by a number of immunochemical procedures." S. Haga et al. 2001. "Overexpression of the p53 gene product in canine mammary tumors." *Oncol Rep.* Nov 1;8(6): 1215–1219. See also G. R. Sahu et al. 2002. "Rearrangement of p53 gene with overexpressed p53 protein in primary cervical cancer." *Oncol Rep.* March 1;9(2): 433–437; V. K. Moudgil et al. 2001. "Hormonal regulation of tumor suppressor proteins in breast cancer cells." *J Steroid Biochem Mol Biol.* Jan–Mar;76(1–5): 105–117.

38 K. J. Chang et al. 1995. "Influences of percutaneous administration of estradiol and progesterone on human breast epithelial cell cycle in vivo." *Fertil Steril.* Apr;63(4): 785–791.

39 S. Shantha et al. 2002. "Natural vaginal progesterone is associated with minimal psychological side effects: a preliminary study." *J Women's Health Gend Based Med.* Dec;10(10): 991–997 and S.Ferrero et al. 2002. "Vaginal micronized progesterone in continuous hormone replacement therapy. A prospective randomized study." *Minerva Gynecol.* Dec;54(6): 519–530.

40 E. Darj et al. 1993. "Liver metabolism during treatment with estradiol and natural progesterone." *Gynecol Endocrinol.* Jun;7(2): 111–114.

41 R. D. Langer. 1999. "Micronized progesterone: a new therapeutic option." *Int J Fertil Women's Med.* Mar–Apr;44(2): 67–73.

42 J. L. Shifren et al. 2000. "Transdermal testosterone treatment in women with impaired sexual function after oophorectomy." *N Engl J Med.* Sep 7;343(10): 682–688.

43 E. Wespes and C. C. Schulman. 2002. "Male andropause: myth, reality, and treatment." *Int J Impot Res.* Feb;14(Suppl 1): S93–S98.

44 J. P. Heaton and A. Morales. 2001. "Andropause—a multisystem disease." *Can J Urol.* Apr;8(2): 1213–1222.

45 Shortly after the introduction of Proscar, Merck and Company introduced Propecia, which is identical to Proscar, only in a 1-mg rather than a 5-mg strength, as a treatment for male pattern baldness.

46 F. Debruyne et al. 2002. "Comparison of a phytotherapeutic agent (Permixon) with an alpha-blocker (tamsulosin) in the treatment of benign prostatic hyperplasia: a 1-year randomized international study." *Eur Urol.* May;41(5): 497–507; G. Campault et al. 1984. "A double-blind trial of an extract of the plant seronoa repens in benign prostatic hyperplasia." *Br. J. Clin. Pharm.* 18: 461.

47 G. S. Gerber. 2000. "Saw palmetto for the treatment of men with lower urinary tract symptoms." *J Urol.* May;163(5): 1408–1412.

48 L. B. Nieuwoudt et al. 1990. "Correlation between the macromolecular effects of estradiol and catecholestradiols and the total prostatic catecholestrogen concentration." *Clin Physiol Biochem.* 8(5): 231–237.

49 A. M. Nakhla et al. 1994. "Estradiol Causes the Rapid Accumulation of cAMP in Human Prostate." *Proc Natl Acad Sci USA.* June 7; 91 (12): 5402–5405.

50 B. T. Ashok et al. 2001. "Abrogation of estrogen-mediated cellular and biochemical effects by indole-3-carbinol." *Nutr Cancer.* 41(1–2): 180–187; J. J. Michnovicz, H. Adlercreutz, and H. L. Bradlow. 1997. "Changes in levels of urinary estrogen metabolites after oral indole-3-carbinol treatment in humans." *J Natl Cancer Inst.* May 21;89(10): 718–723.

51 K. J. Auborn et al. 2003. "Indole-3-carbinol is a negative regulator of estrogen." *J Nutr.* 133(7 Suppl): 2470S–2475S.

52 H. J. Jeong et al. 1999. "Inhibition of aromatase activity by flavonoids." *Arch Pharm Res.* Jun;22(3): 309–312.

53 P. Taxel et al. 2001. "The effect of aromatase inhibition on sex steroids, gonadotropins, and markers of bone turnover in older men." *J Clin Endocrinol Metab.* Jun;86(6): 2869–2874.

54 J. S. Bland. 2002. *Nutritional Endocrinology: Breakthrough Approaches for Improving Adrenal and Thyroid Function.* Gig Harbor, Washington: Metagenics Educational Programs, pp. 141–142.

55 A. T. Guay et al. 2003. "Clomiphene increases free testosterone levels in men with both secondary hypogonadism and erectile dysfunction: who does and does not benefit?" *Int J Impot Res.* Jun;15(3): 156–165.

56 As one study concluded, "epidemiological studies provide no clues that the levels of circulating androgen are correlated with or predict prostate disease. Similarly, androgen replacement studies in men do not suggest that these men suffer in a higher degree from prostate disease than control subjects. It seems a defensible practice to treat aging men with androgens if and when they are testosterone-deficient, but long-term studies including sufficient numbers of men are needed." L. Gooren. 2003. "Androgen deficiency in the aging male: benefits and risks of androgen supplementation." *J Steroid Biochem Mol Biol*. Jun;85(2–5): 349–355. See also A. Morales. 2002. "Androgen replacement therapy and prostate safety." *Eur Urol*. Feb;41(2): 113–120.

57 J. E. Morley. 2003. "The need for a men's health initiative." *J Gerontol A Biol Sci Med Sci*. 58: 614–617.

CHAPTER 21

1 B. N. Ames and P. Wakimoto. 2002. "Are vitamin and mineral deficiencies a major cancer risk?" *Nat Rev Cancer*. Sep;2(9): 694–704.

2 J. E. DaVanzo et al. 2003. "A study of the cost effects of daily multivitamins for older adults." The Lewin Group, Inc., Oct 8.

3 T. S. Church et al. 2003. "Reduction of C-Reactive Protein Levels Through Use of a Multivitamin." *Am J Med*. Dec 15;115(9): 702–707.

4 D. Salisbury. 2004. "Chemists develop antioxidants 100 times more effective than vitamin E." *Exploration*, Jan. 16; http://exploration.vanderbilt.edu/news/news_antioxidant.htm.

5 K. Dean. "Breathing new life into medicine." *Wired News*, July 16, 2003; www.wired.com/news/print/0,1294,59635,00.html.

6 C. Lok. "Smarter drugs." *MIT Technology Review*, March 2004; www.technologyreview.com/articles/print_version/launchpad0304.asp.

7 K. Philipkoski. "Souped-up rice goes against the grain." *Wired News*, June 5, 2003; www.wired.com/news/medtech/0,1286,59117,00.html.

8 D. Shintani and D. DellaPenna. 1998. "Elevating the Vitamin E Content of Plants Through Metabolic Engineering." *Science*. Dec 11;282: 5396.

9 I. Ajjawi and D. Shintani. 2004. "Engineered plants with elevated vitamin E: a nutraceutical success story." *Trends in Biotechnology*. Mar;22(3): 104–107.

10 K. Kleiner. "Biotech researchers create safer soybeans." *NewScientist.com*, September 2002; www.newscientist.com/news/news.jsp?id=ns99992782.

11 AP News Service. "Scientists foresee genetically engineered healthier steak." February 2, 2004; www.usatoday.com/news/health/2004-02-04-healthy-steak_x.htm; S. M. Kitessa et al. 2004. "Supplementation of grazing dairy cows with rumen-protected tuna oil enriches milk fat with n-3 fatty acids without affecting milk production or sensory characteristics." *Br J Nutr*. Feb;91(2): 271–278.

12 B. Demmig-Adams and W. Adams III. 2002. "Antioxidants in photosynthesis and human nutrition." *Science*. Dec 13;298: 2149–2153; I. Raskin et al. 2002. "Plants and human health in the twenty-first century." *Trends in Biotechnology*. 20(12): 522–531.

13 Ibid, p. 2153.

14 The antioxidant enzymes don't have to commit suicide (hara-kiri) when they give up one of their electrons; rather, they can work together as a team. For example, when a molecule of vitamin E gives up an electron to quench a free radical, a molecule of vitamin C often comes along and gives up one of its electrons to restore the vitamin E molecule, and then a glutathione molecule gives one of its electrons to vitamin E.

15 Swiss Institute of Bioinformatics. *ENZYME*. Enzyme nomenclature database. Release 27.0, October 2001, updates up to 1 Feb 2002; www.expasy.ch/enzyme (accessed February 2, 2002).

16 This seminal article by Bruce Ames provides a scientific basis for the use of high-dose vitamin therapy in the treatment of numerous diseases. B. N. Ames, I. Elson-Schwab, and E. A. Silver. 2002. "High-dose vitamin therapy stimulates variant enzymes with decreased coenzyme binding affinity (increased K(m)): relevance to genetic disease and polymorphisms." *Am J Clin Nutr*. Apr;75(4): 616–658.

17 Ibid., p. 1.

18 Much of our DNA does not encode any proteins, so this number of mutations is not quite as worrisome as it first appears. Cells can repair some of the changes to DNA; changes that cannot be repaired are called mutations. Mutations are particularly common when cells divide because the DNA must be replicated at that time. See T. Beardsley. "Mutations galore: humans have high mutation rates. But why worry?" *Sci Am*. April 1999; www.sciam.com/article.cfm?articleID=0004AC33-68BB-1C71-9EB7809EC588F2D7.

"Repeated sequences that do not code for proteins ('junk DNA') make up approximately 98 percent of the human genome. This so-called junk DNA is not junk as it plays a critical role in gene expression. Repetitive sequences shed light on chromosome structure and dynamics. Over time, these repeats reshape the genome by rearranging it, thereby creating entirely new genes or modifying and reshuffling existing genes . . . "

"Humans share most of the same protein families with worms, flies, and plants, but the number of gene family members has expanded in humans, especially in proteins involved in development and immunity. The human genome has a much greater portion (50%) of repeat sequences than the mustard weed (11%), the worm (7%), and the fly (3%)." Human Genome Project Sequence Analysis; www.ornl.gov/TechResources/Human_Genome/project/journals/insights.html.

Humans have between 30,000 to 35,000 genes, and there are about 1.4 million locations where small changes can occur on a gene. The laboratory mouse has 30,000 genes.

19 Favism is a disease in which red blood cells are destroyed after eating fava beans, a common broad bean in the Mediterranean region.

20 A search on the National Library of Medicine Web site www.ncbi.nlm.nih.gov for the keyword *antioxidant* revealed over 89,000 references. A search for *vitamin C* came up with 22,000 articles.

21 The *New England Journal of Medicine* suggests that it is reasonable for most adults to take a multivitamin supplement at the RDA level, with the possibility of higher levels of folic acid, vitamins B_6, B_{12} and D, depending on risk of cardiovascular disease and bone loss. W. C. Willett and M. J. Stampfer. 2001. "Clinical practice. What vitamins should I be taking, doctor?" *N Engl J Med*. Dec 20;345 (25):1819–1824.

In the *JAMA* article, the authors admit, "Most people do not consume an optimal amount of all vitamins by diet alone." R. H. Fletcher and K. M. Fairfield. 2002. "Vitamins for chronic disease prevention in adults: clinical applications." *JAMA*. Jun 19;287(23): 3127–3129.

22 IOM (Institute of Medicine). 2000. "Dietary Reference Intakes. Applications in Dietary Assessment. A Report of the Subcommittee on Interpretation and Uses of Dietary Reference Intakes and the Standing Committee on the Scientific Evaluation of Dietary Reference Intakes. Food and Nutrition Board." National Academy Press: Washington, D.C.

23 For reasons of space, only a brief summary of each nutrient is included. For additional information, see S. Lieberman and N. Bruning. 1997. *The Real Vitamin and Mineral Book*. Garden City Park, New York: Avery Publishing Group; R. Atkins. 1998. *Dr. Atkins' Vita-Nutrient Solution*. New York: Simon and Schuster.

24 In a study published in the *New England Journal of Medicine* in 1998, Melissa K. Thomas, M.D., Ph.D., of Boston's Massachusetts General Hospital, reported that blood levels of vitamin D were deficient in 57 percent of hospitalized patients, and 22 percent were severely deficient.

25 R. P. Heaney. 2003. "Long-latency deficiency disease: insights from calcium and vitamin D." *Am J Clin Nutr*. Nov;78(5): 912–919.

26 R. P. Heaney et al. 2003. "Human serum 25-hydroxycholecalciferol response to extended oral dosing with cholecalciferol." *Am J Clin Nutr*. Jan;77(1): 204–10.

27 G. A. Plotnikoff and J. M. Quigley. 2003. "Prevalence of severe hypovitaminosis D in patients with persistent, nonspecific musculoskeletal pain." *Mayo Clin Proc*. Dec;78(12): 1463–1470.

28 The results of one of the largest studies ever performed involving vitamin E were published in the *New England Journal of Medicine* in 1993. In this study, 87,000 female nurses and 40,000 male health professionals were followed for eight years during which time they regularly filled out questionnaires about their lifestyles and diets. Those who consumed at least 100 IU of vitamin E as a supplement for at least two years had a 36 percent lower risk of major coronary disease than did those with the lowest intake.

29 A recent book by T. E. Levy, M.D., J.D. *Vitamin C, Infectious Diseases, & Toxins*, Xlibris Corp. 2002, contains over 1,200 scientific references regarding the safety and efficacy of vitamin C.

30 G. J. Fosmire. 1990. "Zinc toxicity." *Am J Clin Nutr*. Feb;51(2): 225–227.

31 H. R. Casdorph and M. Walker. 1995. *Toxic Metal Syndrome*. Garden City Park, New York: Avery Publishing, pp. 75–127.

32 F. L. Crane. 2001. "Biochemical functions of coenzyme Q_{10}." *J Am Coll Nutr*. Dec;20(6): 591–598.

33 O. Portakal et al. 2000. "Coenzyme Q_{10} concentrations and antioxidant status in tissues of breast cancer patients." *Clin Biochem*. Jun;33(4): 279–284.

34 "CoQ_{10} functions as an electron carrier in the mitochondrial respiratory chain as well as serving as an important intracellular antioxidant. Lowered blood and tissue concentrations of CoQ_{10} have been reported in a number of diseases, although whether this deficiency is the cause or an effect of the disease remains largely unresolved." I. P. Hargreaves. 2003. "Ubiquinone: cholesterol's reclusive cousin." *Ann Clin*

Biochem. May;40(Pt 3): 207–218. This study also suggests that more work needs to be done to identify CoQ_{10}'s role: "Although a number of studies have reported clinical improvement in congestive heart failure patients after CoQ_{10} supplementation to standard therapy, concerns about the design of these studies coupled to the small number of patients involved have limited their acceptance." According to one review of the literature, CoQ_{10}'s low toxicity warrants its use even before the additional clinical trials are completed. B. Sarter. 2002. "Coenzyme Q_{10} and cardiovascular disease: a review." *J Cardiovasc Nurs.* Jul;16(4): 9–20. Also see S. Greenberg and W. H. Frishman. 1990. "Co-enzyme Q_{10}: a new drug for cardiovascular disease." *J Clin Pharmacol.* Jul;30(7): 596–608.

35 Selective toxicity of GSPE toward breast, lung, and gastric adenocarcinoma has been noted. See D. Bagchi et al. 2002. "Cellular protection with proanthocyanidins derived from grape seeds." *Ann NY Acad Sci.* May;957: 260–270.

36 Ibid.

37 See S. Lamm. 1997. *Younger At Last.* New York: Pocket Books, pp. 132–150.

38 H. Moini, L. Packer, and N. E. Saris. 2002. "Antioxidant and prooxidant activities of alpha-lipoic acid and dihydrolipoic acid." *Toxicol Appl Pharmacol.* Jul 1;182(1): 84–90.

39 One study pointed to the beneficial effects of alpha-lipoic acid mimicking insulin D. Konrad et al. 2001. "The antihyperglycemic drug alpha-lipoic acid stimulated glucose uptake via both GLUT4 translocation and GLUT4 activation." *Diabetes.* 50: 1464–1471. Another study investigating LA's cellular mechanism of action pointed out that alpha-lipoic acid "directly activates lipid, tyrosine and serine/threonine kinases in target cells, which could lead to the stimulation of glucose uptake . . ." Of particular note, according to this study, "these properties are unique among all agents currently used to lower glycaemia in animals and humans with diabetes." K. Yaworsky et al. 2000. "Engagement of the insulin-sensitive pathway in the stimulation of the glucose transport by alpha-lipoic acid in 3T3-L1 adipocytes." *Diabetologia.* Mar 1;43(3): 294–303. See also A. El Midaoui and J. de Champlain. 2002. "Prevention of hypertension, insulin resistance, and oxidative stress by alpha-lipoic acid." *Hypertension.* Feb;39(2): 303–307.

40 A. M. Wang et al. 2000. "Use of carnosine as a natural anti-senescence drug for human beings." *Biochemistry* (Mosc). Jul;65(7): 869–871; C. Brownson and A. R. Hipkiss. 2000. "Carnosine reacts with a glycosylated protein." *Free Radic Biol Med.* May 15;28(10): 1564–1570.

41 Even more important is that the French don't overeat. The portions are smaller in France than in the U.S., and only 7 percent of the French are obese. They don't snack in between meals and are very slim as a result. See "Secrets of slim French revealed." *BBC News,* August 22, 2003; http://news.bbc.co.uk/1/hi/health/3173997.stm.

42 Researchers are still trying to figure out the mechanism by which resveratrol acts. See, for example, A. Sgambato et al. 2001. "Resveratrol, a natural phenolic compound, inhibits cell proliferation and prevents oxidative DNA damage." *Mutat Res.* Sep 20;496(1–2): 171–180; S. Bastianetto, W. H. Zheng, and R. Quirion. 2000. "Neuroprotective abilities of resveratrol and other red wine constituents against nitric oxide-related toxicity in cultured hippocampal neurons." *Br J Pharmacol.* Oct;131(4): 711–720.

43 Green vegetables have the highest concentrations of both lutein and zeaxanthin; while yellow-orange fruits and vegetables, except for butternut squash, have "a much lower level of lutein in comparison to greens but contained a higher concentration of zeaxanthin." J. M. Humphries and F. Khachik. 2003. "Distribution of lutein, zeaxanthin, and related geometrical isomers in fruit, vegetables, wheat, and pasta products." *J Agric Food Chem.* Feb 26;51(5): 1322–1327. See also E. L. Snellen et al. 2002. "Neovascular age-related macular degeneration and its relationship to antioxidant intake." *Acta Ophthalmol Scand.* Aug; 80(4): 368–371.

44 J. A. Mares-Perlman et al. 2001. "Lutein and zeaxanthin in the diet and serum and their relation to age-related maculopathy in the third national health and nutrition examination survey." *Am J Epidemiol.* Mar 1;153(5): 424–432.

45 R. S. Lord et al. 2002. "Estrogen metabolism and the diet-cancer connection: rationale for assessing the ratio of urinary hydroxylated estrogen metabolites." *Altern Med Rev.* Apr;7(2): 112–129.

46 Another mechanism of controlling estrogen risk is by facilitating the conversion of the more powerful (and carcinogenic) estradiol into estrone. Soy products contain genistein, a phytonutrient that assists in this conversion. See R. W. Brueggemeier et al. 2001. "Effects of phytoestrogens and synthetic combinatorial libraries on aromatase, estrogen biosynthesis, and metabolism." *Ann NY Acad Sci.* Dec;948: 51–66.

47 M. S. Brignall. 2001. "Prevention and treatment of cancer with indole-3-carbinol." *Altern Med Rev.* Dec;6(6): 580–589.

48 E. Giovannucci et al.1995. "Intake of carotenoids and retinol in relation to risk of prostate cancer." *J Natl Cancer Inst.* 87 (23): 1767–1776.

49 J. K. Rossinow et al. 2003. "Effects of lycopene and vitamin E on gamma-irradiated prostate cancer cells." *Int J Radiat Oncol Biol Phys.* Oct 1;57(2 Suppl): S348–349; E. Giovannucci et al. 2002. "A prospective study of tomato products, lycopene, and prostate cancer risk." *J Natl Cancer Inst.* Mar 6;94(5): 391–398.

50 T. Wilt, A. Ishani, and R. MacDonald. 2002. "Serenoa repens for benign prostatic hyperplasia." *Cochrane Database Syst Rev.* (3): CD001423; F. Debruyne et al. 2002. "Comparison of a phytotherapeutic agent (Permixon) with an alpha-blocker (Tamsulosin) in the treatment of benign prostatic hyperplasia: a 1-year randomized international study." *Eur Urol.* May;41(5): 497–506.

51 According to one study, "F(2)-isoprostanes are recently described prostaglandin F isomers produced by cyclooxygenase-independent free radical peroxidation of arachidonic acid. Their quantification in plasma and urine is a sensitive and specific indicator of lipid peroxidation and, hence, of oxidative stress in vivo." Dietary supplementation for 14 days with garlic extract "reduced plasma and urine concentrations of 8-iso-PGF(2 alpha) by 29% and 37% in nonsmokers and by 35% and 48% in smokers. Fourteen days after cessation of dietary supplementation, plasma and urine concentrations of 8-iso-PGF(2 alpha) returned to values not different from those before ingestion of AGE in both groups." S. A. Dillon et al. 2002. "Dietary supplementation with aged garlic extract reduces plasma and urine concentrations of 8-iso-prostaglandin F(2 alpha) in smoking and nonsmoking men and women." *J Nutr.* Feb;132(2): 168–171. See also A. Mohamadi and S. T. Jarrell. 2000. "Effects of wild versus cultivated garlic on blood pressure and other parameters in hypertensive rats." *Heart Dis.* Jan–Feb;2(1): 3–9.

52 C. Borek. 2001. "Antioxidant health effects of aged garlic extract." *J Nutr.* Mar;131(3s): 1010S–5S.

53 Institute of Medicine, Food and Nutrition Board. 2000. *Dietary Reference Intakes for Vitamin C, Vitamin E, Selenium, and Carotenoids.* National Academies Press: Washington, D.C. See also www.fiu.edu/~nutreldr/Resources/Resources/DRIs/DRI_Table_%20One_A.pdf.

54 E. Anngard. 1994. "Nitric Oxide: Mediator, Murderer, and Medicine." *Lancet.* 343: 1199–1207.

55 As a naturally occurring amino acid, side effects from arginine are rare. It should, however, be avoided by individuals with a history of herpes outbreaks or cancer. The herpes virus has an affinity for arginine and can be stimulated into reactivation when arginine is taken in large amounts. Blood sugar levels can be raised by arginine, so caution should be used in cases of diabetes.

CHAPTER 22

1 Improved physical fitness is a "pressing" need for Americans because regular physical activity reduces the morbidity, mortality, and costs associated with chronic illness, according to the 2002 U.S. Department of Health and Human Services report "Physical Activity Fundamental to Preventing Disease." This report highlighted a 1993 study, which claims that "14 percent of all deaths in the United States were attributed to activity patterns and diet." (J. M. McGinnis and W. H. Foege. 1993. "Actual causes of death in the United States." *JAMA.* 270(18): 207–212.)

Research summarized in Physical Activity and Health: A Report of the Surgeon General (1996) emphasized that virtually everyone would benefit from physical exercise (U.S. Department of Health and Human Services. Atlanta, Georgia: Centers for Disease Control and Prevention, National Center for Chronic Disease Prevention and Health Promotion).

Another landmark report and goal-setting document is *Healthy People 2010*, chapter 22: "Physical Activity and Fitness" (www.healthypeople.gov/Document/HTML/Volume2/22Physical.htm). According to this report, "The 1990s brought a historic new perspective to exercise, fitness and physical activity by shifting the focus from intensive exercise to a broader range of health-enhancing physical activities." The report goes on to say: "On average, physically active people outlive those who are inactive. Regular physical activity also helps to maintain the functional independence of older adults and enhances the quality of life for people of all ages . . . The role of physical activity in preventing coronary heart disease (CHD) is of particular importance, given that CHD is the leading cause of death and disability in the United States. Physically inactive people are almost twice as likely to develop CHD as persons who engage in regular physical activity."

Because "physical inactivity characterizes most Americans" and "exertion has been systematically engineered out of most occupations and lifestyles," experts from a National Institutes of Health (NIH) panel produced a consensus statement in 1995 on physical activity and cardiovascular health (NIH Consensus Statement, 13(3), Dec 18–20). They emphasized the importance of a "coordinated national campaign involving a consortium of collaborating health organizations to encourage regular health activity." *Healthy People 2010* embodies that recommendation.

For a summary of landmark reports on fitness, statistics, and other resources, see the Web site of the National Coalition for Promoting Physical Activity, www.ncppa.org/landmarkreports.asp.

2 S. N. Blair et al. 1989. "Physical fitness and all-cause mortality." *JAMA.* Nov;262: 2395–2401.

3 C. D. Lee, S. N. Blair, and A. S. Jackson. 1999. "Cardiorespiratory fitness, body composition, and all-cause and cardiovascular disease mortality in men." *Am J Clin Nutr.* Mar;69(3): 373–380.

4 S. N. Blair et al. 1996. "Influences of cardiorespiratory fitness and other precursors on cardiovascular disease and all-cause mortality in men and women." *JAMA*. Jul 17;276(3): 205–210.

5 I. M. Lee. 2003. "Physical activity in women: how much is good enough?" *JAMA*. Sept 10;290(10): 1377–1378.

6 Exercise is one of the "cornerstones" of treatment for raised levels of triglycerides. M. J. Malloy and J. P. Kane. 2001. "A risk factor for atherosclerosis: triglyceride-rich lipoproteins." *Adv Intern Med.* 47: 111–136. See also A. H. Liem, J. W. Jukema, and D. J. van Veldhuisen. 2003. "Secondary prevention in coronary heart disease patients with low HDL: what options do we have?" *Int J Cardiol.* Jul;90(1): 15–21; P. A. Metcalf et al. 2001. "Factors associated with changes in serum total cholesterol levels over 7 years in middle-aged New Zealand men and women: a prospective study." *Nutr Metab Cardiovasc Dis.* Oct;11(5): 298–305.

7 American College of Sports Medicine, Fit Society Page: Exercise for Health, Winter 2003 (www.acsm.org/health+fitness/pdf/fitsociety/fitsc103.pdf), page 5. Also see Mayo Clinic, "How to measure exercise intensity," www.mayoclinic.com/invoke.cfm?objectid=045751A6-C795-4BE8-ADCD591E1DF5ABBA.

8 The AHA identifies the zone as 50–75 percent of maximum heart rate and suggests that, when you start exercising, you aim at the lower end of that zone. "Target Heart Rates," www.americanheart.org/presenter.jhtml?identifier=4736. Other sources, such as the Mayo Clinic, use American College of Sports Medicine (ACSM) guidelines to define 75–85 percent target heart rates. The Mayo Clinic site notes that target heart rates are "a rough guideline that is less reliable as you grow older." "How to measure exercise intensity," www.mayoclinic.com/invoke.cfm?objectid=045751A6-C795-4BE8-ADCD591E1DF5ABBA.

9 G. F. Fletcher et al. 1996. "Statement on exercise: benefits and recommendations for physical activity programs for all Americans." *Circulation.* 94: 857–862; American Council on Exercise (ACE), "Who should have an exercise stress test and how safe is such a test?" (www.acefitness.org/fitfacts/fitbits_display.cfm?itemid=283); D. H. Mahler. 1995. *American College of Sports Medicine Guidelines for Exercise Testing and Prescription.* 5th ed. Baltimore, Maryland: Williams & Wilkins, p. 373.

10 See, for example, "Who needs cardiac evaluation?" *Johns Hopkins Health After 50*, April 2001; www.hopkinsafter50.com/html/newsletter/2001/ha0401_Feature.php; Your Heart Health Record Web site, Bristol-Myers Squibb Medical Imaging, Inc.; www.adifferentheart.com/healthrecord.htm.

11 J. Torr. "Biotech a healthy market for chips." *Computerworld*, August 15, 2003; www.computerworld.com.au/index.php?id=1327715226&fp=16&fpid=0.

12 S. Shoham et al. 2001. "Motor-cortical activity in tetraplegics." *Nature.* 413: 793. For the University of Utah news release, see "An early step toward helping the paralyzed walk." October 24, 2001; www.utah.edu/news/releases/01/oct/spinal.html.

13 R. A. Freitas Jr. "Say Ah." www.kurzweilai.net/meme/frame.html?main=/articles/art0189.html; R. A. Freitas Jr. *Nanomedicine*, Volumes I and IIA. Austin, Texas: Landes Bioscience. Excerpts from volume 1 available on KurzweilAI.net at www.kurzweilai.net/meme/frame.html?main=/articles/art0602.html.

14 "Stretching lengthens muscles and tendons, and thereby improves flexibility. Longer muscles can generate more force around joints, helping a person jump higher, lift heavier weights, run faster, and throw farther . . . There is scant evidence that stretching prevents injuries or delayed-onset muscle soreness, which is caused by muscle fiber damage." S. Jonas. "Preventing injury," chapter 6, "Exercise and Fitness." *The Merck Manual*, 2nd Home Edition; www.merck.com/pubs/mmanual_home2/sec01/ch006/ch006d.htm.

15 M. C. Escher is a 20th-century artist known for his paradoxical paintings and drawings. One of Escher's most famous drawings shows people walking uphill on four connected stairways in which the last stairway leads back to the first stairway, an apparent impossibility.

16 M. Hargrave et al. 2003. "Subtalar pronation does not influence impact forces or rate of loading during a single-leg landing." *J Athletic Training.* Mar;38(1): 18–23.

17 "Runners report average yearly injury rates from 24% to 68%, of which 2% to 11% involve the hip or pelvis." K. H. Browning. "Hip and pelvis injuries in runners." *Physician & Sports Med.* Jan;29(1); www.physsportsmed.com/issues/2001/01_01/browning.htm, referring to W. van Mechelen. 1992. "Running injuries: a review of the epidemiological literature." *Sports Med.* 14(5): 320–335.

According to Browning, "among athletes, females have been reported to be at 1.5 to 3.5 times greater risk of stress fractures than are males. Recent prospective studies suggest that the difference is not related to athletes' sex per se, but to factors such as amenorrhea, bone density, and diet."

18 J. Hartgerink, S. Stupp, and E. Beniash. 2001. "Self-Assembling Materials: Coated Nanofibers Copy What's Bred in the Bone," *Science.* 294: 1635–1637; www.sciencemag.org/cgi/content/full/294/5547/1635a; www.matsci.northwestern.edu/stupp/sisnews.html.

19 A. H. Goldfarb and A. Z. Jamurtas. 1997. "Beta-endorphin response to exercise: an update." *Sports Med.* Jul 1;24(1): 8–16; D. V. Taylor et al. 1994. "Acidosis stimulates beta-endorphin release during exercise." *J Appl Physiol.* 77(4): 1913–1918. See also "Understanding endorphins," *ProTeamPhysicians.com*, www.proteamphysicians.com/patient/perf/endorphins.asp.

20 American College of Sports Medicine Fit Society Page. "Resistance Training." Fall 2002.

21 Ibid., p. 4.

22 National Association for Fitness Certification, "Weight training basics," www.body-basics.com/libtwo.html.

23 "Weight training: How and Why," *OhioHealth*, www.ohiohealth.com/healthreference/reference/ADFA9F13-2B2C-46FF-AF328D2782CAF854.htm?category=5314.

24 American College of Sports Medicine Fit Society Page. "Enhancing your flexibility," p. 5. Spring 2002.

25 Ibid.

26 "A low-fat diet combined with sufficient exercise may be the one-two punch needed to improve blood cholesterol levels for many persons who have had trouble doing so," according to a Stanford online report on a 1998 study conducted at the Stanford Center for Research in Disease Prevention (www.news-service.stanford.edu/news/july29/cholesterol729.html). Marcia Stefanick, the lead author in the study, pointed to the issues facing doctors in an interview for this article. "If you just reduce dietary fat to lose weight without exercising, you often reduce good cholesterol along with the bad, canceling out the benefits." She went on to say, "The challenge [with adding exercise] is not to think you can get away with eating more just because you started exercising." You will not burn off the calories you eat from a cookie by running or walking a mile, for example. See M. L. Stefanick et al. 1998. "Effects of diet and exercise in men and postmenopausal women with low levels of HDL cholesterol and high levels of LDL cholesterol." *N Engl J Med.* Jul 2;339(1): 12–20; T. R. Thomas et al. 2002. "Exercise training does not reduce hyperlipidemia in pigs fed a high-fat diet." *Metabolism.* Dec;51(12): 1587–1595.

CHAPTER 23

1 H. Malmros. 1950. "The relation of nutrition to health: a statistical study of the effect of the wartime on arteriosclerosis, cardiosclerosis, tuberculosis, and diabetes." *Acta Medica Scandinavia.* 246(Suppl): 141–149. Despite its title, this article is surprisingly nontechnical. The dramatic effect of food rationing on heart disease in several European countries during World War II is presented in standard prose and charts.

A. Keys. 1975. "Coronary heart disease: the global picture." *Atherosclerosis.* 22: 153–154. A comprehensive overview on coronary heart disease by the author of the eminent Seven Countries Study, this article cites extensive evidence from around the world, including studies of global peoples; the effects of wartime; social class and occupation; the impact of exercise, stress, and personality type; the role of risk factors, dietary factors, and genetics; and a discussion on the prevention of coronary heart disease.

R. G. Wilkinson. 1996. *Unhealthy Societies: The Afflictions of Inequality*. London: Routledge. In this book, Wilkinson posits a connection between income inequality and mortality. He also claims that life expectancy at birth in England and Wales increased by 6.5 and 6.6 years, respectively, for men and women from 1911–1921, compared with 2.4 and 2.3 years, respectively, from 1921–1931 and 1.5 and 1.2, respectively, for 1931–1940. A similar jump (7.0 and 6.5) was noted for 1940–1951.

2 This trend has been noticed in many countries. In Australia, for example, "suicide rates were higher during periods of draught and lower during WWII." ("More suicides under Conservative rule." *BBC News*, September 18, 2002; http://newswww.bbc.net.uk/1/low/health/2263690.stm.)

In Croatia, "in the areas directly affected by war, the suicide rate was significantly lower than in other areas during the study period 1993–1998." (M. Grubisic-Ilic et al. 2002. "Epidemiological study of suicide in the Republic of Croatia." *Eur Psych.* Sep 1;17(5): 259–264.) In Jaffna, Sri Lanka, between 1980 and 1989, "there was a marked drop in the suicide rate during the war" (D. J. Somasundarum and S. Rajadurai. "War and suicide in northern Sri Lanka." *Acta Psychiatr Scand.* Jan 1;91(1): 1–4.)

3 Stress may lengthen the time triglycerides stay in the blood. "If a person has a high-fat snack or meal during a time of stress, that fat is going to be circulating in the blood for a longer period of time. That means it may be more likely to be deposited in the arteries where it can contribute to heart disease." C. M. Stoney. 2002. Quoted in "Stress causes heart-damaging fats to stay in blood longer." Ohio State Research (www.acs.ohio-state.edu/researchnews/archive/cholblod.htm). C. M. Stoney et al. 2002. "Acute psychological stress reduces plasma triglyceride clearance." *Psychophysiology.* Jan;39(1): 80–85.

4 A. M. Tacon. 2003. "Meditation as a complementary therapy in cancer." *Fam Community Health.* Jan–Mar;26(1): 64–73; M. J. Kreitzer and M. Snyder. 2002. "Healing the heart: integrating complementary therapies and healing practices into the care of cardiovascular patients." *Prog Cardiovasc Nurs.* Spring;17(2): 73–80.

5 A. Elkin. 1999. *Stress Management for Dummies.* New York: Wiley Publishing, Inc.; S. S. Nussey and S. A. Whitehead, eds. 2001. "Catecholamine synthesis and secretion." *Endocrinology: An Integrated Approach.* BIOS Scientific Publishers, Ltd. Online table of contents: www.ncbi.nlm.nih.gov/books/bv.fcgi?call=bv.View..ShowTOC&rid=endocrin.TOC&depth=1; N. Schneiderman. "Behavior, Autonomic Function, and Animal Models of Cardiovascular Pathology," in T. M. Dembroski, T. H. Schmidt, and G. Blümchen, eds. 1983. *Biobehavioral Bases of Coronary Heart Disease.* Basel, Switzerland: Karger, pp. 304–364, 317–322; R. M. Sapolsky. 1990. "Stress in the Wild." *Sci Amer.* Jan: 116–124; R. Williams. 1989. *The Trusting Heart.* New York: Times Books, pp. 75–82.

6 "When demands are physical, as they often were in earlier times, the hormones and fats released during the stress response are rapidly delivered to the muscles by the increased heart rate and blood pressure . . . Meeting mental demands requires lower, but ongoing, levels of stress hormones and fatty fuels . . . When the work to be done is mental, the hormones and fats that have been mobilized for action are not used up. The unnecessarily high heart rate and blood pressure set up a condition of increased turbulence in the bloodstream, which in turn increases the tension on the walls of the arteries." M. Burg. "Stress, behavior, and heart disease." Chapter 8, p. 97, in B. Zaret et al., eds. 1992. *Yale University School of Medicine Heart Book*; http://info.med.yale.edu/library/heartbk/8.pdf.

7 R. Misslin. 2003. "The defense system of fear: behavior and neurocircuitry." *Clin Neurophysiol.* April;33(2): 55–66.

8 Story provided by Joel Miller, M.D.

9 M. Burg, pp. 99–104; I. Kawachi et al. 1998. "Prospective study of a self-report Type A scale and risk of coronary heart disease." *Circulation.* 98: 405.

According to the Kawachi study, one of the reasons is the difference in results gathered by self-report questionnaires versus structured interview approach (VCE). The authors suggest that the former are less likely to capture some aspects of hostile behavior, such as hurried speech and hostile facial expressions. Another reason is that perhaps the questionnaires don't cover some components of the complex. Different questionnaires emphasize different things. The Kawachi study used the MMPI-2 Type A Scale, which incorporates a broader range of components. The study used a cohort of 2280 community-dwelling men from Boston, ages 21–80, at the start. The MMPI-2 was administered by mail in 1986 and participants were followed up for an average of seven years. Higher type A scores were associated with higher average body mass index, more frequent history of heart disease, more smoking, more alcohol consumption. "The MMPI-2 Type A Scale provides a global score based on 3 apparently critical aspects of TAB: time urgency, competitiveness, and hostility. It may be the confluence of these behavior styles, rather than one aspect alone, that increases risk of CHD."

A much-cited study on type A personality and heart disease risk is T. M. Dembroski et al. 1989. "Components of hostility as predictors of sudden death and myocardial infraction in the Multiple Risk Factor Intervention Trial." *Psychosom Med.* 54(5): 514–522.

T. Hallman. 2001. "Psychosocial risk factors for coronary heart disease, their importance compared with other risk factors and gender differences in sensitivity." *J Cardiovasc Risk.* Feb;8(1): 39–49.

The Hallman study showed that women were more sensitive than men with respect to psychosocial risk factors for CHD.

E. R. Greenglass and J. Julkunen. 1991. "Cook-Medley hostility, anger, and the Type A behavior pattern in Finland." *Psychol Rep.* Jun;68(3 Pt 2): 1059–1066. This study supports the points made by Kawachi by pointing to the importance of "specifying the kind of hostility" measured by the scale used, in this case, the Cook-Medley scale. The study looked at 219 university students. When they used a "subscale" measuring cynical distrust, they found a positive correlation between cynicism and CHD.

The following study was also aimed at figuring out what hostility components were important: Y. Gidron and K. Davidson. 1996. "Development and preliminary testing of a brief intervention for modifying CHD-predictive hostility components." *J Behav Med.* Jun;19(3): 203–220.

The presence of depression can alter the association between hostility and CHD but not reduce the risk of the patient. See N. Ravaja, T. Kauppinen, and L. Keltikangas-Jarvinen. 2000. "Relationships between hostility and physiological coronary heart disease risk factors in young adults: the moderating influence of depressive tendencies." *Psychol Med.* Mar;30(2): 381–393. "Despite the established risk factor status of hostility, lack of anger and hostility, when combined with high depressive tendencies, may represent the most severe exhaustion where the individual has given up. Disregard of this fact may explain some null findings in the research on hostility and CHD risk."

G. E. Miller et al. 2003. "Cynical hostility, depressive symptoms, and the expression of inflammatory risk markers for coronary heart disease." *J Behav Med.* Dec;26(6): 501–515. This study emphasizes the importance of looking at both the "independent and interactive relationships among psychosocial characteristics involved in disease." The study looked at 100 adults regarding hostility and depressive symptoms. The study discusses how these factors operate together to influence coronary heart disease.

J. E. Gallacher et al. 2003. "Is type A behavior really a trigger for coronary heart disease events?" *Pyschosom Med.* May–Jun;65(3): 339–346. This study provides another perspective on the role of the type A personality. The study looked at 2,394 men aged 50–64: "The data show Type A is a strong predictor of when incident coronary heart disease (or coronary event) will occur rather than if it will occur. These findings suggest that Type A increases exposure to potential triggers, rather than materially affecting the process of atherosclerosis."

10 M. Burg, pp. 99–104; R. B. Williams Jr. 1987. "Psychological factors in coronary artery disease: epidemiologic evidence." *Circulation.* Jul;76(Suppl I): 1117–1123.

11 I. Kawachi et al. 1998. "Prospective study of a self-report Type A scale and risk of coronary heart disease." *Circulation.* Aug 4;98(5): 405; see also J. E. Muller et al. 1997. "Mechanisms precipitating acute cardiac events." *Circulation.* Nov 4;96(9): 3233–3239.

12 J. C. Barefoot et al. 1989. "The Cook-Medley hostility scale: item content and ability to predict survival." *Psychosom Med.* Jan–Feb;51(1): 46–57.

13 J. C. Barefoot et al. 1983. "Hostility, CHD incidence, and total mortality: a twenty-five-year follow-up study of 255 physicians." *Psychosom Med.* Mar;45(1): 59–63.

14 J. C. Barefoot et al. 1987. "Suspiciousness, health and mortality: a follow-up study of five hundred older adults." *Psychosom Med.* Sep–Oct;49(5): 450–457.

15 S. Segerstrom et al. 1998. "Optimism is associated with mood, coping, and immune change in response to stress." *J Pers Soc Psych.* Jun;74(6): 1646–1655; M. F. Scheier and C. S. Carver. 1987. "Dispositional optimism and physical well-being: the influence of generalized outcome expectancies on health." *J Pers.* Jun;55(2): 169–210.

16 S. E. Taylor et al. 2000. "Behavioral responses to stress in females: tend-and-befriend, not fight-or-flight." *Psych Rev.* Jul;107(3): 411–429.

17 M. Friedman and D. Ulmer. 1984. *Treating Type A Behavior and Your Heart.* New York: Alfred A. Knopf, pp. 175–237.

18 Data from: T. H. Holmes and R. H. Rahe. 1967. "The social readjustment rating scale." *J Psychosom Res.* Aug;11(2): 227–237. Also see Holmes Rahe Social Readjustment Rating Scale; www.markhenri.com/health/stress.html.

19 Per 2002 CDC statistics, "smoking costs Americans over $157 billion annually in medical care." Of the 442,398 deaths per year in the U.S. from smoking-related causes, 33.5 percent are cardiovascular-related. 2003 Heart Disease and Stroke Statistical Update, American Heart Association (www.americanheart.org/presenter.jhtml?identifier=3000090), p. 26. Approximately a quarter of U.S. men and women smoke (p. 25).

20 "Number of deaths and age-adjusted death rates per 100,000 population for categories of alcohol-related (A-R) mortality, United States and States, 1979–1996." National Institute on Alcohol Abuse and Alcoholism (www.niaaa.nih.gov/databases/armort01.htm). The total number of deaths attributable to A-R mortality was 110,000 in 1996, and is only provided to 1996. However, each of the components of this death rate, such as cirrhosis of the liver, has remained stable since this time, having increased by less than 10 percent. Thus it remains the case that over 100,000 deaths in the U.S. are attributable to A-R mortality.

21 U.S. Department of Health and Human Services, Centers for Disease Control. 1990. "Alcohol-related mortality and years of potential life lost: United States, 1987." *Morbidity and Mortality Weekly Report.* Mar;39(11): 173–178; www.cdc.gov/mmwr/preview/mmwrhtml/00001576.htm.

U.S. Department of Health and Human Services. Tenth Special Report to the U.S. Congress on Alcohol and Health from the Secretary of Health and Human Services, June 2000; www.niaaa.nih.gov/publications/10report/intro.pdf.

22 E. B. Rimm et al. 1991. "Prospective study of alcohol consumption and risk of coronary heart disease in men." *Lancet.* Aug;338: 464–468.

23 In 2000, Americans consumed 17.3 gallons of coffee per capita. U.S. Department of Agriculture, Foreign Agricultural Service (data from Davenport & Company LLC). By some popular reports, 80 percent of Americans drink coffee every day. P. McMahon. "'Cause coffees' produce a cup with an agenda." *USA Today*, July 25, 2001.

24 M. J. Klag et al. 2002. "Coffee intake and risk of hypertension." *Arch Int Med.* Mar 25;162: 657–662.

25 "Caffeine is well known to promote anxious behavior in humans and animal models . . ." M. El Yacoubi et al. 2000. "The anxiogenic-like effect of caffeine in two experimental procedures measuring anxiety in the mouse isn't shared by selective A(2A) adenosine receptor antagonists." *Psychopharmacol* (Berl). Feb;148(2): 153–163; J. P. Boulenger et al. 1984. "Increased sensitivity to caffeine in patients with panic disorders." *Arch Gen Psych.* Nov;41: 1067–1071.

26 S. M. Wolfe et al. 1988. *Worst Pills, Best Pills.* Washington, D.C.: Public Citizen Health Research Group, p. 145; C. R. H. Newton. "Benzodiazepine abuse" (www.emedicine.com/aaem/topic42.htm).

27 T. Thiele et al. 1998. "Ethanol consumption and resistance are inversely related to neuropeptide Y levels." *Nature.* 396: 366–369; M. Cockerill. 1998. "Low levels of brain chemical drives mice to drink." *Br Med J.* Dec(317): 1544B.

28 George F. Koob has written hundreds of articles on the topic of drug addiction.

S. B. Caine and G. F. Koob. 1993. "Modulation of cocaine self-administration in the rat through D-3 dopamine receptors." *Science.* 260: 1814–1816; G. F. Koob et al. 1994. "Corticotropin releasing factor, stress and behavior." *Seminars in the Neurosciences.* 6: 221–229; P. Hyytia and G. F. Koob. 1995. "GABAA receptor antagonism in the extended amygdala decreases ethanol self-administration in rats." *European Journal of Pharmacology.* 283: 151–159; G. Schulteis, A. Markou, M. Cole, and G. F. Koob, 1995. "Decreased brain reward produced by ethanol withdrawal." *Proc Natl Acad Sci USA.* 92: 5880–5884; G. F. Koob. 1996. "Hedonic valence, dopamine, and motivation." *Molecular Psychiatry.* 1: 186–189; G. F. Koob. 1996. "Drug addiction: The yin and yang of hedonic homeostasis." *Neuron.* 16: 893–896; M. Spina et al. 1996. "Appetite-suppressing effects of urocortin, a CRF-related neuropeptide." *Science.* 273: 1561–1564; D. A. Finn, R. H. Purdy, and G. F. Koob. "Animal models of anxiety and stress-induced behavior: effects of neuroactive steroids." In S. S. Smith, ed. 2004. *Neurosteroid Effects in the Central Nervous System: The Role of the GABA-A Receptor.* Boca Raton, Florida: CRC Press, pp. 317–338; P. J. Kenny, I. Polis, G. F. Koob, and A. Markou. 2003. "Low dose cocaine self-administration transiently increases but high dose cocaine persistently decreases brain reward function in rats." *European Journal of Neuroscience.* 17: 191–195; G. F. Koob. 2003. "Neuroadaptive mechanisms of addiction: studies on the extended amygdala." *European Neuropsychopharmacology.* 13: 442–452; G. F. Koob. "Drug reward and addiction." In L. R. Squire, F. E. Bloom, S. K. McConnell, J. L. Roberts, N. C. Spitzer, and M. J. Zigmond, eds. 2003. *Fundamental Neuroscience*, 2nd edition. San Diego: Academic Press, pp. 1127–1143; G. F. Koob, A. J. Roberts, B. L. Kieffer, C. J. Heyser, S. N. Katner, R. Ciccocioppo, and F. Weiss. "Animal models of motivation for drinking in rodents with a focus on opioid receptor neuropharmacology." In M. Galanter, ed. 2003. *Research on Alcoholism Treatment* (series title: *Recent Developments in Alcoholism*, vol. 16). New York: Plenum Press, pp. 263–281; G. F. Koob and L. Pulvirenti. "Drug addiction and loss of control: focus on motivation in an allostatic perspective." In M. Massotti and L. Pulvirenti, eds. 2003. *Neuroscience of Drug Addiction: Focus on Neural Plasticity* (series title: *Rapporti ISTISAN 03/7*). Rome: Instituto Superiore di Sanita, pp. 39–48.

29 T. Phillips et al. 1998. "Alcohol preference and sensitivity are markedly reduced in mice lacking dopamine D2 receptors." *Nature Neuroscience.* Nov 1(1): 610–615.

30 University of California. *Berkeley Wellness Letter* 5.7 (Apr 1989).

31 S. Higgens and J. Katz. 1998. *Cocaine Abuse: Behavior, Pharmacology, and Clinical Applications.* New York: Academic Press; T. Madge. 2004. *White Mischief: A Cultural History of Cocaine.* New York: Thunder's Mouth Press; C. Reinarman and H. G. Levine, eds. 1997. *Crack in America: Demon Drugs and Social Justice.* Berkeley: University of California Press; S. Ali, ed. 2000. *The Neurochemistry of Drugs of Abuse: Cocaine, Ibogaine, and Substituted Amphetamines.* New York: New York Academy of Sciences; E. V. Nunes and J. S. Rosecan. "Human neurobiology of cocaine." In H. I. Spitz and J. S. Rosecan, eds. 1987. *Cocaine Abuse: New Directions in Treatment and Research.* New York: Brunner/Mazel Publishers, pp. 48–97; S. Shiffman and T. A. Wills, eds. 1985. *Coping and Substance Use.* San Diego: Academic Press, pp. 41–42.

32 2003 Heart Disease and Stroke Statistical Update, American Heart Association; www.americanheart.org/presenter.jhtml?identifier=3000090, p. 25.

33 "Projections of DSM-III alcohol abuse and alcohol dependence for the U.S. population aged 18 and older, 1990, 1995, 2000." National Institute on Alcohol Abuse and Alcoholism; www.niaaa.nih.gov/databases/abdep2.htm. "An estimated 20 to 40 percent of patients in large urban hospitals are there because of illnesses that have been caused or made worse by their drinking . . . [O]ne in four children under the age of 18 lives in a household with one or more family members who are alcohol dependent . . ." U.S. Department of Health and Human Services, Tenth Special Report to the U.S. Congress on Alcohol and Health from the Secretary of Health and Human Services, June 2000; www.niaaa.nih.gov/publications/10report/intro.pdf, p. ix.

34 Researchers are still debating whether caffeine is addictive.

According to one study, "[the] data correlate well with the known sensitivity of locomotion, mood and sleep to low doses of caffeine. They also show that low doses of caffeine which reflect the usual human level of consumption fail to activate reward circuits in the brain and thus provide functional evidence of the very low addictive potential of caffeine." A. Nehlig and S. Boyet. 2000. "Dose-response study of caffeine effects on cerebral functional activity with a specific focus on dependence." *Brain Res.* Mar 6;858(1): 71–77.

Another study claims "caffeine is an addictive psychoactive substance. Similar to previous findings with other licit and illicit psychoactive drugs, individual differences in caffeine use, intoxication, tolerance, and withdrawal are substantially influenced by genetic factors." K. Kendler and C. Prescott. 1999. "Caffeine intake, tolerance, and withdrawal in women: a population-based twin study." *Am J Psych.* Feb;156: 223–228.

35 Department of Health and Human Services, 2001 National Household Survey on Drug Abuse: Volume 1. Summary of National Findings (www.samhsa.gov/oas/nhsda.htm), chapter 2, p. 11 and chapter 7, p. 2. An estimated 123,000 Americans used heroin in 2001; 1.7 million, cocaine; and 406,000, crack. A "current user" is defined as someone who has used the drug in the month prior to the survey.

36 Two-thirds of U.S. adults report relying on the weekends to catch up on sleep. National Sleep Foundation. "Sleep in America" Poll, March 2002.

According to one recent study, "[s]leep debt has a harmful impact on carbohydrate metabolism and endocrine function. The effects are similar to those seen in normal aging and, therefore, sleep debt may increase the severity of age-related chronic disorders." K. Spiegel et al. 1999. "Impact of sleep debt on metabolic and endocrine function." *Lancet.* Oct 23;354(9188): 1435–1439.

37 Acknowledgments to Dr. Joel Miller who contributed his ideas to this list.

38 H. Benson et al. 1974. "The relaxation response." *Psychiatry.* Feb;37: 37–46; H. Benson et al. 1975. "The relaxation response: psychophysiologic aspects and clinical applications." *Int J Psych Med.* 6: 87–98.

See also Benson's more recent book: *The Wellness Book: The Comprehensive Guide to Maintaining Health and Treating Stress-Related Illness and Timeless Healing.* 1993. New York: Scribner.

Some other recent studies include:

B. H. Chang et al. 2004. "Relaxation response for Veterans Affairs patients with congestive heart failure: results from a qualitative study within a clinical trial." *Prev Cardiol.* Spring;7(2): 64–70. This study suggested value of RR in congestive heart failure health care (with both physical and emotional changes seen.)

R. Bonadonna. 2003. "Meditation's impact on chronic illness." *Holist Nurs Pract.* Nov–Dec;17(6): 309–319. This is an overview of different types of techniques for "mindfulness" and their role in a clinical setting.

T. Esch, G. L. Fricchione, and G. B. Stefano. 2003. "The therapeutic use of the relaxation response in stress-related diseases." *Med Sci Monit.* Feb;9(2): RA23–34. This study points to the therapeutic uses of RR techniques, particularly in mild or early disease states when "a high degree of biological and physiological flexibility may still be possible." Interesting connection to nitric oxide production.

G. D. Jacobs. 2001. "Clinical applications of the relaxation response and mind-body interventions." *J Altern Complement Med.* 7(Suppl 1): S93–101. This study refers to the "several hundred peer-reviewed studies in the past 20 years" that have shown that "the relaxation response and mind-body interventions are clinically effective in the treatment of many health problems that are caused or made worse by stress." It suggests that these techniques are very effective when combined with standard medical care.

C. L. Mandle et al. 1996. "The efficacy of relaxation response interventions with adult patients: a review of the literature." *J Cardiovasc Nurs.* Apr;10(3): 4–26. This study reviewed 37 studies of the efficacy of RR interventions with adult patients. "Consistencies in the results suggest the effectiveness of the relaxation response in reducing hypertension, insomnia, anxiety, pain, and medication use across multiple populations, diagnostic categories, and settings."

39 H. Benson et al. 1974. "Decreased blood pressure in borderline hypertensive subjects who practiced meditation." *J Chron Dis.* 27: 163–89.

40 C. N. Alexander et al. 1989. "Transcendental meditation, mindfulness, and longevity: an experimental study with the elderly." *J Pers Soc Psych.* Dec;57: 950–964.

41 H. Benson with M. Z. Klipper. 1975 [2000]. *The Relaxation Response.* New York: William Morrow, pp. 23–25, 68–74.

42 J. Kabat-Zinn. 1990. *Full Catastrophe Living: Using the Wisdom of Your Body and Mind to Face Stress, Pain, and Illness.* New York: Delacorte Press; J. Kabat-Zinn. 1995. *Wherever You Go, There You Are: Mindfulness Meditation in Everyday Life.* New York: Hyperion.

43 Biofeedback is being proposed for a variety of health issues, such as urinary incontinence (N. M. Shinopulos and J. Jacobson. 1999. "Relationship between health promotion lifestyle profiles and patient outcomes of biofeedback therapy for urinary incontinence." *Urol Nurs.* Dec;19(4): 249–253) and cutaneous problems (P. D. Shenefelt. 2003. "Biofeedback, cognitive-behavioral methods, and hypnosis in dermatology: Is it all in your mind?" *Dermatol Ther.* Jun;16(2): 114–122). See also D. Shapiro and R. S. Surwit. "Biofeedback." In O. F. Pomerleau and J. P. Brady, eds. *Behavioral Medicine: Theory and Practice.* Baltimore: Wilkins & Williams.

44 Although Benson derived his technique largely from transcendental meditation (TM), proponents of TM point out that Benson's technique isn't the same as TM and differs from it in subtle but important ways. There has been extensive research reported on the health benefits of TM, which includes papers in over 100 referenced journals. A summary of this research and a complete listing of references is contained in D. Orme-Johnson and C. N. Alexander, "Summary of research on the transcendental meditation and TM-Sidhi program," available from TM centers.

For information on TM, visit www.maharishi.org/tm/index.html and for centers that teach transcendental meditation in your area, visit www.maharishi.org/locations/locations.html or call 1-888-432-7686. Online courses are also available from the Maharishi Open University (www.mou.org).

45 Benson with Klipper, pp. 158–166. See note 41 on page 441.

INDEX

Boldface page references indicate illustrations. Underscored references indicate boxed text.

A

B

C

G

H

I

K

L

M

N

O

P

Q

R

S

T

U

V

W

X

Y